THE ROUTLEDGE
HANDBOOK OF EMERGENCE

Emergence is often described as the idea that the whole is greater than the sum of the parts: interactions among the components of a system lead to distinctive novel properties. It has been invoked to describe the flocking of birds, the phases of matter and human consciousness, along with many other phenomena. Since the nineteenth century, the notion of emergence has been widely applied in philosophy, particularly in contemporary philosophy of mind, philosophy of science and metaphysics. It has more recently become central to scientists' understanding of phenomena across physics, chemistry, complexity and systems theory, biology and the social sciences.

The Routledge Handbook of Emergence is an outstanding reference source and exploration of the concept of emergence, and is the first collection of its kind. Thirty-two chapters by an international team of contributors are organised into four parts:

- Foundations of emergence
- Emergence and mind
- Emergence and physics
- Emergence and the special sciences

Within these sections important topics and problems in emergence are explained, including the British Emergentists; weak vs. strong emergence; emergence and downward causation; dependence, complexity and mechanisms; mental causation, consciousness and dualism; quantum mechanics, soft matter and chemistry; and evolution, cognitive science and social sciences.

Essential reading for students and researchers in the philosophy of mind, philosophy of science and metaphysics, *The Routledge Handbook of Emergence* will also be of interest to those studying foundational issues in biology, chemistry, physics and psychology.

Sophie Gibb is a Professor and Head of Department in the Department of Philosophy at Durham University, UK.

Robin Findlay Hendry is a Professor and Director of Research in the Department of Philosophy at Durham University, UK.

Tom Lancaster is a Professor in the Department of Physics at Durham University, UK.

Routledge Handbooks in Philosophy

Routledge Handbooks in Philosophy are state-of-the-art surveys of emerging, newly refreshed, and important fields in philosophy, providing accessible yet thorough assessments of key problems, themes, thinkers, and recent developments in research.

All chapters for each volume are specially commissioned and written by leading scholars in the field. Carefully edited and organized, *Routledge Handbooks in Philosophy* provide indispensable reference tools for students and researchers seeking a comprehensive overview of new and exciting topics in philosophy. They are also valuable teaching resources as accompaniments to textbooks, anthologies, and research-orientated publications.

Also available:

For more information about this series, please visit: www.routledge.com/Routledge-Handbooks-in-Philosophy/book-series/RHP

THE ROUTLEDGE
HANDBOOK OF EMERGENCE

*Edited by Sophie Gibb, Robin Findlay Hendry
and Tom Lancaster*

Routledge
Taylor & Francis Group

LONDON AND NEW YORK

First published 2019 by Routledge

2 Park Square, Milton Park, Abingdon, Oxon, OX14 4RN
605 Vanderbilt Avenue, New York, NY 10017

Routledge is an imprint of the Taylor & Francis Group, an informa business

First issued in paperback 2020

British Library Cataloguing-in-Publication Data
A catalogue record for this book is available from the British Library

Library of Congress Cataloging-in-Publication Data
A catalog record for this book has been requested

ISBN: 978-1-138-92508-3 (hbk)
ISBN: 978-0-367-78388-4 (pbk)

Typeset in Bembo
by Apex CoVantage, LLC

CONTENTS

Contents

Contents

NOTES ON CONTRIBUTORS

Lynne Rudder Baker was a Distinguished Professor at the University of Massachusetts Amherst. She published numerous books and articles on the philosophy of mind, metaphysics, the philosophy of religion, and the nature of philosophy itself.

Umut Baysan is a Lecturer in Philosophy at the University of Oxford. Previously, he worked as a University Teacher and a Postdoctoral Researcher at the University of Glasgow, where he obtained his PhD in 2014. He works in the philosophy of mind and metaphysics, specialising in the metaphysics of mind–body relations. His research has appeared in journals such as *Analysis, Australasian Journal of Philosophy, Erkenntnis, Minds and Machines,* and *Philosophical Quarterly.*

John Bickle is Professor of Philosophy and Adjunct Professor of Psychology at Mississippi State University, and Affiliated Faculty in Neurobiology at the University of Mississippi Medical Center. A philosopher of neuroscience especially interested in cellular and molecular neurobiology, he is the author of four books, more than eighty papers and chapters in philosophy and neuroscience journals and volumes, and edited the *Oxford Handbook of Philosophy and Neuroscience* (2009, paperback edition 2012).

William Bechtel, a philosopher of science at the University of California, San Diego, focuses his research on the strategies life scientists employ to develop mechanistic explanations, especially in cell and molecular biology, and the tools they develop to facilitate their endeavours (e.g., experimental instruments and diagrammatic and network representations). He is coauthor of *Discovering Complexity and Connectionism and the Mind* and author of *Discovering Cell Mechanisms and Mental Mechanisms.*

Robert Bishop is Associate Professor of Physics and Philosophy and the John and Madeleine McIntyre Endowed Professor of Philosophy and History of Science at Wheaton College. He has published extensively on reduction and emergence, determinism, and philosophy of the social sciences. His most recent books are *The Philosophy of the Social Sciences* (2007) and *The Physics of Emergence* (forthcoming).

Stephen J. Blundell is a Professor of Physics at the University of Oxford and a Fellow of Mansfield College. His research used muon-spin rotation to study problems in condensed matter physics, particularly relating to magnets and superconductors. His books include *Magnetism in Condensed Matter*, *Concepts in Thermal Physics* (coauthored with his wife, the astrophysicist Katherine Blundell), and *Quantum Field Theory for the Gifted Amateur* (coauthored with Tom Lancaster).

Alex Carruth is Assistant Professor of Philosophy at Durham University. His work focusses on issues in philosophy of mind, metaphysics, and philosophy of science.

Stewart Clark is Professor of Condensed Matter Physics in the University of Durham specialising in quantum theory and computational physics. Some of his particular interests include the development of methods for solving problems in electronic structure to predict the properties of new materials for technological applications. He has also been involved recently in projects discussing the philosophy of science, especially in describing the structure of materials from first principles.

Piers Coleman is Professor of Physics at Rutgers University and holds a part-time position at Royal Holloway, University of London. His research focusses on the emergent properties of materials, such as superconductivity and magnetism, that derive from the many-body physics of electrons within. He is also interested in the history of discovery in physics, which he finds helps inspire him and provide perspective on his research.

George Ellis, FRS, coauthored *The Large Scale Structure of Space Time* with Stephen Hawking. He is Professor Emeritus of Applied Mathematics at the University of Cape Town and has been Visiting Professor at the University of Texas, University of Alberta, Boston University, London University, and Oxford University, and has been Professor of Cosmic Physics at SISSA, Trieste. His recent books include *How Can Physics Underlie the Mind? Top-Down Causation in the Human Context* (2016).

Sophie Gibb is Professor and Head of Department in the Department of Philosophy at Durham University. Her research interests lie within contemporary metaphysics (in particular, ontology) and the philosophy of mind (in particular, the mental causation debate).

Carl Gillett is Professor of Philosophy at Northern Illinois University who works primarily on the philosophy of science and the philosophy of mind, focussing in particular on the metaphysics of science. Gillett has published extensively on issues concerning scientific composition (realisation of properties, part-whole of individuals, implementation of activities), compositional explanation, reduction, and emergence.

John Heil is Professor of Philosophy at Washington University in St Louis, Honorary Professor of Philosophy at Durham University, and Honorary Research Associate at Monash University. His most recent book is *The Universe as We Find It* (2012). He is currently working on a successor monograph in metaphysics tentatively titled *Appearance in Reality*.

Robin Findlay Hendry studied chemistry and philosophy of science at King's College London and the LSE. After completing his PhD, he taught briefly at the University of Edinburgh before coming to Durham University in 1994, where he is now Professor of Philosophy. His research interests include philosophical issues in chemistry, such as the nature of chemical substances, molecular structure, and the chemical bond, and more general issues, such as natural kinds, classification, emergence, reduction, and the unity of science.

Emily Herring studied philosophy at the Sorbonne and is now completing a PhD at the University of Leeds on the reception of French philosopher Henri Bergson's theories among British biologists.

Tuukka Kaidesoja is a Researcher at the Unit of Practical Philosophy of the University of Helsinki. He is the author of *Naturalizing Critical Realist Social Ontology* and articles on philosophy of the social sciences and sociological theory. His current research project explores the relationship between sociology and the cognitive sciences.

Tom Lancaster is Professor of Physics at Durham University. His research focusses on the physics of magnetism in one and two dimensions. He is the coauthor of *Quantum Field Theory for the Gifted Amateur* (with Stephen Blundell).

Cynthia Macdonald is Honorary Senior Research Fellow at the University of Manchester, where she was previously Professor of Philosophy, and Professor Emerita at Queen's University Belfast and the University of Canterbury, New Zealand. She has held Visiting Professorial positions at the Universities of Utrecht, Columbia, Rutgers, and Connecticut. Her research publications are in the areas of philosophy of mind, philosophy of psychology, and metaphysics.

Graham Macdonald is Honorary Senior Research Fellow at the University Manchester and Emeritus Professor, University of Canterbury, New Zealand. He has previously held Visiting Professorial positions at Queen's University Belfast and King's College London. His research and publications have been in the areas of mental causation, teleosemantics, philosophy of social science, and the philosophy of Karl Popper.

Kerry McKenzie is an Associate Professor in Philosophy at the University of California, San Diego. She received her PhD from the University of Leeds in 2012 and then took up postdoctoral fellowships at the University of Calgary and the Rotman Institute, University of Western Ontario. She works on a variety of themes in scientific metaphysics, most saliently fundamentality, modality, structuralism, and naturalism.

Brian P. McLaughlin is Distinguished Professor of Philosophy and Cognitive Science at Rutgers University, and Director of the Rutgers Cognitive Science Center. He has well over one hundred publications, and is the author of numerous papers in the philosophy of mind, the philosophy of psychology, metaphysics, and philosophical logic. He and Vann McGee have a forthcoming book entitled *Terrestrial Logic: Bringing Formal Semantics Down to Earth*.

Tom McLeish, FRS, is Professor of Natural Philosophy in the Department of Physics and in the Centre for Medieval Studies and the Humanities Research Centre at the University of York, UK. His highly collaborative research in soft matter and biological physics has been recognised by the Bingham Medal of the Society of Rheology and the Weissenberg Medal of the European Society of Rheology. He is currently Chair of the Royal Society's Education Committee.

Michel Morange was Professor of Biology at the Sorbonne University and at the Ecole Normale Supérieure in Paris. He worked at the Institut Pasteur and at the Ecole Normale Supérieure in enzymology and later in developmental biology. He also worked on the history and philosophy of the life sciences. He has published many books on the molecular revolution and its actors and the recent transformations of biology linked with the development of systems and synthetic biology.

Denis Noble is Emeritus Professor at the University of Oxford. He created the first biophysically detailed mathematical model of heart rhythm. Since retirement from his chair of Cardiovascular Physiology in 2004 he has returned to his earlier interests in evolutionary biology, where he takes issue with neo-Darwinism. The result is two books, *The Music of Life* (2006) and *Dance to the Tune of Life* (2016), and a series of articles with his brother Raymond Noble.

Raymond Noble is an Honorary Senior Lecturer at the Institute for Women's Health, University College London, where he was previously a graduate tutor and medical ethicist. He graduated in Zoology and has a PhD in neuroscience. He now works on understanding causation in biology.

Paul Noordhof is Anniversary Professor of Philosophy at the University of York. His main work has been in causation, mental causation, self-deception and belief, the imagination, and emergence. He has just completed a full-length book on causation, *A Variety of Causes*.

Mark Pexton has researched scientific explanation (specifically non-causal explanations) and emergence. He has developed a theory of contextual emergence, in part based on computational emergence. He is a research fellow at Durham University, and was formerly a member of the Durham Emergence Project.

David Pines was the founding director of the Institute for Complex Adaptive Matter (ICAM); Distinguished Professor of Physics, University of California, Davis; and Research Professor of Physics and Professor Emeritus of Physics and Electrical and Computer Engineering in the Center for Advanced Study, University of Illinois at Urbana-Champaign (UIUC).

Gregory Radick was educated at Rutgers and Cambridge. Currently he is Professor of History and Philosophy of Science at the University of Leeds. His books include *The Simian Tongue: The Long Debate about Animal Language* (2007) and, as co-editor with Jonathan Hodge, *The Cambridge Companion to Darwin* (2009).

David Robb teaches philosophy at Davidson College. He is interested in the philosophy of mind and metaphysics, especially mental causation, free will, and the nature of properties.

Michael Silberstein is Professor of Philosophy at Elizabethtown College, director of the interdisciplinary cognitive science program at Elizabethtown, and affiliated faculty in the Philosophy Department at the University of Maryland, College Park, where he is also a faculty member in the Foundations of Physics Program and a Fellow on the Committee for Philosophy and the Sciences. His most recent book is *Beyond the Dynamical Universe: Unifying Block Universe Physics and Time as Experienced* (2018).

Susan Stepney is Professor of Computer Science and Director of the York Cross-disciplinary Centre for Systems Analysis at the University of York, UK. She is Vice-president of the International Society for Artificial Life. She has a first degree in Natural Sciences (Theoretical Physics) and a PhD in Theoretical Astrophysics. Her current research interests include the study of emergent properties, agent-based modelling of complex systems, artificial chemistries, and open-ended evolution.

Iorwerth Thomas was raised in Carmarthenshire, Wales, and graduated with an MPhys from Swansea University, where he also obtained a PhD in lattice gauge theory in 2006. His research interests lie in various areas of condensed matter theory, including density functional theory, low-dimensional systems, phonons, strongly correlated systems, and their applications. He is currently working on calculations of the thermal conductivities of transition-metal dichalcodenide nanocomposites using effective medium theories at the University of Exeter.

Robert Van Gulick is Professor of Philosophy and Director of the Cognitive Science Program at Syracuse University. He has published on a wide variety of topics in the philosophy of mind with a special focus on problems of consciousness.

Jessica Wilson is Professor of Philosophy at the University of Toronto (Scarborough and School of Graduate Studies). Her research focuses on metaphysics, philosophical methodology, and epistemology, with applications to philosophy of mind and science. In 2014 she was a co-recipient of the Lebowitz Prize for Philosophical Achievement and Contribution. Wilson's book, *Metaphysical Emergence*, is forthcoming with Oxford University Press.

Jason Winning is a PhD candidate in philosophy and cognitive science at the University of California, San Diego. He completed a BS in computer science and an MA in philosophy at Northern Illinois University. His research focusses on the theoretical foundations and interrelatedness of agency, information, and the structure and organisation of reality.

Hong Yu Wong is Professor of Philosophy and Head of the Philosophy of Neuroscience group at the Department of Philosophy and the Werner Reichardt Centre for Integrative Neuroscience, University of Tübingen. His research interests are in the philosophy of mind and cognitive science, metaphysics, and aesthetics. He has written on aspects of emergence in the special sciences.

Christian Wüthrich is Associate Professor of Philosophy at the University of Geneva. He works in philosophy of physics, general philosophy of science, and metaphysics, with a research focus on the philosophy of quantum gravity and the emergence of spacetime from underlying non-spatiotemporal physics.

Julie Zahle is a Research Fellow in the Department of Philosophy at the University of Bergen. Her research focuses on various topics in the philosophy of the social sciences, including the individualism–holism debate, theories of practice, qualitative methods, values in social science, and the philosophy of anthropology and sociology.

ACKNOWLEDGEMENTS

We would like to express our thanks to the following people for stimulating our interest in the subject of emergence and for their help in assembling this volume. The handbook grew from the activities of the Durham Emergence Project, and we are indebted to all members of the project and to participants in the project's meetings and conferences. The project was funded by The John Templeton Foundation (Grant ID 40485), and we gratefully acknowledge their support. The opinions expressed in this publication are those of the authors and do not necessarily reflect the views of the John Templeton Foundation. Research workshops during the academic year 2014–2015 were supported by the Arts and Humanities Research Council (project title 'Emergence: Where Is the Evidence?' grant reference AH/M006395/1) and Durham University's Institute of Advanced Study. We also thank the following individuals: Adam Johnson from Routledge for his help and understanding; and Stephen Blundell, Alex Carruth and Peter Vickers for their comments on an earlier version of the introduction. Finally, we acknowledge the great effort of each of the authors of the chapters in this book. We thank them for accommodating our many requests and for their patience and support throughout the long process of compiling the handbook.

INTRODUCTION

Robin Findlay Hendry, Sophie Gibb and Tom Lancaster

Emergence is often described in terms of the slogan 'the whole is greater than the sum of the parts'. We assemble a model building from building blocks. If we understand the properties of the building blocks, then we might expect that the finished building possesses properties that are understandable from the things we know about the blocks and how they fit together. However, if it turns out that the building has some new feature that does not follow from what we know about the blocks, then that feature might be called emergent. Many things are at stake when asking questions about emergence. Do we believe that there are barriers to deriving the laws governing complex systems? Do we think that there is more to complex systems than simply the complication of their large size? Do we believe that the mind is only the result of a collection of biological circuitry, or is it something more or different? This volume addresses these questions.

To claim that a thing is emergent involves asserting something about the relationship between that thing and its more fundamental parts. For instance, although a thing is dependent on its parts (that is, it could not exist without them), it is also novel with respect to them; it is something new and distinct. As we will see, exactly how we should elaborate on such claims is a matter of much debate. Another way of describing emergence is as a failure of reduction. Reduction implies that some relatively complex phenomenon can be explained in terms of some simpler phenomenon. New fields of knowledge seem to specify how high-level entities are composed. A reductionist might argue that since high-level entities like molecules are made of more fundamental particles governed by the laws of quantum mechanics, then chemistry should be reducible to quantum mechanics. In contrast to this, the emergentist might hold that an appeal to lower-level phenomena will be inadequate since some phenomena cannot be explained by reduction. The challenge for supporters of emergence is to make clear what is added as we move up through these levels of reality and in what way higher-level phenomena do not merely result from the interactions of lower-level entities.

Not all descriptions of emergence are the same. One way of dividing emergence is between those phenomena that emerge because of the way we represent or gain knowledge about the world (epistemic emergence) versus the emergence of genuinely existing new entities (for example, objects or properties) or new kinds of causes (ontological emergence). Epistemic emergence is relatively uncontroversial: we often accept that it is convenient to represent the world in certain ways, but also that the most convenient representation might involve fictional entities. Ontological emergence is different and implies that the higher level will contain entities that are just as real

1

as those found at the lower level. Their reality might manifest itself by the higher level being able to act downwards to affect the lower level. Of course, one might also object to talk of levels altogether, and instead choose to frame the discussion in terms of emergence occurring at different scales or at different levels of abstraction. Another way of dividing up the subject is into strong and weak emergence. Sometimes this division aligns with the distinction between epistemic and ontological emergence so that strong emergence implies that emergent entities exist, and weak emergence implies that emergent entities arise from the way we gain knowledge and represent it. However, strong and weak emergence can also be thought of as categories that can sub-divide epistemic and ontological emergence themselves.

Emergence, then, is a way of characterising relationships between complex entities and their parts, relationships between the sciences and the place of life and mind in the physical world. In the last ten years it has once again become a major focus of interest among scientists and philosophers, but it remains the subject of entrenched disagreements over several issues. First, emergentists and reductionists disagree on whether composite objects can have properties and behaviour that go beyond those of their basic constituents. Second, they typically disagree over whether advances in science support emergence or reduction. Emergentists see science as revealing the complexity of the world and the diversity of its laws. Reductionists see science as undermining emergence by failing to uncover clear examples, thus supporting a reductionist view of the world in which there is unity underlying the apparent complexity: just a few kinds of things, governed by a few simple laws, which provide the ultimate basis for everything that happens.

This volume aims to investigate philosophical and scientific characterisations of emergence and carefully examine the evidence for its existence in different domains. We will address the following questions: How is emergence defined in different areas of science and philosophy? How *should* it be defined? Is it a coherent possibility? Is it genuinely interesting or a mere triviality? What could count as evidence for it, and which way does the evidence point? We have brought contributors together from metaphysics, philosophy of mind, philosophy of science, physics and other empirical sciences. This allows us to draw on important developments across all these areas.

Our task is complicated by the diversity of intellectual contexts in which the idea of emergence has been applied: it has a long history, and it is employed by both philosophers and scientists of different kinds. It began as a way of characterising the difficult relationship between the mental and the physical, a difficulty which philosophers have engaged with at least since Descartes. Emergence was first conceived as a way of placing the mind *in* nature. Descartes explained the mind and the body as distinct substances, each of which could be conceived to exist without the other. This made the mental appear superadded, which in a sense it was. The early emergentists argued that there is only one kind of stuff (the physical), but that the mental is a distinct way that some physical entities can be or behave (see McLaughlin 1992). Mental properties *arise from* physical stuff. Although some philosophers (including McLaughlin) have argued that the achievements of twentieth-century science (such as, for instance, the unification of physics and chemistry) make such a view scientifically implausible, scientists have (since the 1920s) used the notion themselves to explain the perceived autonomy of the sciences of life and (since the 1970s) of the behaviour of certain complex systems in, for example, condensed matter physics. Does this mean that the philosophers and the scientists are using the same word to mean different things? The sheer variety of different ways in which the term is used supports this view.

In the preamble to a recent book on emergence, Paul Humphreys (2016) notes that the problem is not just that different fields have used the term 'emergence'. The concept of causation, he argues, makes a telling contrast: it, too, is employed by different disciplines, but what afflicts emergence in particular is that 'we do not have a firm pre-theoretical grasp of emergence in the way that we do with something like causation' (xvii). Thus, in an 'intellectually ordered world' the term would be eliminated (xviii), to be replaced by more specific terms tied to the different contexts in which it is used. A less drastic response (adopted in practice by Humphreys himself) is pluralism: even if emergence is a somewhat amorphous notion, one might continue to use the term for importantly different kinds of phenomena while recognising that no single definition will do justice to all its uses. The hope is that one might still be able to say something informative about emergence in the abstract, balanced with critical examination of particular examples from different fields.[1] This handbook proceeds with that hope in mind.

Foundations of emergence

The first section of the book, comprising twelve chapters, concerns the foundations of emergence. How should emergence be characterised? One must answer that question before undertaking any investigation of whether emergence exists in nature. Moreover, this foundational task only becomes more pressing if there is widespread disagreement about how to characterise emergence. One plausible approach is to go to the historical source, by considering the theories advanced by those who first introduced the idea of emergence. Brian McLaughlin does just that in a classic article (McLaughlin 1992) and also in his contribution to this volume. We then turn to some key concepts or issues that have been associated with particular characterisations of emergence, to each of which is devoted a chapter. The first six of these address emergence in the abstract, discussing concepts or issues which are widely employed in both the philosophical and the scientific traditions. First comes dependence (discussed by Paul Noordhof), then fundamentality (Kerry McKenzie), reduction (John Bickle), and functional realisation (Umut Baysan). Next come three chapters concerning the connections between causal autonomy and emergence, with discussions of Alexander's Dictum by Alex Carruth; downward causation by Carl Gillett; and the causal completeness of the physical, by Sophie Gibb. We then turn to three chapters devoted to emergence as it arises within some sciences, but not all of them: these are Mark Pexton's critical examination of information-theoretic approaches to emergence, Jason Winning and William Bechtel's argument that a focus on patterns allows a distinct kind of emergence to be identified and Robert Bishop and Michael Silberstein's discussion of how the notions of complexity and feedback can be used to characterise emergence. The relationship between the different approaches to emergence taken by philosophers and scientists is a theme that runs through the whole volume: we end the section with Jessica Wilson addressing this issue directly.

As already noted, the notion of emergence was first introduced into philosophy as a way of placing the mind in nature. Brian McLaughlin's contribution to his volume examines British Emergentism, a philosophical tradition that began with John Stuart Mill and ended with C.D. Broad, whose book *The Mind and Its Place in Nature* (Broad 1925) is 'the last major work' (30) in the movement.[2] The British Emergentists, argues McLaughlin, proposed what is now called a 'layered view of reality' to explain the apparent disunity of the sciences, which arises from 'the way the natural world itself is' (24). This contrasts with the reductionist explanation of disunity, according to which autonomous special sciences arise as a result of cognitive, conceptual or mathematical limitations that prevent scientists from perceiving or exposing the metaphysical relationships that (in their view) connect higher-level entities, facts or laws to the more fundamental entities, facts or laws from which they arise. For the British Emergentists, special sciences exist because in some complex physical situations new properties arise which are subject to laws that cannot *even in principle* be deduced from the laws governing the fundamental physical entities whose interactions (ultimately) give rise to those situations. The laws connecting these new properties with the complex physical situations in which they arise have no deeper explanation and so must be considered fundamental. It is worth noting here that Broad distinguished between two kinds of emergent law (see Broad 1925, 50–52; Hendry 2006, 177–180). 'Trans-physical' laws concern essentially macroscopic properties, such as colours and temperatures. 'Intra-physical' laws concern emergent properties which might in principle be instantiated at lower levels, but which are in fact mechanistically inexplicable: Broad's example is breathing, which is a 'particular kind of movement' (1925, 80). The fact that trans-physical laws are emergent is in some important sense necessary; while the fact that intra-physical laws are emergent is correspondingly contingent.

McLaughlin gives us a detailed and sympathetic account of British Emergentism, defending it against accusations of incoherence levelled by such prominent contributors to the philosophy of mind as Jaegwon Kim. McLaughlin does, however, argue that British Emergentism fails, though for different kinds of reasons in different candidate cases of emergence. The claim that there is emergence in chemistry fails, according to McLaughlin, because quantum mechanics provides general physical laws – as expressed, for instance, by the Schrödinger equation – which it is plausible to think *do* entail every chemical fact, even if scientists cannot in practice perform the corresponding derivations. In other cases, there may be no such physical 'theory of everything' on the horizon, but it may be possible to provide functional reductions of special-science properties. A law would then link such a special-science property with its physical realiser, doing the same job as Broad's trans-physical laws but without being fundamental. What of qualia, the subjective qualitative character associated with experiences of certain kinds? Some philosophers who are sympathetic to reductionism (for instance, Kim 2005), see them as presenting insurmountable difficulties to the reductionist project. This makes them the likeliest application of key British Emergentist claims, but even here McLaughlin seeks to explain their anomalous nature by appealing to our conceptual limitations rather than the metaphysics of emergence.

Emergence is widely associated with dependence, and not just by the British Emergentists. This might be because, in a broad sense, emergence involves something 'new' arising out of something else, perhaps showing some degree of independence or autonomy from that from which it emerged (its emergence base), or even the power to affect or constrain it (sometimes called 'top-down' or 'downward' causation). The emergent depends on that from which it emerges in the sense that it cannot exist without it. Emergence in this view might simply be characterised as a combination of a dependence relation of some kind with the *failure* of a dependence relation of another kind. The detailed work is then to identify the kinds of dependence and independence involved, which might vary in different cases of emergence. Thus, for instance, Broad held that emergent properties (which, he argued, included both the chemical and the mental) supervene on physical properties, in the sense that once the latter are fixed, so are the former (see McLaughlin 1992; Hendry 2006). The Latin etymology of the term 'supervenience' suggests that it should be a term whose meaning is close to that of 'emergence', a 'coming along on top of' something more basic. Since the early twentieth century, however, the term has taken on a rather specific technical meaning for philosophers: it is a kind of consistency constraint between groups of properties, which may hold across difference possible worlds.[3] But this kind of covariation might be produced in at least two ways: on a 'mechanist' (or reductionist) view, fundamental physical laws bring about instantiations of (for instance) mental properties, while for the emergentist, emergent properties are connected to their physical emergence bases by autonomous laws of nature which are not determined to hold by the more basic laws governing the more fundamental physical parts of the systems which bear such emergent properties.

Paul Noordhof provides an overview of different attempts to characterise dependence and its role in understanding emergence and some of the difficulties they face. On a purely modal account, for instance, one might say that X depends on Y just in case X could not exist unless Y did (or if X and Y are properties, X could not be instantiated unless Y is). Such accounts face well-known difficulties posed by an example due to Kit Fine (1995) and discussed by Noordhof:

in any possible world in which Socrates exists, the singleton set containing him also exists *and vice versa,* but then Socrates and his singleton set depend on each other symmetrically, if 'dependence' is defined as noted earlier. Hence, the problem: for it is also plausible to suppose that in some important sense the singleton set depends on Socrates, but not the other way round. Noordhof also presents counterexamples to the simplest attempts to characterise supervenience in purely modal terms. One might respond to these difficulties by adding to purely modal accounts further conditions involving other metaphysical notions such as identity, essence or grounding. Such 'modality plus' approaches solve the problems which they were designed to address, but raise new ones of their own. Moreover, even if the dependence between Socrates and his singleton set cannot be captured in purely modal terms, such cases 'don't undermine the status of a purely modal account as one kind of dependence relation' (48). Noordhof concludes by defending purely modal accounts of some dependence relations, such as that between the mental and the physical in various positions in the philosophy of mind, and advocating a focus on construction (broadly, how a thing is made or constituted) in others.

Closely related to dependence is fundamentality, which, as Kerry McKenzie notes, seems to be involved in understanding emergence in two ways. On the one hand, emergents are said to be non-fundamental because they arise out of or depend on something else. On the other hand, in the respects in which they represent novelty or autonomy, emergents are fundamental: we have already seen that Broad held his trans-physical laws to be fundamental, in the sense that they are not entailed by the laws governing their emergence base. But what is fundamentality? In the philosophy of science it has long been associated with Oppenheim and Putnam's hierarchical view of the sciences, ordered by part–whole relationships (Oppenheim and Putnam 1958). According to McKenzie this is 'at best, a heuristic and a metaphor' (56) and of dubious completeness: the place of cosmology in this hierarchy, for instance, is rather unclear because the structures it studies are not obviously composed of structures at smaller scales in the way that (for instance) molecular or cellular structures are. It also fails to capture many debates concerning ontological priority: metaphysicians are often concerned with priority relations between categories (such as objects and relations), while Oppenheim and Putnam's hierarchy concerns only '*intra-*categorical' (56) relations of priority (that is, between the entities and laws of different sciences); philosophers of physics debate the relative merits of relationism and substantivalism, but spatiotemporal relations and space-time must exist '*wherever in the levels hierarchy spacetime is*' (56). One might doubt whether a 'one-size-fits-all' (57) relation of priority is what is involved in all these debates. This worry deepens when considering the debates in *a priori* metaphysics concerning the formal features of fundamentality and its alignment with such closely related notions as dependence, determination and grounding. One issue that reveals the different approaches of metaphysicians and philosophers of science most starkly concerns the existence of the fundamental and of the non-fundamental. On the one hand, it is 'a surprisingly popular view among metaphysicians that no non-fundamental entity could ever have the status of *existent*' (61), while on the other hand, 'the more fundamental an entity is, the more remote from our experience it is likely to be, and – the history of science counsels us – the less confident we should correspondingly be about its existence and nature' (61).

It is a commonplace of both scientific and philosophical traditions that emergence and reduction are opposed: emergence is even sometimes just non-reducibility,[4] so no discussion of emergence would be complete without an exploration of different conceptions of reduction. This is provided by John Bickle, who argues that models of reduction should resonate with 'actual current scientific practice' (70). Against the current received wisdom, he argues that two stalwarts of twentieth-century philosophy are more nuanced and typically admit weaker (and hence more plausible) formulations than is often supposed. These are 'term/sentence' reduction, associated with positivism, in which, for instance, discourse about the mental is replaceable by neutral (and physicalistically acceptable) language, and Ernest Nagel's 'theory reduction' in which the laws of a reduced theory (or the observational data they explain) are derivable from the reducing theory.[5] Nagel distinguished between homogeneous and heterogeneous reductions: in the former, one theory is deduced from another theory that is couched in the same vocabulary; in the latter, the two theories are not expressed in the same language, and so the deduction requires 'bridge laws' relating the vocabularies of the two theories. Nagel's view of reduction allows more subtlety than is often supposed, because there is some leeway in how bridge laws are interpreted. If they are mere conventions, then the fact that a reduction can be achieved may not be very significant from the point of view of either science or metaphysics. Bridge laws whose truth reflect metaphysically more interesting identities allow metaphysically more interesting reductions.[6] The usual candidates for such interesting cases are 'temperature is mean kinetic energy', 'water is H_2O' and 'pain is c-fibres firing', though each of these is highly controversial. Bickle argues that term/sentence reduction and Nagelian reduction both find counterparts in real scientific projects. He is less sanguine about 'functional reduction', which is the starting point of many anti-reductionist arguments concerning qualia: this is a 'philosopher's fiction' (70). More permissive models that focus on explanation and mechanisms are, he argues, more in tune with actual research programmes in science.

Although it clearly constitutes a form of autonomy, multiple realisation has traditionally been associated with non-reductive physicalism (NRP) rather than emergentism. NRP is the claim that everything is in some sense physical but that the subject matter of special sciences (including the mind) is in some important sense irreducible to that of physics. Unlike strong emergentism, NRP is widely understood to be, at best, agnostic on whether higher-level sciences introduce novel causal powers, where a causal power is the ability a thing has to bring about an effect in other things. Umut Baysan therefore sets out to provide a detailed analysis of the differences between realisation and emergence, though he acknowledges that the two kinds of positions 'can overlap' (79). He notes four dimensions of difference:

1   Strong emergentism is widely understood to be incompatible with the physicalist claim that nothing exists 'over and above' the physical. In contrast, realised properties are often taken to be 'ontologically innocent' (79): no addition to being over and above their realisers.[7]

2   If X realises Y, X must be more fundamental than Y. In contrast, emergentists see higher-level properties as fundamental.

3   While NRP is committed to the *metaphysical* supervenience of higher-level properties on fundamental physical properties, emergentists typically require only *nomological* supervenience as expressed, for instance, in Broad's trans-physical laws.

4   Baysan acknowledges that it is controversial whether or not novel causal powers can be associated with realised properties: Kim's 'causal inheritance principle' assumes not, but temperamental anti-reductionists have argued that the truth of this principle is not intrinsic to the idea of realisation.

NRP itself is also open to criticism: if it does not allow higher-level properties a distinct causal role, isn't it just reductionism in all but name?

The next three chapters turn to emergence and causation, the connection between which we have already seen expressed in three central claims: that strong emergence requires novel (or distinct) causal powers, that emergent wholes exert 'downward causation' on their parts and that strong emergence is incompatible with the causal closure of the physical. Alex Carruth critically examines the role of 'Alexander's Dictum' (Alexander 1920) in articulating the disagreement between strong emergentists and their reductionist opponents. According to Alexander's

Dictum, exerting novel causal powers is a necessary condition for emergents to constitute distinct existences: something over and above that from which they emerge. Making a contribution to the causal structure of the world may not be necessary in every case of existence: if abstract entities lack spatiotemporal location, they can hardly be expected to bump into lamp posts. Yet it seems at least possible that one might have reasons to believe that some abstract entities exist, for instance, via Quinean indispensability arguments for the existence of mathematical objects. However, causal efficacy is a plausible requirement when restricted to concrete entities, Carruth argues. Thus, Alexander's Dictum, when carefully formulated, can be a focus for the debate over strong emergence, especially because, as a criterion for ontological commitment, it can be a neutral point of agreement between emergentists and their reductionist opponents.

The idea that some parts can come together and form a whole which in some important way modifies or constrains the behaviour of those very same parts is associated both with British Emergentism and the more recent emergentist tradition in physics and elsewhere (Ellis 2012). Carl Gillett casts a sceptical eye over various attempts to articulate this idea. He rejects the term 'downward causation', arguing that using the *intra*-level relation of causation to explicate an *inter*-level relation is a strategic mistake: causation is supposed to hold only between *distinct* individuals (this was certainly Hume's assumption). Wholes and their parts are 'compositionally related, rather than being wholly distinct' (100), so to understand what he calls the 'foundational determinative relation' (99) involved in emergence, philosophers and scientists should look elsewhere. What is sought is a 'sui generis relation different from both causal, and compositional relations', but 'substantive work' (109) needs to be done in showing that such a relation can be instantiated – that it is a coherent possibility – and that it is instantiated in concrete cases.

The focus of Sophie Gibb's chapter is the causal closure of the physical. Strong emergentism is committed to two claims: first that emergent properties are distinct from physical properties (and, hence, that emergent causes are distinct from physical causes), and second that emergent properties are causally relevant in the physical domain. Put together, these commitments bring the position into conflict with the supposed principle of the causal closure of the physical. How could emergent entities have (non-redundant) physical effects if the causal closure principle is correct? After setting out the causal closure argument, Gibb considers various formulations of the causal closure principle, of which there are many, exhibiting different strengths. Some formulations of the principle simply appeal to the notion of causal *completeness*: roughly, the notion that every physical effect has a sufficient physical cause. Hence, nonphysical entities never *need* to appear in a complete causal account of the physical world: they would be redundant. However, stronger versions of the causal closure principle, such as the principle that 'No physical effect has a non-physical cause' directly exclude the very possibility of non-physical causes in the physical domain. On the one hand, a causal closure principle must not be so weak that it renders the causal closure argument invalid. On the other hand, it must not be so strong that it has no empirical support or begs the question against emergentism and other forms of interactive dualism. Whether proponents of the causal closure principle can meet this challenge is the focus of Gibb's chapter. Her conclusion is that they cannot.

The next three chapters in this section concern the relation between the concept of emergence and various notions which find applicability in some of the empirical domains in which the term 'emergence' has come to be used, but not necessarily in all of them: information, pattern and complexity. In the pluralist spirit of this volume we assume that we can learn something from the domain-specific conceptions of emergence that are developed using these notions, without trying to apply what we have learned to all candidate cases of emergence.

There is a long association between prediction and emergence. For the British Emergentists, the behaviour of an emergent could not *in principle* be predicted even by a being with unlimited cognitive abilities to make predictions and who can access all the relevant information.[8] This kind of non-predictability has continued to be associated with a kind of non-reducibility, but one foundational question is how far such considerations can motivate a 'merely epistemic' or weak emergence that applies only to the perspective of epistemically limited beings. First, there are objective limits to the deductive strength of axiomatic systems (expressed, for instance, in Gödel's incompleteness theorems). Even if laws of nature exist governing the most fundamental parts of reality which, in some sense, determine the course of every event, these laws, and the information needed to apply them, may not be recursively axiomatisable. Second, deduction takes time: physical laws may not allow that a computer could calculate, within the lifetime of the universe, the course of events using information only at the most fundamental level. In either case empirical knowledge of non-fundamental entities and processes, which are typically studied by special sciences (and also by non-fundamental parts of physics), would become indispensable in predicting the course of events. Moreover, it would seem too quick to bracket this kind of emergence with 'merely epistemic' or weak emergence. Mark Pexton's chapter offers a critical overview of just these issues, also considering the prospects of 'pancomputational emergence', which 'identifies higher-level aggregates as necessary for the universe to "calculate" the behaviour of complex systems' (121). The scare quotes around the word 'calculate' indicate the metaphorical nature of the suggestion as proposed, and one central task in articulating pancomputational emergence is to unpack that metaphor.

Jason Winning and William Bechtel offer a distinction between 'being emergence' and 'pattern emergence'. Being emergence has dominated discussions of emergence in the philosophical tradition, including those of the British Emergentists. Assuming a mechanistic perspective in which less fundamental entities are related to more fundamental entities by constitution or composition, debates concerning being emergence (and its alternatives, such as ontological reducibility) concern how far, and in what sense, more fundamental entities fully explain or determine the being of the less fundamental entities they ground. In contrast, pattern emergence concerns how far patterns of behaviour, which may be considered in abstraction from their material constitution, are generated and explained by more fundamental patterns. Seeking a more detailed way of developing this idea that would explicate the emergence of control and goal directedness in the biological sciences, they settle on Howard Pattee's account in terms of the complementarity of rate-dependent and rate-independent features of biological processes.

According to the British Emergentists, emergent properties can arise only when the systems that bear them reach a certain level of complexity. Their discussions, however, were innocent of the technical sense of the term 'complex system' which became widespread across a number of empirical sciences in the second half of the twentieth century. Robert Bishop and Michael Silberstein's chapter explores this technical notion through close attention to scientific examples. Sometimes, complexity is primarily a feature of a system's behaviour (like Winning and Bechtel's pattern emergence), a complexity which may be produced by a relatively simple classical system such as three billiard balls on a torus. Sometimes, in contrast, complex interactions among the parts of a system give rise to relatively simple (i.e. predictable) behaviour, as for instance, in the case of Rayleigh-Bénard convection (RBC), which figures throughout their discussion. The emphasis on organisation and dynamical interaction, it should be noted, is far removed from the talk of 'arrangements' appearing in discussions of composition and constitution by philosophers who do not engage with empirical science. Bishop and Silberstein elect to stay close to the scientific examples, which include RBC, quantum entanglement, Belousov-Zhabotinsky reactions and various kinds of feedback and control in the biological sciences. On this kind of descriptive approach, 'complexity' may track a number of different features that give rise to emergence in quite different ways.

The last chapter of this section, by Jessica Wilson, addresses explicitly a theme running implicitly throughout this book and arising particularly in the last three chapters: the way that discussions of emergence typically involve both detailed consideration of scientific examples and the attempt to extract general lessons from them. Wilson addresses the contrasting advantages and disadvantages of two corresponding strategies one might take in characterising a metaphysically interesting notion of emergence: scientism and abstractionism. On a scientistic approach, we commit to 'extracting features of emergence from close consideration of scientific case studies' (160). In contrast to scientism is abstractionism, which characterises emergence in 'terms floating free of scientific notions' (158). Each approach tends to fail one of the two requirements Wilson proposes for accounts of metaphysical emergence: that being emergent should rule out reduction and that it should do so in an illuminating way, explaining how failures of reduction are possible. Wilson goes on to argue that 'substantive' approaches, characterizing emergence in powers-based terms which are both scientifically informed and metaphysically contentful, can jointly meet these constraints.

Emergence and mind

The thought that beliefs and desires can give rise to bodily movements is central to our pre-theoretical notion of human agency. My desire to raise my hand, for example, seems to bring about my hand's raising. And, had I not had this desire, in normal circumstances, I would not have raised my hand. However, identifying a plausible relationship between mental and physical entities that is consistent with psychophysical causation is problematic.

Until recently, the general assumption in the philosophy of mind was that for mental entities to cause physical entities, some sort of physicalism – the doctrine that all entities are identical with or, in some sense 'nothing over and above' physical entities – must be correct. This is primarily because of the generally accepted claim that the physical domain is causally closed – that is, the claim that every physical effect has a direct, sufficient physical cause. The causal closure principle appears to render *sui generis* mental entities causally redundant in the physical domain, and thus any form of interactive dualism ultimately implausible. However, in light of issues regarding the tenability of both reductive and non-reductive physicalism, together with more critical examinations of the causal closure principle, this consensus is now being challenged. As a consequence, emergent dualism – a form of interactive dualism – is gaining increasing attention in the philosophy of mind. According to emergent dualism, mental entities are dependent yet novel entities. Mental entities emerge from – they arise out of – more fundamental physical entities, and consequently they are dependent entities – their existence depends on the existence of physical entities. Yet despite being dependent on physical entities, they are novel with regard to them. They are novel because they have novel causal powers – that is, full-blooded causal powers that cannot be reduced to the causal powers of the physical entities from which they emerge. Because emergent mental entities arise out of physical entities, a naturalistic account of mental entities can be accepted. It is therefore considered to be a more attractive form of interactive dualism than the type of interactive substance dualism typically associated with Descartes.

If emergent dualism is correct, then either the causal closure principle must be rejected or a version of emergent dualism must be advanced that is harmonious with a causally closed physical world. How should the emergent dualist respond to this challenge? This issue is but one of several interrelated issues facing an emergent theory of mental entities. Are the characteristics of emergent dualism compatible? Are certain versions of emergent dualism nothing other than disguised versions of physicalism? What evidence is there for emergent dualism? Does emergent dualism allow one to explain central features of the mind? Is the framework of the debate in which emergent dualism is located – that is, the mental causation debate – actually misconceived? These are but some of the questions explored in the following chapters.

Hong Yu Wong's chapter sets out the framework of the emergence debate in the philosophy of mind. Distinguishing between epistemological (weak) emergence and ontological (strong) emergence, he explains that the concern of his chapter is with the latter – emergent dualism is supposed to be in opposition to physicalism, but the epistemological variety is not. Wong goes on to distinguish and discuss two forms of ontological emergence found in the contemporary mental causation debate: supervenience emergentism and causal emergentism. Common to both positions is the view that the relation between emergent properties and basal properties holds as a matter of nomological (i.e. lawful) necessity. According to supervenience emergentism, this is because there are fundamental emergent laws that guarantee supervenience. According to causal emergentism, this is because emergent entities are caused by basal properties. As he explains, one of the central problems with supervenience emergentism is whether it is in fact in opposition to physicalism, because supervenience emergentism and non-reductive physicalism arguably share the same basic commitments. Wong's chapter concludes with a discussion of three challenges facing emergent dualism: the causal closure of the physical domain, the lack of positive empirical evidence for emergent dualism and the availability of rival positions in the mental causation debate that undercut the motivation for emergent dualism.

David Robb uses Sperry's characterisation of emergent mental causation as a framework for his discussion (Sperry 1980), according to which emergent mental causation is novel, dependent, harmonious and unexceptional. Mental causation is novel because, as Robb puts it, mental entities 'have novel *causal powers*, powers that cannot be found among (or reduced to) powers' of the entities from which they emerge (p. 187 this volume). Despite mental entities having novel causal powers, mental entities are dependent on the entities from which they emerge. Emergent mental causation is harmonious because it does not disturb or intervene in lower-level causal processes. It is unexceptional, because emergent mental causation is just one instance of emergent causation. What evidence is there for such emergent mental causation? Robb argues that there is a model of emergent mental causation, grounded in the plausible thesis of psychological autonomy, that displays these four characteristics and, hence, which provides evidence for emergent mental causation. Pursuing the question of what further evidence there is for emergent mental causation, Robb proceeds to a discussion of the 'spaces of reasons' (Sellars 1956) and consciousness. However, throughout his discussion of the evidence for emergent mental causation, Robb's central concern is whether the four characteristics of emergent mental causation can be made compatible – in particular, whether novelty can be reconciled with harmony. For, of course, even if there is apparent evidence for emergent mental causation, if the four characteristics of emergence are incompatible, then it is an untenable position. Hence, Robb's chapter concludes with a discussion of the ways in which one might attempt to reconcile novelty and harmony.

Cynthia Macdonald and Graham Macdonald's chapter focuses on non-reductive physicalism – the orthodox position in the mental causation debate – and why tension within non-reductive physicalism has led to increased interest in emergent positions in current philosophy of mind. Non-reductive physicalism can, Macdonald and Macdonald explain, be formulated in terms of the combination of a token identity claim (each individual mental event or phenomenon is identical with a physical one) and a type distinctness claim (mental properties or kinds are distinct from physical ones). The non-reductive physicalist's primary motivation for maintaining a type dualism is the argument from multiple realisability – the argument that mental properties are multiply realised by and, hence, cannot be identical with, physical properties. The tension inherent in non-reductive physicalism results from the combination of its commitment to physicalism, irreducibility and psychophysical causation. Macdonald and Macdonald's chapter first sets out the main objections to non-reductive physicalism and the various responses that proponents of non-reductive physicalism have offered. The chapter then turns to a discussion of a weaker version of non-reductive physicalism which is not committed to a token identity thesis, but which instead relies on a notion of physical realisation. However, as Macdonald and Macdonald argue, these versions of non-reductive physicalism are no less vulnerable to charges of epiphenomenalism. The refusal to abandon neither irreducibility nor psychophysical causation has caused some of those in the mental causation debate to turn their attention to emergence.

Turning from discussions of the framework of the emergence debate in the philosophy of mind, Lynne Rudder Baker's focus is on intentionality and the ways in which emergence and intentionality might be related. Intentionality is a central feature of the mental and is 'the property of being about, or representing, or being directed toward something. For example, my promising to call you tomorrow is about calling you tomorrow. My remembering that Vienna is in Austria is about Vienna's being in Austria' (p. 206 this volume). After a discussion of the notion of intentionality and the features that individuate it, Baker addresses the question of whether intentional properties are emergent. Baker considers four possible responses to this question. First, intentional properties are not emergent because intentional properties do not exist. Second, there are emergent properties, but they are only weakly emergent – that is, they are not ontologically significant, introducing no new causal powers. Third, intentional properties are strongly emergent – they carry ontological weight and introduce new causal powers. The fourth response is to maintain what Baker refers to as 'ontological dualism'. Baker defends the third response.

In contemporary philosophy of mind, much of the focus on emergence concerns consciousness. How could an explanation of consciousness be given that is consistent with physicalism? Robert Van Gulick's chapter is specifically concerned with the question of whether the notion of emergence enables us to understand the nature and basis of consciousness. It is widely agreed that the specific problem of how to explain consciousness as part of the physical world is a particularly difficult one in the philosophy of mind. Does an appeal to emergence help? And if one maintains an emergent account of consciousness, what are the implications for physicalism? Van Gulick observes that this depends on how the concepts of 'consciousness', 'emergence' and 'physicalism' are defined. Consequently, the first section of Van Gulick's chapter is devoted to a discussion of consciousness, the second to a discussion of emergence and the third to a discussion of physicalism. Van Gulick distinguishes between different strengths of both epistemological emergence and ontological emergence. His conclusion is that while most forms of consciousness are weakly ontologically emergent, most forms of consciousness (with the possible and debatable exception of phenomenal consciousness) are not strongly ontologically emergent. Regarding epistemological emergence, most forms of consciousness are to some degree epistemically emergent, and some forms of consciousness, especially phenomenal consciousness, might also be strongly epistemically emergent. However, to claim that some forms of consciousness are strongly epistemically emergent is not thereby to reject physicalism, because this is consistent with a non-reductive physicalist's framework.

Finally, John Heil's chapter addresses the question of how to reconcile conscious experiences with physics, with a particular focus on emergentism and panpsychism. According to emergentism, consciousness is an emergent phenomenon – a non-physical add-on to the physical universe. Contrary to this, according to panpsychism, consciousness is not an add-on to the physical universe, but is instead present within the fundamental physical domain. Heil considers neither the option of emergence nor the option of panpsychism to be plausible, arguing that the puzzle of consciousness is of our own making and is reliant on implicit, but often implausible, assumptions. The most important of these assumptions is that the mental-physical distinction is a distinction, not merely in conception, but a distinction in reality. Upon metaphysical scrutiny, there are good reasons to abandon this assumption, and once we do so, the contemporary discussion in which emergentism is located should be reassessed and the puzzle of consciousness dissolves.

Emergence and physics

A milestone of the recent interest in emergence in science is the much-discussed article by Philip Anderson 'More Is Different' (Anderson 1972), which includes the statement that 'at each level of complexity entirely new properties appear, and the understanding of the new behaviors requires research which I think is as fundamental in its nature as any other'. It is these new properties and behaviours that form the basis of many claims for emergence in physics and beyond. Several of the ideas in the article were developed further in the context of the sub-discipline of condensed matter physics (CMP) in Anderson's influential book *Basic Notions of Condensed Matter Physics* (Anderson 1984), which provides a useful framework for understanding the conceptual basis of many of the articles in this section of the Handbook.

The field of CMP grew from the older study of the physics of solids. The foundations for this solid-state physics, involving the application of quantum mechanics and statistical physics to crystalline metals and insulators, were laid through the work of many physicists in the early twentieth century and were set out by Frederick Seitz in an influential book (Seitz 1940). The more modern version of the subject, CMP (which arguably also owes its name to Anderson), encompasses the general study of solids and liquids, including quantum-mechanical phenomena such as magnetism, superconductivity and heavy fermion physics, through to the classical realm of soft matter, which encompasses systems such as flexible polymer chains and, more recently, extends towards biological phenomena. Despite its breadth, the subject of CMP is founded on a set of unifying concepts and theoretical tools. Many of the latter originate from the quantum field theory of high-energy particle physics, which have been absorbed into CMP since the 1950s (Anderson 1984). These deal with quantum-mechanical properties of matter and, perhaps surprisingly, also the classical statistical mechanics of soft matter. In comparison to high-energy particle physics, condensed matter phenomena occur at the limit of relatively low energies (or long length scales), but, as we shall see, emergence gives us grounds to regard both high-energy physics and CMP as fundamental in Anderson's sense noted earlier.

In the four sections of *Basic Notions*, Anderson identifies the concepts around which discussions of emergence in (condensed matter) physics has largely been organised. These are (i) broken symmetry, (ii) adiabatic continuity, (iii) quantum matter and (iv) the renormalisation group. The raw material for these central concepts can be traced back to the Russian physicist Lev Landau and his followers, working in Moscow in the post-war years.[9]

Considering the first of these, it is notable that very many arguments in physics are based around the notion of symmetry: the way that it is possible to transform a system such that its description remains unchanged. (A well-known example is the rotation of an equilateral triangle about its centre through 120 or 240 degrees, which results in an identical figure.) For solid matter the relevant symmetries might include rotational, and also translational, symmetry (the idea that the behaviour probed at some point is the same as the behaviour at some other point). The point here is that symmetry is connected to the thermodynamic phases of matter.

Different phases (solid, liquid, magnetically ordered, superconducting, etc.) are stable in particular environments, that is, at different temperatures, pressures etc., and these phases are separated by phase transitions (generalising behaviour such as melting or boiling, etc.). One special variety of phase transition separates matter with different symmetries, and it is here that we encounter a claim for emergent behaviour and properties. A liquid, on average, has complete translational symmetry: sample a liquid at different points, and you are just as likely to find the same average density of particles. A solid does not have this symmetry. The atoms in a solid are regularly placed, and so we have translational symmetry only through integer numbers of special translations known as lattice vectors. The complete translational symmetry has been broken or lowered and replaced with a more restrictive one. Anderson teaches us to look for four properties when a phase transition takes place, and it is here, again, that emergence comes into play. (i) Phase transitions are sharp: they take place discontinuously and separate regions of different behaviour. (ii) After a symmetry is broken, *new* forces emerge that act to prevent deformations of the ordered state. This effect is known as rigidity. (iii) *New* particle-like excitations emerge in the ordered state. (iv) The ordered state hosts another species of *new* particle-like state: defects, which separate spatial regions where symmetry has been broken in slightly different ways. The emergence of new forces and particles, as well as the existence of the transitions themselves, lend themselves to a description based on emergentist ideas. We can then, for example, identify rigidity as a property of a system emerging at a phase transition, or phonon excitations as emergent particles.

The physical basis of broken symmetry is explored in Stephen Blundell's chapter in the context of scaling and the renormalisation group. When we approach a phase transition, our usual notion of physics occurring at a particular scale breaks down and, at a phase transition, fluctuations occur on all length scales. Blundell describes the statistical mechanical analysis of symmetry breaking along with Anderson's argument about its stability, related also to Hund's paradox (Hund 1927). Another intriguing property of many systems at phase transitions is universality: the idea that at phase transitions, the macroscopic, thermodynamic behaviour of very different systems is described by the same mathematical parameters. The apparent independence of the macroscopic behaviour from the low-level constituents has led many to look for emergence in these cases. Blundell goes on to describe the renormalisation group: a systematic method of examining a system at different length scales. This can be seen as providing a method of coarse-graining: moving from a low-level description of a system to a higher-level one by averaging out fluctuations to identify more slowly varying quantities.

While renormalisation reveals that some details of the interactions of a system are irrelevant to its large-scale behaviour, some are revealed to be relevant. Classifying systems in terms of the relevant interactions reveals deep connections between disparate models and goes some way to elucidating the similarities between, for example, high-energy physics and CMP. Renormalisation group analysis therefore teaches us that the structure of physical field theories describing so-called fundamental physics is shared with physics on far larger length scales. We can then exploit the fact that an emergent, effective (that is, low-energy) description of the physics of solids, for example, shares the same mathematical form as is seen in the description of an often quite different phenomenon in higher-energy physics. This is, perhaps, the source of the promise of emergence in the sciences, where, in Anderson's sense, at each scale, new physics emerges that can equally well be regarded as fundamental.

In the same spirit, Tom McLeish presents the case of soft-matter physics. This is a discipline studying the classical statistical mechanics of polymeric fluids, colloidal suspensions, gels, foams and, most recently, biological systems. McLeish describes how soft matter makes use of the same models and techniques as quantum field theory, a use which is made possible by a mathematical technique known as the Wick rotation (Wick 1954), which transforms quantum models into statistical ones by transforming between a time variable and a temperature one. McLeish also examines the constraints on a system encoded in notions of topology and explores how these relate to downward causation and the possibility of strongly emergent properties. McLeish goes on to describe successes in the more recently founded discipline of biophysics and the new possibilities it presents of causal relationships between variables at quite different scales.

In the chapter on quantum-mechanical matter, Stewart Clark and Iorwerth Thomas examine how the quantum-mechanical Schrödinger equation makes links to emergent structure in physical systems. The authors' take on this is an operational one that exploits the current state of the art in computational quantum mechanics. The techniques they describe rely on a method of tackling the multi-particle Schrödinger equation known as *density functional theory*. Clark and Thomas express the view that many of the macroscopic properties of materials that derive from basic quantum processes are amenable, or will shortly be amenable, to a first-principles treatment, where only the constituent particles in a piece of matter are specified, without additional data. The success of such a project would wipe out claims to a strong form of emergence in physics.

Adiabatic continuity is Landau's other key insight into the interactions of constituents within condensed matter systems. This is an argument that shows how the properties of particles result from particle–particle interactions. The resulting emergence of particle properties is taken up in Tom Lancaster's chapter. The adiabatic continuity argument relies on a thought experiment that starts by considering a collection of non-interacting particles. We then imagine turning on the interactions and we watch how the particle properties evolve. In the fully interacting system we call the excitations *quasiparticles*. In many cases, turning on interactions does not perturb the system too dramatically, and we say that the particle properties evolve adiabatically. The changes to the properties of the particles might be regarded as being emergent, albeit in a rather weak sense. However, if the turning on of interactions forces the system to break a symmetry and experience a phase transition, then it becomes impossible to track the properties of a single particle across the phase transition. This breakdown of adiabatic continuity results in new particles with a claim to emergence in a stronger sense.

David Pines traces the recent history of the study of emergence in condensed matter physics, and beyond, through the founding of the Institute for Complex Adaptive Matter (ICAM). Pines' article *The Theory of Everything* (Laughlin and Pines 2000), written with R.B. Laughlin and identifying the quantum protectorate, re-energised the subject of emergence in mainstream physics research in the early 2000s, and the activities of ICAM were instrumental in making emergence a driving force in the methodology of modern condensed matter physics. Pines goes on to describe how the emergentist approaches routinely used in scientific research are now being applied to education, with emergence at the heart of educating young people in how to think like scientists.

Piers Coleman continues this theme, discussing the relationship between emergence and reductionism in terms of the methodology of the condensed matter physicist. In Coleman's conception, reductionist practice in CMP involves the study of the interactions between the microscopic, electronic and atomic units in the system. In contrast, emergentist practice often involves a phenomenological description of macroscopic properties and their (effective) interactions. Coleman sees these two approaches as playing an 'intertwined' role in the work of the physicist, arguing that they complement one another, forming what he sees as an 'awkward alliance', but that ultimately unites experimental and theoretical analysis. Coleman's flavour of emergence can then be seen as a consequence of reductionism, where understanding the consequences of microscopic interactions often requires both phenomenology and experiment. His thesis is illustrated with a number of profound examples from current research topics, including exotic states of matter such as topological insulators and strange metals.

Our current picture of the large-scale structure of space-time (Hawking and Ellis 1973) is based on the geometrical notion of a fundamental space-time which has a metric structure described by the general theory of relativity. In his chapter, Christian Wüthrich discusses the emergence of space and time themselves. He takes recent research in quantum gravity as suggesting that the universe might not, in fact, be fundamentally spatiotemporal, but instead that space-time emerges from a non-spatiotemporal structure. This raises serious concerns about the coherence and adequacy of theories in the field of quantum gravity. Wüthrich argues, however, that if it can be established that space-time emerges in non-spatiotemporal terms (and also how its relevant aspects are explained), then the challenge is met. To meet the challenge, Wüthrich describes a form of space-time functionalism that might offer a promising template for such a project.

Emergence and the special sciences

As we have seen, the British Emergentists saw emergence as a way of placing mind in the natural world in a manner consistent both with a broad materialism and the causal efficacy of the mental. They took as their model an emergentist conception of the relationship between physics and the special sciences. As McLaughlin notes in his contribution to this volume (see Chapter 1), they assumed (i) that emergence provides a good metaphysical explanation of the relationship between physics and the special sciences and (ii) that the available empirical evidence supports this model, against alternative (for instance, reductionist) explanations. These are both highly controversial assumptions, and it is the purpose of the chapters in this section to examine them critically in the cases of computation, chemistry and the biological and social sciences.

Susan Stepney returns to Anderson's 'More Is Different' in examining emergence in the digital, computational domain, in particular, Anderson's picture of emergence being the result of 'a quantitative change becoming a qualitative one'. In classical computers, *More* comes from the addition of *space* (memory) or *time* (number of operations). Stepney first investigates how space can contribute to emergence through the amount of memory and how that memory can be 'chunked' into higher-level concepts from simple data structures to entire virtual machine layers. She describes how some of the properties and functionality of computation can be straightforwardly reduced to the underlying structures, but how others appear to be stronger candidates for emergence, by virtue of their being global properties of the system. Stepney then turns to the concept of time by considering the effect of iteration, which involves repetition of an operation from a different starting point at each step. She shows how this can give rise to a whole zoology of candidates for emergence.

Robin Findlay Hendry's contribution takes up the case of chemistry, a discipline whose concepts and theories many philosophers consider to be prime candidates for reduction to physics, even though Broad used it as a central example of emergence. The first group of issues concerns the relationship between chemical substances and their structure at the molecular scale. Theoretical identities, such as 'water is H_2O', are sometimes taken to express an identity between chemical and physical properties, which would rule out the distinctness and autonomy assumed by emergentists. Hendry argues that liquid water is complex and heterogeneous at the molecular scale, and scientific explanations of its macroscopic causal powers (e.g. to conduct electricity) invoke statistical facts about populations of molecules and concerted processes involving organised systems of molecules. The second group of issues concerns the fact that substances display different structures at different scales of energy, length and time: it cannot be assumed that entities and processes at higher-energy scales and shorter scales of length and time causally exclude entities and processes at lower energy scales and longer scales of length and time. The third group of issues concern the chemical bond: for reductionists, the application of quantum mechanics to atoms and molecules from the 1920s onwards affected the reduction of structure. Hendry argues that explanations of molecular structure appeal to non-trivial assumptions concerning dynamic interactions between electrons and nuclei. Such explanations presume specific kinds of dynamic organisation in quantum-mechanical systems of electrons and nuclei, which can be regarded as the necessary conditions for the emergence of structure.

The next four chapters turn toward emergence and emergentism in the biological sciences. During the 1920s a distinct tradition of emergentist thinking developed within the biological sciences that overlaps with British Emergentism both doctrinally and in the person of C. Lloyd Morgan. Emily Herring and Gregory Radick provide a historical overview and critical examination of these developments. The main doctrinal connections with British Emergentism concern the motivation of the position: it saw itself as a 'third way' (353) between mechanism, according to which living organisms are nothing more than 'complex machinery' (352), and vitalism, according to which life is somehow superadded, much like a Cartesian mental substance. The empirically evident autonomy of the biological sciences would be explained in terms of a holism of a kind that is so characteristic of emergence. Yet emergentism in biology was more than simply an offshoot of British Emergentism: on the one hand, it was influenced by the process philosophy of Henri Bergson and Alfred North Whitehead; on the other hand, it developed into systems biology, which offered an organised way to resist the widespread reductionism which had been encouraged by Crick and Watson's identification of DNA as 'the secret of life'.

The focus of Michel Morange's contribution is emergence in the cell. He begins by noting the 'widespread and non-rigorous' (363) use of the term 'emergence' by biologists: sometimes 'the verb "appear" could easily replace "emerge" (363)', in which case the 'emergence' discussed will have no specific relevance to the contents of this volume. In other cases, the relevance is clear and substantial. The kind of emergence associated with systems theory (as outlined in Herring and Radick's chapter) is, he argues, an illuminating way to approach at least three kinds of cellular processes which have been investigated in detail by biologists since the 1960s: the oscillations in the concentrations of components that are central to understanding basic metabolic pathways such as the glycolytic process; Peter Mitchell's chemiosmotic model of the synthesis of adenosine triphosphate (ATP), for which he was awarded the 1978 Nobel Prize in Chemistry; and the idea of a 'cell decision' to describe 'the existence of checkpoints in (364) the cell cycle'. For Morange, the key point is that to understand these processes biologists must build models at the level of the entire cell, or supracellular structures such as organs (citing work on the heart by Denis Noble, another contributor to this volume), or even whole organisms. He concludes by tracing the influence of this work on synthetic biology.

Drawing together themes that have come up in a number of other chapters in this volume, George Ellis writes about the link between evolution, information and emergence. For Ellis, the emergence of life is based on processes that accumulate *information*, and it is the use of information that is key to how life functions. The complexity seen in life is based, Ellis argues, in modular, hierarchical structures, which use information differently at each level. The physically realised information needed for evolution comes into being over time and therefore emergence takes place with genuine and novel causal powers coming into being at each new level in a hierarchy, allowing the processes appropriate to that level to occur largely independently of the lower-level structure. The origin of information itself is proposed to come from processes of adaptive selection. Ellis concludes that the processes of information origin and usage are only possible via top-down realisation of higher-level requirements at lower levels.

Raymond Noble and Denis Noble develop a framework for thinking about emergentist explanation in science that makes explicit a number of features that have been noted in other contributions to this volume: first, they argue that emergence is real and available for empirical study by scientists; second, in emergent systems causation is 'circular' (387), with structures at higher levels constraining processes at lower levels just as structures at lower levels constrain processes at higher levels; furthermore 'the upward and downward forms of causation do not occur in sequence, they occur in parallel (i.e. simultaneously)' (387). Noble and Noble introduce the term 'a-mergence' to emphasise the 'lack of causal directionality' (387), which they then apply to the explanatory relationship between genotype and phenotype. Among philosophers, they cite Nancy Cartwright and John Dupré as having developed metaphysical accounts of relationships between the sciences, and of scientific explanation, that are congenial to their proposals and hope that biology might one day 'escape the limited metaphysical straitjackets of purely gene-centric interpretations' (398) of biological processes.

In the final chapter in this volume, Julie Zahle and Tuukka Kaidesoja explore emergence in the social sciences. Emergentist ideas in social theory usually invoke the emergence of social facts, properties or relationships of the kind exemplified by 'universities, states, traffic jams, wealth distributions, declarations of war, firms' firing of employees, and norms' (400). The idea of emergence can, they argue, be traced at least back to the work of Émile Durkheim, though without the use of the term 'emergence', which appears only 'sporadically' (400) in subsequent social theory. Zahle and Kaidesoja begin by distinguishing between 'an epistemic notion of emergence', which is 'a feature that social phenomena have relative to our limited knowledge of them', and 'an ontological notion of emergence', which is independent of any state of knowledge (400). They concentrate on the work of two influential critical realists, Roy Bhaskar and Dave Elder-Vass, in both of whose work emergence is quite prominent and ontological in conception. They nevertheless differ in how they see (for instance) social roles as generating novel causal powers with respect to the individuals who occupy them.

We would like to conclude this introduction by remembering three people who contributed to this volume, either directly or indirectly, but who passed away before it was completed: E. J. (Jonathan) Lowe, Lynne Rudder Baker and David Pines. We are honoured to be able to present the chapters by David Pines and Lynne Rudder Baker in this volume.

Jonathan Lowe (1950–2014) was a leading contributor to metaphysics, philosophical logic and the philosophy of mind, and also a friend and mentor to both Sophie Gibb and Robin Findlay Hendry. His notable achievements included the formulation of a new 'four-category ontology', proposed as a metaphysical foundation for all empirical scientific thought, and distinctive accounts of the ontological status of the person and of human agency. His many books include *Kinds of Being: A Study of Individuation, Identity and the Logic of Sortal Terms* (1989); *Subjects of Experience* (1996); *The Possibility of Metaphysics: Substance, Identity and Time* (1998); *The Four-Category Ontology: A Metaphysical Foundation for Natural Science* (2006); and *Forms of Thought: A Study in Philosophical Logic* (2013). He was heavily involved in the conception and planning of the Durham Emergence Project, for which he was due to write a book-length articulation and defence of his non-Cartesian emergence substance dualism.[10] His absence was a source of sadness throughout the project.

Lynne Rudder Baker (1944–2017) was another key figure in the philosophy of mind and metaphysics. In particular, she was one of the central opponents of reductionist and physicalist conceptions of the universe and one of the most important proponents of the constitution view of human persons – the view that human persons are 'constituted by', but not identical with, their bodies. She is the author of several books, including *Saving Belief: A Critique of Physicalism* (1987), *Explaining Attitudes: A Practical Approach to the Mind* (1995), *Persons and Bodies: A Constitution View* (2000), and *The Metaphysics of Everyday Life: An Essay in Practical Realism* (2007). With other scholars, Baker also delivered the 2001 Gifford Lectures in Natural Theology at the University of Glasgow. These were published as *The Nature and Limits of Human Understanding* (ed. Anthony Sanford, T & T Clark, 2003). Her chapter in this Handbook explores the ways in which emergence and intentionality might be related.

David Pines (1924–2018) was one of the pioneers of understanding electronic properties and excitations in condensed matter physics. His discoveries in the field of the electronic properties of matter laid the foundations for the modern treatment of the subject. His early work with David Bohm on the electron gas led to the development of the celebrated random phase approximation, while his work on superconductivity was a direct antecedent of the famous BCS theory. He is well known across physics for his influential books on many-body physics and on quantum liquids, which include *The Many-Body Problem* (1961), *Elementary Excitations in Solids* (1963) and *The Theory of Quantum Liquids, Vol. I: Normal Fermi Liquids.* (1966) and *The Theory of Quantum Liquids Vol. II: Superfluid Bose Liquids* (with P. Nozières) (1990). More recently, he was noted for his work on high-temperature superconductivity and heavy fermion physics. Pines was a leading proponent of emergentist thinking in physics. His agenda-setting article 'The Theory of Everything', co-authored with Robert Laughlin, introduced a generation of scientists to the use of emergence as a way of understanding, and also of doing, science. The Institute of Complex Adaptive Matter, which he cofounded, has played a key role in developing the idea that emergence represents an organising principle of nature. Professor Pines' chapter in this Handbook gives his personal history of the subject and the development of the Institute, along with his hopes for the use of emergence in education practice.

Notes

1 For further discussion of two different approaches to understanding emergence, which she calls 'abstractionism' and 'scientism', see Jessica Wilson's chapter in this volume, and later.
2 McLaughlin points out that Mill did not himself use the term 'emergence'; the term in its modern philosophical usage derives from George Henry Lewes' distinction between 'resultants' and 'emergents'.
3 Note that once philosophers came to see supervenience as a mere trans-world consistency relation, they often rejected it as offering any kind of *explanation* of the relationship between the mental and the physical (see for instance Horgan 1993).
4 The opposition is not a universal assumption: see for instance Butterfield (2011).
5 For Nagel's model see Nagel (1979). See also Needham (2010) and Dizadji-Bahmani, Frigg and Hartmann (2010), who also defend Nagelian reduction as a viable analysis, albeit from different sides of the debate over *reductionism*, the view that different sciences bear reductive relationships to each other, or would do in an ideal, completed science.
6 Note the parallel between Broad's 'trans-physical' laws and Nagel's bridge laws.
7 One might wish to distinguish realisation *simpliciter* (the making real of something) from the particular analyses of it which have been developed under the constraint of making NRP work.
8 Broad used the term 'mathematical archangel' (1925, 71) for this kind of hypothetical predictor, a similar kind of being to Laplace's demon.
9 Landau's arguments are profound and penetrating, but tend to be brief and rather sketchy, although they are given, in something like their original form, in the textbook series that Landau co-authored with Evgeny Lifshitz (Landau and Lifshitz 1980).
10 This position is critically examined by Alex Carruth and Sophie Gibb in a memorial volume for Jonathan, where they explore in detail how his conception of mental causation might fit into his four-category ontology (see Carruth and Gibb 2019).

References

Alexander, S. 1920 *Space, Time and Deity* London: Macmillan.
Anderson, P.W. 1972 'More Is Different' *Science* 177, 393.
Anderson, P.W. 1984 *Basic Notions of Condensed Matter Physics* Boulder, CO: Westview Press.
Broad, C.D. 1925 *The Mind and Its Place in Nature* London: Kegan Paul, Trench and Trubner.
Butterfield, J. 2011 'Emergence, Reduction and Supervenience: A Varied Landscape' *Foundations of Physics* 41, 920–959.
Carruth, A. and S. Gibb 2019 'The Ontology of E. J. Lowe's Substance Dualism' in A. Carruth, S. Gibb and J. Heil (eds.) *Ontology, Modality, and Mind: Themes from the Metaphysics of E. J. Lowe* Oxford: Oxford University Press.

Dizadji-Bahmani, F., R. Frigg and S. Hartmann 2010 'Who's Afraid of Nagelian Reduction?' *Erkenntnis* 73, 393–412.

Ellis, G.F.R. 2012 'Top-down causation and emergence: some comments on mechanisms' *Interface Focus* 2, 126–140.

Fine, K. 1995 'Ontological Dependence' *Proceedings of the Aristotelian Society* 95, 269–290.

Hawking, S.W. and G.F.R. Ellis 1973 *The Large Scale Structure of Space-Time* Cambridge: Cambridge University Press.

Hendry, R.F. 2006 'Is There Downward Causation in Chemistry?' in D. Baird, L. McIntyre and E. Scerri (eds.) *Philosophy of Chemistry: Synthesis of a New Discipline* Dordrecht: Springer, 173–189.

Horgan, T. 1993 'From Supervenience to Superdupervenience: Meeting the Demands of a MaterialWorld' *Mind* 102, 555–586.

Humphreys, P. 2016 *Emergence: A Philosophical Account* Oxford: Oxford University Press.

Hund, F. 1927 'On the interpretation of molecular spectra, III' *Zeitschrift für Physik* 43, 805.

Kim, J. 2005 *Physicalism, or Something Near Enough* Princeton, NJ: Princeton University Press.

Landau, L.D. and E.M. Lifshitz 1980 *Statistical Physics*, Vol. 2 Oxford, UK: Butterworth-Heinemann.

Laughlin, R.B. and D. Pines 2000 'The Theory of Everything' *Proceedings of the National Academy of Sciences of the United States of America* 97, 28–31.

McLaughlin, B. 1992 'The Rise and Fall of British Emergentism' in A. Beckermann, H. Flohr, and J. Kim (eds.) *Emergence or Reduction? Essays on the Prospects for Non-Reductive Physicalism* Berlin: Walter de Gruyter, 49–93.

Nagel, E. 1979 *The Structure of Science: Problems in the Logic of Scientific Explanation*, Second Edition Indianapolis: Hackett.

Needham, P. 2010 'Nagel's Analysis of Reduction: Comments in Defence as Well as Critique' *Studies in History and Philosophy of Modern Physics* 41, 163–70.

Oppenheim, P. and H. Putnam 1958 'Unity of Science as a Working Hypothesis' in H. Feigl, M. Scriven, and G. Maxwell (eds.) *Minnesota Studies in the Philosophoy of Science*, Vol. 2 Minneapolis: University of Minnesota Press, 3–36.

Seitz, F. 1940 *Modern Theory of Solids* New York: McGraw-Hill.

Sellars, W. 1956 'Empiricism and the Philosophy of Mind' in H. Feigl and M. Scriven (eds.) *Minnesota Studies in the Philosophy of Science*, Vol. 1 Minneapolis: University of Minnesota Press.

Sperry, R.W. 1980 'Mind-Brain Interaction: Mentalism, Yes; Dualism, No' *Neuroscience* 5, 195–206.

Wick, G.C. 1954 'Properties of Bethe-Salpeter Wave Functions' *Physical Review* 96, 1124–1134.

PART 1

Foundations of emergence

1

BRITISH EMERGENTISM

Brian P. McLaughlin

The endeavor to understand the natural world through scientific enquiry, while having proved enormously successful, has resulted in numerous sciences. Taking the widest divisions, there is physics and then there are the special sciences, which include, among others, chemistry, biology, psychology, linguistics, sociology, and economics. There are, moreover, many divisions within these broad cuts, including within physics. To be sure, one of the main aims of scientific theorizing is unification, but though science unifies, it also diversifies. As Jerry Fodor once remarked, "[T]he development of science has witnessed the proliferation of specialized disciplines at least as often as it has witnessed their elimination" (1975, 9–10) or, we may add, their unification. This bodes well for employment in science, at least if adequate funding is available. Still, though, the Milesian longing for a comprehensive, systematically unified, final scientific theory of the natural world persists, and indeed is a driving force for some physicists.

Let us step back from this situation, draw a circle around it, and ask: Given that there is one and only one natural world that the enterprise of science seeks to understand, why are there many sciences rather than just one science? There is no received answer. The factors that immediately come to mind are many and varied. To name just a few: there are, arguably, scientific unifications yet to be achieved; but also, different sciences serve different specific purposes; they can deploy different methods; different concepts can cross-classify the same phenomena; there is computational intractability; and there are limitations on our ability to theorize imposed by our cognitive architecture, including, arguably, gaps in our conceptual schemes that we are constitutionally unable to build conceptual bridges to close. However, our scientific to-do lists, our aims, our methods, our built-in cognitive limitations, and other factors concerning us aside, does the natural world at least in principle admit of a comprehensive, systematically unified, final scientific theory?

In this chapter, I discuss the British Emergentist movement, a movement that presented a view of the natural world according to which the answer is "no." Although it has ancient roots (Caston 1997), the movement began around the mid-nineteenth century and flourished in the first quarter of the twentieth century (McLaughlin 1992). The truly major works in the movement are John Stuart Mill's *System of Logic* (the first edition of which was published in 1843, and the eighth edition of which was published in 1872), Samuel Alexander's two-volume *Space, Time and Deity* (1920), Lloyd Morgan's *Emergent Evolution* (1923), and C.D. Broad's *The Mind and Its Place in Nature* (1925), but other notable works include Alexander Bain's *Logic* (1870) and George Henry Lewes's two-volume *Problems of Life and Mind* (1875).

A view can be found in these works according to which there are many sciences rather than just one science, because of the way the natural world itself is. In its mature form in Alexander (1920), Morgan (1923), and Broad (1925), the view is that the natural world is layered: it has a hierarchical structure in which higher tiers are dependent on, but are not reducible to, lower tiers. The elements of higher tiers are wholes or systems entirely composed of elements of lower tiers, but possessed of the kinds of properties not possessed by any of their constituents, properties that emerge from their constituents being propertied and related in certain ways. The properties of wholes that so emerge are thereby emergent properties. Such emergent properties figure in fundamental laws of nature.

On this view of the nomological structure of the natural world, there is a vast collage of fundamental laws. The natural world is thus such that the Milesian longing for a comprehensive, small group of systematically integrated fundamental laws cannot be satisfied. Broad quipped, adding a spoonful of sweetness to help the medicine go down, that if there is indeed such a lack of unity in the natural world, it "must simply be swallowed whole with that philosophical jam that Professor Alexander calls 'natural piety'" (1925, 55).

In what follows, I first lightly sketch the history of the British Emergentist movement from Mill to Broad, with a focus just on the issue of why there are many sciences.[1] Then, I briefly address the issue of whether there are emergent properties in the sense in question. I conclude with a few remarks about appeals in contemporary physics to a different notion of emergence.

In *A System of Logic* (Book III, Ch. VI), Mill distinguishes "two modes of the conjoint action of causes, the mechanical and the chemical" (1868, Seventh Edition, xvi). He tells us that causes combine in the mechanical mode to produce an effect just in case that effect is the result of their conjoint action and the sum of what would have been the effects of each of the causes had they acted alone. He illustrates this mode with the example of forces acting jointly to produce a certain movement. The resulting movement is the vector sum of what would have been the effects of each of the component forces had they acted alone. Citing "the principle of the Composition of Forces in mechanics," Mill says that "in imitation of that well-chosen name," he gives "the name of the Composition of Causes to the principle which is exemplified in all cases in which the joint effect of several causes is identical with the sum of their separate effects" (406). The principle of the Composition of Causes, he tells us, "by no means prevails in all departments of the field of nature" (406). Often, causes combine instead in the chemical mode, so-called because it is exhibited in chemical interactions, though by no means exclusively in such interactions. Causes combine in the chemical mode to produce an effect just in case the effect is the result of their conjoint action and not the sum of what would have been the effect of each of the causes had they acted alone. The product of a chemical process is in no sense the sum of the effects of each reactant. Combining methane and oxygen, for instance, produces carbon dioxide and water, which is in no sense the sum of what would have been the effects of methane and oxygen acting alone. Given that an effect of the causes that act together to produce it either will be the sum of what would have been the effects of each cause acting alone or it will not be, the distinction is exhaustive: causes combine either in the mechanical mode or in the chemical mode with respect to any effect that results just from their conjoint action. This distinction, Mill tells us, is "one of the most fundamental distinctions in nature" (409).

According to Mill, sciences strive to offer deductive explanations of phenomena in terms of laws, and so deductive nomological explanations. But there is no single science, since sometimes when the principle of the Composition of Causes fails, "the concurrence of causes is such as to determine a change in the properties of the body generally, and render it subject to new laws, more or less dissimilar to those to which it conformed in its previous state" (413). On his view, we have special sciences because at some particular points in the transition from separate to united action, the laws change, and an entirely new set of effects are either added to, or take the place of, those which arise from the separate agency of the same causes: the laws of these new effects being again susceptible of composition, to an indefinite extent, like the laws which they superseded.

(411)[2]

Consider, for instance, organic bodies. They are wholly composed of kinds of ingredients that also figure as ingredients of inorganic matter, but causal factors have brought entities of these kinds together into an organization, a whole or system or complex body, that exhibits new properties, properties that figure in laws of physiology. Those laws are not deducible from the laws concerning the ingredients as they occur in inorganic matter. Moreover, the laws of physiology in question supersede them. As concerns bodies that are ingredients of inorganic matter and come together to make up organic bodies, he says: "Those bodies continue, as before, to obey mechanical and chemical laws, in so far as the operation of those laws is not counteracted by the new laws that govern them as organized beings" (409).

In Mill's view, sciences are nomothetic, but the natural world is not governed by a small group of systematically, well-integrated fundamental laws. It is governed by a collage of fundamental laws, with laws concerning complex organizations superseding laws concerning their constituents in isolation. The various departments of science are concerned with the various compartments of nature.

Mill's distinction between the mechanical and the chemical modes of the conjoint action of causes ignited the British Emergentist movement. Mill himself, however, never used the term "emergence." He called effects of causes acting conjointly in the mechanical mode, "homogeneous" effects (412); those of causes acting conjointly in the chemical mode, "heterogeneous" effects; and the laws governing the latter causal transactions, "heteropathic laws" (409). George Henry Lewes (1875) called Mill's heterogeneous effects "emergents" and his homogeneous effects "resultants." Effects are either resultants or emergents. An emergent, Lewes tells us, "is unlike its components insofar as these are incommensurable, and it cannot be reduced to their sum or their difference" (1875, 412). Given that, the first occurrence of each kind of emergent is taken to introduce genuine novelty into the world. Lewes's talk of emergents led to talk of emergence in the work of Alexander, and Morgan and Broad followed Alexander in this. These theorists took chemical substances and organic bodies to be wholly composed of atoms and subatomic particles (they knew about electrons and protons), and so took changes to involve rearrangements of atoms and more fundamental particles. But they held that new configurations of them can possess genuinely novel, and indeed irreducible, properties. Their works inspired a large, international literature, both supportive (see, e.g., Lovejoy 1927) and critical (see, e.g., Pepper 1926).[3]

The idea that the natural world has a hierarchical structure is first explicitly articulated in the British Emergentist literature in Alexander's *Space, Time, and Deity*. Alexander writes of the emergence of new qualities from the complexity of organization, telling us:

> The emergence of a new quality from any level of existence means that at that level there comes into being a certain constellation or collocation of the motions belonging to that level, and this collocation possesses a new quality distinctive of the higher-complex. . . . The higher-quality emerges from the lower level of existence and has is roots therein, but it emerges therefrom, and it does not belong to that lower level, but constitutes its possessor a new order of existent with its special laws of behavior. The existence of emergent qualities thus described is something to be noted, as some

would say, under the compulsion of brute empirical fact, or, as I should prefer to say in less harsh terms, to be accepted with the "natural piety" of the investigator. It admits of no explanation.

(1920, 45–47)

Although existents at higher levels are dependent on existents at lower levels in that they are wholly composed of such existents, the higher-level existents have emergent qualities and are governed in part by autonomous laws of behavior that cite those qualities.

If Alexander's hierarchical view of the natural world is correct, then the goal of finding a small group of systematically, well-integrated fundamental laws of nature that govern the entire natural world is a pipe dream. The cure for the Milesian longing is a large dose of natural piety.

In *Emergent Evolution*, Morgan embraced Alexander's view of ascending levels of reality and proposed an evolutionary cosmology inspired by Alexander's claim that Darwin's principle of adaptation extends "below the level of life" (Alexander 1920, Vol. 2, 310). As concerns the ascending levels or grades of reality, Morgan says there are

> physical and chemical events in progressively ascending grades. Later in evolutionary sequence life emerges – a new "quality" of certain material or physico-chemical systems with supervenient[4] vital relations hitherto not in being. Here again there are some progressively ascending grades. Then within this organic matrix, or some highly differentiated part thereof, already "qualified" . . . by life, there emerges the higher quality of consciousness or mind.
>
> *(1923, 9–10)*

As concerns the evolution of the ascending grades, Morgan says:

> At any emergent stage of evolutionary progress is a new kind of relatedness . . . hitherto not in being. In virtue of such new kinds of relatedness, not only have natural entities new qualities within their own proper being, but new properties in relation to other entities. The higher entities are not only different in themselves; but they act differently in the presence of others.
>
> *(19)*

He draws a useful distinction between two kinds of relatedness:

> I speak of the relatedness which obtains wholly within a given system as *intrinsic* [to the system]; and I shall distinguish the relatedness of this system to some other system, or systems, as *extrinsic*. A system of intrinsic relatedness I shall provisionally call an entity.
>
> *(19)*

He tells us:

> At each ascending step, there is a new entity in virtue of some new kind of relation, or sets of relations within it, or as I phrase it, intrinsic to it. Each exhibits new ways of acting, and reacting to, other entities.
>
> *(64)*

As concerns such new ways of acting and reacting to other entities, he says:

> when some new kind of relatedness is supervenient (say at the level of life), *the way in which the physical events which are involved run their course is different in virtue of its presence — different from what it would have been if life had been absent.*
>
> *(16, emphasis his)*

The ascending levels or grades are not levels or grades of scale. They are mereological levels, levels organized by part–whole relations. To be sure, a part of an entity or system will be at a smaller scale than the entity or system, and that entity or system in turn will be at a smaller scale than any entities of which it is a part, but the parts of an entity can be at very different scales, and the relevant ways of dividing the parts of entities or systems are (in Morgan's terminology) a matter of different kinds of intrinsic relatedness, not scale. The ascending hierarchy is a matter of the intrinsic relatedness of systems from which new qualities emerge. These emergent qualities determine some extrinsic relations, including causal relations. Entities at one scale can, of course, causally influence entities at another scale. A star and an atom will exert a gravitational attraction on each other. But gravitational attraction is a kind of extrinsic relatedness that, in Morgan's view, is at the lowest grade of the hierarchy, since all entities have mass. Gravitational attraction is thus a same grade kind of extrinsic relatedness. Following Alexander, Morgan maintains that there can be causation from higher grades to lower grades in the hierarchy. Indeed, he holds that higher-grade events can affect (to use his phrase) "the go of events" at lower grades in ways unanticipated by the laws of lower grades. There is thus, in his view, top-down causation, causation from higher grades or levels to lower grades or levels.[5]

The sciences are concerned with respective emergent orders of the ascending hierarchy. Thus, Morgan tells us:

> On this understanding we distinguish mind, life, and matter. Within each, of course, there are many emergent sub-orders of relatedness. It is for science to work out the details for psychology, for biology, for chemistry and for physics.
>
> *(22)*

There are many sciences rather than just one because the natural world consists of emergent orders of an ascending hierarchy. The special sciences are concerned with the laws governing higher grades of the hierarchy.

This emergentist view of the structure of the natural world receives its most careful and detailed formulation in Broad's *Mind and Its Place in Nature*. I will henceforth focus on that formulation, rather than highlighting differences between the views of Alexander, Morgan, and Broad (though such there be[6]).

Broad examines two views of "the relations between the various sciences," "Mechanism" and "Emergence" (1925, 76). He tells us that according to Mechanism in its purest form,

> the external world has the greatest amount of unity which is conceivable. There is really only one science and the various "special sciences" are just particular cases of it.
>
> *(76)*

(This passage might well call to mind the quip, typically attributed to Ernest Rutherford, that "all sciences are either physics or stamp collecting.") Broad states that in contrast to the mechanist view, in the emergentist view,

[w]e have to reconcile ourselves to much less unity in the external world and a much less intimate connexion between the various sciences. At best the external world and the various sciences that deal with it form a hierarchy. . . . [W]e should have to recognize aggregates of various orders. And there would be two fundamentally different types of law, which might be called "intraordinal" and "trans-ordinal" respectively. A trans-ordinal law would be one which connects the properties of aggregates of adjacent orders. A and B would be adjacent, and in ascending order, if every aggregate of order B is composed of aggregates of order A, and if it has certain properties which no aggregate of order A possesses and which cannot be deduced from the A properties and the structure of the B-complex by any law of composition which has manifested itself at lower-levels. An intra-ordinal law would be one which connects the properties of aggregates of the same order. A trans-ordinal law would be a statement of the irreducible fact that an aggregate composed of aggregates of the next lower order in such and such proportions and arrangements has such and such characteristic and non-deducible properties.

(77–78)

According to Broad, the aggregates of a given order (or level or grade) have three kinds of properties: ordinally neutral ones, reducible ones, and ultimate ones. The ordinally neutral properties of aggregates of a certain order are properties that are also possessed by aggregates of lower orders. (He cites inertial and gravitational mass as examples [79].) The reducible ones reduce either to other properties of the same order or to properties of lower orders. The ultimate properties of an order are specific to aggregates of the order and are, moreover, irreducible. They figure in trans-ordinal laws and also intra-ordinal laws of the order in question.

In Broad's terminology, trans-ordinal laws are, by stipulation, emergent laws. His notion of an emergent property is explained in terms of the notion of a trans-ordinal law. So the key to understanding his notion of an emergent property is understanding his notion of a trans-ordinal law.

Trans-ordinal laws link properties of wholes with (what I will call) microstructural properties of the wholes. The notion of a microstructural property can be explicated as follows. MS is a microstructural property of an object or system S just in case (a) S is decomposable into some set of nonoverlapping constituents $C_1 \ldots C_n$, (b) MS is the property of consisting of $C_1 \ldots C_n$ being respectfully propertied in such-and-such ways and related so propertied to each other in so-and-so ways, and (c) S has MS. A trans-ordinal law will state that whatever has a certain microstructural property MS has a certain other property E, where E is not possessed by any of the constituents of MS. A property E is an emergent property just in case necessarily if something has E, there is some microstructural property MS such that it is a trans-ordinal law that whatever has MS has E. (That is a kind of strong, nomological supervenience thesis [see McLaughlin 1994].) E emerges from a microstructural property MS only when there is such a trans-ordinal law. E may emerge from more than one microstructural property, but each such microstructural property is linked to E by a trans-ordinal law. A trans-ordinal law will be contingent and only a posteriori knowable (65). It is logically possible for something to have the microstructural property without having E. But possession of the microstructural property will nomologically necessitate, and so in that sense determine, that the whole has E.[7]

Trans-ordinal laws are contingent, only a posteriori knowable, and state that whatever has a certain microstructural property has a certain other property, a property not possessed by any constituent of the microstructural property. But not all laws that meet those conditions are trans-ordinal laws, for a law can meet those conditions and yet be reducible. A trans-ordinal law is a statement of the irreducible fact that whatever has a certain microstructural property MS has a property E (78). Trans-ordinal laws are, by stipulation, irreducible. If a law is reducible, it is not a trans-ordinal law.

Broad takes law reduction to require deduction of the law from other laws and statements of conditions, either intra-ordinal laws and conditions (some intra-ordinal laws are reducible to other intra-ordinal laws) or lower-ordinal laws and conditions. (I here count compositional principles, such as, e.g., the principle of the additivity of as mass, as laws, and ones that have instances at lower levels as lower-ordinal laws.) By deducible, he means deducible *"in theory"* (70, italics his), that is, deducible in principle. He readily acknowledges that in some cases of deducibility, "the mathematical difficulties might be overwhelming in practice" (70). But our inability to carry out the deduction, even because of built-in limitations in our cognitive architecture, would not suffice for nondeducibility in his intended sense. Nor would even the physical impossibility of constructing a computer that could carry out the deduction. He says that even "a mathematical archangel, gifted with the further power of perceiving the microscopic structure of atoms as easily as we can perceive hay-stacks" (70) could not deduce a trans-ordinal law from laws and statements concerning the microstructure of atoms.

Broad appeals to various conditions in characterizing a trans-ordinal law, including that such a law not be deducible from other laws and conditions, that it not be "a special case which arises by combining two or more laws" (65), and that it not be "a special case which arises through substituting certain determinate values for determinate variables in a given law" (65). These are all in the service of explicating the claim that a trans-ordinal law is a *"unique* and *ultimate* law"* (65, emphases his), a lawful statement of an irreducible fact (78). Given that a trans-ordinal law is supposed to be such a law, I take it that a trans-ordinal law must be a fundamental law. The property of being a fundamental law is a global property of a law: it depends on what other laws and conditions hold (McLaughlin 2017). Trans-ordinal laws are fundamental laws and so admit of no explanation.

A trans-ordinal law is a statement. Statements are made using sentences. Sentences are composed of words. A trans-ordinal law will be stated using a term for the emergent property in question. That term will not be a term of any minimal vocabulary for any lower-order theory. It will not appear in any lower-order laws or lower-order statements of conditions. The kind of deducibility required for reduction of one theory to another must allow premises that contain terms both in the reducing theory and in the theory to be reduced. Ernest Nagel (1961) tried to explicate the notion of theory reduction in terms of bridge laws that contain terms of both theories and which can figure as premises in the relevant deductions. He allowed such laws to be of this form: whenever anything is F, then it is G. Trans-ordinal laws have that form and contain both higher-order and lower-order terms. Should they, then, count as bridge laws? In a word, "no." The reason is that if trans-ordinal laws count as bridge laws, then Nagel fails to state a sufficient condition for theory reduction (McLaughlin 1992). Ignoring fine points, for a theory T to reduce to a theory T*, the true statements of T must be entailed by the true statements of T* alone. Other premises may be required to deduce the true statements of T, but such premises must be entailed by the true statements of T* alone (or in conjunction with a second-order statement that they are all the true statements of T*). It is thus legitimate to appeal to other premises in the deduction if those premises are a priori and necessary. But trans-ordinal laws are a posteriori and contingent. If trans-ordinal laws are needed in addition to a set of statements couched in the vocabulary of T* to deduce the true statements of T, then Broad would claim, rightly, that we have a case of emergence, not reduction.[8]

Reduction requires deducibility. Deduction requires entailment: a statement cannot be deduced from a set of premises unless the premises in the set entail it. Trans-ordinal laws are (by stipulation) irreducible, and so must not be deducible from other laws and statements of conditions, even with the help of a priori necessary truths. Broad holds trans-ordinal laws are not deducible from lower-level laws and statements of conditions because they are not entailed by such. Given that, he could appeal directly to non-entailment, rather than to non-deducibility in characterizing trans-ordinal laws.[9] Failure of entailment requires no appeal to a mathematical archangel or to a kind of logical system. It takes time and resources to carry out a deduction, how much time and what resources depends on, among other things, the number of steps in the deduction in the logical system in question. Entailments require no time at all and no resources whatsoever, and they are not carried out in any logical system since they are not carried out at all.

There is a distinction between two kinds of entailment, one that Broad did not know was available. There is epistemic entailment and semantic entailment. P epistemically entails Q just in case the material conditional if P then Q is a priori; and P semantically entails Q just in case the material conditional if P then Q is necessary (true in every possible world) (Chalmers and Jackson 2001). There can be epistemic entailment without semantic entailment and semantic entailment without epistemic entailment. The reason is that, as Kripke (1980) showed us, there are contingent truths that are a priori and necessary truths that are a posteriori. Consider contingent a priority. It is a priori yet contingent that if Benjamin Franklin is the actual inventor of bifocals, then he is the inventor of bifocals. The reason it is contingent is that it is a necessary truth that Benjamin Franklin is the actual inventor of bifocals, but a contingent truth that Benjamin Franklin is the inventor of bifocals.[10] The antecedent of the conditional is necessary, while the consequent is contingent, and so the conditional itself is contingent. The statement that Benjamin Franklin is the actual inventor of bifocals and the statement that Benjamin Franklin is the inventor of bifocals epistemically entail each other. But there is semantic entailment only in one direction. Benjamin Franklin is the inventor of bifocals semantically entails that Benjamin Franklin is the actual inventor of bifocals; indeed, every statement semantically entails the latter statement since it is a necessary truth. But the statement that Benjamin Franklin is the actual inventor of bifocals does not semantically entail that Benjamin Franklin is the inventor of bifocals. Turn to a posteriori necessity. Given that water is H_2O, it is necessary that if there is water in the sink, then there is H_2O in the sink. The statement that there is water in the sink thus semantically entails that there is H_2O in the sink. But the entailment is not a priori, since it is only a posteriori knowable that water is H_2O.

Reduction is a kind of explanation and so must satisfy epistemic conditions. Although the terminology is not Broad's and he did not recognize that there are a priori contingent truths and a posteriori necessary truths, it is nonetheless clear that he took trans-ordinal laws to be such that they fail to be either epistemically or semantically entailed by other laws and conditions. Trans-ordinal laws are, in that sense, fundamental laws. Emergent properties are properties linked to microstructural properties via such fundamental laws.

Trans-ordinal laws can figure in the explanation of why there are many sciences rather than just one, or indeed the explanation of anything, only if there are such laws. Broad readily acknowledges that it is an empirical question whether there any trans-ordinal laws connecting subatomic microstructural properties with chemical properties, or any trans-ordinal laws connecting microstructural chemical properties with biological properties. He thus acknowledges that it is an empirical question whether there are any chemical or biological emergent properties. His speculation that there may indeed be such emergent properties was reasonable at the time he was writing. But our epistemic situation has changed.

Mind and Its Place in Nature, published in 1925, was the last major work in the British Emergentist movement. I have speculated elsewhere that the movement lost its momentum because of a series of truly revolutionary developments in science (McLaughlin 1992). Erwin Schrödinger stated his famous equation of nonrelativistic quantum mechanics in 1926, which was followed several years later by the result that quantum mechanics can in principle at least explain chemical bonding, which was later followed by the development of organic chemistry and of molecular biology. There is good reason to think that all of the fundamental forces of nature are ones that are exerted at the subatomic level: the gravitational force, the electro-magnetic weak force, and the strong force; and further unification may be possible. Moreover, on the conceptual front, the notion of reduction via functional analysis has been developed. A functional property is a second-order property of having some property or other instances of which occupy certain roles as causes and effects. Functional properties are realized by properties instances of which occupy the relevant causal roles. A law linking a microstructural property with a functional property that it realizes will be contingent and a posteriori. But it will not be a fundamental law, since it will hold in virtue of laws and conditions concerning the microstructural property, namely the laws and conditions that determine that the microstructural property occupies the relevant causal role. It will thus not be a trans-ordinal law. A number of would-be examples of candidate emergent properties in the British Emergentist literature, such as the property of being able to reproduce, and dispositional properties of various sorts are, arguably, functional properties that can be reduced via functional analyses (see, e.g., Chalmers 1996). Although the issue is indeed an empirical one, there seem to be no emergent chemical or biological properties in Broad's sense. We must look elsewhere for an explanation of why there are, in addition to physics, the sciences of chemistry and biology.[11]

There remains a serious issue whether qualia, the what-it-is-like for the subject aspects of subjective experiences, are emergent in Broad's sense (see, e.g., van Cleve 1990; Chalmers 1996; Kim 2005). I myself think that qualia do not admit of reduction by functional analyses. But I also think that the problem of the place of qualia in nature arises because of gaps in our conceptual scheme that we are constitutionally unable to build conceptual bridges to close, rather than because qualia are emergent properties in his sense. I cannot, however, pursue that matter here and so will say no more about qualia.

The term "emergence" is used today in a very wide range of fields. Some theorists use the term in a way that at least takes its inspiration from the British Emergentist literature (to name just a few: van Cleve 1990; Beckerman 1992; Kim 1992; McLaughlin 1992, 1997; Stephan 1992; Chalmers 1996, 2006; Crane 2001; Gillett 2002; O'Connor and Wong 2005; Shoemaker 2007; Vision 2011; Wilson 2016). However, many theorists use "emergence" and "emerge" in ways that do not. Nonetheless, I speculate that even in such cases there are historical chains of usage of "emergence" and "emerge" tracing back to their use in Alexander (1920), which traces back to the use of "emergents" in Lewes (1875).

Despite such historical chains of usage, there are contemporary uses of "emerge" and "emergence" that differ, often quite markedly, from their use in Alexander (1920), Morgan (1923), and Broad (1925). For example, Jeremy Butterfeld (2011) articulates a notion of emergence that he maintains is useful in the metaphysics of physics. He uses "emergence" in a sense in which emergence is compatible with reduction. Assuming that "reduction" is being used in the same sense as in Alexander's, Morgan's, and Broad's work (something I will not explore here), emergence in Butterfield's sense is, then, incompatible with emergence in the sense of the term in that literature. Nothing can be emergent in both senses, since nothing can be both reducible and irreducible (in the same sense). There is, however, no genuine conflict here. "Emergence" is used differently. There are just two homophones. "Emergence" is a term of art. The early bird doesn't get to keep the worm. No one owns the term. Definitions of "emergence" in science and metaphysics are stipulative. They should be judged just by their theoretical fruits. There is no issue of coming up with the correct definition of "emergence," unless one is just trying to offer a definition that captures how it happens to be used by certain theorists (as I tried to capture Broad's use). Given a notion of emergence, the interesting questions are whether anything is emergent in the sense in question and, if so, whether facts of such emergence can explain anything. If Butterfield's notion of emergence is indeed useful in the metaphysics of science, then that is all to the good. But there is simply no real issue about whether Butterfield or instead Broad-inspired emergentists are right about the relation between emergence and reduction.

P.W. Anderson's seminal "More Is Different: Broken Symmetry and the Nature of the Hierarchical Structure of Science" sparked interest in a notion of emergence in physicists working in condensed matter physics, currently the largest subfield of physics. Anderson underscored the fact that there are symmetry-breaking phase transitions.

Emergence in the sense now used in discussions of condensed matter physics is different from emergence in Broad's sense. Anderson would, I believe, agree. Consider that, to avert possible misunderstanding of his view, he says:

> [W]hen I speak of scale change causing fundamental change I don't mean the rather well-understood idea that phenomena at a new scale may actually obey different fundamental laws – as, for example, general relativity is required on the cosmological scale and quantum mechanics on the atomic.[12] I think that it will be accepted that all ordinary matter obeys simple electrodynamics and quantum theory, and that really covers most of what I shall discuss. (As I said, we must all start with reductionism, which I fully accept.)
>
> *(1972, 222–223; Bedau and Humphreys 2008)*

Anderson thus tells us that he accepts reductionism. Although he denies, for instance, that psychology is applied biology and that biology is applied chemistry (222), I take it that given his embrace of what he calls "reductionism," he would also deny that there are trans-ordinal psychobiological and biochemical laws. Moreover, he doesn't think there are fundamental psychological or biological laws. His main point is that

> the reductionist hypothesis does not by any means imply a "constructionist" one: The ability to reduce everything to simple fundamental laws does not imply the ability to start from those laws and reconstruct the universe.
>
> *(222)*

Condensed matter physics provides a testing ground for quantum mechanics, but it also, in addition, makes enormous contributions to our understanding of solids and liquids. We have made discoveries about ordinary matter that, so far as we know, could not have been made other than by doing condensed matter physics. More can thus indeed be different in ways that are ripe for new and exciting science. But unless with more sometimes comes trans-ordinal laws, and so new fundamental laws of nature, that offers no vindication of the claim that there is emergence in Broad's sense. It is my understanding that the prospects for such vindication are taken to be truly dim indeed.

I noted earlier that quantum mechanics can in principle at least explain chemical bonding. But as concerns reduction without construction, it should be noted that although Schrödinger's equation, the fundamental equation of nonrelativistic quantum mechanics, can be written down quite simply and, on the evidence, Hamiltonians can be specified by a small number of types of quantities, it is nonetheless the case that save for the hydrogen atom, approximation techniques must be used to solve the equation, the most innocent of which is the Born-Oppenheimer approximation, which makes the simplifying assumption that the nucleus of the atom does not move. It would be physically impossible to build a computer that could accurately solve the equation for a system of as few as, say, 200 particles. It is thus physically impossible to construct solutions for no end of cases. It might be claimed that chemistry nevertheless reduces to quantum mechanics, but it is physically impossible to deduce all chemical truths from truths of quantum mechanics. The issue of reduction aside, however, that is no reason whatsoever to think there are trans-ordinal physicochemical laws. Entailment takes no time and requires no resources. On the evidence, the equation together with a specification of a small number of kinds of quantities will entail a unique solution to the equation for any chemical phenomenon.

The British Emergentists tried to offer an explanation of why there are many sciences rather than just one. The explanation, in its most developed form, appealed to the thesis that nature has a hierarchical structure, with higher levels dependent on lower levels, but governed in part by autonomous laws of nature so that there is a vast collage of fundamental laws of nature, rather than a single fundamental law of symmetry or even a small group of systematically well-integrated fundamental laws. Later scientific advances, I maintain, undermined that emergentist view, casting serious doubt indeed on the claim that there are trans-ordinal laws (save, perhaps, ones linking microstructural properties with qualia). The question remains, however, whether, as many physicists hope, a systematically unified, final scientific theory of the natural world can be found. A point to underscore in closing is that even if there is such a theory and we someday formulate it, the question remains whether we would be able to use it to "construct" all correct scientific theories from that fundamental theory. We might very well be unable to do that, even with the help of ingenious approximation techniques and the best computers the laws of physics allow. Discovery of The Theory of Everything, if such there be, would, I think, still leave plenty of work not only for poets and literary criticism theorists, but also for condensed matter physicists, chemists, biologists, psychologists, linguists, sociologists, and economists.

Notes

1 Thus, I don't, for instance, discuss the role of Deity in Alexander's and Morgan's emergentist metaphysics.
2 For an explanation of how a law can supersede another in Mill's sense without contravening it, see McLaughlin (1992, 61).
3 See McLaughlin (1992) and Stephan (1992) for historical discussion.
4 It should be cautioned that Morgan uses "supervenient" in its vernacular sense, not in its contemporary philosophical sense (see McLaughlin 1997).
5 Jaegwon Kim (1999) has challenged the coherence of this kind of view. He has argued that "higher-level properties can serve as causes in downward causal relations only if they are reducible to lower-level properties" (150) and adds: "The paradox is that if they are so reducible, they are not really 'higher-level' any longer" (150). Given space constraints, the issue is too complex to address properly here, but for a defense of the conceptual possibility of top-down causation in the British Emergentist sense both in Newtonian mechanics and in quantum mechanics, see McLaughlin (1992). It is an empirical question whether there is top-down causation in that sense.
6 One difference is that, unlike Alexander and Morgan, Broad does not build it into the very notion of an emergent property that emergent properties are causally efficacious.
7 Kim (2010) argues that Broad fails to formulate a coherent notion of emergence. Let it suffice for me to note that in making that case, he tacitly relies on the mistaken assumption that Broad uses "determines" to mean something like metaphysically necessitates. Broad instead uses it just to mean nomologically necessitates and takes laws to be metaphysically contingent. A microstructural property MS will determine an emergent property E just in that it will be a contingent law of nature, a trans-ordinal law, that whatever has MS has E. Kim fails to identify any incoherence in Broad's notion of an emergent property.
8 It is this consideration that leads Kim (2005) to deny that "Nagelian reduction" is in fact a kind of reduction.
9 Chalmers (2006) stipulates two notions of emergence by appealing to nonentailment.
10 Identity is necessary: if A is identical with B, then necessarily A is identical with B. But there are contingent statements of identity. "Benjamin Franklin" and "the actual inventor of bifocals" are rigid

designators, while "the inventor of bifocals" is not. "Benjamin Franklin is the actual inventor of bio-focals" is a noncontingent statement of identity since the identity sign is flanked by ridged designators, whereas "Benjamin Franklin is the inventor of bifocals" is a contingent statement of identity.

11 For further discussion, see McLaughlin (1992, 1997).

12 There is, of course, as yet no quantum theory of gravity.

References

Alexander, S. 1920. *Space, Time, and Deity*, 2 Vols. London: Macmillan.

Anderson, P.W. 1972. More Is Different. In *The American Society for the Advancement of Science*. (Reprinted in *Emergence: Contemporary Readings in Philosophy of Science*, eds. M.A. Bedau and P. Humphreys, 2008, 221–230. Cambridge: MIT Press).

Bain, A. 1887. *Logic*, new and revised edition. New York: Appleton.

Beckerman, A. 1992. Supervenience, Emergence, and Reduction. In *Emergence or Reduction?*, eds. A. Beckermann, H. Flohr, and J. Kim, 94–118. Berlin: Walter de Gruyter.

Bedau, M., and P. Humphreys, eds. 2008. *Emergence: Contemporary Readings in Philosophy and Science*. Cambridge, MA: MIT Press.

Broad, C.D. 1925. *The Mind and Its Place in Nature*. London: Routledge & Kegan Paul.

Butterfeld, J. 2011. Less Is Different: Emergence and Reduction Reconciled. *Foundations of Physics* 4(6): 1065–1135.

Caston, V. 1997. Epiphenomenalism, Ancient and Modern. *The Philosophical Review* 106(3): 309–363.

Chalmers, D.J. 1996. *The Conscious Mind*. New York City, NY: Oxford University Press.

Chalmers, D.J. 2006. Strong and Weak Emergence. In *Re-Emergence of Emergence*, eds. P. Clayton and P. Davies, 244–256. New York City, NY: Oxford University Press.

Chalmers, D.J., and F. Jackson. 2001. Conceptual Analysis and Reductive Explanation. *Philosophical Review* 110: 315–361.

Crane, T. 2001. The Significance of Emergence. In *Physicalism and its Discontents*, eds. C. Gillett and B. Loewer, 207–224. Cambridge: Cambridge University Press.

Fodor, J. 1975. *The Language of Thought*. Cambridge, MA: Harvard University Press.

Gillett, C. 2002. The Varieties of Emergence: Their Natures, Purposes, and Obligations. *Grazer Philosophische Studien* 65(1): 95–121.

Kim, J. 1992. 'Downward Causation' in Emergence and Non-Reductive Physicalism. In *Emergence or Reduction?*, eds. A. Beckermann, H. Flohr, and J. Kim, 119–138. Berlin: Walter de Gruyter.

Kim, J. 1999. Making Sense of Emergence. *Philosophical Studies* 95(1–2): 3–36. (Reprinted in Reprinted in *Emergence: Contemporary Readings in Philosophy of Science*, eds. M.A. Bedau and P. Humphreys, 2008, 127–154. Cambridge: MIT Press).

Kim, J. 2005. *Physicalism or Something Near Enough*. Princeton, NJ: Princeton University Press.

Kim, J. 2010. 'Supervenient and Yet Not Deducible': Is There Are Coherent Concept of Ontological Emergence? In *Essays in the Metaphysics of Mind*, ed. J. Kim. Oxford: Oxford University Press.

Kripke, S. 1980. *Naming and Necessity*. Cambridge: Harvard University Press.

Lewes, G.H. 1875. *The Problems of Life and Mind*, Vol. 2. London: Trübner.

Lovejoy, A. 1927. The Meanings of 'Emergence' and Their Modes. *Journal of Philosophical Studies* 2(6): 167–181.

McLaughlin, B.P. 1992. The Rise and Fall of British Emergentism. In *Emergence or Reduction?*, eds. A. Beckermann, H. Flohr, and J. Kim, 49–93. Berlin: Walter de Gruyter. (Reprinted in *Emergence: Contemporary Readings in Philosophy of Science*, eds. M.A. Bedau and P. Humphreys, 2008, 19–60. Cambridge: MIT Press).

McLaughlin, B.P. 1994. Varieties of Supervenience. In *Supervenience: New Essays*, eds. E.E. Savellos and Ü.D. Yalcin, 16–59. Cambridge: Cambridge University Press.

McLaughlin, B.P. 1997. Emergence and Supervenience. *Intellectica* 25: 25–53. (Reprinted in *Emergence: Contemporary Readings in Philosophy of Science*, eds. M.A. Bedau and P. Humphreys, 2008, 81–98. Cambridge: MIT Press).

McLaughlin, B.P. 2017. Mind-Dust, Magic, or a Conceptual Gap Only? In *Panpsychism: Contemporary Perspectives*, eds. G. Brüntrup and L. Jaskolla, 305–333. New York City: Oxford University Press.

Morgan, C.L. 1923. *Emergent Evolution*. London: Williams & Norgate.

Nagel, E. 1961. *The Structure of Science: Problems in the Logic of Scientific Explanation*. New York, Chicago, and Burlingame: Harcourt, Brace & World, Inc.

O'Connor, T., and H.Y. Wong. 2005. The Metaphysics of Emergence. *Nous* 39: 658–678.

Pepper, S. 1926. Emergence. *Journal of Philosophy* 23: 241–245.

Schrödinger, E. 1926. An Undulatory Theory of the Mechanics of Atoms and Molecules. *The Physical Review* 28(6): 1049–1070.

Shoemaker, S. 2007. *Physical Realization*. Oxford: Oxford University Press.

Stephan, A. 1992. The Historical Facets of Emergence. In *Emergence or Reduction?*, eds. A. Beckermann, H. Flohr, and J. Kim, 25–48. Berlin: Walter de Gruyter.

Van Cleve, J. 1990. Mind-Dust or Magic? Pansychism Versus Emergence. *Philosophical Perspectives* 4: 215–226.

Vision, G. 2011. *Re-Emergence: Locating Conscious Properties in a Material World*. Cambridge, MA: MIT Press.

Wilson, J. 2016. Metaphysical Emergence: Weak and Strong. In *Metaphysics in Contemporary Physics*, eds. T. Bijaj and C. Wüthrich, 345–402. Leiden: Brill Rodopi.

2

DEPENDENCE

Paul Noordhof

Dependence is the most general notion under which a host of familiar metaphysical relations between entities – causation, supervenience, grounding, realisation, etc. – fall. In the first section of this chapter, I will offer some preliminary clarifications to outline the territory in a little more detail. Some years back, this would have primarily involved differentiating kinds of dependence in terms of the strength of the modal operators used and the other details of an analysis deploying them. Now, there has been a proliferation of non-purely modal accounts of dependence. The second section identifies the various reasons that have been offered for this proliferation. The third section discusses a notion of ontological dependence and grounding, each of which draws on an appeal to the essence of the depending, or depended upon, entities. In spite of their popularity, we will see that such notions are of little assistance in capturing a central case of interest to us: the proper understanding of emergence. In the light of this, the fourth section defends a purely modal treatment of some of the problem cases outlined in the first section and also discusses a non-modal notion of construction. I close with a hypothesis that the combination of three features, a non-dependence account of fundamentality, various notions of construction and purely modal properties, remove the motivation for appeal to an independent account of grounding in this, and perhaps any, area.

Clarification of the territory

Three preliminary clarifications are needed. First there is a distinction between necessary dependency and sufficient dependency. For entities F and G, G is *sufficient dependent* on F if and only if the existence or instantiation of F is sufficient in a to-be-further-specified respect for the existence or instantiation of G. In many cases, this will mean that the existence or instantiation of G is necessary in the same to-be-further-specified respect for the existence or instantiation of F. In which case, F is *necessary dependent* upon G. In recognising this connection between sufficient dependency and necessary dependency, we are not assuming that dependency is a non-symmetric relation against those who hold it is an asymmetric one. The point is just that there are these distinct types of dependency. Many familiar characterisations of dependency relation – for example, supervenience and grounding – have focused on sufficient dependence relations (e.g. see Fine (2012), pp. 37–40). Nevertheless, necessary dependence relations are important manifestations of dependence too: an entity upon which something, an F, is necessary dependent is something whose absence is sufficient for the absence of F. E. J. Lowe, amongst others, has focused on this idea of dependence (Lowe (1998), pp. 136–141).

The second clarification concerns dependency as a general, neutral term and dependency as a particular minimal kind of relationship. Dependency relations include grounding, supervenience and so forth. The use of 'dependency' in the last sentence is the general neutral term for these different kinds of relations. However, dependency is also used to characterise the minimal way in which there is a connection between the instantiation of two properties, for example. Often, the latter is characterised modally. Other notions of dependency – such as supervenience or grounding – are constructed from, or given additional characteristics to, this more minimal notion. The use of 'dependence' here is not neutral about what dependency relation is being referred to. For example, whereas it would be incorrect to characterise dependency in the neutral sense earlier as non-symmetric because grounding, for example, is taken to be an asymmetric relation, the second minimal sense of dependency is *non-symmetric*. If F depends upon G, it does not follow that either G does not depend upon F or that G does depend upon F.

A third preliminary clarification relates to the strength of the dependence relation involved. Some philosophers distinguish at the outset between a particular kind of dependence relation often dubbed ontological or metaphysical dependence from sometimes nomic, sometimes causal dependence. They take our present topic to be metaphysical dependence (e.g. Fine (1995), pp. 270–271; Lowe [1998], pp. 136–137). The sharper focus may be the outcome of theorising, but it should not be a starting point. After all, others have begun with causation as the obvious model for dependence relations (e.g. Kim (1984), p. 53). The exact relationship between metaphysical dependence and nomic dependence is open for debate – especially if, for example, you are drawn to a powers ontology, as we shall discuss later. So it is better to consider dependence to be a more general category under which these potentially different kinds of dependence relationships may hold.

There may be different sets of sufficient conditions for the target entity. For a property G, say, possible sufficient conditions are F_1, F_2, F_3. Suppose that the F properties are complex properties: $F_1 = M_1$ & M_2, $F_2 = M_3$ & M_4, $F_3 = M_5$ & M_6. Then a necessary condition of one of these sufficient conditions will not be necessary for the target entity. Relative to a particular sufficient condition holding, though, the target entity will be necessary dependent upon particular components of these nomic sufficient conditions. Call this notion *relative necessary dependence*. In this sense, for example, G is relative necessary dependent upon M_1.

Metaphysical and nomic dependence relations keep the conditions backing these dependence relations fixed. For example, in the case of nomic dependence, the same laws are assumed to hold. To cover the case of nomic dependence arising from how things are arranged at the start of the universe, the initial conditions may be assumed to be fixed too. It is plausible that this will cover the case of dependence between effects of a common cause. However, the dependence relation between cause and effect is one for which it is not the case that the dependence conditions are kept fixed, certainly not in a deterministic universe (Lewis (1973), pp. 13–19). To consider whether a candidate effect is present in the absence of a cause, there will be at least some cases in which we are evaluating circumstances in which the laws are a little different (or the particular matters of fact are) in order for the cause to be absent. Thus, Peter Simons was incorrect to suggest that causal dependence just involved a weaker sense of must than metaphysical dependence, although, setting aside the powers ontology, this is more plausible for the case of nomic dependence (Simons (1987), p. 295). So, there is what we may call *variable condition dependence* of which causation is the most familiar example.

As a sketch of the territory ahead, within the context of the clarifications already made, I provide the following diagram (Figure 2.1). I have left off relative necessary dependence and kindred notions because they will not detain us in the discussion ahead.

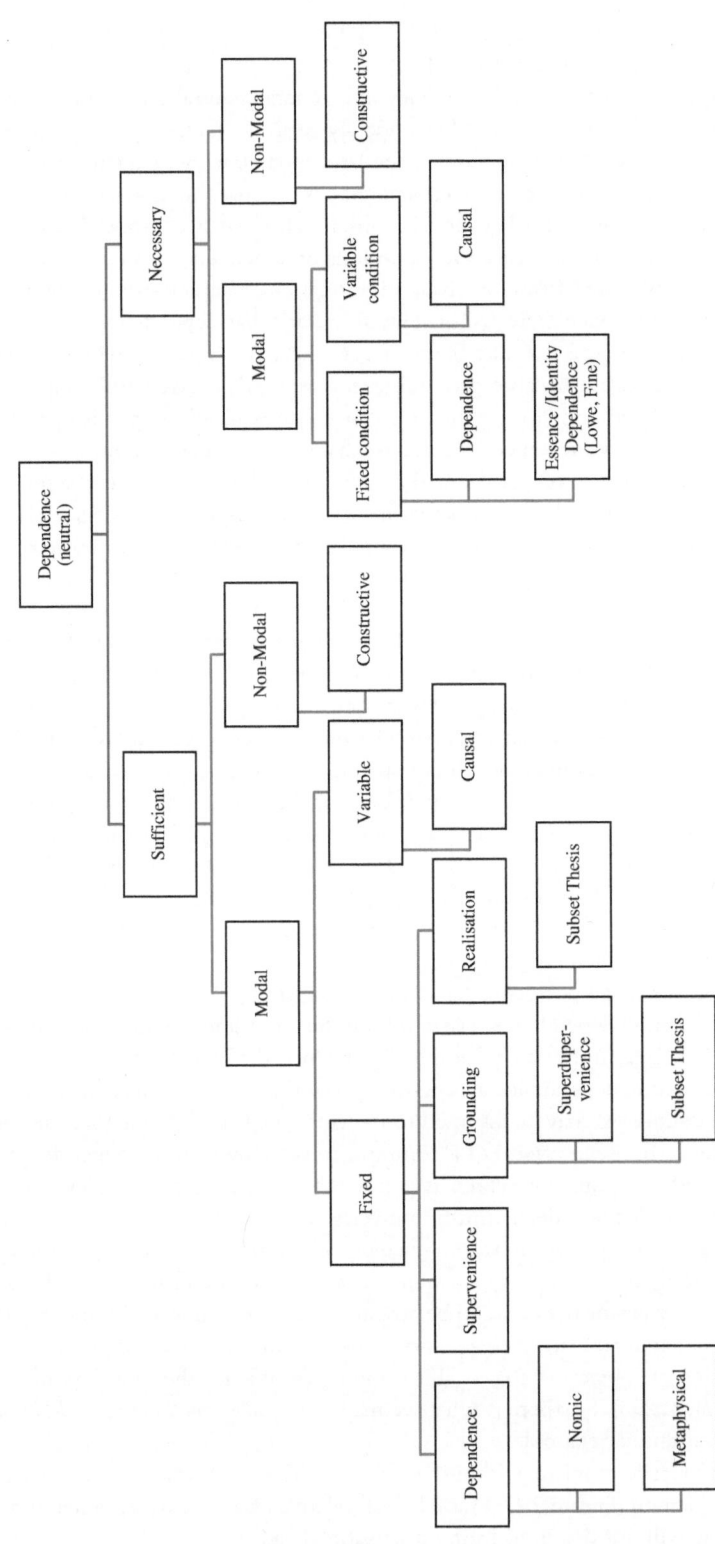

Figure 2.1 Various kinds of dependence.

The issue of non-modal accounts of dependence will be discussed later. They may require some adjustment to the diagram. Before that, I am going to focus on modal accounts and the issue that has received significant attention as of late, namely, whether a pure modal analysis can be given. This will be relevant to the issue of how we should formulate the distinction between doctrines of emergence and non-reductive physicalism.

The case against pure modal accounts of dependence

An important recent theme in the discussion of dependence has been the putative failure of the modal analysis of dependence. Figure 2.1 shows how the varieties of dependence have proliferated. This is largely due to the failure of the modal analysis. According to this analysis,

> X is necessary dependent upon Y if and only if necessarily, if X exists/is instantiated, then Y exists/is instantiated (see e.g. Simons (1987), p. 295).

> X is sufficient dependent upon Y if and only if necessarily, if Y exists/is instantiated, then X exists/is instantiated.

Where X and Y are property instances, replace 'exists' on the right-hand side with 'is instantiated'. For the purposes of our present discussion, the strength of necessity in these formulations should be taken to be metaphysical. A number of cases have been presented to demonstrate the problematic character of these analyses and in favour of a hyperintensional notion of dependence: grounding and its kin. A context is hyperintensional if and only if substituting logically equivalent propositions (or necessarily co-extensive terms) fails to be truth preserving.

Consider the relationship between Socrates and the singleton set {Socrates}. By the modal analysis noted earlier, each is necessary dependent upon the other. If {Socrates} exists, then Socrates exists and if Socrates exists, then {Socrates} exists. Each is also sufficient dependent upon the other as a result. Yet, the claim runs, {Socrates} is dependent upon Socrates but not vice versa (e.g. Fine (1995), p. 271; Lowe (2006), pp. 34–35).

A second type of case concerns metaphysically necessary existents (e.g. the number 2). By the modal analysis, my existence is necessary dependent upon 2's existence and 2's existence is sufficient dependent upon me. Yet it may be argued that these dependency claims should not fall out of the fact that 2 is a necessary existent alone and are, intuitively, false (Fine (1995), p. 271).

A third type of case concerns the essential properties of substances. The particular property of *Socrates' humanity* is essential to Socrates and can only be possessed by him. In which case, by the modal analysis of dependence, Socrates is necessary and sufficient dependent upon his humanity. From which it follows that, given substances are independent existences, Socrates is not a substance.

A related fourth case concerns events like that of being Socrates' life. Socrates' life couldn't fail to imply the involvement, and hence existence, of Socrates. Equally, Socrates couldn't exist without having his life, although its exact features may vary. Therefore, Socrates existence is both necessary and sufficient dependent on his life. Nevertheless, it is plausible that the event of being Socrates life is dependent on the existence of Socrates but not vice versa. To rule out another option, Socrates and his life cannot be identified because the life has a certain duration that Socrates does not, and Socrates has a certain weight that his life does not (Lowe (1998), pp. 143–145).

Modal analyses of dependence have been at the heart of analyses of the more complex notion of supervenience. Dissatisfaction with the utility of this notion has compounded concern about the appeal to modality. Supervenience is primarily a complex relation of sufficient dependence. A preliminary characterisation in terms of families of properties, A properties and B properties, is

> A properties supervene upon B properties if and only if there cannot be a difference of A-properties without a difference of B-properties.

or, as it is more normally put,

> it cannot be the case that there is indiscernibility with respect to B-properties without indiscernibility with respect to A-properties (e.g. see the first explicit modern use in Hare (1952), p. 145).

The basic idea differs from the simple case of dependence described earlier because it takes a family of properties, the B-properties, to be exclusively determinative of the A-properties. By contrast, if X is sufficient dependent upon Y, it is still possible that X may also occur without Y, either just by itself or because of some Z upon which it is also sufficient dependent. Nevertheless, like modal dependence, the envisaged relation is taken to be

> *Reflexive:* If A = B, then A supervenes upon B and B supervenes upon A.
> *Non-symmetric:* If A-properties supervene upon B-properties, it is possible that B-properties supervene upon A properties.
> *Transitive:* If A-properties supervene upon B-properties and B-properties supervene upon C-properties, then A-properties supervene upon C-properties.
> *(Kim (1984), p. 67)*

The preliminary understanding of supervenience is inadequate. As McLaughlin and Bennett point out, it has the implication that F properties supervene upon not-F properties. Something cannot fail to differ in F without differing in not-F. Yet it is not plausible to suppose that this relationship is the one intended by those who appeal to supervenience (McLaughlin and Bennett (2011), 3.5).

More developed accounts of supervenience avoid this consequence. Here are two illustrations. First, there is a version of Jaegwon Kim's notion of strong supervenience.

> A *strongly supervenes* on B just in case, necessarily, for each x and each property F in A, if x has F, then there is a property G in B such that x has G, and *necessarily* if any y has G, it has F. i.e. $\Box(x)(F)(Fx \& F\ e\ A \supset (\exists G)\ (G\ e\ B\ \&\ Gx\ \&\ \Box(y)(Gy \supset Fy)))$.
> *(cf. Kim 1984), p. 65*

If x has F, it does not have not-F, and it is not the case that, necessarily, if x has not-F, then it has F. There is also Frank Jackson's account of global supervenience developed from work by John Haugeland (1982), David Lewis (1983) and Terence Horgan (1982). Jackson holds that

> Any world which is a minimal physical duplicate of our world is a duplicate simpliciter of our world.
> *(Jackson (1998), p. 12)*

In more general terms,

> A properties supervene upon B properties if and only if, any world which is a B-duplicate, is a duplicate simpliciter (and, thus, an A duplicate).
>
> (cf. Kim 1984, p. 68)

Kim's characterisation allows for various readings of the modal operators. For example, in the case of nomological dependence, the second occurrence will be that of nomological necessity as will, plausibly, the first. The model for a supervenience formulation of metaphysical or ontological dependence is plausibly

> A *strongly supervenes* on B just in case, necessarily$_n$, for each x and each property F in A, if x has F, then there is a property G in B such that x has G, and *necessarily$_m$* if any y has G, it has F. i.e. $\Box_n(x)(F)(Fx \ \& \ F \ e \ A \supset (\equiv G) \ (G \ e \ B \ \& \ Gx \ \& \ \Box_m \ (y)(Gy \supset Fy)))$.
>
> (e.g. Noordhof (1999a), p. 295)

Here the subscripts 'n' and 'm' represent nomological and metaphysical necessity, respectively. The crucial point is that the instantiation of the G property is taken to metaphysically necessitate the instantiation of the F property. Kim suggests the opposite ordering, but this would mean that the supervening properties necessarily had a supervenience base but were just a nomological consequence of the supervenience-base properties (Kim (1984), p. 66).

As I noted earlier, supervenience adds to pure modal dependence relations the idea of exhaustive determination within a world. It is for this reason that supervenience formulations of physicalism and the relationship between the evaluative and the natural have been popular. Nevertheless, it has been argued that inadequacies in such formulations add to the case for impure modal, or non-modal, accounts of dependence. Some alleged inadequacies relate to specific features of global supervenience but, in fact, rest upon assumptions about what we should take as the supervenience-base of mental properties that should not be accepted. For example, consider two simple worlds. In w_1, a has F and G and b has G; in w_2, a has G but not F and b doesn't have G (where F is a mental property, G a narrowly physical one). Since the two worlds differ in G properties, they are compatible with the global supervenience of F properties on G properties. Although they are incompatible with the strong supervenience of F on G, they are not incompatible with the strong supervenience of F on an object having G and occurring in a world with another object having G. The latter might be a surprising supervenience-base for F, but if it holds in all possible worlds, then there is no reason to suppose that the instantiation of F is incompatible with physicalism, contrary to what Kim implies (Kim (1987), pp. 319–326); the original case is Petrie's (1987, p. 121), though he does not draw Kim's conclusion.

The more substantial inadequacies of formulation are brought out by comparison with other doctrines such as dualism within an occasionalist framework, emergence and ethical non-naturalism. I go through them in turn.

A version of Malebranche's occasionalism holds that God takes arrangements of narrowly physical properties as the occasion to bring about the instantiation of non-physical mental properties. If God is a consistent necessary existent, then in all possible worlds, corresponding to each arrangement of physical properties, he has a distinct non-physical mental property instantiated. Thus, there is a metaphysically necessary connection between arrangements of physical properties and avowedly non-physical properties (Wilson (2005); Wilson (2014), p. 543).

We can deal with this case now. One response is to distinguish between intra-world and inter-world consistency. While a consistent God would have a care to ensure the same narrowly physical properties caused the same non-physical mental properties within a world, there is no reason to think that consistency requires the same non-physical mental properties across all possible worlds. A second is to note that God cannot falsify necessary truths and specifically, necessary truths about possibilities (assuming S5 modal logic). Thus, if the distinct natures of P_1 and M_1 are such that it is necessarily true that it is possible that P_1 and not-M_1, then God cannot make it otherwise. To claim he can is question begging against the proposed definition of non-reductive physicalism. A third is to count this counterexample as a special case. The characterisation of non-reductive physicalism would have a 'no divine power' external condition. Finally, of course, if it turns out that a putative necessary God is not possible, then there is no need to deal with this case.

The second case concerns the proper characterisation of emergentism if a powers ontology is true. A distinctive feature of the latter is that fundamental properties are powers that have their causal profiles essentially. Suppose that one part of G_1's causal profile is that, in circumstances C (as specified by the instantiation of an arrangement of powers), G_1 causes the instantiation of F_1 (a candidate non-physical mental property). Then the connection between C, G_1 and F_1 is that of metaphysical necessitation. So, appeal to such a connection cannot capture what is characteristic of non-reductive physicalism. Another example with the same conclusion is the case of ethical non-naturalism which I have discussed further elsewhere (Noordhof (2003)).

In the next section, we will examine whether accounts that are not purely modal succeed in analysing these cases.

Modality plus accounts: grounding and ontological dependence

In this section, I am going to consider accounts that take necessitation to be a necessary but insufficient condition for the relevant connection. There is an additional element.

Identity or essence dependence

One approach that has found favour with many with regard to initial cases described is to develop a notion of identity or existence dependence appealing to essence. For example, E. J. Lowe and Tuomas Tahko have suggested that

> x depends for its identity upon y $=_{df}$ There is two-place predicate 'F' such that it is part of the essence of x that x is related by F to y (Tahko and Lowe (2015), 4.2) (a slight variant is Koslicki's talk of y being a constituent of x's essence).
>
> *(see Koslicki (2012), p. 190)*

Similarly Fine has argued that

> x (essence) depends upon y $=_{df}$ It is part of the essence of x that it exists only if y exists.
>
> *(Fine (1995), pp. 272–273)*

To apply the former to the case of Socrates and {Socrates}, the two-place predicate is '– has as a member –', x is {Socrates} and y is Socrates (see also Lowe (1998), p. 149). It is plausible that my existence is not identity dependent upon the number 2. Part of the essence of Socrates' life is that it is lived by Socrates, and part of the essence of Socrates' humanity is that it is possessed by Socrates. Thus, Lowe's account of identity dependence deals with many of the cases listed earlier. Similar claims about the essence of these things will be the basis for corresponding claims about essence dependence in the formulation given by Fine. Both of these are examples of necessary dependence because they talk of *part* of the essence of x being such and such. So we don't as yet have an account of sufficient dependency appealing to essence. As we shall shortly see, this proves more problematic.

The definitions also have some surprising consequences. Many non-reductive physicalists who are proponents of a powers ontology hold that for each mental property M_i, its causal powers are a subset of the causal powers of the specific arrangement of the narrowly physical properties, $A_i(p_1, p_2, \ldots p_n)$, that realise them. Although we shall see later that the subset claim is problematic, the point for now is that it follows that physical properties are identity and essence dependent upon the mental properties they putatively realise. This is the opposite of what was required for the non-reductive physicalist. More generally, suppose that the instantiation of M_i is part of the causal profile of $A_i(p_1, p_2, \ldots p_n)$, then $A_i(p_1, p_2, \ldots p_n)$ is *identity* dependent upon M_i. Let $A_i(t, p_1, p_2, \ldots p_n)$ be the causal basis for M_i, where t is the trigger for $A_i(p_1, p_2, \ldots p_n)$ to instantiate M_i. It is part of the essence of $A_i(t, p_1, p_2, \ldots p_n)$ that M_i is instantiated. In which case, the causal basis of M_i is essence dependent upon M_i. Again, this is the opposite of what is required.

So while these notions may capture the idea that water depends for its existence and identity on hydrogen, say, because part of the essence of water includes hydrogen, these notions do not help with the characterisation of non-reductive physicalism in a powers ontology.

Grounding

The grounding relation has typically been taken to include two elements. First, the grounding entities are somehow explanatory of the grounded entities (Rosen (2010), pp. 122–126; Dasgupta (2015), p. 558). Indeed, Paul Audi argues for the existence of the grounding relation from the observation that there are *non-causal explanations* of why certain facts are the case; e.g. the fact his shirt is maroon grounds the fact his shirt is red (Audi (2012a), pp. 688, 693). Second, the grounding entities are ontologically more fundamental ('prior' to the grounded) (e.g. see Schaffer (2012), p. 122; Barnes (2012), pp. 875–876; Audi (2012a), p. 686). The grounded entities are derivative. However, Elizabeth Barnes explicitly dissociates fundamentality from explanatoriness, so these two elements are not to be taken as an inevitable package (Barnes (2012), pp. 897–899).

As it is standardly envisaged, grounding is a case of sufficient dependency. Thus, Gideon Rose writes that suppose the fact that p, written $[p]$, is grounded, then typically several facts together ground $[p]$. The general form of the grounding claim is $[g_1]$ and $[g_2]$ and $[g_3] \ldots$ ground $[p]$. The notion of a partial ground is of a member of the set that grounds $[p]$ (Rosen (2010), p. 115). Grounding is taken to imply metaphysical necessitation but involve either or both of the additional features mentioned earlier (Rosen (2010), p. 118).

In contrast to supervenience, grounding is generally taken to be asymmetric and irreflexive (e.g. Audi (2012b), p. 102). A third plausible feature is transitivity. If $[p]$ grounds $[q]$ and $[q]$ grounds $[r]$, then $[p]$ grounds $[r]$. Jonathan Schaffer offers counterexamples to transitivity, but his counterexamples appeal to partial grounding rather than grounding, whereas those, like Rosen, asserting transitivity have full grounding in mind (Schaffer (2012); Rosen (2010), p. 116).

What in addition to metaphysical necessitation is required for grounding? Just as in the case of identity or essence dependence characterised earlier, there has been an appeal to essence. Fine argues that the following will be *true of the essence* of grounded properties, say the grounded property A:

$B_1(x_1, x_2 \ldots), B_2(x_1, x_2 \ldots)$ is a ground for the truth C whenever $A(x_1, x_2 \ldots), B_1(x_1, x_2 \ldots),$
$B_2(x_1, x_2 \ldots)$ (where $A(x_1, x_2 \ldots), B_1(x_1, x_2 \ldots), B_2(x_1, x_2 \ldots)$ hold of an object a, say) (Fine

(2012), p. 75, a similar view is adopted in Audi (2012b), p. 109, except the latter also allows such truths to be part of the essence of grounding property where Fine does not).

Here the truth C will be either attributing A to an object or the claim that something is A. Since facts involving property A may be grounded in facts involving different properties – for example, something is coloured may be grounded in facts about objects being red, green, blue, etc. – there will be a host of these grounding truths that are part of the essence of A. Because they hold of the essence of colour, it is appropriate to say that something is coloured in virtue of having one colour or another (Fine (2012), p. 77).

This relates grounding to the essence of properties by brute force. With essence dependence, a property was dependent on another property if that property was part of the essence. Dependence was analysed by reference to essence. This relationship is not available in the case of grounding because, due to variable realisation, no particular ground is part of the essence of a property. So, instead, it is suggested that it is part of the essence of a property that it should be grounded in such and such a way, depending upon what holds. This does not explain grounding in terms of essence. It takes grounding to be part of what characterises the essence. Mention of grounding cannot be excised. If, instead of the envisaged formula, we just appealed to a conditional relating the putative grounds to the putative grounded, then it is just as plausible to say that this kind of conditional is part of the essence of the grounding properties too, in which case, we would have no basis for the asymmetry of grounding. If part of the nature of $B_1(x_1, x_2 \ldots)$, $B_2(x_1, x_2 \ldots)$ is the fact that $A(x_1, x_2 \ldots)$, then the former would depend on the latter. That's why, as Fine acknowledges, there must be explicit reference to grounding (Fine (2012), p. 78).

Fine, then, is committed to claiming that part of the essence of pain is that, if a subject has c-fibre firing, then c-fibre firing is a *ground* for pain. But there is little evidence that this is so (Rosen (2010), p. 132). While it might be part of the essence of colour that there are various ways of being colour-wise that ground it, it does not seem to be true of pain that there are different physical ways of being in pain that are part of the essence of pain – for example, that it is c-fibre firing. Of course, this may be offered as a conjecture. An alternative is that it is part of the essence of various different physical conditions that they should be grounds of pain. Paul Audi envisages it is both part of the essence of the grounding *and* the grounded properties that the former ground the latter (Audi (2012b), p. 109). A cost of either of these other options is that the grounding properties then become essence or identity dependent upon the grounded properties in the sense specified in the previous section. They are instantiated in virtue of grounding certain facts about the mental, in which case, the alignment is lost between the fact that *p is grounded in* the fact that q and the fact that q does not hold *in virtue of* the fact that *p* but vice versa.

Although Fine's position avoids this last consequence, it has a further cost. Consider

(P) The fact that S's brain instantiates $A(p_1, p_2, p_3 \ldots p_n)$ grounds S's being in pain.

Is (P) itself grounded in arrangements of narrowly physical properties or not? Fine's preferred approach suggests not. The explanation at least partly involves a fact about the essence of being in pain (Dasgupta (2015) develops the point generally, pp. 565–576). In this case, how does grounding serve to differentiate itself from other connections between the mental and the physical, for example, nomological necessitation, metaphysical necessitation and the like, in attributing *priority* to arrangements of physical properties? All appeal to explanatory principles relating facts at different levels and grounding appeals to facts about mental properties such as being in pain. It seems that the proponents of grounding must argue that their relation is special. Although appeal might be made to the essence of pain, since this essence involves various ways in which pain is *grounded* in arrangements of narrowly physical, such an appeal is unproblematic. It is not sufficient to argue, as Shamik Dasgupta does, that principles based on essences (or real definitions) are not apt to be grounded because the issue is whether the essence of pain (for example) to which the putatively ungroundable (P) appeals is of an acceptably physical thing (cf. Dasgupta (2015), pp. 577–580). So grounding plays a primitive role in the characterisation of essence and, it must be assumed, a primitive role in the explanation of why inter-level grounding claims are acceptable to the physicalist without claiming that this is because the essence of pain is grounded in the physical. Truly, the sui generis is coming in aid of the theoretically problematic!

For some, grounding is just a certain kind of non-causal explanatory relation between worldly facts. It is not meant to capture the idea that the grounded involve *nothing over and above* what they are grounded in (Audi (2012a), pp. 708–710). However, this threatens the motivation for considering grounding to be distinct from metaphysical necessitation and, thus, its putative role in distinguishing between non-reductive physicalism and emergentism. More, thus, are tempted to see the matter as follows. In the case of non-reductive physicalism, facts about mental properties are *grounded* in facts about arrangements of narrowly physical properties, but not vice versa (e.g. Bliss and Trogdon (2014), 6.1; Dasgupta (2015)). In the case of emergentism, they are not. Instead the relationship is either nomological or metaphysical necessity depending upon whether the emergent properties are non-physical or physical but with emergent causal powers.

Nevertheless, the characterisation has a substantial drawback. It mistakenly classifies a weak version of non-reductive physicalism as a form of emergentism. Without the aid of God to ensure that one property metaphysically necessitates another, the claim that one property metaphysically necessitates another indicates that the nature of the former is intimately connected with, and exhausts in character, the latter. Otherwise, there would still be a way the world could fail to be, given the instantiation of the former. Consider a position that makes the following three claims; first, that arrangements of narrowly physical properties metaphysically necessitate mental properties; second, that there are no emergent causal powers; and, third, we should not recognise a distinction between fundamental and derivative properties. This deserves to be classified as a version of non-reductive physicalism, which we might call harmony physicalism (cf. Noordhof (2003), p. 106). It recognises a layered world in the sense given earlier, from the subject matter of physics upwards, with the requirement of harmony between the layers characterised by metaphysical necessitation. I don't deny that many non-reductive physicalists want the additional fundamentality claim. However, we should not allow their preferences to dictate the most general characterisation of the doctrine of non-reductive physicalism.

This suggests the following picture of various kinds of non-emergent physicalism and emergence (Figure 2.2).

A related approach put forward by Elizabeth Barnes holds that emergent entities are *dependent* fundamental entities. She characterises dependence as follows:

> (OD) An entity *x* is dependent if for all possible worlds w and times t at which a duplicate of *x* exists, that duplicate is accompanied by other concrete, contingent objects in w at t.
> *(Barnes (2012), p. 880)*

Independent entities are capable of lonely existence, that is, existence in possible worlds without other objects. The notion of a fundamental entity is taken to be a primitive. An entity is fundamental if its existence is not derivative from any other entities (Barnes (2012), p. 876). Or, as one might say, its existence is ungrounded.

(OD) is a purely modal characterisation of dependence. It is not meant to have any implications about what is fundamental and what is not. We have already noted that a standard

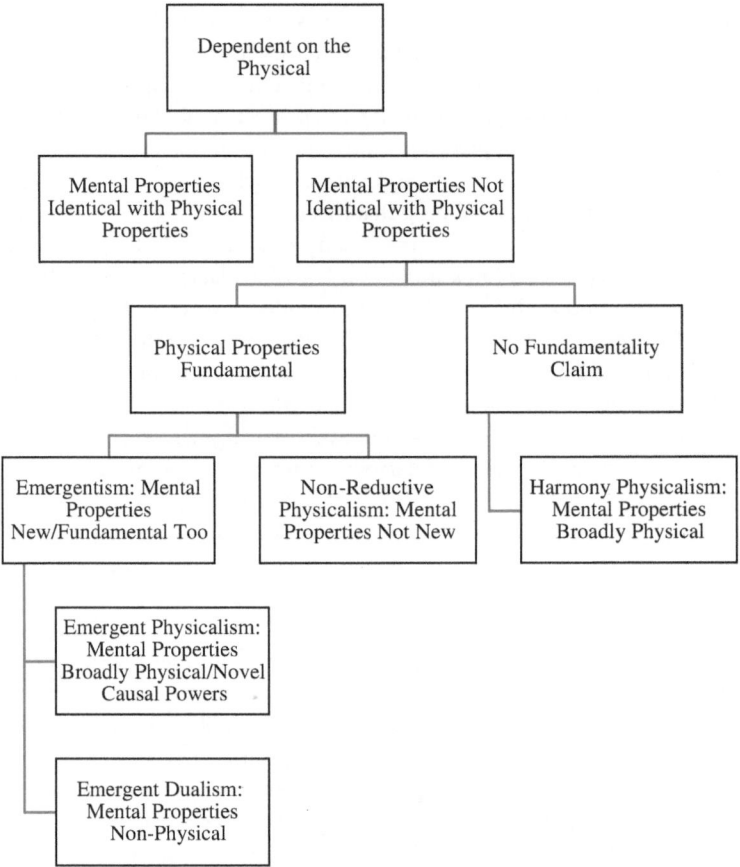

Figure 2.2 Various kinds of mental dependence upon the physical.

understanding of emergence is that arrangements of narrowly physical properties nomologically, but not metaphysically, necessitate non-physical mental properties. Do the mental properties come out as dependent on this picture?

The answer might seem to be straightforward. If the connection is only one of nomological necessity, then there will be possible worlds in which it does not hold. In which case, mental properties turn out not to be dependent by (OD), which requires that there must be some entity or other in addition to the instances of mental properties from which these mental properties emerge. The standard understanding of emergentism fails to be accommodated.[1] However, that's not quite right. Although the nomological necessitation of mental properties may be contingent, that there is some nomologically necessary relationship or another may be metaphysically necessary. If that is correct, the immediate problem is avoided.

Instead, the answer would seem to turn on whether the brain, or the arrangements of narrowly physical properties that constitute it, is a distinct concrete contingent object accompanying the instances of mental properties. Suppose the brain (to fix ideas) possesses both the arrangements of narrowly physical properties and the non-physical mental properties, which are nomically related to them. Then it isn't a distinct concrete contingent object. In the absence of an alternative viable candidate, the non-physical mental properties would fail to be classified as emergent, which is counterintuitive.

Perhaps Barnes takes non-physical mental properties to be instantiated in an object distinct from the brain. If that is the case, then there seems no reason why there couldn't be a world in which instances of the mental properties are instantiated without the brain. Mental properties would fail the analysis of dependence and so not count as emergent, which is also a counterintuitive result.

Barnes might respond that she is providing a characterisation of emergence, not defending it as a doctrine in the mental case. However, she seems to lose a distinctive feature of emergence. It doesn't have to be claimed that emergent entities are dependent in all possible worlds. It is simply that, in this world (for example), they are so dependent.[2]

Suppose, instead, that arrangements of physical properties metaphysically necessitate mental properties. If mental property instances are distinct objects from the arrangements of physical property instances, then this would make these arrangements of physical property instances dependent entities. Moreover, if these arrangements of physical properties are also fundamental, then the resulting position seems to be classified as a case of narrowly physical properties emergent from mental properties. In contrast with typical cases of emergence, these arrangements of physical properties would be dependent upon entities that might not, themselves, be fundamental, viz the mental properties. So not only does a standard characterisation of non-reductive physicalism get classified as physical emergence, but, in addition, we have a new bifurcation in the characterisation of emergence turning upon whether or not the entities upon which target emergent entities are dependent are fundamental.

The alternative is for Barnes to deny that mental property instances are objects distinct from the arrangements of narrowly physical properties that metaphysically necessitate them. But this would undermine the motivation for her position. Metaphysical necessitation would rule out the necessitated properties being something over and above the necessitating properties, just as proponents of a purely modal account require.

It is time to take a step back. Proponents of grounding accounts take themselves to be postulating a tighter, more explanatory, relationship between entities than metaphysical necessitation. An alternative is to deny that there is a tighter relationship but place the emphasis instead on characterising entities on the necessitating side as fundamental. If the latter works, the motivation for grounding is significantly undermined. [p] holds in virtue of [g₁] and [g₂] not because of the nature of the grounding relation between [p] and *[g₁] and [g₂]* but because *[g₁] and [g₂]* are fundamental and metaphysically necessitate [p]. Thus, for example, non-reductive physicalism with the fundamentality claim will be characterised in terms of supervenience, plus the claim that narrowly physical properties are fundamental (see Bricker (2006), pp. 270–272).

Given that proponents of grounding take grounding to be a primitive, it is within the rights of this alternative position to, as Philip Bricker recommends, take *fundamentality* as a primitive notion. However, there are features to which we might appeal to provide some flesh to the notion of fundamentality.

First, suppose, for some target property, M, P metaphysically necessitates M but not vice versa. Instead, at best, there is some disjunctive property P_1 or P_2 or P_3 . . . or P_n such that, metaphysically necessarily, if M is instantiated, then P_1 or P_2 or P_3 . . . or P_n is instantiated. The fact that M fails to settle the detail of the world with respect to which P_i necessitates it but that some P_i does settle the detail of the world M-wise indicates a way in which the P properties are more fundamental. Call this *differential determination*. Second, and relatedly, suppose laws relating the P properties will explain why Ms are instantiated, given Ps metaphysically necessitate Ms, but the reverse does not hold. That doesn't mean that emergence is impossible. I have described conditions under which it holds earlier. However, these require special claims about which laws are fundamental, etc. In their absence, the default will be that P properties have a certain priority for the reason just given. Call this *complete coverage*. Third, although there are some P properties that necessitate Ms, there are other P properties that do not, whereas the reverse is not the case. There are no M properties without the P properties that necessitate them. Thus, P properties constitute an additional level of detail about the world. Call this *extra specificity*. Fourth, if there are relationships of the following sort

$$\Box_m (x)(A(g_1, g_2)\ x \leftrightarrow (Fx))$$

then the left side is fundamental seems more fundamental because it is made up of an arrangement of properties, for at least some of which properties such relationships don't hold, but not vice versa. Call this *decompositional extra specificity*. Fourth, the P properties may be identified as purely natural properties that capture facts about resemblance and duplication in objects and, as a result of which, figure in laws of nature. Let's dub this *naturalness*.

All of these features support the idea that the target properties, or facts about them, hold *in virtue of* properties with these features. However, there is one way in which the relationship between the target properties and the necessitating properties may be made tighter without abandoning the guiding idea of a purely modal account. That is, to identify the minimal metaphysical necessitation base of the target properties rather than any old necessitation base, however supplemented (see e.g. Noordhof (1999a), p. 307). The minimal necessitation base will have no redundant elements. The target properties should have relative necessary dependence upon them (to deal with the issue identified by Fine (2012), p. 57). With this final adjustment in place, the issue is whether there is any further need for an appeal to grounding. We shall examine this in two ways in the final section.

Modality and construction

The first few cases discussed in the second section show that there are additional dependence relations to one characterised purely modally. But they needn't show that these are appropriate for the characterisation of the distinction between non-reductive physicalism and emergence, and they don't undermine the status of a purely modal account as one kind of dependence relation. Let me go through the cases in turn.

In the case of Socrates and {Socrates}, it is clear that the latter essence depends upon the former. It is part of the essence of {Socrates} that it has Socrates as a member but not part of the essence of Socrates that he figure in the unit set of Socrates. By contrast, because my essence does not include the number 2, or vice versa, neither I nor the number 2 essence depends upon the other. In the case of Socrates' humanity or life, each essence depends upon Socrates, but it is not part of the essence of Socrates that it is characterised by a particular case of humanity or living a certain life (Lowe (1998), p. 153).

As we have already seen, essence dependence is not the proper way to characterise the dependence of mental properties on physical properties. Indeed, it threatens to get the priority relations the wrong way round in the case of a powers ontology. Here purely modal accounts fare better. They also capture an important notion of dependence in themselves, namely the strength of correlations between the existence and/or features of different entities.

Some limit this dependency to contingent existences (e.g. Simons (1987), p. 295). However, we might take purely modal dependencies between, say my existence and the number 2, to be a merely trivial but harmless case, because 2 is a necessary existent. Another option is to deny that necessary existents are independent existents. Rather, they are existents that depend upon the conditions for the existence of any possible world. So, while I am not sufficient dependent for my existence on the number 2, the number 2 is sufficient dependent upon my existence (as part of the world in general) from which its existence derives.

This brings me to the case of distinguishing between non-reductive physicalism and emergentism given a powers ontology. There are at least two relevant kinds of emergentism. A non-physical emergentism – emergentist dualism – holds that arrangements of narrowly physical properties generate novel non-physical qualities. A physicalist emergentism holds that no non-physical qualities are generated. Instead, certain highly distinctive arrangements of narrowly physical properties give rise to novel causal powers.

It is not obvious that a powers ontology has to recognise the first kind of emergentism. A typical motivation for emergentist dualism is that the felt qualities of experience – qualia – are neither physical properties nor functional properties realised by physical properties. Many proponents of a powers ontology hold a functionalist view of all properties – taking their nature to be exhausted by their causal profile – and thus a functionalist view of qualia too, which is compatible with physicalism. Even those who do allow an internal, qualitative aspect to powers do not view this as specifically non-physical. It is said to be identical to, or entail, the causal profile (e.g. Martin (1997), pp. 215–217). So we can just set this case aside. Nevertheless, it is also worth noting that the move I'm about to make regarding the second kind of emergentism could be developed with respect to the first.

How are we to make sense of the idea of novel causal powers in the context of physicalist emergentism? The *collapse objection* is that this is not possible either. If physicalist emergentism is true, then M (an emergent non-physical mental property) has a novel causal power F. $A_1(P_1, P_2, P_3)$ metaphysically necessitates M. This is conceded even by those who deny that metaphysical necessitation is sufficient for physicalism. So $A_1(P_1, P_2, P_3)$ has F. Properties, or arrangements of properties, have the causal powers of the properties they metaphysically necessitate. A variant of this argument would say that P_1 has F in circumstances $A_1(P_1, P_2, P_3)$. Either way, F is not a novel causal power of M after all.

The only answer that seems available is that the causal powers relating to M are relatively isolated and exceptional when compared with the rest of the causal framework. The laws responsible for them are independent of the laws relating to the rest of the framework. In this case, it is reasonable to argue that narrowly physical properties may lack this exceptional part of their causal profile in other possible worlds and still be instantiated. It is not essential to their essence in the way powers related to the broader network of narrowly physical properties are. So the very considerations that favour emergence undermine the motivation for insisting that the relevant arrangements of narrowly physical properties *metaphysically necessitate* novel causal powers. In this case, the same way of differentiating between dualistic emergentism and non-reductive physicalism can be applied to the difference between the latter and physicalist emergentism. The connection between the arrangements of narrowly physical properties and their emergent causal powers is nomologically rather than metaphysically necessary. The mistake is to suppose that physicalist emergentism must be characterised in terms of causal powers $A_1(P_1, P_2, P_3)$ does not have, as opposed to having them but via independent fundamental laws.

Other responses to the collapse objection provide further support for the move I have just made. One response is to argue that characterisation of powers reflects the forces in which they are grounded. Many of the powers of $A_1(P_1, P_2, P_3)$ are reflections of fundamental narrowly physical forces like electromagnetic interaction or gravitation (Wilson (2002)). If there are novel emergent powers, these will be grounded in novel forces. However, this reinforces the point that there may be worlds in which $A_1(P_1, P_2, P_3)$ fails to have the powers associated with M. If there are genuinely novel forces, they will be unrelated to the other forces and, hence, these other forces, and the powers they ground, may occur independently.

A second response is to argue that emergent properties must be instantiated in new objects and thus that the emergent causal powers are not possessed by $A_1(P_1, P_2, P_3)$ but by this new object. As Umut Baysan and Jessica Wilson note, there are various ways in which this claim may be motivated (Baysan and Wilson (2017)). The remaining question is how these objects may be characterised as new – that is something over and above $A_1(P_1, P_2, P_3)$ from which they arise – and yet dependent upon $A_1(P_1, P_2, P_3)$ or other arrangements of narrowly physical properties so that they cannot occur without some such arrangement.

Tensions in the position seem exacerbated by Baysan's, at least, preferred motivation for postulating a new object, namely that a property's causal powers are derived from their possession by objects. If the new object is dependent upon $A_1(P_1, P_2, P_3)$, the latter metaphysically necessitating the new object having the causal power F, then how can it be denied that F is derived from an object's possessing $A_1(P_1, P_2, P_3)$ too? What extra strict notion of *derivation* is the basis for the distinction? On the other hand, any weaker connection between $A_1(P_1, P_2, P_3)$ and the new object allows us to draw the distinction between non-reductive physicalism and physicalist emergentism in terms of the difference between nomological and metaphysical necessity once more.

In place of an appeal to the modal difference I have defended, many proponents of a powers ontology appeal to what has been dubbed the *subset thesis*. It holds that

> An instance of P realises an instance of M if and only if the token causal powers of m on a given occasion are a non-empty subset of the token causal powers of p (where 'm', 'p' are instances of M, P).
>
> *(Wilson (2011), p. 128)*

If the subset thesis holds for all mental properties, non-reductive physicalism is true. If the causal powers of some mental properties are not a subset of any narrowly physical properties or arrangements thereof, then emergent physicalism is true. The treatment of the collapse objection demonstration already indicates that this proposal is problematic. However, my argument will be that it is problematic even if the objection can be met.

The subset thesis rests upon the idea of one kind of entity being constructed from, and nothing over and above, another entity or entities without any assumption about or appeal to modal consequences. There are a number of other potential cases:

> *Set Formation*: For any objects, x, y, z . .., there is a set $\{x, y, z \ldots \}$
> *Part-Whole*: A whole is the mereological sum of its parts.
> *Constitution*: Matter and form together constitute an object (Fine (1999))
> *Micro-basing*: A property P by O instantiated as a result of properties of entities that are a part of O (Kim (1998), pp. 113–114)
> *Event Constitution*: An event occurs if O has P at t (Kim (1976), pp. 160–161)
> *Object Formation*: Objects are *bundles* of properties at times (all mentioned by Bennett (2011), pp. 83–84)

Modal consequences are not ruled out, but if they hold, they hold as a result of the nature of the constructed entity and not as a result of the construction. Consider the case of realisation given earlier. Suppose that an instance of M has causal powers that are a subset of the causal powers of an instance of P. It doesn't follow that metaphysically necessarily, if P then M unless, in every possible world, M has the same causal powers or its causal powers are always a subset of P's causal powers. This may be so, but it doesn't fall out from the construction relation.

The characterisation of event constitution given a moment ago is another illustration. According to Jaegwon Kim, events are constituted from objects having properties at times. Nevertheless, he denied that the constitutive object, property or time of an event is among its essential properties (Kim (1976), pp. 171–173). Others have disagreed, arguing that events need some or all of these as essential properties in order for reference to events to play the appropriate role in causal explanations (Yablo (1992), pp. 414–419). I think this is mistaken, but it is not a point I want to emphasise here (Noordhof (1998c)). Rather, the point is that the modal properties of events are not argued to follow from how they are constructed. Similarly, opinions differ over constitution. Lynne Rudder Baker emphasises that it is a contingent matter whether those things that constitute an object entail its existence, Kit Fine would, I presume, claim it is necessary (Baker (2000), p. 34; Fine (1999), pp. 73–74).

Although these various forms of construction are a way in which one entity may be dependent upon other entities, it is not immediately helpful to the characterisation of the difference between emergence and non-reductive physicalism. The relevant notion of construction – the subset thesis – fails to provide the required demarcation.

Here is one way of developing the argument I advertised. Suppose that a particular type of pain, P, is variably realised in creatures with different biologies – humans and some rather strange creatures from Alpha Centauri VI – and it is also realised in robots: by $A_1(N_1, N_2, N_3)$, $A_2(P_1, P_2, P_3)$ and $A_3(S_1, S_2, S_3)$, respectively. Each of these states gives rise to a certain distinctive kind of pain behaviour – B_N, B_P and B_S – corresponding to the differences in constitution of the creatures in question. The non-reductive physicalist is committed to instances of P being capable of causing each of these types of pain behaviour when realised in these different ways. Their causal powers include, then, causing in different circumstances B_N, B_P and B_S. However, the *realisations* of P aren't capable of causing the pain behaviours of the different types but just their own type. Therefore, the causal powers of P exceed those of its realisers (Noordhof (1997), p. 246, repeated in Noordhof (1999b), pp. 113–114, and developed at greater length in Noordhof (2013), pp. 98–106). Some proponents of grounding have made a similar, but more general, point by saying that a grounded property's commitments about the nature of fundamental reality are not less than, but indeed exceed, those of its grounds. This is because the commitments of the grounded property derive from each of its grounds and not just one of them (Trogdon (2013), p. 114).

This is not to deny that the subset thesis may hold for certain properties. For example, pain in humans may be realised by $A_1(N_1, N_2, N_3)$. The former's causal powers don't outstrip those of its realisation base because it does not have distinct realisation bases. It fails to be identical with its realisation base because the neurophysiological base has additional causal powers not associated with pain. Non-reductive physicalism, though, is supposed to take the variable realisation of mental properties seriously. It holds that these properties are not reducible to other properties but are genuine properties in their own right. These features of the non-reductive physicalist position are responsible for the failure of the subset thesis.

By contrast, the purely modal characterisation of the relationship between mental properties and arrangements of narrowly physical properties is well placed to characterise the relationship between their causal powers. According to the non-reductive physicalist, the causal powers of mental properties are the union of a subset of the causal powers of each of their minimal supervenience-bases. Thus, failure for one property to be locally constructed from other properties is insufficient for emergence to be true, as it is generally understood. It is an interesting fact that non-reductive physicalists seem to be committed to the failure of certain local constructive relations holding as a result of a more global modal relationship.

Concluding remarks

We have seen that dependency relations come in various sorts. First, there are purely modal dependency relations. We saw how they may be used to characterise emergence. Second, there are constructive relations. These are of significant interest but appear insufficient to characterise emergence because they group non-reductive physicalism along with it.

Many of the classic cases cited to motivate talk of grounding only required a notion of necessary dependence rooted in essences. The sufficient dependence notion of grounding faced technical difficulties relating to how grounding drew on an appeal to essence and appeared unnecessary to the formulation of the difference between emergence and non-reductive physicalism. Indeed, a combination of a purely modal characterisation of the connection plus a characterisation of entities on one side of the dependency as fundamental had a better claim to capture what some folk have in mind.

Although essence dependence appears to capture a distinctive kind of necessary dependence relation even here, it is unclear whether the appeal to essence is doing significant work. An alternative is that this kind of dependence is a consequence of a particular combination of construction and a purely modal account. Thus, to illustrate, {Socrates} is constructed from Socrates by the set membership construction relation. The connection between Socrates and {Socrates} has modal force because it is metaphysically necessary to {Socrates} that it is constructed in this way. If that's right, modality is not to be understood in terms of essence, but rather essences are ways it is metaphysically necessary for certain entities to be constructed. If constructive relations and modality have priority, then the case of non-reductive physicalism shows that we must look to modal differences to understand emergence.

Notes

1 A paper I discovered subsequent to developing the argument of this part of the chapter makes this point (Pearson (2017), p. 8).
2 See Pearson (2017, p. 7) for the same objection developed independently.

References

Alexander, Samuel (1920), *Space, Time and Deity*, 2 vols. (London, Macmillan).
Audi, Paul (2012a), 'Grounding: Toward a Theory of the *In-Virtue-Of* Relation', *Journal of Philosophy*, 109, no. 12, pp. 685–711.
Audi, Paul (2012b), 'A Clarification and Defense of the Notion of Grounding', Fabrice Correia and Benjamin Schneider (eds.), *Metaphysical Grounding* (Cambridge, Cambridge University Press), pp. 101–121.
Baker, Lynne Rudder (2000), *Persons and Bodies* (Cambridge, Cambridge University Press).
Barnes, Elizabeth (2012), 'Emergence and Fundamentality', *Mind*, 121, no. 484, pp. 873–901.
Baysan, Umut and Jessica Wilson (2017), 'Must Strong Emergence Collapse?', *Philosophica*, 91: 49–104.
Bennett, Karen (2011), 'Construction Area (No Hard Hat Required)', *Philosophical Studies*, 154, pp. 79–104.
Bliss, Ricki and Kelly Trogdon (2014), 'Metaphysical Grounding', Edward N. Zalta (ed.), *The Stanford Encyclopedia of Philosophy* (Winter 2016 Edition), https://plato.stanford.edu/archives/win2016/entries/grounding
Bricker, Phillip (2006), 'The Relation between General and Particular: Entailment vs. Supervenience', *Oxford Studies in Metaphysics*, Vol. 2 (Oxford, Oxford University Press), pp. 251–287.
Dasgupta, Shamik (2015), 'The Possibility of Physicalism', *The Journal of Philosophy*, 111, pp. 557–592.
Fine, Kit (1995), 'Ontological Dependence', *Proceedings of the Aristotelian Society*, 95, pp. 269–290.
Fine, Kit (1999), 'Things and Their Parts', Peter A. French and Howard K. Wettstein (eds.), *Midwest Studies in Philosophy*, Vol. 23 (Boston, Blackwell), pp. 61–74.
Fine, Kit (2012), 'Guide to Ground', Fabrice Correia and Benjamin Schneider (eds.), *Metaphysical Grounding* (Cambridge, Cambridge University Press), pp. 37–80.

Hare, R. M. (1952), *The Language of Morals* (Oxford, Oxford University Press).

Haugeland, John (1982), 'Weak Supervenience', *American Philosophical Quarterly*, 19, pp. 93–103.

Horgan, Terence (1982), 'Supervenience and Microphysics', *Pacific Philosophical Quarterly*, 63, pp. 29–43.

Jackson, Frank (1998), *From Metaphysics to Ethics* (Oxford, Oxford University Press).

Kim, Jaegwon (1976), 'Events as Property Exemplifications', M. Brand and D. Walton (eds.), *Action Theory* (Dordrecht, Reidel), pp. 159–177.

Kim, Jaegwon (1984), 'Concepts of Supervenience', *Philosophy and Phenomenological Research*, 45, pp. 153–176, reprinted in his (1993), Supervenience and Mind (Cambridge, Cambridge University Press), pp. 53–78 [page references in text to latter].

Kim, Jaegwon (1987), '"Strong" and "Global" Supervenience Revisited', *Philosophy and Phenomenological Research*, 48, pp. 315–326.

Kim, Jaegwon (1998), *Mind in a Physical World* (Cambridge, MA, The MIT Press).

Koslicki, Kathrin (2012), 'Varieties of Ontological Independence', Fabrice Correia and Benjamin Schneider (eds.), *Metaphysical Grounding* (Cambridge, Cambridge University Press), pp. 186–213.

Lewis, David (1973), *Counterfactuals* (Oxford, Blackwell).

Lewis, David (1983), 'New Work for a Theory of Universals', *Australasian Journal of Philosophy*, 61, pp. 343–377.

Lowe, E. J. (1998), *The Possibility of Metaphysics* (Oxford, Oxford University Press).

Lowe, E. J. (2006), *The Four-Category Ontology* (Oxford, Oxford University Press).

Martin, C. B. (1997), 'On the Need for Properties: The Road to Pythagoreanism and Back', *Synthese*, 112, no. 2, pp. 193–231.

McLaughlin, Brian and Karen Bennett (2011), 'Supervenience', Edward N. Zalta (ed.) *The Stanford Encyclopedia of Philosophy* (Spring 2018 Edition), <https://plato.stanford.edu/archives/win2011/entries/supervenience/>.

Noordhof, Paul (1997), 'Making the Change, the Functionalist's Way', *British Journal for the Philosophy of Science*, 48, no. 2, pp. 233–250.

Noordhof, Paul (1999a), 'Causation by Content?', *Mind and Language*, 14, no. 3, pp. 291–320.

Noordhof, Paul (1999b), 'Micro-Based Properties and the Supervenience Argument: A Response to Kim', *Proceedings of the Aristotelian Society*, 99, Part 1, pp. 109–114.

Noordhof, Paul (1998c), 'The Overdetermination Argument versus the Cause-and-Essence Principle: No Contest', *Mind*, April 1999, 108, no. 430, pp. 367–375.

Noordhof, Paul (2003), 'Not Old . . . But Not That New Either: Explicability, Emergence and the Characterisation of Materialism', Sven Walter and Heinz-Dieter Heckman (eds.), *Physicalism and Mental Causation: The Metaphysics of Mind and Action* (Charlottesville, Imprint Academic), pp. 85–108 (ISBN 0-907-84547-9).

Noordhof, Paul (2013), 'Mental Causation: Ontology and Patterns of Variation', Sophie Gibb, E. J. Lowe and R. D. Ingthorsson (eds.), *Mental Causation and Ontology* (Oxford, Oxford University Press), pp. 88–125.

Pearson, Olley (2017), 'Emergence, Dependence, and Fundamentality', *Erkenntnis*, published online, 8.4.17.

Petrie, B. (1987), 'Global Supervenience and Reduction', *Philosophy and Phenomenological Research*, 48, pp. 119–130.

Rosen, Gideon (2010), 'Metaphysical Dependence: Grounding and Reducton', Bob Hale and Aviv Hoffmann (eds.), *Modality* (Oxford, Oxford University Press), pp. 109–135.

Schaffer, Jonathan (2012), 'Grounding, Transitivity and Contrastivity', Fabrice Correia and Benjamin Schneider (eds.), *Metaphysical Grounding* (Cambridge, Cambridge University Press), pp. 122–138.

Simons, Peter (1987), *Parts* (Oxford, Oxford University Press).

Tahko, Tuomas E. and E. J. Lowe (2015), 'Ontological Dependence', Edward N. Zalta (ed.), *The Stanford Encyclopedia of Philosophy* (Spring 2015 Edition), https://plato.stanford.edu/archives/spr2015/entries/dependence-ontological/.

Trogdon, Kelly (2013), 'An Introduction to Grounding', M. Hoeltje, B. Schneider and A. Steinberg (eds.), *Varieties of Dependence: Ontological Dependence, Grounding, Supervenenience, Response Dependence* (München, Philosophia, Basic Philosophical Concepts), pp. 97–122.

Wilson, Jessica (2002), 'Powers, Forces, and Superdupervenience', *Grazer Philosophische Studien*, 63, pp. 53–78.

Wilson, Jessica (2005), 'Supervenience-Based Formulations of Physicalism', *Noûs*, 39, pp. 426–459.

Wilson, Jessica (2011), 'Non-Reductive Realisation and the Powers-Based Subset Strategy', *The Monist*, 94, no. 1, pp. 121–154.

Wilson, Jessica (2014), 'No Work for a Theory of Grounding', *Inquiry*, 57, pp. 535–579.

Yablo, Stephen (1992), 'Cause and Essence', *Synthese*, 93, pp. 403–449.

3

FUNDAMENTALITY

Kerry McKenzie

1. Introduction

The concept of metaphysical emergence is intimately tied up with our concept of fundamentality. Whether it is unpredictability, irreducibility, or metaphysical or dynamical autonomy that are taken as its hallmarks, it seems that that which characterizes the metaphysically emergent could equally characterize the fundamental. But the idea of the emergent as something arising out of complexity suggests that the concept involves the non-fundamental just as essentially. In order to understand emergence, then, it seems we need to get a grip on what we mean by fundamentality and how it is that we should understand the relation between the fundamental and that with which it is contrasted.

Fortunately, however, by now there exists an extensive literature on how to conceptualize fundamentality. For if anything characterizes the last decade in metaphysics, it is surely the explosion of interest metaphysicians have shown in precisely this question. Although the reasons for this foregrounding of fundamentality issues are not easy to determine, in the last few years a number of philosophers in the *a priori* tradition have gone as far as to claim that metaphysics just is the study of the fundamental and its relation to the non-fundamental – some even more restrictively that it is that of the fundamental alone.[1] While the idea that there is something novel about taking fundamentality to be of prime metaphysical significance has at times been over-stated, and while the idea that metaphysics is about the fundamental exclusively can be criticized on a number of grounds, it can hardly be denied that fundamentality considerations do indeed do a huge amount of work in canonical metaphysical projects.[2] As has been pressed by Schaffer, a swathe of the paradigm questions of metaphysics – such as the questions of Platonism vs. nominalism, idealism vs. realism, and substratum vs. bundle theories of objects – are arguably best construed not so much as questions concerning *what there is*, but rather of *what is more fundamental than what* – in other words, of what is *ontologically prior*.[3] Thus, it seems anyone interested in metaphysical issues – not just that of emergence – needs to think about what we are talking about when we talk about ontological priority. But just as fundamentality considerations feature centrally in *a priori* metaphysics, the same can largely be said for many projects in the philosophy of science. The debates over ontic structural realism, for example, or those between relationalists and substantivalists about general relativistic spacetime, are likewise naturally viewed as primarily debates over what we should take as fundamental. And what gives many questions concerning the special sciences their point is precisely the fact that their structurally complex subject matter seems to have non-fundamental status. Whatever team we play for in metaphysics, then, it seems we need to think carefully about conceptual questions concerning fundamentality; about ontological questions concerning what, if anything, we should take its extension to be; about 'metametaphysical' questions such as what it is about metaphysics that makes fundamentality so central to it; and − in the metaphysics of science especially − about methodological questions such as how the sciences themselves can assist us in answering the others.

This brief chapter will have something to say about all of these questions, and the labour will be divided as follows. In Section 2, we will introduce some of the idioms pertaining to *ontological priority* that are intended to help us get a grip on the notion of fundamentality. In Section 3, we will discuss whether priority ought to be conceptualized in terms of some kind of determination relation or rather in a form ontological dependence − a matter over which there seems to be a surprising amount of confusion in the literature. In Section 4, we will consider some issues pertaining to the level of 'grain' at which relations of priority ought to be conceptualized. In Section 5, we will consider the issue of whether we should believe that anything fundamental exists at all.

2. The 'levels hierarchy': science and metaphysics

Talk of the fundamental connotes a domain of entities somehow distinguished − and 'distinguished' in the sense of *privileged* − with respect to everything else. One says that the fundamental is *ontologically prior* to those less distinguished entities. Some take the fundamental to be defined in terms of the relation of priority it bears to non-fundamental stuff, with ontological priority regarded as the analysing primitive; others take fundamentality to be an absolute notion that must ultimately resist analysis.[4] But even if one adopts the latter relational conception, it would be hard to deny that part of the job description we at least associate with the fundamental involves its bearing of a special relation to the remainder of what exists.[5] Indeed, discussions of fundamentality typically get off the ground through the depiction of reality as equipped with a 'layered structure', in which the fundamental occupies a unique and especially exalted position at the root of it. As Schaffer puts it, the basic world-view that often motivates fundamentality talk is of

> a *hierarchical view of reality* ordered by *priority in nature*. The primary entities form the sparse structure of being, while the grounding relations generate an abundant superstructure of posterior entities. The primary is (as it were) all God would need to create. The posterior is grounded in, dependent on, and derivative from it. The task of metaphysics is to limn this structure.[6]

Such a hierarchically ordered vision of reality is often said to be fractioned into 'levels'. When confronted with this picture, philosophers of science will be apt to recall some descendent of the 'system of reductive levels' laid out in Oppenheim and Putnam's vision of 'unified science' over half a century ago.[7] The purpose of this work was to arrange the diverse sciences into one connected explanatory picture, and what was supposed to effect these explanatory connections was that idea that the subject matters of less fundamental sciences were *mereologically composed* of those of the more fundamental. Thus, subatomic particles were conceived of as sitting at the bottom of the hierarchy; from there we progress through atomic physics through molecular chemistry and cell biology, all the way up to wherever in the social sciences we are willing to stop. Whatever its status in philosophy of science, Schaffer refers to this mereological conception as still the 'central connotation' of the levels metaphor as it is understood in fundamentality metaphysics.[8]

This pyramidic, monolithic, building-block picture of the sciences is surely, at best, a heuristic and a metaphor. (Where, for example, do we place the subject matter of cosmology, and how could the Higgs field – a field pervading all of space and time – be classed as fundamental on this view?) The 'levels hierarchy' surely has to be articulated in less crude – and likely more piecemeal and contextual – terms than this.[9] Nevertheless, for all that we may only have a murky idea of how the relations of priority between any two domains of science ought to be understood, it remains that almost everyone – notable exceptions aside – seems to accept that such ontologies generally do stand in relations of priority of some sort.[10] We will have more to say about the structure of these relations momentarily. But for now, note that if one takes the ontology of the different sciences to (by and large) stand in priority relations (however they are best articulated) and that it is the resultant ordering that defines the 'levels hierarchy', then it seems that this latter hierarchy *cannot* be identified with the ordered structure that Schaffer takes metaphysics to be tasked with limning. The reason is that questions of how the ontologies of the sciences relate to each other seem to be distinct from those questions Schaffer takes to be paradigmatic of metaphysics. Consider, for example, the debate in philosophy of physics over relationalism vs. substantivalism about spacetime. Surely no one is proposing that spacetime points and the relations between them stand on different 'levels' of the hierarchy, where different levels are conceived of (by and large) as populated by the ontology of different sciences. Rather, both the points and the relations exist *wherever in the levels hierarchy spacetime is.* (To put it another way, it is not as if the relationalist takes the spacetime points to be located somewhere 'north' of fundamental physics, assuming that spacetime physics indeed qualifies as fundamental physics.[11]) Or to take another example, suppose we are considering whether the laws of fundamental physics are primitive constituents of nature or rather, as (for example) Humeans will hold, a derived construction from fundamental properties. Then it seems that the laws of fundamental physics belong in, and *only* in, those domains *where fundamental properties are found.* For at macropscopic scales where colour charge and weak isospin, etc., are no longer predicates of objects, it seems the laws of the standard model will not just be impractical in application, but simply undefined. Rather, a different set of laws will be definable at that level. As such, it seems those who regard laws as non-fundamental entities are not committing to the idea that laws inhabit a higher 'level' than that inhabited by the fundamental stuff: their fundamentality is not about where they are on the hierarchy.

For these reasons, it seems that while many questions about how the ontologies of the different sciences relate are focused on their relative priority, and while many canonical questions in metaphysics are also centrally concerned with matters of priority, we cannot take it that the priority relations involved in each case align on a single hierarchical structure. On the contrary, it seems that the layered structure that (those who identify as) metaphysicians are primarily concerned with is a structure that is oriented in some sense 'orthogonally' from that relating the ontologies of the different sciences. What seems ultimately responsible for this two-dimensional priority structure is the fact that the paradigmatically metaphysical debates seem primarily to be over the priority of metaphysical *categories:* for example, over which of the category of objects or that of relations is the ontologically prior category (as in debates over structuralism), or whether the category of objects is ontologically prior to the category of properties (as in debates over nominalism). Questions about the structure of the scientific levels hierarchy, by contrast, seem primarily *intra*-categorical, consisting of considerations over whether the laws of the biological sciences are less fundamental than the laws of physics, or whether the properties of the cell are less fundamental than those of its constituents (and so on). Given that it seems there are two distinct, though connected, levels structures involved in each case, it should be clear that one cannot blithely assume that the priority relations defining each axis are identical. Whether there is a one-size-fits-all relation of priority that may be usefully deployed in any area of fundamentality metaphysics is a matter to which we will return. But for now, let us remain at an abstract level and consider what features, if any, of logical form are shared by relations of priority in general.

3. Characterizing the fundamental

Largely responsible for the recent explosion of literature on fundamentality is the idea that supervenience and other purely modal concepts are insufficiently discriminating to do the work of characterizing the relations of priority defining the layered structure (or structures) of reality. This in itself is not a new observation, and why it has only recently taken on axiomatic status is a matter over which I can only speculate.[12] But regardless of whether it is new or what the reasons for its recent uptake are, I take the basic idea that mere modal correlations are too coarse to capture priority structure to be fundamentally correct.[13] In *a priori* metaphysics at least, the relation of 'grounding' has become the heir of supervenience in characterizing priority structure, with the result that the lion's share of the contemporary discussion seems to concern arguments around the correct analysis and logical form of this putative relation.

Of course, as the 'distinctive structuring relation of metaphysics', grounding needs to be understood more determinately than simply as the hyperintensional relation that relates more to less fundamental stuff.[14] Looking at the literature, however, one is struck by considerable divergence on even the most basic aspects of how grounding is understood. For example, grounding is often identified with a hyperintensional relation of *ontological dependence*, such as that which has historically been taken to exist between attribute and that in which it inheres. In this view, the fundamental would be characterized as that without which the non-fundamental could not exist. Other times, however, grounding is presented as a relation of *ontological determination*, with the non-fundamental being thought of as that somehow *brought about, entailed by*, or *derivative of* the more fundamental stuff.[15] But even before we get into the details of how these relations are to be defined, it would seem that these two relations differ both intensionally and in terms of their logical form, with the former intuitively presenting the fundamental as a *necessary*, and the latter a *sufficient*, condition on the non-fundamental. Moreover, it is obvious that these relations are not just conceptually quite different, but extensionally distinct as well. Thus, to take some examples, entangled states in quantum mechanics seem to give us a case of dependence without a corresponding determination: for while (most would argue that) such relations cannot exist without their relata, the intrinsic properties of those relata famously do not allow for the reconstruction of the overall state. Conversely, a multiply realized but non-fundamental entity plausibly provides us with an example of determination without a corresponding dependence. For since a multiply realized special-science feature could, by definition, be brought about by any number of more fundamental goings-on, it seems right to say that it is determined by any such realizer and yet wrong to say that it is ontologically dependent on any one of them, in the sense of being unable to exist without it.[16] Thus, whether we regard grounding as a relation of dependence or one of determination could drastically change what we are apt to regard as fundamental.

What, then, are we to do, given that the literature is as we find it? One option is to adopt a pluralism about priority and say that both relations track facts about priority. Were we to take this approach, consistency will demand that there are no facts about what is prior to what *simpliciter*, but only with respect to one or other relation. Such a move finds an analogy in the philosophy of biology and the dispute over the species concept.[17] However, given that species pluralism has largely given way to *eliminativism* about the species concept – leaving only facts about what is descended from what, what can reproduce with what, and what is morphologically similar to what in its wake – one might take the fact that non-coextensive relations between things have all been thought to track priority has similarly eliminativist morals. Now, of course, in one sense this could be taken to be an entirely unproblematic consequence. After all, knowing what determines what and what depends on what helps us to develop an intelligible picture of the world and of the way that the different sciences fit together; what, one might ask, is the value added in going further and saying what is 'fundamental'? Indeed, those of a more cynical persuasion might even go as far as to say that the recent explosion of fundamentality talk expresses only metaphysicians' proclivity to pepper their work with a term that has the effect of demanding that it be regarded as significant – look up the word 'fundamental', after all, and you will find that it has synonyms like 'important'. Such a view recalls certain episodes in the history of physics in which fundamentality rhetoric surfaced somewhat in synch with perceived threats on funding.[18] If this is your stance, then banishing the word might help us view metaphysics in a more sober light.

On the other hand, however, we might view all this as much too cynical, holding that fundamentality is a distinction that the world admits whatever one's take on the value of metaphysics (or the insecurities of metaphysicians).[19] Therefore, let us think more carefully about which of determination or dependence has a better claim on characterizing the fundamental. To do so, consider again the intuition pumps that Schaffer (and others) use to help us get a handle on the notion – in particular, the 'all God had to do' metaphor alluded to in the quote earlier. Plausibly, if we are to understand the fundamental in terms of this metaphor, then we understand it via a relation of *determination*; for what it connotes is that, by making the fundamental, everything else was settled, taken care of – or, in other words, *determined*. But it seems the same cannot be said about dependence: for knowing what everything depends on will not suffice to establish what else there is, should there be dependent entities not fully determined by that on which they depend.

It follows from all this that only if the fundamental is conceptualized in determination terms would knowledge of the fundamental suffice, at least in principle, for knowing everything else about the world. Thus, it seems that determination not only captures our initial motivating intuitions but furthermore accounts for the centre-stage role played by the fundamental in metaphysics. For at the heart of metaphysics is ontology, and ontology concerns what there is: if the fundamental is that which *determines everything whatsoever*, then it determines everything we are interested in, and so of course there is ample justification for focusing on it. Moreover, conceptualizing fundamentality in this way would mean that non-determined but ontologically dependent entities would qualify as fundamental, but occupy a sort of half-way house conceptually insofar as dependence has been taken, perhaps erroneously, to capture something relevant to fundamentality. Indeed, Barnes (2013) takes being non-determined but dependent to be the hallmark of emergence, holding that this half-way house character is precisely what emergentist intuitions have been struggling to capture all along (emergence seemingly involving, as mentioned at the outset, both the fundamental and the non-fundamental simultaneously). By contrast, multiply realized entities – that is, entities that are determined by but not dependent on their realizer in a given instance – may continue to pose problems for theses of reductionism in science consistently with saying they are not metaphysically fundamental. For the inevitably disjunctive nature of that upon which they arguably do depend makes it difficult to *identify* them, for all the usual reasons, with whatever realizes them in a given instance; similarly, it will problematize the idea that the details of any given realizer provide the *best explanation* of any token instance of the realized property. Thus, it seems that, on the determination-based conception, debates over (certain versions of) reductionism can flourish even though we should all be able to agree that the property we are supposing to be non-reduced is metaphysically non-fundamental.

4. Getting specific

So far, so *a priori*. What we wanted to understand was the levels hierarchy, and what we've done so far is identify something about the logical form of the relations the fundamental stands in to the non-fundamental – namely, that it is their job to entail its existence – and waved at the idea that it transcends that form (insofar as the relation is more fine-grained than such mere entailment). What more can we say? One question we should think about is whether a single relation – sometimes referred to as the 'big-G' relation of grounding – suffices for expressing the work that the fundamental does in determining the remainder of ontology. The fact that so much effort in the *a priori* metaphysics literature has gone into characterizing this relation suggests that many take this question to be answered in the affirmative, even if they do not all explicitly say so.

Others have argued, however, that all this is deeply misguided. They hold that if we want to develop a metaphysics of priority more contentful and illuminating than we have managed so far, then we need to go about the project piecewise, focusing on the *sui generis* relations that pertain to entities of different types.[20] A principal reason stated in favour of this is that establishing the mere fact that an entity is grounded leaves open many critical metaphysical questions about that entity – supposedly including even that of whether the grounded entity should be taken to exist.[21] Now, one might of course question how this could possibly be, for surely to say that an entity stands in any real relation or that it has a place in reality (layered or otherwise) is *ipso facto* to commit to the existence of that thing.[22] Nevertheless, given that we can agree that non-fundamental scientific theories exist, that their subject matter is whatever it is, and be very much uncertain as to whether that subject matter really *exists*, it seems that one can make at least an analogous point about the priorities between scientific ontologies. There at least it seems clear that merely stating that an entity may be 'derived' from a more fundamental one does indeed leave open many crucial metaphysical questions about it and that it will do so until we provide much more detail about how it is that the derivation goes.

To see this, let us retreat from *a priori* metaphysics and consider more highly specified, scientific ontologies: that is, ontologies that we understand in the context of particular, well-developed scientific *theories*. In particular, consider how it is that the ontology of relativistic quantum theory – the most fundamental theory produced to date, at least that we know how to submit to empirical test – determines that of non-relativistic quantum theory and of how that in turn determines the ontology of the still less fundamental theory of classical physics. Taking the latter transition first, the process by which classical ontology may be recovered from that non-relativistic quantum theory is by now well understood, due to the development of the theory of *environmentally induced decoherence* over about the last quarter-century. In a nutshell, one can show – and utilizing only *quantum theory itself* – that the effect of the environmental interactions on a physical system is to suppress, over extremely short timescales, the phenomena most characteristic of quantum regimes. In particular, suppose that one takes as a target system a tiny dust mote and considers it in an environment consisting of the microwave background, solar photons and neutrinos, and air molecules at standard atmospheric pressure. Then one can show that the quantum interference associated with the mote becomes negligible virtually instantaneously; that in any world in which such a mote occurs, it occurs approximately localized in both position and momentum, achieving the smallest spread consistent with Heisenberg uncertainty; that in any world the trajectory followed by the peak of its probability distribution will closely approximate, compared with other physically relevant scales, Newtonian trajectories; and that the probability distributions appropriate to these systems may be interpreted as they would be classically. In other words, though we start with a quantum system in an environment (which is likewise modelled quantum-mechanically), what we get back in short order is a system that we can with every justification regard as *approximately classical*. Now to be sure, because (it turns out) one is left at the end of this 'decoherence' analysis with (what is most naturally interpreted as) a *multitude* of worlds, each containing a classical-like dust mote – not simply the one mote that we took ourselves to start out with – decoherence does not in itself solve the 'measurement problem' lying at the heart of the interpretational difficulties of quantum theory. As such, one still needs an interpretation of quantum mechanics – Bohmian, many-worlds, or whatever one prefers – in order to tell a story about why only one tranche of the resulting approximately classical ontology appears to us to be observed. But it is widely agreed that through decoherence the problem of *emergent classicality* has been solved dynamically and that it need not be solved afresh for each interpretation.[23] The problem, by contrast, is that there is something of an 'embarrassment of riches' when it comes to deriving classical stuff!

Consider now how it is that the ontology of non-relativistic (and hence non-fundamental) quantum theory is determined from that of its relativistic successor, namely quantum field theory. Recent work by Myrvold (2015) has shown in detail how wavefunctions – that is, solutions of the (non-relativistic) Schrödinger equation featuring a definite number of particle quanta – may be derived, in some circumstances, from an underlying ontology of quantum fields in the non-relativistic limit. This is the limit in which the effects of the relativity of simultaneity may be taken to be negligible. Such effects will be negligible so long as 'we are not dealing with processes that are spread out too far in space' – a spread which will, of course, be frame-relative – and so long as we are happy to record the properties of fields to a truncated temporal resolution.[24] As such, the ontology of non-relativistic quantum theory may be shown to emerge from that of more fundamental theory in a limit defined by a restriction on the accuracy with which we measure the properties of the underlying field systems, and in the circumstance that our interest is limited to regions that are spatially sufficiently small.

It seems, then, that we can by now tell intricate stories about how it is that the ontology of non-relativistic quantum theory is derived from that of its more fundamental relativistic counterpart and how classical ontology may in turn be derived from that. But it seems that the stories involved in each case raise different issues from a metaphysical perspective. For the story in the quantum-classical transition is exclusively *dynamical*: in order to produce classical systems, one need only take quantum systems (embedded – as they always will be in practice – in an ambient environment) and let them run forward in time. By contrast, the derivation of non-relativistic from relativistic systems can crucially involve frame-dependent assumptions, which we may of course expect to raise questions about the objectivity of the derived non-relativistic entities. While not wishing to prejudge the outcome of a pursuit of those questions here, the point for present purposes is simply that the manner in which the ontology of non-fundamental theories is derived can have significant implications for whether or not we will feel justified in regarding the derived entity as genuinely existent or real.

These developments in the understanding of inter-theory relations make salient the following morals. It reminds us, first of all, that showing how the ontology of non-fundamental scientific theories correspond to those of more fundamental theories represents a scientific achievement, the details of which insipid claims that the former is 'grounded' in the latter might easily obscure. It also reminds us that there is no one prescription on the sort of limiting relations that take us from more to less fundamental theories.[25] But it also reminds us that establishing that the ontology of a given non-fundamental scientific theory may be regarded as real in the first place is going to be a function of the nature of the derivation relations in which that ontology partakes. Assuming that a minimum condition on an entity's having a place in the layered structure of reality is that the entity concerned has some measure of objective existence, it seems clear that we are going to have to attend carefully to the nature of these derivations before we can start thinking about whether it is 'grounded' or not. But by that point, however, it becomes difficult to see what work the 'grounded' predication is doing in enhancing our understanding of the hierarchy of scientific ontologies beyond that which has already been achieved through attending to the details of the derivations themselves.

As such, it is unclear to me at least how much of the discussion in *a priori* metaphysics over 'big-G grounding' has to offer apropos the task of understanding the world as it is described in our best scientific theories. Rather, it seems to me that we need to do a great deal of work in order to even get to the point at which we can begin to talk about grounding, but also that by that same point it is unclear what still remains to be said. Less sceptically, however, it was noted earlier that the distinctively metaphysical applications of priority talk seem to pertain to a structure in some sense 'orthogonal' to that most naturally taken to relate the ontologies of scientific theories. Perhaps it is in that context that metaphysicians' discussions over grounding in the abstract will prove their usefulness – at the very least, by providing useful tools and resources that can be adapted for a more *sui generis* treatment.[26]

5. The existence of the fundamental

It was just argued that oftentimes, before we can begin to think about matters of grounding, we have to work to establish that a given candidate for being grounded even exists in the first place. For presumably only if an entity exists can it be said to be grounded and hence non-fundamental. It is, however, a surprisingly popular view among metaphysicians that no non-fundamental entity could ever have the status of *existent*. After all, arguments based on causal exclusion and a variety of principles of parsimony are intended to militate against the view that there is anything non-fundamental at all.[27] But given that clearly *something* exists, the view that the non-fundamental is non-existent is plausible only to the extent that we are confident that reality bottoms out in a realm of fundamental entities. This, of course, seems to place us in a rather dangerous predicament, given that the more fundamental an entity is, the more remote from our experience it is likely to be and – the history of science counsels us – the less confident we should correspondingly be about its existence and nature. To be sure, many have asserted that the existence of the fundamental is *a priori* necessary for there to exist anything whatsoever – usually by expressing some variant of the view that without the fundamental 'reality would be infinitely deferred, never achieved'.[28] But this intuition, for all that it seems to be widely shared, has proven remarkably resistant to further defence.

For example, an argument often given for the intuition is that if there were no fundamental entities, then the world would fall victim to a vicious existential regress – a regress taken to be somehow more problematic than the cosmological argument's causal analogue. But as Bliss (2013) argues, the only compelling interpretation of what makes a regress 'vicious' seems to result in the regresses envisioned in many infinite-descent worlds being more appropriately classified as benign. Cameron (2008) argues similarly that the best reason to believe in fundamentalia is a virtue-based argument to the effect that worlds with fundamentalia are more 'unified' than those without – an argument which he claims makes their existence a contingent feature.[29] And not only is it the case that arguments for the necessity of fundamentalia have been called into question, but arguments outright denying the existence of a fundamental level have been marshaled as well. For example, Schaffer (2003) argues that we have good inductive grounds for denying that there is anything fundamental, since the particles we have previously regarded as fundamental have often been found to not be so after all. And while Callender (2001) is probably right to regard this as 'an inductive leap of the wildest sort', McKenzie (2011) argues that it is at least conceivable that scientific theories could have anti-fundamentalist consequences. Indeed, the S-matrix theory – the theory which, through its application to strong interactions, played a crucial role in the emergence of string theory – may here be said to represent a prototype, since it enjoyed at least a modicum of success and had a natural anti-fundamentalist interpretation.

Given the failures suffered by *a priori* justifications for the necessity of fundamentalia, perhaps the right way to approach the issue is in broadly empirical terms. Taking this route commits us to believing in fundamental entities if and only if our best scientific theories imply that there are such things. As noted, this need not commit us to the existence of fundamental particles (or perhaps even fundamental laws), since it is conceivable that real physical theories could furnish us with positive reasons to deny that there in fact are any (just as, arguably, did the S-matrix theory). But if we go this way, it seems that we will inevitably be committed to another sort of 'half-way house' fundamentality. The reason is that if we take this naturalistic approach, certain theoretical principles must be at least *treated* as axiomatic in order to deduce any anti-fundamentalist conclusions. In this sense, it seems we can never get around a commitment to fundamental *principles*, even if we can use those same principles to deny the existence of fundamental *particles*. While philosopher-scientists such as Bohm and Popper have harboured more radical anti-fundamentalist visions, it is entirely unclear from a naturalistic perspective how we could compellingly argue for them.[30] Still, what does seem clear is that the world could admit more recherché possibilities than those insisted on by the *a priori* fundamentalist.

Notes

1 See e.g. Schaffer (2009), Dorr (2008), and Paul (2012) for statements of the former view; Sider (2011, p. 1) for a statement of the latter.
2 For example, Stebbing 1932 lays out a conception of analytic metaphysics closely analogous to that of Paul and Schaffer.
3 Schaffer (2009, Sec. 2.2).
4 Schaffer (2009), Rosen (2010), and Bennett (2017) take a relational approach. Wilson (2014), Sider (2011), and Fine (2015) take fundamentality to be primitive.
5 Or, if one denies that there exists anything non-fundamental at all, one should hold that the fundamental has an essential role to play in determining the correctness or incorrectness of statements that putatively refer to non-fundamental stuff, even if they are not taken to be defined by any such role. (Note that there is room for comparison here with the notion of perfectly natural properties: while it seems these resist analysis, we get a grip on what they are through appreciating the work that they do in canonical metaphysical projects.)
6 Schaffer (2009), p. 351.
7 Oppenheim and Putnam (1958).
8 Schaffer (2003), p. 500.
9 See Craver (2015) for a defence of the levels hierarchy conceived of in a more piecemeal context of enquiry-dependent terms.
10 See Cartwright (1999) for a notable such exception.
11 For discussion of the possibility that spacetime is non-fundamental, see Huggett and Wüthrich (2013).
12 The basic observation is in Kim's criticisms of supervenience (see eg. Kim 1993).
13 For example, approaches to priority based purely on supervenience render all necessary features of the world, such as whether the continuum hypothesis is true, non-fundamental features; nor – perhaps more importantly – can they adjudicate on relations of priority between necessary co-existents.
14 To say that grounding is hyperintensional is to commit to it being a relation stronger than necessitation: the grounded and its grounds could co-exist in all the same possible worlds and yet the grounding relations between them not thereby be settled.
15 For example, Rosen (2010), Trogdon (2013), Cameron (2008), and Bliss (2013) all present dependence as synonymous with grounding. For examples of work in which grounding is taken to be a relation of determination, see Fine (2015) and Dasgupta (2015).

16 These cases can be a little tricky to construct, but the easiest cases are where the realizers cannot co-exist in the same world. Since laws are global entities, different fundamental laws presumably cannot co-exist at the same world; thus, if special-science laws can be multiply realized, as many believe, this would constitute just such a case of determination without dependence.

17 Ereshefsky (1998) is the classic reference here.

18 See Martin (2015) for an illuminating discussion.

19 As Craver (*op cit.*, p. 2) puts it, 'The suggestion that we might be better off abandoning the levels metaphor is about as likely to win converts as the suggestion that we should abandon metaphors involving weight or spatial inclusion. These metaphors are too basic to how we organize the world to seriously recommend that they could or should be stricken from thought and expression'.

20 Jessica Wilson stands at the forefront of this debate (see eg Wilson *op cit.*). Of course, this point is consonant with that made in Section 2 about intra-vs. – inter-categorical notions of priority, and at the end of the last section.

21 See Wilson *op cit.*, Section II.i.

22 See *ibid*, footnote 33.

23 See Rosaler (2015) for the defence of this claim and a much more careful discussion of decoherence and the quantum-classical relation.

24 Myrvold (2015, p. 3246).

25 To recall a point made in Section 2, note that rather than one pyramidic 'levels hierarchy' of physical ontology, the 'cGh cube' depicts the three dimensions of limiting relations that we expect to extend out from a fundamental theory of quantum gravity.

26 See French and McKenzie (2015) for a general defence of the value of *a priori* metaphysics along these lines.

27 See Thomasson (2007) for a survey of many of these arguments and extended defence against them.

28 Schaffer (2003).

29 Note that Orilia (2009) argues that Cameron's argument for contingent fundamentality on the basis of unity considerations does not succeed (independently of the perennial problem of associating 'likeliness' with 'lovely' features such as unity).

30 See Bohm (1957), Chapter 5; Popper (1972, pp. 194–196).

Works cited

Barnes, Elizabeth (2013). 'Emergence and Fundamentality', *Mind* 121 (484) (2012): 873–901 first published online April 10, 2013 doi:10.1093/mind/fzt001

Bennett, Karen (2017). *Making Things Up*. Oxford: Oxford University Press.

Bliss, Ricki (2013). 'Viciousness and the Structure of Reality', *Philosophical Studies* 166: 399–418.

Bohm, David (1957). *Causality and Chance in Modern Physics*. Abingdon, UK: Routledge & Kegan Paul Ltd.

Callender, Craig (2001). 'Why Be a Fundamentalist? Reply to Schaffer', paper presented at the Pacific APA, San Francisco 2001, available at PhilSci Archive paper 215.

Cameron, Ross (2008). 'Turtles All the Way Down: Regress, Priority and Fundamentality', *Philosophical Quarterly* 58 (230): 1–14.

Cartwright, Nancy (1999). *The Dappled World: A Study of the Boundaries of Science*. Cambridge: Cambridge University Press.

Craver, C.F. (2015). 'Levels', in *OpenMIND: 8(T)*, eds. T. Metzinger and J.M. Windt. Frankfurt am Main: MIND Group. doi:10.15502/9783958570498

Dasgupta, S. (2015). 'The Possibility of Physicalism', *Journal of Philosophy* 111 (9/10): 557–592.

Dorr, Cian (2008). 'There Are No Abstract Objects', in *Contemporary Debates in Metaphysics*, eds. Theodore Sider, John Hawthorne and Dean W. Zimmerman. Oxford: Blackwell.

Ereshefsky, Marc (1998). 'Species Pluralism and Anti-Realism', *Philosophy of Science* 65 (1): 103–120.

Fine, Kit (2015). 'Unified Foundations for Essence and Ground', *Journal of the American Philosophical Association* 1 (2): 296–311.

French, Steven and Kerry McKenzie (2015). 'Rethinking Outside the Toolbox', in *Metaphysics in Contemporary Physics (Poznan Studies in Philosophy of Science (Poznan Studies in the Philosophy of the Sciences and the Humanities)*, eds. Tomasz Bigaj and Christian Wüthrich, Vol. 104, pp. 25–54. Rodolpi: Brill.

Huggett, Nick and Christian Wüthrich (2013). 'Emergent Spacetime and Empirical Coherence', *Studies in History and Philosophy of Science Part B: Studies in History and Philosophy of Modern Physics* 44 (3): 276–285.

Kim, Jaegwon (1993). 'Supervenience as a Philosophical Concept', in *Supervenience and Mind: Selected Philosophical Essays, Cambridge Studies in Philosophy*. Cambridge: Cambridge University Press.

Martin, Joseph D. (2015). 'Fundamental Disputations: The Philosophical Debates That Governed American Physics, 1939–1993', *Historical Studies in the Natural Sciences* 45 (5), November: 703–757. doi:10.1525/hsns.2015.45.5.703

McKenzie, Kerry (2011). 'Arguing against Fundamentality', *Studies in History and Philosophy of Science Part B* 42 (4): 244–255.

Myrvold, Wayne C. (2015). 'What Is a Wavefunction?', *Synthese* 192 (10): 3247–3274.

Oppenheim, Paul and Hilary Putnam (1958). 'Unity of Science as a Working Hypothesis', in *Concepts, Theories, and the Mind-Body Problem*, Minnesota Studies in the Philosophy of Science, eds. H. Feigl, M. Scriven and G. Maxwell, Vol. 2, pp. 3–36. Minneapolis: University of Minnesota Press.

Orilia, Francesco (2009). 'Bradley's Regress and Ungrounded Dependence Chains: A Reply to Cameron', *Dialectica* 63 (3): 333–341.

Paul, L.A. (2012). 'Building the World from Its Fundamental Constituents', *Philosophical Studies* 158 (2): 221–256.

Popper, Karl (1972). *Objective Knowledge*. Oxford: Oxford University Press.

Rosaler, Joshua (2015). 'Interpretation Neutrality in the Classical Domain of Quantum Theory', *Studies In History and Philosophy of Science Part B Studies In History and Philosophy of Modern Physics* 53.

Rosen, Gideon (2010). 'Metaphysical Dependence: Grounding and Reduction', in *Modality: Metaphysics, Logic, and Epistemology*, eds. R. Hale and A. Hoffman, pp. 109–136. Oxford: Oxford University Press.

Schaffer, Jonathan (2003). 'Is There a Fundamental Level?', *Nous* 37 (3): 498–517.

Schaffer, Jonathan (2009). 'On What Grounds What', in *Metametaphysics: New Essays on the Foundations of Ontology*, eds. David Manley, David J. Chalmers and Ryan Wasserman, pp. 347–383. Oxford: Oxford University Press.

Sider, Theodore (2011). *Writing the Book of the World*. Oxford: Oxford University Press.

Stebbing, L.S. (1932). 'The Method of Analysis in Metaphysics', *Proceedings of the Aristotelian Society* 33 (1932–1933): 65–94.

Thomasson, Amie L. (2007). *Ordinary Objects*. Oxford: Oxford University Press.

Trogdon, Kelly (2013). 'An Introduction to Grounding', in *Varieties of Dependence: Ontological Dependence, Grounding, Supervenience, Response-Dependence (Basic Philosophical Concepts)*, eds. Miguel Hoeltje, Benjamin Schnieder and Alex Steinberg, pp. 97–122. Philosophia: Verlag.

Wilson, Jessica M. (2014). 'No Work for a Theory of Grounding', *Inquiry* 57 (5–6): 535–579.

4

REDUCTION

John Bickle

Why include an entry on reduction in a handbook on emergence? Historically, reduction has been the antithesis of emergence. Their antagonism was noted by Ernest Nagel in his classic chapter on theory reduction (1961) (much more on which below), which ends with a detailed discussion of emergence. We can often learn interesting things about a concept by investigating its contradictory.

I will restrict my discussion to reduction as characterized in philosophy of science. Even so restricted, "reduction" is vastly equivocal. Nevertheless, all varieties share "nothing-but"-ism: a reduction shows that the reduced kind (whatever kind it might be) is thereby "nothing but" the reducing; no reduced content is left out or over. This is the feature that sets reduction in opposition to emergence.

Term/sentence reduction

Term or sentence reductionism was popular through the mid-20th century. It was one of two "dogmas" Quine (1951) urged fellow empiricists to abandon. Its brief popularity in philosophy of mind rested on attempts to translate psychological terms or sentences into "topic-neutral" synonyms, which in turn were hypothesized to be contingently identifiable with physical terms or sentences: those of future neuroscience, according to central state materialists (Smart 1959). The proposed definability of psychological expressions into "topic-neutral" synonyms, for example, "something is going on in me which . . .," and the predicted future contingent identities of those "somethings" with physical (e.g., brain) goings-on, captured reductionism's "nothing-but"-ism.

This approach to reduction has zero popularity now. Yet it is worth dwelling on here briefly, because the most philosophically sophisticated advocate of term/sentence reduction, Rudolph Carnap, held a much more nuanced view, as far back as the mid-1930s, than is typically realized. Logical positivism is widely acknowledged, and roundly criticized, for advocating a definitional reductionism based on a "verificationist" theory of meaning. Carnap (1932) held such a view. However, by his 1936 work, he realized the limits of verificationism. There he shifted explicitly to confirmation, a much weaker notion, adopted an explicitly pragmatic interpretation of observation languages for science . . . and then gave up on definability. Deeply theoretical terms from physics weren't the only problem; simple disposition terms like "soluable" "cannot be defined by means of the terms by which conditions and reactions are described" (1936, 440). Reduction by definability, for the most prosaic of sciences, had to go.

Instead of being defined by observation terms, a new scientific term, say, Q_3, can be introduced initially by a "reduction pair":

(R1) $Q_1 \supset (Q_2 \supset Q_3)$
(R2) $Q_4 \supset (Q_5 \supset {\sim}Q_3)$,

where Q_1 and Q_4 describe experimental conditions, and Q_2 and Q_5 experimental results (1936. 441). A single reduction pair (typically) does not establish the "complete" meaning of a new term. It only does so for cases where one of the test conditions, $Q_1 \vee Q_4$, is fulfilled; for all other cases Q_3 remains indeterminate, "meaningless" in Carnap's (1936) parlance. But scientists can further diminish this "region of indeterminateness" by adding additional reductive pairs, to specify additional conditions and related experimental results that determine whether Q_3 or ${\sim}Q_3$ applies. (The most straightforward example is a term for a physical magnitude that can be determined by different methods.) The term will then have meaning for the disjunction of all the experimental conditions in its set of reduction pairs, and this set expands with each new "decision" about its usage (1936, 448–449).

Carnap's target in his 1936 work are those who insist on reduction via definability. The problem is that "a definition determines the meaning of a new term once and for all," a practice "not in accordance with the intentions of the scientist concerning the use of the predicate" (1936, 449). For the indeterminate cases not (yet) governed by a reduction pair, Carnap insists, "the scientist wishes to leave these questions open until the results of further investigations suggest the statement of a new reduction pair" (1936, 449). Reduction pairs thus constitute a "partial definition of meaning only and can therefore not be replaced by a definition" (1936, 449–450). This shift raises a fascinating puzzle for even the brief history of analytic philosophy. W.V.O. Quine at least acknowledges that Carnap had "long since" given up on the "radical reductionism" of term/sentence definability, but then famously goes on in the remainder of that essay to level his "Duhemian holism" challenge to "dogmatic" empiricists, Carnap included. Yet seventeen years prior Carnap had written: "Science is a system of statements based on direct experience, and controlled by experimental verification. Verification in science is not, however, of single statements but of the entire system or subsystem of such statements" (1934). It is difficult to find a more straightforward statement of "Duhemian holism." It is also worth noting that Carnap's appeals in his 1936 work to "scientists' intentions" suggest that he had given up on "rational reconstructions" as early as then. It can be baffling how credit for influential ideas gets distorted over even short histories in philosophy.

Aside from the historical interest just noted, why is all this worth repeating? Because Carnap's nuanced account of term introduction via sets of reduction pairs may not be so widely implausible about actual scientific practice than logical positivism is so often accused of being. Morton and Bickle (2005) argue that the scientific term, "memory consolidation," looks to have been introduced and developed in present-day molecular neurobiology in a fashion resembling Carnap's reduction pairs.

Theory reduction

We turn next to the most prominent account of the reduction relation, theory reduction. The classic account is Ernest Nagel's (1961, chapter 11); Schaffner (2012) provides the best recent overview of Nagel's account, its numerous proposed revisions and replacements, and its remarkable resiliency. For Nagel, reduction was *deduction*: of the sentences constituting the reduced theory, from those of the reducing serving as premises. Logical derivability captured reductionism's "nothing-but"-ism: no content of the reduced theory remains beyond the scope of the reducing.

In many actual scientific cases, the reducing theories correct the reduced. Real planets do not travel in ellipses; real bodies do not fall with uniform vertical acceleration over any finite interval near the surface of the earth. Kepler's theory of planetary motion and Galileo's mechanics are false vis-à-vis Newton's mechanics, yet these are textbook scientific reductions. How can a false theory validly be derived from a true one (presumably, at the time of the reduction)? Nagel captured this feature by adding various limiting assumptions and boundary conditions on the applicability of the reducing theory, possibly counterfactual, to the premises of the derivation. He also noted that many scientific reductions are "heterogeneous": the reduced theory contains terms not part of the descriptive vocabulary of the reducing. "Heat" and "pressure" of classical equilibrium thermodynamics, for example, nowhere occur in statistical mechanics and the kinetic/corpuscular theory of matter. For Nagel, various "conditions of connectability" must also be included among the premises of the derivation. These "bridge laws" or "correspondence rules," as they soon came to be called, had to have at least the logical strength of conditionals, with sentences containing terms of the reducing theory as antecedents, and others containing terms of the reduced theory which did not occur in the reducing as consequents. Unfortunately, in his famous illustration, the reduction of the ideal gas laws of classical thermodynamics to statistical mechanics, these connections are stated as biconditionals. This perhaps rationalizes the otherwise puzzling fact that biconditional conditions of cross-theoretic connectability were so often assumed to be required on Nagel's account. He explicitly states that only conditionals are required, and the logic of first-order derivability clearly only requires conditionals. This misunderstanding misinformed some anti-reductionist "multiple realization" arguments in philosophy of mind for decades. If "[brain sentence]⊃[psychological sentence]" is all the cross-theory connectability that a Nagel reduction (deduction) of psychology via neuroscience requires, it matters not at all that [psychological sentence] is true of other creatures lacking brains. Nagel discussed some difficulties interpreting these conditions as either conventional definitions or synthetic statements. Their status remained a constant source of discussion within the Nagel-inspired reduction literature.

Nagel's work (1961) was not the first account of intertheoretic reduction to appear. Schaffner (2012) provides a helpful history of Nagel's account, tracing it back a dozen years prior to 1961. Other philosophers were working on intertheoretic reduction, too. John Kemeny and Paul Oppenheim (1956) offered a logically weaker account, in which the reducing theory explains all the "observational data" that the reduced theory explains, and typically more. The reducing theory's increased explanatory scope captures reductionism's "nothing-but"-ism; the reduced theory thereby plays no ineliminable explanatory role. Additionally, for Kemeny and Oppenheim, the reducing theory must be at least as "well systematized" as the reduced. Systematization balances simplicity and explanatory power. It is a measure of how well the increased complexity of the reducing theory is compensated for by its increased explanatory power. A simpler theory thus can be reduced to a more complex one if the more complex theory is significantly stronger in explanatory power by enough to offset its increased complexity. Unfortunately, Kemeny and Oppenheim leave systematicity intuitive. Their account was quickly rejected as being logically too weak: presumably, it counts as reductions cases which are not. Despite this popular challenge, Theurer and Bickle (2013) argue that some apparent reductions in the recent neurobiological field of "molecular and cellular cognition" appear to follow a modified Kemeny–Oppenheim pattern.

Patrick Suppes (1967) suggested an isomorphic "sameness of structure" condition on intertheoretic reduction within a broader framework which conceived of theories as ordered sets of models. Building explicitly on Suppes' suggestion, Balzer, Moulines, and Sneed (1987) specified additional set-theoretic conditions, including analogues of Nagel's connectability and derivability conditions, and "ontological reduction links" between components of reducing and reduced models, such that every "confirmed empirical application" of the reduced theory is related to some model of the reducing. Logically stronger than mere isomorphism, these conditions expressed a model-theoretic analog of derivability, and thereby captured reductionism's characteristic "nothing-but" feature. With great formal ingenuity, Balzer, Moulines, and Sneed (1987, chapter VI) reconstruct historical cases of scientific reductions using their set-theoretic apparatus.

Paul Feyerabend (1962) leveled one of the first responses to Nagel's work (1961). Denying that any scientific examples actually meet Nagel's connectability and derivability conditions, Feyerabend proposed a radical "ontological replacement" account of reduction: the ontology of the reducing theory replaces without remainder that of its "incommensurable" reduced analog in all contexts of scientific usage, including the observational. Most philosophers of science rejected Feyerabend's radical alternative, but both Kenneth Schaffner (1967, 1992) and Clifford Hooker (1981) saw reason to accommodate some of his views. For Schaffner, reduction remains deduction, with the reducing theory serving as premises, but not of the reduced theory, but rather, of a *corrected version* of the reduced theory. The deduced structure is still in the theoretical language of the reduced theory (perhaps altered by corrections necessitated by the reducing), and so something akin to Nagelian "bridge laws" remain required in heterogeneous reductions. Schaffer dubbed these "reduction functions." His "General Reduction Paradigm" (later renamed the "General Reduction-Replacement Paradigm") also insisted that an "analog relation" held between the actual reduced theory and the corrected version actually deduced in the reduction. He never specified this analog relation in any formal detail. Over the years Schaffner has increasingly characterized his account as more closely aligned with Nagel's (see most recently Schaffner (2012)).

Hooker (1981) accommodated some Feyerabendian features also while still keeping reduction as deduction, but replacing the actual reduced theory with its *image*, already formulated within the descriptive vocabulary of the reducing theory. The constructed image is *explanatorily equipotent* to the actual reduced theory, with the strength of logical derivation and this explanatory equipotence capturing reductionism's "nothing-but"-ism. In cases of reduction which correct, and perhaps even falsify significantly, the actual reduced theory, various counterfactual limiting assumptions and boundary conditions must also be added as premises from which this image is derived. But since the derived image is already within the vocabulary of the reducing theory, no "conditions of connectability" or "reduction functions" are required to affect its derivation. The number and counterfactual extent of the limiting assumptions and boundary conditions needed to derive the explanatory equipotent image of the reduced theory suggests a measure of the relative "smoothness" of a given intertheoretic reduction, although Hooker admits that he leaves this notion in a frustratingly intuitive state (1981, 223–224). Still, working with a number of actual scientific examples displaying varying amounts of Feyerbendian "incommensurability," Hooker suggests that a given case's (intuitive) location on the "smoothness/bumpiness of intertheoretic reduction" spectrum nicely matches its location on the "ontological retention/replacement" spectrum for the reduced theory's postulated kinds. So we learn something about cross-theoretic ontology – identity, to revision, to outright elimination – when we investigate a given intertheoretic reduction for its relative "smoothness."

Hooker's account influenced Paul and Patricia Churchland's neuroscience-based eliminative materialism (P.M. Churchland 1985; P.S. Churchland 1986). Bickle (1998) exploited Hooker's insights within a "semantic" (model-theoretic) account of theory structure, borrowing resources from Balzer, Moulines, and Sneed (1987). He offers a detailed model of the isomorphic location of scientific cases across the intertheretic reduction and ontological consequences spectra; develops a detailed Hooker-inspired account of the reduction of the classical gas laws of equilibrium thermodynamics to statistical mechanics and the kinetic theory of gases, including the derivation of an image of the gas laws within the reducing theory; develops a quasi-formal account of "smoothness of reduction" (based on set cardinality); develops in detail a response to the multiple realizability challenge to reduction inspired by Hooker's remarks in his work (1981, Part III); and argues for a revisionary ontological outcome for the propositional attitudes of folk psychology vis-à-vis a then-current model of synaptic plasticity in cellular neurobiology. Endicott (1998) subjects Hooker's, the Churchlands', and Bickle's argument to searching evaluation and criticism. Schaffner (2012) recently concurs with Endicott's criticism.

Well into the 1990s, "reduction" in serious philosophy typically meant intertheretic reduction, and that usually meant Nagel's account. But over the past twenty years some new accounts have developed, and we close with some of those.

Functional reduction

One alternative to intertheoretic reduction emerged from consciousness studies: functional reduction. Joseph Levine (1993) was an early advocate of an "explanatory gap" between the subjective qualitative features of conscious experiences, that "what-it-is-like-ness" (Thomas Nagel 1974) of, for example, the vivid redness of the visual experience of a polished fire truck or the taste of a ripe peach, and the physical features of the object experienced and the conscious experiencer. Levine (1993) defended this gap by way of an account of "reductive explanation," one quickly adopted by fellow "qualiaphiles" such as David Chalmers (1996) and even Jaegwon Kim (2005). Reduction, according to Levine, is a two-stage process. "Stage 1 involves the (relatively? quasi?) *a priori* process of working the concept of the property to be reduced 'into shape' for reduction by identifying the causal role for which we are seeking the underlying mechanisms" (1993, 132). This is to "functionalize" the concept – hence the name of this account. "Stage 2 involves the empirical work of discovering just what those underlying mechanisms are" (1993, 132), that is, doing the science to discover which kinds in the natural world play, or at least approximate, that functional role.

Levine fleshes out this skeleton with an example: the reduction of water to collections of H_2O molecules in liquid state. "Our very concept of water is of a substance that plays such-and-such a causal role" (1993, 131): for example, boils at 212°F at sea level. Empirical-cum-theoretical investigations of collections of H_2O molecules, including their varying speeds and locations near the liquid's surface, intermolecular attractive forces between them, and their increasing kinetic energy with increased heat and its relation to atmospheric pressure, explain the 212°F boiling point. Those are the empirically discovered features of the world that realize this aspect of water's full functional profile. That the behavior of the physical aggregates matches these causes and effects "without remainder" captures reductionism's "nothing-but"-ism. If there is "nothing more" to being water than its possessing that complete set of causes and effects, and if the empirically discovered behavior of collections of H_2O molecules in liquid state realize that full functional profile, then water is "nothing but" H_2O molecules in liquid state.

With so much serious work on intertheretic reduction already in existence, why would qualiaphiles bother to articulate another account? Conveniently, at least for the anti-reductionists among them, adopting functional reduction yielded a straightforward argument for the *irreducibility* of qualia. A central conclusion of qualiaphiles is that qualia cannot be functionalized: cannot be "fully elaborated without remainder" into a profile of causes and effects. Such treatments inevitably "leave out" their qualitative features, the "what-it-is-like-ness" of these conscious experiences, as thought experiments involving inverted spectra, absent qualia, what Mary the future neuroscientist doesn't know, and philosophical zombies purport to show. Chalmers (1996) is a key source for many of these famous puzzles, but the entire consciousness studies literature wallows in them and their seemingly endless variations. But according to functional reduction, if no functionalization of a concept is possible . . . then neither is any (functional) reduction of it! If Stage 1 of a reduction is a nonstarter, then no Stage 2 empirical work will find the mechanisms which exhaust the functional profile, which exhausts that concept's content – because no functional profile does. Viola! Qualia are irreducible to physical mechanisms!

The problem with functional reduction, at least in philosophy of science, is that it is as much a philosopher's fiction as the famous thought experiments purporting to show that qualia can't be functionalized. Levine's (1993) "scientific" example, which Chalmers (1996), Kim (2005), and others adopt, draws neither from contemporary science nor its recent history – as advocates of intertheoretic reduction attempted. The case study comes instead from elementary-school science education; it's an example we use to instruct children about our basic scientific world view. One may be excused for questioning whether real reductionism in real scientific practice circa now resembles an analysis derived from cartoon "science."

This counter to functional reduction (first offered in Bickle 2012) amounts to a challenge: take the Methods section from a published experimental report appearing in any reputable current science journal – *Cell, Science, Nature, Proceedings of the National Academy of Sciences* – and show how the experiments described there can be interpreted reasonably as "functionalizing" some concept, even partially. Help yourself to the Supplementary Materials now routinely published online with all major scientific publications. Prediction: you will not succeed. That is because "functional reduction" is no part of actual current scientific practice.

Explanatory, metascientific, and mechanistic reduction

Reduction in philosophy of science has experienced a recent renaissance. This final section documents just three examples. Michael Strevens' "explanatory reduction" is particularly apt to include in a volume on emergence. He aims to provide an account that is genuinely reductionist, but that also accommodates concerns that have motivated anti-reductionists, including emergentists. He builds his account explicitly on a doctrine of ontological physicalism: everything that exists is made up entirely of physical stuff, and everything that occurs does so because of the effects of physical laws. From this he concludes that all events, no matter how high level (above that of fundamental physics) or abstract, can be derived from fundamental physical facts and laws. Some of these derivations, though not all and in fact very few, are also explanations: namely, those successful derivations of the explanandum limited exclusively to the components of the causal network that "make a difference" to its occurrence. This is Strevens' broader "kairetic" account of explanation, detailed most extensively in his 2011 work; its name is drawn from the Greek "kairos," meaning "decisive point." Strevens even provides a "recipe" for finding these reductive explanations, via a process of abstracting away from the details of a "complete" specification of the causal network generating the explanandum until a derivation of the event remains, but no further abstracting away of causal details will maintain that logical relation. Explanatory reductionism is then one further conjunct: for any higher-level phenomenon, there is at least one derivation of it from the most fundamental physical level that is (kairetic) explanatory, and that one is its best explanation. The general reductionist appeal to derivation is familiar: it was the logical relation Nagel (1961) required between reduced and reducing theories. Here, however, "derivation" is part of the apparatus for representing the operation of a causal process, not merely a relation between theories qua syntactic structures. It is also the key relation in Strevens' account that captures reductionism's "nothing-but"-ism.

How then does Strevens save anti-reductionist motivations, like the methodological autonomy of the special sciences? In his 2016 work responding to Phillip Kitcher's famous appeal to the autonomy of classical genetics from molecular biology, Strevens distinguishes between two senses of explanatory irrelevance. "Objectively" irrelevant causal details make no difference to the occurrence of the explanandum. "Contextually" irrelevant causal details can be, and often are, objectively relevant to a complete explanation of it. But for practical purposes of pursuing particular scientific endeavors, these details can be, and often must be, ignored: "black boxed," as Strevens puts it, for the efficient special-scientific investigation into the explanandum to proceed. This scientific division of labor can be either "compartmental," at a single level of investigation, or "stratified," across levels. And while kairetic explanatory reductionism requires that eventually all the objectively relevant difference-making causal details must get included in our best (fundamental physical) explanation, the scientific practices involved in achieving such explanations recognize a demand for the contextual autonomy of special-scientific pursuits.

Strevens (2016) remains uncommitted about "whether it is worth our explanatory while" actually to trace particular lines of implementation down to their fundamental physics (karietic) explanation. The ensuing reductive explanation will be better than any which stops partway down, but practically speaking achieving it may be too expensive . . . or too boring. These practicalities depend on the specific case at hand. But what about entire "reductionistic" scientific endeavors that seek explanations of phenomena down many recognized levels at once, in a "single bound"? The autonomy Strevens (2016) seeks to salvage within his reductionism might be no part of such scientific endeavors; these sciences might violate even his practical norms of "contextual irrelevance." His shift to context and pragmatics affords him wiggle room; perhaps "compartmentalizing" norms typically operative in single-level special sciences get suspended in these "single-bound" endeavors.[1] It is worth reflecting on such cases, however, because a different sense of reduction has been built directly upon them.

If our overarching goal in investigating reduction is to understand a prominent (yet hardly universal) feature of scientific practice, that goal suggests a strategy. Find some field that both its practitioners and scientists working in related fields label "reductionist," and investigate practices – experimental, explanatory – specific to it. Conduct this analysis as bereft of philosophical/epistemological assumptions about "what reduction has to do or be" as one can render oneself. The result should be a "metascientific" account of actual scientific reductionism in actual scientific practice. With such an account in hand, one can follow up investigating whether other scientific fields routinely labeled "reductionistic" employ similar practices and whether the account delivers what philosophers have wanted (or targeted) in "an account of reduction."

John Bickle (2006, 2009, 2012) has sought such an account. The first step is choosing a scientific field to investigate. Bickle chooses "molecular and cellular cognition" (MCC), a field that raises the worry suggested earlier for Strevens' account. MCC started in the early 1990s with the application of gene targeting techniques from developmental molecular biology to neurobiology and has expanded to encompass more recent developments such as optogenetics (Bickle 2016). The Molecular and Cellular Cognition Society (www.molcellcog.org) now has more than 2300 members worldwide. Its proponents contrast their approaches and goals explicitly with those of "cognitive neuroscience" in their focus on cellular and molecular mechanisms of cognitive functions and their use of animal models (http://molcellcog.org/index.asp?page=about). But MCC is hardly some fringe field in current neuroscience. Its centrality to the discipline for nearly two decades is apparent in a quote from the introductory chapter to one of the principal textbooks in the field, the edition from fifteen years ago (since supplanted by a fifth edition):

> This book . . . describes how neural science is attempting to link molecules to mind – how proteins responsible for the activities of individual nerve cells are related to the complexity of neural processes. Today it is possible to link the molecular dynamics of

individual nerve cells to representations of perceptual and motor acts in the brain and to relate these internal mechanisms to observable behavior.

(Kandel, Schwartz, and Jessel 2000, 3–4; my emphases)

These "links" are nothing less than "single-bound" mind-to-molecular pathway reductions, introduced fifteen years ago as "textbook" neuroscience. Learning and memory, including "declarative" or "explicit" forms, constitute MCC's most notable achievements to date (Sweatt 2009; Silva, Landreth, and Bickle 2014).

What does a metascientific analysis of landmark MCC results reveal actual scientific reductionism to be? First, it is part of the more general search for causal connections between neuroscientific kinds. Three kinds of experiments are crucial to establish any causal hypothesis A ➔ B with a strong degree of scientific confidence. *Negative interventions* manipulate the hypothesized cause (A) by decreasing its probability or intensity and measure changes in the hypothesized effect (B). For hypothesized excitatory causes, under widely accepted conditions of experimental control, a decrease in the hypothesized effect shows that the hypothesized cause is necessary for the effect. (For hypothesized inhibitory causes, the change to the effect will be exactly the opposite, i.e., an increase. Similarly for positive intervention experiments, to follow.) But even the most scrupulously controlled negative intervention experiments can never show the sufficiency of the hypothesized cause. *Positive interventions*, especially ones integrated with successful negative interventions, can. Here experimenters manipulate the hypothesized cause (A) by increasing its probability or intensity and measure changes to the hypothesized effect (B). Increases in the effect show the sufficiency of the hypothesized cause. Now, however, the problem of experimental artifact looms, since in most positive interventions the hypothesized cause is increased beyond its normal biological limits. The integration of results from a third kind of experiment is required: *non-intervention experiments*, in which the hypothesized cause (A) is not manipulated experimentally, but merely measured in correlation with the occurrence of the effect (B) in as biologically realistic circumstances that permit the required measurements. For any specific causal hypothesis, scientific confidence that it has been established experimentally increases with the successful integration of results from all three types of experiments.[2]

So far this metascientific analysis of landmark MCC experimental results does not distinguish the field from causal-mechanistic sciences more broadly, not all of which are reductionist. What additional features makes MCC reductionist? The first concerns particular experimental practices. Specific cognitive functions are operationalized into behavioral protocols and measures for use with animal models, typically rodents. The molecular-biological and gene targeting experimental tools then routinely employed to manipulate the hypothesized causes manipulate components of the biological system generating the operationalized behavioral measures – and most often components of the behaving system's components (of its components): specific proteins involved in intra- and inter-cellular signaling pathways. These are the hypothesized mechanisms of the experimentally operationalized cognitive functions under direct experimental test. The second reductionist feature is another form of multiexperiment integration, mediation analysis (Silva, Landreth, and Bickle 2014). Even when a neurobiological causal hypothesis has been established by the integration of all three negative manipulation, positive manipulation, and nonintervention experiments, MCC investigations do not stop. In particular, MCC scientists next investigate causes mediating the established connection A ➔ B: those causes by which A causes B. To do this, MCC scientists inevitably "look down" into the components that constitute A and B. This is what their experimental tools of choice, drawn from molecular biology and increasingly biochemistry, permit. The now quarter-century track record of this approach far exceeds anything that less reductionistic fields of science can offer, at least for the causal mechanisms of learning and memory.

Or so holds metascientific, "ruthless" reductionism. The underlying picture is straightforward, even if the MCC scientific details can be daunting. Real reduction in real scientific practice is a matter of intervening experimentally into increasingly lower levels of biological organization, from specific neurons now down to specific configured proteins in intra- and inter-neuronal signaling pathways, then tracking the effects of these interventions on the system's behavior in experimental protocols well accepted as indicators of specific cognitive functions. When these experiments are individually successful *and* successfully integrated, a reduction is claimed: of the specific cognitive function directly to the lowest level of biological organization for which the full set of integrated interventionist experimental results has been achieved. This is explanation "in a single bound," of cognition directly to its cellular or molecular mechanisms. Nowadays, at least for many aspects of learning and memory, the operative mechanistic level is specific genes and proteins. This account wears reductionism's "nothing-but"-ism on its sleeve. The cognitive function is the cellular/molecular activity experimentally intervened into and integrated successfully, delivered via the system's neuroanatomy to the behavioral periphery, to drive muscle contractions against the skeletal frame and produce the behaviors that operationalize the specific cognitive function for experimental investigation. Ruthless reductionism indeed!

Both Strevens' and Bickle's talk of causes and mechanisms will ring familiar to recent philosophers of science. The last fifteen years has seen the rise and increasing dominance of a "new mechanism" (Machamer, Darden, and Craver 2000, with a precursor in Bechtel and Richardson (1993) and before that numerous papers of William Wimsatt's throughout the 1970s and 1980s). New mechanists are not univocal about reduction, but one view has been developed by William Bechtel (2009). Bechtel's "mechanistic reduction" draws directly on the basic new mechanist account of explanation. To explain the action of some system S's Ψing, one specifies its mechanism: the individual components of S $x_1, \ldots, x_n$, those components' activities $\phi_1, \ldots, \phi_m$, and the causal organization of the x_is' $\phi_i$ings which produces S's Ψing (Craver 2007, 7, Fig. 1.1). Bechtel illustrates his account using ruthless reductionists' favorite case studies: the consolidation of long-term memory. But he does not find reduction "in a single-bound" lurking here, from behavior to molecular pathways. Instead, he finds "nested mechanisms within mechanisms," organized like Russian nesting dolls. The molecular pathways intervened into in MCC experiments are not mechanisms of the behaving animal (e.g., the rat navigating the Morris water maze). Rather, those molecular activities are the mechanisms of specific cells in the rat's hippocampus and cortex inducing long-term potentiation (LTP), a form of activity-driven synaptic plasticity (enhancement). And those specific neurons inducing LTP are a mechanism of place cell spatial map formation across the rat's hippocampus. And place cell spatial map activity is the mechanism of the rat navigating the water maze. A complete mechanistic explanation requires specifying the full mechanism for each component of the system at the next level down. Animal experimental psychology and cognitive neuroscience attend to the mechanisms at higher levels of organization, with cellular and molecular biology the lower levels. No level is explanatorily privileged; all are required for a full mechanistic reduction of the cognitive function. The ruthless reductionist's alternative account, of seeking explanation "in a single bound," risks various methodological errors that cognitive neuroscientists routinely impress upon their MCC colleagues (Bechtel 2009; see also Craver 2007, chapter 7). Nevertheless reductionism's "nothing-but"-ism still obtains: the cognitive function is the full nested collection of mechanisms within mechanisms, with nothing "left over."

This dispute between ruthless and mechanistic reductionists appears resolvable: Which account gets the actual science right? That debate continues, with new case studies continually on offer. But two lines of possible rapprochement have also emerged. Perhaps mechanistic reduction correctly describes the vision of cognitive/systems neuroscience, while ruthless reductionism that of cellular and molecular neurobiology? Bechtel (2009) first suggested this rapprochement, describing the history of neuroscience's mid-20th-century split into the Society for Neuroscience and the cognitive/systems/neuropsychology camps. Neuroscience has been a vastly interdisciplinary enterprise since its inception. Why assume that it speaks with one voice about what reduction is?

Second, new mechanists David Kaplan and Carl Craver (2011) have addressed a challenge raised by "dynamic systems theorists," that purely mechanistic explanations in neuroscience must be supplemented with another form of explanation. Dynamicists insist that most neural systems involve thousands of parts or more, organized and interacting in myriad complex ways, with multiple redundancy and feedback. As this complexity increases, it becomes increasingly implausible to decompose the system into its discrete, interacting parts. New concepts must appear in explanations, including systems variables and order parameters, and the appropriate explanatory tools are turning out increasingly to be ones from nonlinear dynamical systems theory. These explanations violate the strictures of mechanistic explanation.

Kaplan and Craver's (2011) response is radical. To the extent that dynamical systems "explanations" violate the principles of correct mechanistic explanations, such "explanations" aren't genuine explanations at all, but rather just descriptions of the phenomenon to be (mechanistically) explained. They examine in some detail a landmark case of dynamicist neuroscience, the Haken, Kelso, and Bunz (1985) model of human bimanual finger-movement coordination. They point out that these modelers only intended for their dynamical systems model to be a mathematically compact description of the temporal evolution of a purely behavioral dependent variable. None of the model's variables or parameters were interpreted to map onto components or operations of any hypothetical mechanism generating the behavioral data; none of the mathematical relations or dependencies between variables in the model were intended to map onto hypothesized causal interactions between components or activities of any mechanisms. Furthermore, subsequent to developing and publishing their dynamicist model, Kaplan and Craver point out that these modelers themselves began to investigate how the behavioral regularities their model described might be produced by neural motor system components, activities, and organization. This suggests that the modelers themselves were taking their dynamicist model as a heuristic and sought to move towards a "how-possibly," and ultimately a "how-actually," mechanistic explanation.

Kaplan and Craver's (2011) last argument is reminiscent of Silva, Landreth, and Bickle's appeal to mediation analysis in MCC, to "looking down" to components of established causal hypotheses to find the causes mediating them. Mediation Analysis is part of what makes MCC (metascientifically/ruthlessly) reductionistic. So in addition to the challenge Kaplan and Craver (2011) raise for dynamicists, a comparison of one of their arguments with Mediation Analysis yields a significant potential comparison between an explicit kind of reductionism and a new mechanism. It doesn't resolve the dispute about explanations in a "single bound," from behavior directly to molecular pathways, versus nested hierarchies of mechanisms within mechanisms, but it does reveal more clearly what is at dispute. It is the scientific status of "higher-level" causal mechanisms when lower-level ones have been established experimentally with strong scientific confidence. Ruthless reductionists then eschew the higher level as mechanistic; mechanistic reductionists continue to embrace them as such.

Notes

1 Strevens (personal communication) suggested this reply, along with numerous helpful comments that improved my discussion of his views.
2 The terminology for these three kinds of experiments is from Silva, Landreth, and Bickle (2014). Sweatt (2009) refers to them, respectively, as "blockages," "mimicries," and "correlations." There are subtle differences between Silva, Landreth, and Bickle's "positive interventions" and Sweatt's "mimicries," but there is not space to explore those here.

References

Balzer, W., Moulines, C.-U., and Sneed, J. (1987). *An Architectonic for Science.* Dordrecht: Reidel.

Bechtel, W. (2009). "Molecules, systems, and behavior: Another view of memory consolidation." In Bickle, J. (ed.), *The Oxford Handbook of Philosophy and Neuroscience.* New York: Oxford University Press, 3–40.

Bechtel, W. and Richardson, R. (1993). *Discovering Complexity: Decomposition and Localization as Strategies in Scientific Research.* Princeton, NJ: Princeton University Press.

Bickle, J. (1998). *Psychoneural Reduction: The New Wave.* Cambridge, MA: MIT Press.

Bickle, J. (2006). "Reducing mind to molecular pathways: Explicating the reductionism implicit in current cellular and molecular neuroscience." *Synthese* 151 (3): 411–434.

Bickle, J. (2009). "Real reductionism in real neuroscience: Metascience, not philosophy of science (and certainly not metaphysics!)." In Hohwy, J. and Kallestrup, J. (eds.), *Being Reduced: New Essays on Reduction, Explanation, and Causation.* New York: Oxford University Press, 34–51.

Bickle, J. (2012). "A brief history of neuroscience's actual influences on mind-brain reductionism." In Gozzano, S. and Hill, C. (eds.), *New Perspectives on Type Identity Theory.* Cambridge: Cambridge University Press, 88–109.

Bickle, J. (2016). "Revolutions in neuroscience: Tool development." *Frontiers I Systems Neuroscience* (March): http://journal.frontiersin.org/article/10.3389/fnsys.2016.00024/full

Carnap, R. (1932). "The elimination of metaphysics through the logical analysis of language." (translation by Arthur Pap), *Erkenntnis* 2: 60–81.

Carnap, R. (1934). *The Unity of Science.* London: K. Paul, Trench, Trubner and Co. Ltd.

Carnap, R. (1936). "Testability and meaning." *Philosophy of Science* 3 (4): 419–471.

Chalmers, D. (1996). *The Conscious Mind.* New York: Oxford University Press.

Churchland, P.M. (1985). "Reduction, qualia, and the direct introspection of brain states." *Journal of Philosophy* 82 (1): 8–28.

Churchland, P.S. (1986). *Neurophilosophy.* Cambridge, MA: MIT Press.

Craver, C.F. (2007). *Explaining the Brain.* New York: Oxford University Press.

Endicott, R. (1998). "Collapse of the new wave." *Journal of Philosophy* 95 (2): 53–72.

Feyerabend, P.K. (1962). "Explanation, reduction and empiricism." In Feigl, H. and Maxwell, G. (eds.), *Minnesota Studies in Philosophy of Science* 3, Minneapolis: University of Minnesota Press, 28–96.

Haken, H., Kelso, S., and Bunz, H. (1985). "A theoretical model of phase transitions in human hand movements." *Biological Cybernetics* 51 (5): 347–356.

Hooker, C.A. (1981). "Towards a general theory of reduction, Part I: Historical and scientific setting, Part II: Identity in reduction, Part III: Cross-categorial reduction." *Dialogue* 20: 38–59, 201–236, 496–529.

Kandel, E.R., Schwartz, J., and Jessel, T. (eds.) (2000). *Principles of Neural Science*, 4th Ed. New York: McGraw-Hill.

Kaplan, D.M. and Craver, C.F. (2011). "The explanatory force of dynamical and mathematical models in neuroscience: A mechanistic perspective." *Philosophy of Science* 78 (4): 601–627.

Kemeny, J. and Oppenheim, P. (1956). "On reduction." *Philosophical Studies* 7 (1–2): 6–19.

Kim, J. (2005). *Physicalism, or Something Near Enough.* Princeton, NJ: Princeton University Press.

Levine, J. (1993). "On leaving out what it is like." In Davis, M. and Humphreys, G.W. (eds.), *Consciousness: Psychological and Philosophical Essays.* London: Blackwell, 121–136.

Machamer, P., Darden, L., and Craver, C.F. (2000). "Thinking about mechanisms." *Philosophy of Science* 67 (1): 1–25.

Morton, A.L. and Bickle, J. (2005). "Re-examining logical positivism: Testability and meaning in contemporary neuroscience." *Journal of Contemporary Philosophy* 25: 3–11.

Nagel, E. (1961). *The Structure of Science.* New York: Harcourt, Brace, and World.

Nagel, T. (1974). "What is it like to be a bat?" *Philosophical Review* 83 (4): 435–450.

Quine, W.V.O. (1951). "Two dogmas of empiricism." *Philosophical Review* 60 (1): 20–43.

Schaffner, K. (1967). "Approaches to reduction." *Philosophy of Science* 34 (2): 137–147.

Schaffner, K. (1992). "Philosophy of medicine." In Salmon, M., Earman, J., Glymour, C., Lennox, J., Machamer, P., McGuire, J., Salmon, W., and Schaffner, K. (eds.), *Introduction to the Philosophy of Science.* Englewood Cliffs, NJ: Prentice Hall, 310–344.

Schaffner, K. (2012). "Ernest Nagel and reduction." *Journal of Philosophy* 109 (8–9): 534–565.

Silva, A.J., Landreth, A., and Bickle, J. (2014). *Engineering the Next Revolution in Neuroscience.* New York: Oxford University Press.

Smart, J.J.C. (1959). "Sensations and brain processes." *Philosophical Review* 68: 141–156.

Strevens, M. (2011). *Depth: An Account of Scientific Explanation.* Cambridge, MA: Harvard University Press.

Strevens, M. (2016). "Special science autonomy and the division of labor." In Couch, M. and Pfeifer, J. (eds.), *The Philosophy of Phillip Kitcher.* New York: Oxford University Press.

Suppes, P. (1967). "What is a scientific theory?" In Morgenbesser, S. (ed.), *Philosophy of Science Today.* New York: Basic Books, 55–67.

Sweatt, D. (2009). *Mechanisms of Memory*, 2nd Ed. San Diego: Academic Press.

Theruer, K. and Bickle, J. (2013). "What's old is new again: Kemeny-Oppenheim reduction at work in current molecular neuroscience." *Philosophia Scientia* 17 (2): 89–113.

5

EMERGENCE, FUNCTION AND REALIZATION

Umut Baysan

"Realization" and "emergence" are two concepts that are sometimes used to describe same or similar phenomena in philosophy of mind and the special sciences, where such phenomena involve the synchronic dependence of some higher-level states of affairs on the lower-level ones. According to a popular line of thought, higher-level properties that are invoked in the special sciences are *realized* by and/or *emergent* from lower-level, broadly physical properties. So these two concepts are taken to refer to relations between properties from different levels where the lower-level ones somehow bring about the higher-level ones. However, for those who specialize in inter-level relations, there are important differences between these two concepts – especially if emergence is understood as *strong* emergence. The purpose of this chapter is to highlight these differences.

Realizing a function

Realization as an inter-level relation is often thought to be tightly related to the notion of a *function*, and this is arguably due to the fact that the notion of realization was imported into the contemporary philosophy of mind literature with a defence of functionalism alongside the refutation of the view that mental properties (e.g. *being in pain*) are identical with physical properties (e.g. *having C-fibre stimulation*). It was suggested by Putnam (1967) that mental properties are multiply realizable by different physical properties in different organisms, such that it is possible for two organisms to instantiate the same mental property without having any physical properties in common; therefore, a given mental property cannot be identified with any particular physical property. After all, if it is possible to be in pain without having C-fibre stimulation, *being in pain* and *having C-fibre stimulation* cannot be identical. This argument has come to be known as "the multiple realizability argument", and it is sometimes interpreted to support functionalism – the view that mental properties are functional properties. For those who hold this view, just as the very same functional property (e.g. *being a vending machine*) can be instantiated in different physical ways, a mental property can be instantiated in different physical ways – as long as it carries out its constitutive function, whatever that function might be.

Because of the association of the multiple realizability argument with functionalism, the idea that some higher-level property *M* is realized by physical properties is often interpreted as an endorsement of functionalism about *M* – to the extent that one occasionally sees interchangeable uses of "functionalism" and "realizationism" (e.g. Polger & Shapiro 2016). In line with this, there is a widely held view that for some lower-level property to realize some higher-level property, the former must play the role that the latter is associated with, where the said role is understood causally or functionally. As Polger (2004) remarks, we do not need to have a definitive account of what a function is just for the purposes of explaining what realization is. Different philosophers will have different conceptions of what a function is, and for each conception, there may be a different putative realization relation that involves how a function of that sort can be physically realized.[1] For example, one might agree with the early function-alists about the mind that mental properties are computational properties (e.g. Putnam 1967), and thereby think that how a mental property is brought about by the instantiation of some physical properties and relations in one's brain is analogous to, or perhaps identical in type with, how the processes that are constitutive of the software of a computer are brought about by the physical processes in the computer's hardware. Or one might think that having a function is a just matter of playing some causal role (i.e. having characteristic causes and effects), where the causal role in question needn't be computational. Suppose that *being a heart* is individuated by the causal role of pumping blood in an organism's circulatory system through blood vessels. Then, for a physical system, realizing the property of *being a heart* would be a matter of having the right sort of physical properties that can play the causal roles that contribute to the pumping of blood through blood vessels.

More generally, like those who have defended the "role-playing" account of realization, one might understand functional realization along the following lines:

Functional Realization (FR): A property P functionally realizes a property F if and only if, for some causal role R, (i) F is individuated by R and (ii) P is a property that plays R.[2]

In FR, we have three key elements: *first*, a property F to be realized; *second*, a causal/functional role R that individuates F; and *third*, a property that can play R. "Role playing" is really the right metaphor here. Let's say in a story we have a character (e.g. Batman), a role that is asso-ciated with that character (e.g. throwing batarangs at criminals), and someone who plays the role (e.g. whoever plays the Batman role in that particular Batman story). Just as in some stories, Bruce Wayne plays the Batman role (partly) by throwing batarangs at criminals, phys-ical properties realize functional properties in virtue of playing the causal/functional roles that characterize the latter – if functionalism is true.[3] If properties in some domain D are individuated by causal/functional roles and, as a matter of fact, such causal/functional roles are occupied by physical properties, then properties in D are functionally realized by physical properties.[4]

Even though not all theories of realization give a central role to functions in their formula-tions of the realization relation (see Baysan 2015), in nearly all accounts of realization, there is an emphasis on the causal profiles of realized and realizer properties, and moreover in these theories, it is suggested that the causal powers of a realized property are *fully accounted for* in terms of the causal powers of its realizer properties. Such claims of fully accounting-for come in two incom-patible forms. First, there are those who agree with versions of Kim's (1992) *causal inheritance principle* that if a higher-level property Q is instantiated in virtue of some lower-level property P, then all causal powers of Q are also causal powers of P (e.g. Wilson 1999; Clapp 2001; Shoemaker 2001; Baysan 2016). Second, there are those who disagree with the causal inheritance principle but still hold that the causal powers of realized properties are conferred on their bearers in virtue of the causal powers of their realizers (e.g. Pereboom 2002; Gillett 2003). We will revisit the causal inheritance principle in the section "Realization and causal powers" later.

Realization, emergence, and physicalism

There is a sense in which claims of emergence and claims of realization can overlap, in particular if emergence is understood as *weak* emergence. For example, an emergent property can be physically realized if it is a functional property in one sense and its function is performed by some physical property. In fact, Wilson (2015) uses the term "weak emergence" to refer to the relation that I call "realization" here. In Wilson's use, weak emergence and strong emergence are *incompatible* relations: if a higher-level property Q is strongly emergent from a lower-level property P, then Q is *not* weakly emergent from P (and vice versa). The reason is that whereas weak emergence (i.e. physical realization) of Q entails non-reductive physicalism about Q, its strong emergence is incompatible with physicalism about it.[5] (Here, non-reductive physicalism is to be understood as the conjunction of physicalism and the rejection of the identity of higher-level properties with the lower-level ones due to multiple realizability.) So the main difference is this: whereas realization is a relation that relates higher-level properties to physical properties only if non-reductive physicalism is true, the existence of strongly emergent higher-level properties is incompatible with non-reductive physicalism.[6]

If these observations are correct, then we can explain the differences between the relations of realization and strong emergence (henceforth simply emergence) by tracing the differences between the commitments of (non-reductive) physicalist views about higher-level properties and anti-physicalist views about them. These are:

1   Physicalists who put forward realization claims hold that higher-level properties are "nothing over and above" physical properties. Their (anti-physicalist) emergentist counterparts – those who put forward emergence claims – hold that some higher-level properties are "over and above" physical properties.
2   Physicalists take only physical properties to be fundamental, and hence hold that realized properties are non-fundamental. Emergentists attribute fundamentality to some higher-level properties.
3   Physicalists hold that the distribution of higher-level properties (i.e. physically realized properties) supervenes on the distribution of physical properties with metaphysical necessity. Emergentists typically reject this metaphysical supervenience claim.
4   Physicalists typically hold that the causal powers of higher-level properties are fully accounted for in terms of the causal powers of their base properties. Emergentists reject this fully accounting-for claim.

As we shall see, these four differences between non-reductive physicalism and emergentism will point us towards four differences between the relations of realization and emergence.

Over-and-aboveness

Physicalism about higher-level properties (e.g. mental properties) is typically understood as the view that they are "nothing over and above" physical properties (Smart 1959). This way of formulating physicalism might have a reductive flavour, but there are reasons to think that all physicalists about the mind, reductive or non-reductive, should endorse the nothing-over-and-aboveness of the mental vis-à-vis the physical. The following analogy might illuminate the possibility of nothing-over-and-aboveness without identity: suppose that (mereological) composition is ontologically innocent but that composition is not identity; under this supposition, there are instances of nothing-over-and-aboveness without identity: a whole would be nothing over and above its parts, yet it would not be identical with them.

Assuming then that non-reductive physicalists will take realized properties to be nothing over and above their realizers, we can propose the following necessary condition on realization:

> *Realization-OA* (ROA): A property P realizes a property Q only if Q is nothing over and above P.

Whatever the other features of the realization relation might be, if ROA is true, then it will never be the case that a property realizes another one whereby the latter is something over and above the former. And this is something we cannot say for the relationship between emergent properties and their bases. So in this sense, emergent properties are not realized by their physical bases. For emergence, the following seems to be the case:

> *Emergence-OA* (EOA): A property Q is emergent from a property P only if Q is over and above P.

Again, the analogy from composition can be illustrative. Emergence is sometimes presented as the failure of exactly the aforementioned claim about the ontological innocence of composition: an emergent whole is supposed to be more than the sum of its parts. (More precisely, a whole with emergent properties is more than the sum of its parts.) So it is reasonable to think that the emergent properties of a whole are not merely "resultant" properties of the properties of the parts of the whole.

Admittedly the "over and above" talk is somewhat metaphorical, but some have filled in the metaphor in interesting ways, one of which we will cover in the section entitled "Realization and causal powers".

Fundamentality

Let us now see how we can use the fundamentality criterion to highlight another difference between the realization and emergence relations. Non-reductive physicalists hold that mental properties are non-fundamental properties. So a non-reductive physicalist, unlike her eliminativist counterparts, does not eliminate mental properties from her ontology; but unlike her reductive physicalist counterparts, she does not identify mental properties with broadly physical properties, which are supposed to be relatively fundamental properties.

This point about fundamentality is helpful in making progress in terms of a well-known problem about the formulation of physicalism. The problem is the difficulty of responding to the following dilemma: if we formulate physicalism with reference to current physics, physicalism will be false (because current physics is not complete); if we formulate physicalism with reference to an ideal future physics, we will not know what physicalism is (because we do not know what future physics will be like). One way to get around this problem is to restrict the formulation of non-reductive physicalism to certain domains of higher-level properties and then propose that physicalism about that domain of properties is the view that those properties are real but not fundamental properties, whereas physical properties – whatever they might turn out to be like – are fundamental properties.[7] Now in order to capture the physicalist idea that higher-level properties are dependent on physical properties, a non-reductive physicalist should also hold that these non-fundamental higher-level properties are instantiated in virtue of the instantiations of some fundamental (physical) properties. Given that realization is supposed to be the relation that relates these non-fundamental properties to their fundamental (physical) bases, we have the following observation:

> *Realization-F* (RF): A property P realizes a property Q only if P is more fundamental than Q.[8]

Note that RF is not committed to there being multiple degrees or levels of fundamentality. If there is only one level of fundamentality such that anything is either strictly fundamental or just non-fundamental, there can still be instances of RF: realizer properties could be strictly fundamental, and realized properties could be just non-fundamental.

Now given that emergentists hold that some higher-level entities are fundamental despite being dependent on lower-level physical entities, then the following is a plausible conditional:

> *Emergence-F* (EF): A property Q is emergent from a property P only if Q is at least as fundamental as P.

EF entertains the controversial idea that there could be dependent-yet-fundamental entities. Could there be such entities? There are ways of thinking about such possibility. One example is the early emergentists' thought that some higher-level causal powers are fundamental (Broad 1925; McLaughlin 1992). Surely, in suggesting that some higher-level causal powers are fundamental, early emergentists were not committed to such causal powers being ontologically independent. Also, Barnes (2012) surveys a number of metaphysical views in which some entities are both fundamental and dependent; for example, some philosophers think that *persons* are fundamental entities but are dependent on the parts of their physical bodies.

The fundamentality criterion to distinguish between realization and emergence is also helpful in resolving a problem that comes up in the realization literature: *the problem of conjunctive realizers*. RF helps us see why conjunctive properties – if there are any – are not realizers of their conjuncts (as is widely accepted).[9] For example, the conjunctive property of *being red and spherical* is not a realizer of *being red*. Intuitively, this is because although the (conjunctive) property of *being red and spherical* necessitates *being red*, the former does not bring about the latter. The problem of conjunctive realizers is the difficulty of getting this result without stipulating from the outset that no conjunctive property realizes any conjuncts. Now, note that RF already has the resources to get this result. Conjunctive properties – if there are any – are typically less fundamental than their conjuncts, as they are constructions out of their conjuncts. So a conjunctive property and its conjuncts typically will not satisfy RF. So although it might strike some as a truism that realization relates fundamental properties to non-fundamental properties and that emergence does not, there are interesting consequences one can discover by exploring the implications of this suggestion.

Supervenience

Many have thought that there is a difference between realization and emergence because non-reductive physicalism is committed to the metaphysical supervenience of (realized) higher-level properties on the physical ones, whereas emergentists reject this metaphysical supervenience claim and are committed to only the nomological supervenience of the higher level on the physical. Among others, van Cleve (1990), McLaughlin (1997), Noordhof (2003) and (arguably) Chalmers (1996, 2006) are those who have endorsed this difference between non-reductive physicalism and emergentism. So, for realization, the following seems to be true:

> *Realization-S* (RS): A property P realizes a property Q only if P metaphysically necessitates Q.

And for emergence, the following:

> *Emergence-S* (ES): A property Q is emergent from a property P only if Q is nomologically, but not metaphysically, necessitated by P.

Let me offer two important clarifications at this point. The *first* clarification is that, in RS, realizer properties must involve background conditions in order to enable metaphysical necessitation of the realized property by the realizer (Shoemaker 1981), where such background conditions may include even the laws of physics (but not inter-level laws, if there are any).[10] If a higher-level property is not metaphysically necessitated by some physical property that includes such rich background conditions, then it cannot be realized by that physical property. Suppose that a phenomenal property M is a higher-level property of this sort. What might guarantee the instantiation of M if such an enriched physical base may not? In addition to the laws of physics, we may have to include "trans-ordinal" (Broad 1925) or psychophysical laws (Chalmers 1996) in the background conditions of the base property, in which case we cannot be physicalists about M. And this brings us to the second clarification: the reason why ES requires no more than nomological necessitation is that unless other laws of nature (including the trans-ordinal or psychophysical laws) are included in the base, the instantiation of the base does not guarantee, in the metaphysical sense, the instantiation of an emergent property. That is, P and the background conditions alone can only nomologically necessitate M. But P, the background conditions, and all the laws of nature (or a proper subset of them that include the inter-level laws), together, metaphysically necessitate M.

This criterion is controversial, however. This way of explaining the difference between realization and emergence (and likewise non-reductive physicalism and emergentism) is committed to there being a distinction between nomological and metaphysical necessity, and such a distinction is sometimes rejected for reasons that do not primarily concern us here (see Shoemaker 1998). But because of this, some may think that RS and ES are not successful in highlighting the difference between realization and emergence.[11]

Realization and causal powers

So far we have covered three putative ways of explaining the difference between realization and emergence: over-and-aboveness, fundamentality, and supervenience. Now we will revisit Kim's (1992) causal inheritance principle introduced earlier and explore whether we can use it to provide a fourth one.

Recall that the causal inheritance principle suggests that higher-level properties inherit their causal powers from the lower-level base properties they depend on. As Wilson (1999) suggests, there is a connection between the causal inheritance principle and the nothing-over-and-aboveness claims we discussed in the section titled "Over-and-aboveness". She argues that

> [p]hysicalists [about mental properties] . . . cannot . . . allow that mental properties have any causal powers that are different from those of their physicalistically acceptable base properties, for this violates the physicalist thesis that mental properties are "nothing over and above" their base properties.
>
> *(ibid., p. 41)*

Here, the idea is that we can understand the over-and-aboveness of a higher-level property in terms of the novelty of the causal powers it confers on its bearers, and thereby understand the *failure* of over-and-aboveness as the *failure* of such causal novelty. Then, it must be the case that whereas realized properties inherit all of their causal powers from their base properties, emergent properties do not; the latter must have novel causal powers, and the former must not. So for realization we should say the following:

> *Realization-CP* (RCP): A property *P* realizes a property *Q* only if every causal power of *Q* is also a causal power of *P*.

And for emergence, the following:

> *Emergence-CP* (ECP): A property *Q* is emergent from a property *P* only if some causal power of *Q* is not also a causal power of *P*.

Wilson (2015) argues that these two conditionals are true and any other putative distinction between realization and emergence is either entailed by these two conditionals or inadequate to do the job.[12] Note that even if one were to agree with Wilson that these two conditionals are true, one might still reject the further claim that other criteria are either redundant or inadequate.

There is plenty of literature to back up ECP as a conditional, as emergentists are characteristically committed to there being novel causal powers that are associated with some higher-level properties (e.g. Broad 1925; O'Connor 1994; Crane 2001), unless they want to take emergent properties to be epiphenomenal. But RCP is more controversial, and that is what we shall look at more closely.

What might be the reasons to think that RCP is true? There are numerous arguments in favour of RCP, but for reasons of space, here I shall focus on only one of these arguments.[13] Consider the following two principles:

A   A property *P* has a causal power *C* if and only if, as a matter of nomological necessity, all bearers of *P* have *C*.

B   If a property *P* realizes a property *Q* as a matter of (at least) nomological necessity, all bearers of *P* are also bearers of *Q*.

Suppose that *having C-fibre stimulation in a suitably functioning nervous system* (*C-fs*, for short) realizes *being in pain* (*pain* for short) and that cp_1 is a causal power of *being in pain*. (Exactly what cp_1 is does not matter for the purposes of this argument.) If (A) is true, it will be a matter of nomological necessity that all bearers of *pain* have cp_1. Note that this does not mean that all bearers of *pain* must manifest cp_1, as it is shared wisdom that causal powers may exist unmanifested. And if (B) is true, it will be a matter of (at least) nomological necessity that all bearers of *C-fs* are also bearers of *pain*. From these two observations, it follows that it is a matter of nomological necessity that all bearers of *C-fs* have cp_1. This last observation and (A) entail that cp_1 is a causal power of *C-fs*. Note that what is true of cp_1 must be true for all causal powers of *pain*. So any causal power *pain* is a causal power of *C-fs*. More generally, any causal power that we can attribute to a realized property must also be attributed to its realizer, which suggests that RCP is true.

Is this argument persuasive? The argument makes use of two principles: (A) and (B). The second of these is not controversial, and it is entailed by RS discussed earlier. However, the principle stated in (A) is controversial, and admittedly the argument's success hinges on it.[14] Nevertheless, the argument shows that explaining the relationship between properties and the causal powers they confer on their bearers is important for defending RCP and hence understanding the realization relation.[15]

RCP has received much criticism, but due to space considerations, I have handpicked three objections. The first one is from Noordhof (1997, 2013),[16] and it suggests that a realized property may confer different causal powers on its bearers depending on how it is realized. Pains realized in human beings and pains realized in robots may cause different pain behaviours. So the causal powers of a realized property outstrip the causal powers of its individual realizers. Hence, RCP is false.

The second objection is from Pereboom (2002) and is based on the assumption that there is a constitution relation between the causal powers of a realized property and the causal powers of its realizers. Assuming that constitution is not identity, Pereboom argues, a causal power of a realized property cannot be identical with any of the causal powers of its realizers. So RCP must be false.

The third objection is from Gillett (2003), who motivates his objection by giving the following example: imagine a hard diamond which has the causal power to cut glass. In this diamond, *hardness* is realized by the properties (and relations) of the small bits that constitute the diamond. Although the causal power to cut glass is conferred on the diamond by *hardness*, no realizer property (i.e. properties of the small bits of the diamond) can confer this power on the diamond, as those bits are too small to cut glass! Therefore, RCP should be rejected.

What can the proponents of RCP say in response to these objections? Regarding the first objection, they could reject Noordhof's claim that realizer-dependent causal powers are causal powers of realized properties; instead, such causal powers can be naturally said to belong to the realizer properties. Against the second objection, they could say that Pereboom is explaining a different relation: realization of causal powers by (other) causal powers, not realization of properties by (other) properties. Against the third objection, they could point out that they take realization to be a relation between properties of the very same object – not properties of a whole and the properties of its parts as Gillett does.[17]

Nevertheless, responding to an objection is one thing, and persuading one's opponent is another; whether these responses will be persuasive remains to be seen.

Conclusion

Realization and emergence are two inter-level relations that have been invoked by some philosophers of mind and the special sciences to explain the dependence of higher-level properties on the lower-level ones. It is often suggested that higher-level properties are realized by and/ or emergent from lower-level properties. Despite the similarity of realization and emergence claims, there are important differences between these two relations, in particular if emergence is understood as strong emergence. After briefly focusing on a functional theory of realization, I offered four different (but partially overlapping) ways of explaining the differences between these two relations. These differences are due to the fact that whereas realization claims are made as part of physicalist frameworks about higher-level properties, emergence claims are made as part of anti-physicalist frameworks.

Notes

1 See Polger's (2004) Chapters 3 and 4 for a discussion.
2 See Melnyk (2003) for an account of realization along these lines.
3 Here, I am assuming that "Batman" is not a proper name, but refers to a vigilante via a role.
4 The discussion earlier does not cover the difference between *realizer* functionalism and *role* functionalism (about some property *P*). Roughly, the difference between these two views is that the former identifies *P* with the property that occupies *P*'s causal/functional role, whereas the latter does not. So a realizer functionalist about mental properties may identify mental properties with physical properties because physical properties occupy the causal/functional roles that individuate mental properties (see Lewis 1994).

5 In some usages, "weak emergence" and "strong emergence" do not refer to incompatible relations. For example, Chalmers (2006) takes strong emergence of x to be a matter of strict unexplainability of x (i.e. unexplainability, even in principle), and weak emergence of x to be a matter of non-strict unexplainability of x (i.e. unexplainability due to technical or practical difficulties). Assuming that something's strict unexplainability entails, but is not entailed by, its unexplainability in practice, in Chalmers's usages, the relations picked out by these two terms are not incompatible.

6 It is important to note that strong emergence *simpliciter* does not falsify physicalism *simpliciter*. If there are strongly emergent phenomena within physics (e.g. quantum entanglement), we should not be expected to be non-physicalists about physics! But if higher-level properties turned out to be strongly emergent, then physicalism about such higher-level properties should be falsified.

7 Relatedly, see Wilson (2009) for a characterization of "physical" that involves a "not fundamentally mental" clause.

8 See Bennett (2011) for the view that such asymmetry of fundamentality should be the case in all in-virtue-of relations. I am not committed to this stronger claim, and in fact I think the (possible) cases of emergence are counterexamples to it.

9 See Yablo (1992, p. 253), Shoemaker (2007, p. 23), and Wilson (2009, p. 152).

10 If the laws of physics are metaphysically necessary, then they do not need to be included in the background conditions. Also, if there are inter-level laws, physicalism suggests that they are derivable from the laws of physics.

11 See O'Connor (1994) and Wilson (2005) for such criticisms.

12 I should remind the reader that in Wilson's terminology, this distinction is between "weak emergence" and "strong emergence".

13 See Wilson (1999, 2011), Clapp (2001), Shoemaker (2001, 2007), and Baysan (2016) for arguments for RCP. The one we shall look at now is from Baysan (2016).

14 This principle is defended in Baysan (2018).

15 An interesting consequence of this argument is that it appears to apply to strong emergence too, if strong emergence is to be understood simply as a same-subject nomological necessitation relation. If true, that would be bad news for emergentists, as it would be impossible for emergent properties to have novel causal powers. However, emergentists have resources to deal with this problem. See Baysan & Wilson (2017) for a number of ways out of this problem.

16 Noordhof's (1997) objection is actually against Kim's causal inheritance principle, but in (2013) he presents it as an objection against Shoemaker (2007).

17 I should note that Gillett foresees such a response and thinks that a scientifically interesting realization relation should not be a same-subject necessitation relation but should trace part–whole relations.

Bibliography

Barnes, E. (2012). Emergence and fundamentality. *Mind*, 121, 873–901.

Baysan, U. (2015). Realization relations in metaphysics. *Minds and Machines*, 25, 247–260.

Baysan, U. (2016). An argument for power inheritance. *Philosophical Quarterly*, 66, 383–390.

Baysan, U. (2018). Epiphenomenal Properties. *Australasian Journal of Philosophy*, 96, 419–431.

Baysan, U. & Wilson, J. (2017). Must strong emergence collapse? *Philosophica*, 91: 49–104.

Bennett, K. (2011). Construction area (no hard hat required). *Philosophical Studies*, 154, 79–104.

Broad, C.D. (1925). *The mind and its place in nature*. London: Routledge.

Chalmers, D. (1996). *The conscious mind*. Oxford: Oxford University Press.

Chalmers, D. (2006). Strong and weak emergence. In P. Clayton & P. Davies (eds.), *The re-emergence of emergence* (pp. 244–255). Oxford: Oxford University Press.

Clapp, L. (2001). Disjunctive properties. *Journal of Philosophy*, 98, 111–136.

Crane, T. (2001). The significance of emergence. In C. Gillett & B. Loewer (eds.), *Physicalism and its discontents* (pp. 207–224). Cambridge: Cambridge University Press.

Gillett, C. (2003). The metaphysics of realization, multiple realizability, and the special sciences. *Journal of Philosophy*, 100, 591–603.

Kim, J. (1992). Multiple realization and the metaphysics of reduction. *Philosophy and Phenomenological Research*, 52, 1–26.

Lewis, D. (1994). Reduction of mind. In S. Guttenplan (ed.), *A companion to the philosophy of mind* (pp. 412–431). Oxford: Blackwell.

McLaughlin, B. (1992). The rise and fall of British emergentism. In A. Beckerman, H. Flohr & J. Kim (eds.), *Emergence or reduction? Essays on the prospects of non-reductive physicalism* (pp. 49–93). Berlin: De Gruyter.

McLaughlin, B. (1997). Emergence and supervenience. *Intellectica*, 2, 25–43.

Melnyk, A. (2003). *A Physicalist manifesto: Thoroughly modern materialism.* New York, NY: Cambridge University Press.

Noordhof, P. (1997). Making the change, the functionalist's way. *The British Journal for the Philosophy of Science*, 48, 233–250.

Noordhof, P. (2003). Not old . . . but not that new either: Explicability, emergence and the characterisation of materialism. In S. Walter & H. Heckman (eds.), *Physicalism and mental causation* (pp. 85–108). Exeter: Imprint Academic.

Noordhof, P. (2013). Mental causation: Ontology and patterns of variation. In S. Gibb, E.J. Lowe & R.D. Ingthorsson (eds.), *Mental causation and ontology* (pp. 88–125). Oxford: Oxford University Press.

O'Connor, T. (1994). Emergent properties. *American Philosophical Quarterly*, 31, 91–104.

Pereboom, D. (2002). Robust nonreductive materialism. *Journal of Philosophy*, 99, 499–531.

Polger, T. (2004). *Natural minds.* Cambridge, MA: MIT Press.

Polger, T. & Shapiro, L. (2016). *The multiple realization book.* Oxford: Oxford University Press.

Putnam, H. (1967). Psychological predicates. In W.H. Capitan & D.D. Merrill (eds.), *Art, mind and religion* (pp. 37–48). Pittsburgh: University of Pittsburgh Press.

Shoemaker, S. (1981). Some varieties of functionalism. *Philosophical Topics*, 12, 93–119.

Shoemaker, S. (1998). Causal and metaphysical necessity. *Pacific Philosophical Quarterly*, 79, 59–77.

Shoemaker, S. (2001). Realization and mental causation. In C. Gillett & B. Loewer (eds.), *Physicalism and its discontents* (pp. 74–98). Cambridge: Cambridge University Press.

Shoemaker, S. (2007). *Physical realization.* Oxford: Oxford University Press.

Smart, J.C.C. (1959). Sensations and brain processes. *Philosophical Review*, 68, 141–156.

van Cleve, J. (1990). Emergence vs. panpsychism: Magic or mind dust? *Philosophical Perspectives*, 4, 215–226.

Wilson, J. (1999). How superduper does a physicalist supervenience need to be? *Philosophical Quarterly*, 49, 33–52.

Wilson, J. (2005). Supervenience-based formulations of physicalism. *Noûs*, 39, 426–459.

Wilson, J. (2006). On characterizing the physical. *Philosophical Studies*, 131, 61–99.

Wilson, J. (2009). Determination, realization, and mental causation. *Philosophical Studies*, 145, 149–169.

Wilson, J. (2011). Non-reductive realization and the powers-based subset strategy. *The Monist*, 94, 121–154.

Wilson, J. (2015). Metaphysical emergence: Weak and strong. In T. Bigaj & C. Wuthrich (eds.), *Metaphysics in contemporary physics* (pp. 345–402). Leiden: Brill.

Yablo, S. (1992). Mental causation. *Philosophical Review*, 101, 245–280.

6

STRONG EMERGENCE AND ALEXANDER'S DICTUM

Alex Carruth

Introduction

Emergentists hold that higher-level phenomena are something 'over and above' the sum of their most basic parts. This usually involves the emergent phenomena being taken to be both *distinct from* and *novel with respect to* the base phenomena from which they emerge, whilst nevertheless being dependent upon the base phenomena. How *distinctness* and *novelty* should be understood depends on the kind of emergence being proposed: epistemically emergent higher-level phenomena are indispensable features of certain explanatory or predictive practices, whereas with metaphysically emergent phenomena, their 'over and above-ness' is a matter of ontology. This division of kinds of emergence into epistemic and metaphysical is neither exhaustive nor maximally specific, but it should be sufficient for the purposes of this chapter. For those interested in a more nuanced division, there has been a lot of recent work on varieties of emergence; see for instance: Chalmers (2006), Silberstein (2001), van Gulick (2001) and Wilson (2015).

One popular way to characterise a strong form of metaphysical emergence is to say that emergent entities must possess *novel causal powers*. For instance, Jaegwon Kim asserts that if emergentism is to be a coherent position, then emergent entities must have distinctive causal powers (1999). Timothy O'Connor and Hong Yu Wong characterise emergent properties as basic properties had by composite entities, where 'basicness' is at least in part a matter of conferring novel causal powers (2005). Jessica Wilson takes strongly metaphysically emergent entities to have "fundamentally novel powers" (2015, p. 356). There isn't space in this chapter to address these various accounts in detail, but it should be clear that characterising a strong form of metaphysical emergence in terms of novel causal powers is a common feature of the current literature.

One strand of thought which underlies this trend is a commitment to 'Alexander's dictum' (also sometimes called the 'Eleatic Principle' or the 'Causal Criterion'). Alexander's dictum is typically roughly glossed as follows (see for instance Cargile, 2003, p. 1):

> **AD:** To exist is to have causal powers

Adherents of **AD** thus hold that we ought to allow into our ontology only those entities which are capable of engaging causally with other entities, which can make a difference to what happens in the world. This formulation is loose and sloganistic, and one of the aims of this chapter is to offer some clarification on how the principle ought to be interpreted – presenting such criteria in this sloganistic manner is somewhat commonplace. Compare the earlier text, for instance, with Quine's "to be is to be the value of a variable" (e.g., 1980, p. 15).

If **AD** is true, then the motivation for outlining metaphysical emergence in causal terms should be clear: the senses of *distinctness* and *novelty* which really matter in formulating metaphysical emergence will be those which refer to causal powers; it is by possessing or conferring *distinct*, *novel* causal powers that emergent entities gain ontic status, and from which this form of emergentism gains its metaphysical 'strength'. It might initially seem that accepting something like **AD** might automatically commit one to accepting the contentious claim that causal powers are real, irreducible features of the world. This is not, however, the case: those who take a broadly anti-realist stance on powers – as many neo-Humean metaphysicians do – can simply read the term 'causal power' as shorthand for whatever they think properly analyses or reduces this concept, be that the truth of certain conditional statements (see, for instance, Lewis, 1997) or the possession of certain non-dispositional, categorical bases (see, for instance, Prior, Pargetter and Jackson, 1982).

AD, however, does not enjoy universal assent. Nor is it clear exactly how the rough gloss of the principle given earlier ought to be finessed. This chapter examines the role **AD** plays in the debate between emergentists and reductionists, the motivations for endorsing **AD** and some criticisms which **AD** faces, as well as some responses to these criticisms. It argues that whilst these criticisms might be taken to show that **AD** cannot be endorsed as a fully general principle of ontological commitment, nevertheless, the principle can be formulated in a manner that makes it suitable for use in the emergence debate. Finally, some wider consequences of giving causal powers a central role in the debate between emergentists and reductionists are examined.

Alexander's dictum

In *Space, Time and Deity*, British emergentist Samuel Alexander writes of epiphenomenalism – the view that mental states do not cause anything – that:

> The doctrine is not simply to be rejected because it supposes something to exist in nature which has nothing to do, no purpose to serve, a species of noblesse which depends on the work of its inferiors, but is kept for show and might as well, and undoubtedly would in time be abolished.
>
> *(Alexander, 1920, p. 8)*

Alexander goes on to suggest that the strongest grounds for the rejection of epiphenomenalism are, in fact, empirical. Whilst the details of his claim need not detain us here, the passage quoted is suggestive of a general principle concerning ontology: in order for some posited entity to be admitted to an ontological system, that entity ought to play some distinctive role or serve some distinctive purpose in that system, one that is distinctive in the sense that that role or purpose is not played or served by some more basic or fundamental entity or entities – Alexander's 'inferiors'. If the entity cannot be given any such role, then it ought to be abolished in favour of the more basic or fundamental entity or entities which in fact play that role or serve that purpose.

Alexander was certainly not the first to suggest this line of thought, which goes back to antiquity. In the *Sophist*, Plato's Eleatic Stranger, when discussing the nature of being, says:

> I suggest that everything which possesses any power of any kind, either to produce a change in anything of any nature or to be affected even in the least degree by the

slightest cause, though it be only on one occasion, has real existence. For I set up as a definition which defines being, that it is nothing else but power.

(Soph. 247d-e)

The Eleatic Stranger identifies real existence with the potential to engage in causal interactions. This implies that those putative entities which cannot engage in such interactions, that are neither causes nor effects, lack real existence. They ought not to be included in our ontology, which should only contain entities which *do* have the potential to be changed, as effects, or to produce change, as causes.

This leads us to something like **AD**, the claim that to exist is to have causal powers. At this point, it is worth reflecting on a couple of general points concerning this principle. First, it should be noted that whilst being committed to **AD** may often go hand in hand with the acceptance of views such as dispositional essentialism or a powers account of properties – a family of views which may very roughly be characterised as holding that properties are essentially causal in nature and do not derive their causal relevance from, for instance, laws of nature – it is perfectly consistent to accept **AD** even if one rejects these views. David Armstrong rejects the powers account of properties (see, for instance, 1996, ch.5) but is an advocate of **AD** (1978, 1989).

Second, for the purposes of the investigation at hand, **AD** should properly be understood as providing a *criterion of ontological commitment* – that is, a methodological philosophical principle that provides a standard by which to adjudicate which putative entities we take a realist attitude towards. A stronger reading of **AD** is possible: one which *equates* existence with the possession of causal power – this sort of reading is perhaps suggested by the passage from the *Sophist* quoted earlier.

Third, the term 'have' as it appears in **AD** must be read in a fairly loose sense if the principle is to be generally applicable to entities of a variety of different ontological categories. An object, for instance, might 'have' a power in virtue of possessing a particular property, but arguably a property will not 'have' a power in virtue of possessing some further property. Rather, depending on one's views concerning the relationship between properties and causation, the property will 'have' a power, say, in virtue of its essential nature, or because it falls under some law, and will *confer* this power on objects by which the property in question is instantiated. That properties do not literally *have* powers, however, should not be taken to mean that, according to **AD** they therefore do not exist, and thus that nominalism is true. Thus, it is crucial that **AD** be read loosely enough that it can apply to properties, objects and perhaps also to entities belonging to other ontological categories, such as events and processes.

Emergence, reduction and Alexander's dictum

Having briefly outlined the nature and origins of **AD**, it is worth spending a little time thinking about the role the principle has come to play in the debate between emergentists and reductionists. There are two distinct intellectual directions from which one can approach this issue. First, prior to considering the debate between emergentists and reductionists, one might have an antecedent commitment to **AD**. Coming at the issue from this direction, it would seem only natural to apply the principle within the debate. On the other hand, one's primary concern might be with the debate, prior to one having any particular commitment to **AD**. Coming at the issue from this direction, however, raises the question of why one ought to adopt this particular criterion and accompanying characterisation of metaphysically emergent entities as being those higher-level entities which possess or confer novel causal powers.

There are a number of compelling reasons to adopt **AD** within the emergence debate, and thus to frame strong metaphysical emergence in terms of the possession of distinct, novel causal powers. First, **AD** seems to provide at least a sufficient criterion of existence. Suppose, for sake of argument, that E is a higher-level object which is dependent on some set of lower-level objects, the Bs, but which possesses causal powers that are genuinely distinct from and novel with respect to the causal powers possessed by the Bs. Remember that the issue at stake, in terms of strong metaphysical emergence, is whether there exist any higher-level entities which are properly characterised as 'over-and-above' the lower-level entities upon which they depend. E's having its own causal efficacy which cannot be attributed to the Bs looks like evidence *par excellence* that E is something over and above the Bs, that E ought to be accorded genuine irreducible ontological status. If it were not, then it seems that there would be no entity to which to attribute this causal efficacy, and this seems absurd.

Second, having a distinctive causal role seems the appropriate criterion for picking out *strong* forms of metaphysical emergence. There might well be reasons to hold that some higher-level entity E is metaphysically, and not just epistemically, emergent from the Bs even if E does not have distinct, novel causal powers. That is to say, whilst we have seen already that **AD** provides a sufficient condition for metaphysical emergence, it is at least prima facie conceivable that it does not provide a necessary condition. Suppose this is the case and that there could be grounds for thinking that some higher-level entity E_1 was in some sense metaphysically, and not merely epistemically – that is, merely with regard to explanatory or predictive practices – distinct from and novel with regard to the lower-level entities, the Bs, upon which it depends. E_1 might well count as metaphysically emergent, then. But compare E_1 to a higher-level entity E_2 which *does* have its own novel causal powers – it seems very natural to see the latter as emergent in a *stronger* sense than the former. This kind of distinction between strong and weak metaphysical emergence can be found, for instance, in Wilson (2015).

Finally, to the extent to which one has confidence in science's ability to trace causal goings-on, adopting **AD** promises to help make the debate between emergentists and reductionists empirically tractable – at least in part. This is something that all parties to the debate ought to agree is desirable. For arguments in favour of the claim that one of the central roles of science, especially the physical sciences, is to trace causal goings-on, see for instance, Blackburn (1990) or Hawthorne (2001).

It is also worth noting that adopting **AD** will not only be attractive to emergentists, as its combination with various forms of causal exclusion argument (see e.g. Kim, 1999) provides a clear framework and procedure for reductionism (this point is nicely elaborated in Elder, 2003). Having a criterion of ontological commitment that both emergentists and reductionists can agree upon is crucial for the debate. If Smith is an emergentist concerning entities in some domain, but Jones is reductionist about them, and Smith and Jones subscribe to differing criteria of ontological commitment, then there is a very strong chance that there is only the appearance of genuine disagreement between them – Smith and Jones may well simply be talking past one another, as those entities count as real according to Smith's criterion but not according to Jones's. The form of reduction that stems from the acceptance of **AD** might also be thought attractive because it is clear to see how it can be distinguished from eliminativism – if E reduces to the Bs, as all of E's causal powers can be identified with causal powers of the Bs, then E still *exists* according to **AD** – we have just been ascribing causal powers to E! E does not, however, exist as *distinct from* or *novel with respect to* the Bs: E is nothing *over and above* the Bs.

Adopting **AD** thus gives a clear sense of what both strong metaphysical emergence and the relevant form of reduction would consist in, provides a clear and robust conceptual framework in which to situate the debate and may help make the debate empirically tractable. These considerations suggest that if **AD** can be suitably motivated, finessed and defended from the criticisms that have been raised against it, then its adoption in the debate – already widespread in the literature, as noted earlier – is both desirable and justifiable.

Arguments in favour of Alexander's dictum

We have seen that there are reasons in favour of employing **AD** in the debate between emergentists and reductionists. It is now time to examine some of the arguments that have been put forward in support of the principle independently of any role that the principle might play in the debate between emergentists and reductionists.

Some arguments in support of **AD** appeal to epistemological concerns. For instance, talking of properties in *A Theory of Universals*, Armstrong argues as follows:

> [E]very property bestows some active and/or passive power upon the particulars which it is a property of . . . it seems possible to conceive of a property of a thing which bestows neither active nor passive power of any sort. But if there are any such properties, then we can have absolutely no reason to suspect their existence. For it is only in so far as properties bestow powers that they can be detected by the sensory apparatus or other mental faculty.
>
> *(1978, pp. 44–45)*

Armstrong's central point is that in order for an entity – it is worth noting at this juncture that whilst Armstrong is discussing properties specifically, the point can be extended to other sorts of entities too – to make itself known to us, be that through featuring directly in sense perception or less directly by registering on a measuring instrument of some sort, it must be endowed with some sort of causal power. If it were not, then it could not affect our sense organs or our measuring devices, for it would lack the powers necessary to so do. Even if causally inert entities are logically possible, there cannot, of necessity, be any empirical evidence for their existence. And on this basis, suggests Armstrong, we should refrain from supposing that there are any such entities. This argument moves from an epistemic claim – that we could never have evidence of causally inert entities – to a metaphysical one: that we should exclude such entities from our ontology.

Colyvan (1998) identifies a potential inductive argument in favour of **AD**. The argument begins with the claim that there is an intuitively plausible account of where the divide between real and fictional/instrumental entities ought to lie, which will "include physical objects, including theoretical entities, perhaps fields and hence waves as disturbances in these fields, amongst the real entities, but should not include (concrete) possible worlds and frictionless planes" (ibid., p. 3). It can then be noted that the intuitively real entities are all entities which possess causal powers and engage in causal interactions (in the actual world). The intuitively non-real entities, on the other hand, seem to lack this feature. This observation can then be used as an intuitive inductive basis for the plausibility of something like **AD**.

A third form of argument for **AD**, which, like the first, has been put forward by Armstrong, is explanatory in nature. It begins by considering the question of the existence of abstracta – entities that exist outside of space and time, and which can be contrasted with concreta, entities that exist within space and time. Suppose for the sake of argument that there are real abstract entities – numbers, say, or propositions. We can then ask whether or not these abstract entities engage in causal transactions with particular, concrete entities. If they do, Armstrong notes that this will be an exceptional form of causation – abstracta are typically taken to be unchanging, for they exist outside of time, and change seems to require at the very least temporal succession (1978,

p. 129). Nevertheless, if we can accept this form of causation, then real abstract entities pose no challenge to **AD**. If, however, we answer the question in the negative but nevertheless maintain the existence of abstracta, then **AD** seems to be faced with a counter-example.

However, it is open to the proponent of **AD** to question what grounds we have to take a realistic attitude towards abstracta. Abstract entities are, by their nature, not observable – neither directly through the senses nor via scientific instruments. But the realist is typically happy to countenance, alongside observables, unobservable entities that play the right sort of *explanatory* role. However, Armstrong argues that in the case of abstracta, the prospects on this front are dim, if, as in the case currently under consideration, they are taken to lack causal power completely:

> But if the entities postulated lie beyond our world, and in addition have no causal or nomic connections with it, then the postulation has no explanatory value. Hence (a further step of course) we ought to deny the existence of such entities.
>
> *(Armstrong, 1989, pp. 7–8)*

Armstrong's key claim is that if abstracta lacking causal power thus also lack explanatory credentials, then we ought not to take a realist attitude towards them. Thus, the putative counter-example to **AD** is undermined.

The three arguments briefly outlined here are all taken to support something like **AD**. There are objections that can be raised to each argument – see Colyvan (1998) for details – but these shall not be subjected to scrutiny here. Instead, the next section shall focus on criticisms which seek to challenge **AD** directly, rather than those which seek to undermine the thesis by weakening the arguments put forward in its favour.

Criticisms of Alexander's dictum

For a criterion of ontological commitment to be successful, it ought to be neither too narrow nor too broad. A criterion that was too narrow would fail to count as existent some entity or entities that we have independent, compelling reasons to consider to genuinely exist. A criterion that was too broad would count as existent some putative entity or entities that we have independent, compelling reasons to consider *not* to genuinely exist. **AD** faces criticisms from both directions: some have found the criterion it provides so broad as to be totally uninformative, whilst others have held that there are plausible counter-examples to the principle in the form of entities that we have good reason to think exist but which nevertheless lack causal powers. The trick for a successful criterion is to meet what one might think of as *Goldilocks' standard* – not too hot, not too cold, but *just right*! This section will outline these criticisms, and in the next section it is argued that there are plausible modifications to **AD** which allow it to meet these objections – at least insofar as it is employed in the debate between emergentists and reductionists.

Cargile complains that **AD** "is not useful because having causal power can be stretched so broadly that its coincidence with existence is trivial" (2003, p. 144). The worry runs something like this: Cargile invites us to consider the disagreement between a dualist and an eliminative physicalist of some stripe concerning the reality of mental images – the former holding that there really are such things; the latter denying that there are. The dualist might make an appeal along the following lines: a particular instance of a particular mental image might occur as the effect of a thinker being asked a certain question and might itself be the cause of some further mental event – say a recollection that the thinker associates with the image.

Prima facie, it seems, it is possible to ascribe causal powers to mental images. But mental images could only ever be candidates for the ascription of causal powers if one has an antecedent reason to believe them to be existent things – as non-entities cannot be the bearers of causal powers (*ibid.*). It seems like the dualist may be in danger of begging the question here. The eliminative physicalist may resist further by claiming that the prima facie plausibility of ascribing causal powers to mental images can be overturned by appealing to some relevant restriction on the notion of causal power – thus, **AD** will rule in their favour. Again, Cargile argues, we are entering question-begging territory: it is highly likely that any restriction which is strong enough to do the work the physicalist requires it to do will itself presuppose in some way the truth of physicalism. If Cargile is right, then appeal to something like **AD** is of no use in settling the ontological dispute concerning the existence of mental images.

The core worry Cargile raises is that whilst it might in fact be true that all and only those entities that really exist are bearers of causal powers, and so **AD** is not false, the criterion cannot fulfil any useful philosophical role: we cannot appeal to **AD** to adjudicate, for instance, between our imaginary dualist and physicalist. It is only when we have independent grounds for taking something to exist that we can ascribe causal powers to it, and it is illegitimate to restrict the notion of causation in such a way that it can be used to discriminate between the existent and the non-existent. This is a problem that will generalise: Cargile spends much of the rest of the paper arguing that similar issues arise, for instance, concerning the existence of abstract entities. Similar thoughts are echoed by Elder, who says of **AD** that it is "so bland as to be scarcely worth stating" (2003, p. 170).

Another form of objection claims that there are some putative entities which both lack causal powers and plausibly exist, and **AD** fails to accommodate their existence. If this is correct, then **AD** cannot provide a satisfactory criterion of ontological commitment. The rest of this section will outline two potential counter-examples to **AD**: abstract objects and epiphenomenal properties.

Consider the following claim: abstracta such as numbers, sets and propositions are not in any way integrated into the causal network of the world – nothing caused them to exist; nothing has ever or will ever cause them to cease to exist or to change in any other way, and they themselves do not cause anything whatsoever. Given that causation is often considered to be a spatio-temporal phenomenon and that abstracta are, by definition, non-spatio-temporal, this claim enjoys a high-degree of plausibility – although, as we have seen earlier, it may be disputed. Suppose for the sake of argument that the claim is true, and that, therefore, abstracta such as numbers, sets and propositions lack causal powers completely. If this is the case, and if **AD** provides a satisfactory criterion of ontological commitment, then it follows that we ought to embrace nominalism and reject the existence of abstracta.

However, there are independent arguments that support Platonism, the view that at least some abstract entities exist. These are most commonly attributed to Quine and Putnam separately: the 'indispensability argument' proceeds from the claims that (i) we ought to take a realist stance towards those entities over which our best theories seem to quantify and (ii) that our best theories seem to quantify over numbers to the conclusion that we ought to be ontologically committed to numbers, and therefore, to at least some abstract entities (see, for instance, Putnam, 1979 or Quine, 1980, 1981). More recently, shifting the focus from quantification, Colyvan has argued – *contra* Armstrong – that abstract entities play an indispensable explanatory role in many of our best theories and, on this basis, ought to be considered to genuinely exist (2001). The foregoing is only the roughest outline of such an argument, but it ought to be sufficient for our purposes here. If there are good reasons to believe that abstracta exist, even though they lack causal power, then on pain of counter-example **AD** cannot provide a satisfactory criterion of ontological commitment. See Colyvan (1998), Marcus (2015) and Oddie (1981) for further discussion of **AD** and abstract objects.

Another class of potential counter-example to **AD** are epiphenomenal properties. Epiphenomenal properties are non-causal properties, properties which do not bestow their bearers with any causal powers. If we have good reason to think that at least some epiphenomenal properties really exist, then it appears that **AD** ought to be rejected. Epiphenomenalism has been most popularly appealed to as a form of property dualism about the mental. Classic arguments such as the conceivability (see, for instance, Chalmers, 1996, ch.4) and knowledge (see for instance Jackson, 1982) arguments suggest that those properties responsible for the intrinsic, phenomenal nature of experiential states are genuinely distinct from any underlying physical basis they may have, but nevertheless make no causal difference whatsoever. If there are good reasons to believe that such properties exist, even though they bestow their bearers with no causal powers whatsoever, then on pain of counter-example **AD** cannot provide a satisfactory criterion of ontological commitment. For an in-depth discussion of epiphenomena and **AD**, see Sabates (2003).

AD faces challenges on two fronts: first, that it is not sufficiently precise as to be a useful or informative principle that can be deployed in order to help settle ontological disputes; and second, that it incorrectly classes as non-existent certain kinds of entities that we have good reason to take seriously, ontologically speaking. In the next section, we shall see that **AD** can be amended such that it can avoid these worries and be suitable for use in the debate between emergentists and reductionists, although this comes at the price of general applicability.

Formulating a satisfactory version of Alexander's dictum (for the purposes of the emergence–reduction debate, anyway . . .)

Consider first the challenge posed to **AD** by abstract entities. There is a simple modification to the principle which can avoid any concerns raised by those who hold that we have independent, compelling grounds to believe that abstracta genuinely exists, namely:

> **AD-1**: For concrete entities, to exist is to have causal powers.

Adopting **AD-1** as opposed to **AD** comes at the cost of generality – **AD-1** does not provide a universal criterion of ontological commitment but is restricted to a particular ontological regime: the spatio-temporal. Is this a cost that participants in the debate concerning emergence and reduction can happily incur? If the only reason that participants in the debate had for adopting something like **AD** was that principle's claim to being a fully general, universal criterion of ontological commitment, then this would be a serious problem. However, as has already been shown in an earlier section, this is not the case: there are a number of compelling reasons for both emergentists and reductionists to accept something like **AD**, and so **AD-1** is not undermined – at least in its application to the relevant debates – by being less than fully general.

Another way in which **AD-1**'s restricted applicability could be problematic would be if emergentists and reductionists were specifically engaged in debate concerning the ontological status of abstract objects. Typically, however, this is not the case: what is at stake in the debate is, for instance, whether there are emergent condensed or soft matter systems (see, for instance, Lancaster and Pexton, 2015 or McLeish, 2017); whether chemistry is emergent from physics (see, for instance, Hendry, 2017); whether the mind or self exists over and above the body (see, for instance, Hasker, 1999); and so on. All these debates concern the ontological status of higher-level *concrete* entities, not abstracta. Thus, for the purposes of the debate between emergentists and reductionists, the restricted scope of **AD-1** is perfectly acceptable.

The concern that **AD** (and equally, **AD-1**) is too broad – almost trivial – can, with a further amendment to the principle, also be met. The core worry here is that the notion of 'causal power' is so broad that it can be applied very liberally and that any proposed restriction on the notion will ultimately beg the question against the existence of the class of entities whose existence it is introduced to exclude. In order to meet this worry, **AD-1** should be amended as follows:

AD-2: For concrete entities, to exist is to have irreducible, non-redundant causal powers.

Unlike **AD, AD-2** – which is very close to something proposed by Merricks (2001, p. 115) – can be appealed to in order to adjudicate between debates such as the one Cargile outlines between the dualist and eliminative physicalist concerning the reality of mental images. Whilst the concept of causal power may, in general, be as broadly applicable as Cargile suggests, the concept of irreducible, non-redundant causal power is not. Furthermore, restricting the criterion of ontological commitment to irreducible, non-redundant causal powers does not beg the question in favour of either party: it remains an open question whether or not mental images, for instance, are possessed of irreducible, non-redundant causal powers. **AD-2** therefore avoids the charge of being so broad as to be metaphysically uninformative.

In the second section it was stated that one of the advantages of adopting something like **AD** in the debate between emergentists and reductionists was that it provides grounds for distinguishing between reductionism and eliminativism. Adopting **AD-2**, however, will rob the principle of this advantage, for if only those entities which have irreducible, non-redundant powers exist, then a putative higher-level entity E possessed only of *reducible, redundant* powers will not simply be reduced to the Bs upon which it depends – it will be eliminated in favour of them. Thus, in order to maintain this advantage, our principle needs to be complicated a little further:

AD-3: For concrete entities, to exist is to have causal powers; to be a fundamental existent is to have irreducible, non-redundant causal powers.

Unlike **AD-2**, **AD-3** allows that higher-level entities which lack irreducible, non-redundant causal powers exist, but only non-fundamentally. One consequence of **AD-3** might strike some readers as odd. In characterising strong emergence at the outset of this chapter, we said that strongly emergent entities are *dependent* entities, and it is natural to some to think of fundamentality as a matter of *independence*. Thus, **AD-3** runs the risk of mis-characterising strongly emergent entities. However, the notions of fundamentality and independence ought not to be run together in this way – to see why this is the case, consider the fact that there doesn't seem to be any contradiction in the following two claims: (i) *properties* as an ontological kind are dependent on the substances by which they are instantiated; and (ii) there are some fundamental properties. These claims may, of course, be false, but they do not seem to be contradictory – for more developed arguments on the conceptual separation of fundamentality and independence, see Barnes (2012).

We have seen that **AD** can be finessed such as to accommodate the concern that it is too broad and to avoid the challenge posed by abstract entities. The final objection outlined in the previous section was that epiphenomenal properties are a potential counter-example to **AD**. The proponent of **AD** can respond to this challenge in (at least) two ways. First, echoing the claims made by Cargile and Elder, one could hold that the sense of 'causal power' employed in the first clause of **AD-3** can be taken to be so broad that even epiphenomenal properties can be ascribed causal powers, albeit reducible, redundant causal powers. It might be objected that this move is illegitimate, as epiphenomena are meant to lack such powers *by definition*. It isn't clear, however, that this complaint is appropriate. Plausibly, what it is crucial to maintain concerning the causal status of epiphenomena is that they *make no causal contribution*, that the causal run of things would be not be affected in the slightest by their absence. This could certainly be the case whether or not one ascribed epiphenomena causal powers in this very liberal sense. Second, taking one's lead from Plato's Eleatic Stranger, one could hold that so long as epiphenomena have what are sometimes called 'backwards-facing' causal powers – that is to say, as long as they can be *caused* to exist, even if they never then go on to cause anything themselves – then this will be sufficient for satisfying the first clause of **AD-3**, and the putative counter-example can be accommodated.

Whilst in a simple form and as a fully general criterion of ontological commitment, **AD** faces serious challenges, it can be finessed – for instance, as **AD-3**, although there are almost certainly other available glosses – such as to make its deployment in the debate concerning strong metaphysical emergence both appealing and justifiable. The arguments in this section have taken these challenges seriously and assumed that the notion of 'causal power' can be stretched as broadly as Cargile and Elder suggest and that there are compelling, independent reasons to think that abstracta and epiphenomena exist. There is, of course, another general strategy of response available to the proponent of **AD**, which is to argue, for instance, that the indispensability argument, or the arguments in favour of epiphenomenal dualism, are not sufficiently strong to motivate these putative counter-examples. To do so, however, would involve taking stances on a variety of issues in meta-ontology, philosophy of science, philosophy of mathematics and philosophy of mind.

Wider consequences

Taking strong metaphysical emergence to be a matter of possessing or conferring distinctive, novel causal powers is widespread in the debate. In part, this is due to a more or less explicit commitment to Alexander's dictum, which has remained largely unexamined. The preceding sections have examined this principle in more depth, detailed the role it plays in the debate between emergentists and reductionists, outlined arguments for and against the principle as a general criterion of ontological commitment and argued that a finessed version of the principle is available which (i) meets major objections and (ii) is suitable for use in the debate concerning strong metaphysical emergence. One consequence of adopting a finessed version of **AD** such as **AD-3** in this debate is that it puts questions concerning the nature of (irreducible, non-redundant) causal powers centre stage.

Such questions include, but probably aren't limited to, how and when simple powers might combine to form complex powers; whether powers are single- or multi-track – that is, whether each power has only one manifestation type, or whether a single power can be directed towards a number of distinct manifestations – and *how* powers operate; whether a lone power manifests when triggered by the presence of a suitable stimulus; or whether powers operate mutually such that several powers must 'work together' to bring about a particular manifestation. That questions concerning strong emergence may well be crucially sensitive to questions in the metaphysics of powers is something that has been largely overlooked in the current literature. Plausibly, implicitly assuming certain answers to the questions outlined earlier may prejudice the debate: in some cases in favour of the reductionist, in others in favour of the emergentist. This means that for the debate to continue in good order, it is essential that these potential prejudices are made explicit and open for assessment. These points are argued in much more detail in Carruth (forthcoming).

Conclusion

Alexander's dictum provides an attractive standard by which to assess putative cases of reduction or emergence: it sets a standard which both parties in the debate can accept and makes clear the sense in which putative strongly emergent entities are supposed to be something 'over and above' the more basic entities from which they emerge, and it promises to help the debate maintain some empirical tractability. It has been argued that the principle can be formulated in such a way as to avoid major objections, and thus that employing the principle in the debate is not only desirable but also justifiable. A consequence of the adoption of Alexander's dictum has been outlined: parties to the debate may need to engage explicitly with questions concerning the nature of powers in order to avoid prejudicing the debate or talking past one another.

Are there strongly metaphysically emergent entities in the sense **AD-3** prescribes? That is, are there any higher-level entities E which possess or confer non-redundant causal powers, powers which are *distinct from* but *novel with respect to* the causal powers of the lower-level entities – the Bs upon which E depends? Nothing said in this chapter entails any answer to this question, and this is as it should be – specific questions concerning whether or not some putatively strongly emergent entity really is emergent or whether it can be reduced to the lower-level entities upon which it depends ought to be settled by detailed, empirically informed analyses of the relevant cases.

Acknowledgements

The research for this publication was conducted during a postdoctoral fellowship attached to the Durham Emergence Project and was made possible through the support of a grant from the John Templeton Foundation. The opinions expressed in this publication are those of the author and do not necessarily reflect the views of the John Templeton Foundation. Many thanks to Stuart Clark, Kim Davies, Robin Findlay Hendry, Tom Lancaster and Galen Strawson for insightful comments by which this chapter was much improved.

References

Alexander, S., 1920, *Space, Time and Deity*, New York: The Humanities Press.
Armstrong, D. M., 1978, *Universals and Scientific Realism*, Cambridge: Cambridge University Press.
Armstrong, D. M., 1989, *A Combinatorial Theory of Possibility*, Cambridge: Cambridge University Press.
Armstrong, D. M., 1996, *A World of States of Affairs*, Cambridge: Cambridge University Press.
Barnes, E., 2012, "Emergence and Fundamentality", *Mind*, 121:873–901.
Blackburn, S. W., 1990, "Filling in Space", *Analysis*, 50:62–65.
Cargile, J., 2003, "On Alexander's Dictum", *Topoi*, 22:143–149.
Carruth, A. D., forthcoming, "Emergence, Reduction and the Identity and Individuation of Causal Powers", *Topoi*.
Chalmers, D., 1996, *The Conscious Mind*, Oxford: Oxford University Press.
Chalmers, D., 2006, "Strong and Weak Emergence", in P. Davies and P. Clayton (eds.) *The Re-Emergence of Emergence*, Oxford: Oxford University Press.
Colyvan, M., 1998, "Can the Eleatic Principle be Justified?", *Canadian Journal of Philosophy*, 28:313–336.
Colyvan, M., 2001, *The Indispensability of Mathematics*, Oxford: Oxford University Press.
Elder, C. L., 2003, "Alexander's Dictum and the Reality of Familiar Objects", *Topoi*, 22:163–171.
Hasker, W., 1999, *The Emergent Self*, Ithaca: Cornell University Press.
Hawthorne, J., 2001, "Causal Structuralism", *Philosophical Perspectives*, 15:361–378.
Hendry, R. F., 2017, "Prospects for Strong Emergence in Chemistry", in F. Orilia and M. P. Paoletti (eds.) *Philosophical and Scientific Perspectives on Downward Causation*, New York: Routledge:146–163.
Jackson, F., 1982, "Epiphenomenal Qualia", *The Philosophical Quarterly*, 32:127–136.
Kim, J., 1999, "Making Sense of Emergence", *Philosophical Studies*, 95:3–36.
Lancaster, T. and Pexton, M., 2015, "Reduction and Emergence in the Fractional Quantum Hall State", *Studies in History and Philosophy of Science, Part B: Studies in History and Philosophy of Modern Physics*, 52(B):343–357.
Lewis, D., 1997, "Finkish Dispositions", *The Philosophical Quarterly*, 47:143–158.
Marcus, R., 2015, "The Eleatic and the Indispensabilist", *Theoria*, 30:415–429.
McLeish, T., 2017, "Strong Emergence and Downward Causation in Biological Physics", *Philosophica*, 92:113–138.
Merricks, T., 2001, *Objects and Persons*, Oxford: Clarendon Press.
O'Connor, T. and Wong, H. Y., 2005, "The Metaphysics of Emergence", *Noûs*, 39:659–678.

Oddie, G., 1981, "Armstrong on the Eleatic Principle and Abstract Objects", *Philosophical Studies*, 41:285–295.

Prior, E. W., Pargetter, R. and Jackson, F., 1982, "Three Theses about Dispositions", *The American Philosophical Quarterly*, 19:251–257.

Putnam, H., 1979, "Philosophy of Logic", reprinted in *Mathematics Matter and Method: Philosophical Papers*, Vol. 1, 2nd edition, Cambridge: Cambridge University Press:323–357.

Quine, W. V. O., 1980, "On What There Is", reprinted in *From a Logical Point of View*, 2nd edition, Cambridge, MA: Harvard University Press:1–19.

Quine, W. V. O., 1981, "Things and Their Place in Theories", in *Theories and Things*, Cambridge, MA: Harvard University Press:1–23.

Sabates, M., 2003, "Being without Doing", *Topoi*, 22:111–125.

Silberstein, M., 2001, "Converging on Emergence: Consciousness, Causation and Explanation", *The Journal of Consciousness Studies*, 8:61–98.

van Gulick, R., 2001, "Reduction, Emergence and Other Recent Options on the Mind/Body Problem: A Philosophic Overview", *Synthese*, 8:1–34.

Wilson, J., 2015, "Metaphysical Emergence: Weak and Strong", in T. Bigaj and C. Wuthrich (eds.) *Metaphysics in Contemporary Physics: Poznan Studies in the Philosophy of the Sciences and the Humanities*, Leiden: Brill.

7

EMERGENCE, DOWNWARD CAUSATION AND ITS ALTERNATIVES

Critically surveying a foundational issue

Carl Gillett

Contemporary scientific emergentism, defended by physicists like Philip Anderson and Robert Laughlin, neuroscientists like Walter Freeman, and many others, takes as its starting point our present scientific evidence that includes "horizontal" causal explanations/models but also ubiquitous "vertical" compositional explanations/models.[1] Crucially, the latter explanations show how almost all higher-level scientific entities, whether individuals, or their activities and properties, are composed by lower level entities.

Scientific reductionists like Steven Weinberg (1994), have long argued that when we have such compositional explanations then this suffices to show that "Wholes are nothing but their parts" and hence that such lower-level parts are the only determinative entities. But inspired by concrete examples from superconductors to slime mold, scientific emergentists have challenged such reasoning by articulating a contrasting picture where we have "emergent" composed entities. The core of this position is depicted in Figure 7.1 by a prominent emergentist in complexity science.

As the figure highlights, scientific emergentists accept that emergent wholes are composed – hence the upward arrow of composition from the parts to the whole – but the scientific emergentist also claims the emergent whole to be "downwardly" determinative upon its parts. As a result, the scientific emergentist now routinely claims, "Parts behave differently in wholes," and argues that this supports her further contention, contra the reductionist, that emergent entities are determinative and that "Wholes are more than the sum of their parts". The diagram thus highlights the core commitment of scientific emergentism to what I will term the "foundational determinative relation" (FDR) by which emergent wholes downwardly determine their parts and around which many of the novel claims of scientific emergentism are founded.

A key task for scientific emergentism is providing a theoretical account of FDR that coheres with the position's other commitments. My focus in this chapter is to critically survey differing treatments of FDR, and my discussion is therefore solely about the species of "emergence" found in the situation framed by Figure 7.1 and endorsed by contemporary scientific emergentists. There are obviously many *other* species of emergence not committed to anything like FDR.

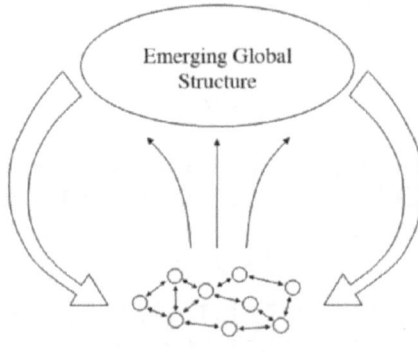

Figure 7.1 Chris Langton's famous diagram of scientific emergence with compositional relations going upwards from the parts to the emergent whole and the foundational determinative relation coming down from the whole to the parts.

[From Lewin (1992), Figure 10, p. 189]

For example, those who take emergent entities to be uncomposed, or who take emergence to do purely epistemic work, endorse nothing like FDR. To make my narrow focus clear, I will use the term "scientific emergence" to refer to the kind endorsed by scientific emergentists.[2]

Philosophers and scientists have offered a number of suggestions about the character of FDR, including that it is:[3]

> *A Boundary Condition* – Michael Polanyi (1968) famously suggested that the emergent whole is a boundary condition on its parts;
>
> *Control* or *Constraint* – the theoretical biologist Pattee (1973), and others following him, suggests that emergent wholes constrain, or bear relations of control to, their parts;
>
> *Reduction in Degrees of Freedom* – Pattee (1973) and others press the related suggestion that emergent wholes result in reductions in the degrees of freedom of their parts.

However, all of these accounts plausibly appear to frame characteristics of FDR, rather than providing an account of the deeper character of this relation. And there appear to be just two competing families of views about the ontological nature of FDR.

By far the most popular approach takes the familiar determinative relation posited in "horizontal" causal explanations and then claims it also holds "downwardly" from the emergent whole to its parts:

> *Downward Causation* – FDR is the relation, or same kind of relation, that we find posited in causal models/explanations between the properties or activities of distinct individuals.

The key question for accounts of FDR as downward causation is whether causation has features that can fit with the characteristics we find in cases of scientific emergence like that in Figure 7.1. For example, the entities in such scientific examples are compositionally related, rather than being wholly distinct individuals, so can causation hold downwardly between the emergent whole and its lower-level parts?

In contrast, a smaller group of scientists and philosophers has explicitly argued that FDR is best understood as a *sui generis* relation different from both causal and compositional relations.

Elsewhere I have coined the term "machresis" (Gillett (2016)) for this new type of relation. But a range of philosophers espouse such a relation, whatever we call it, including Van Gulick (1993), Gillett (2016), and Stump (2012), and also scientists like Freeman (2000) or Laughlin (2005), among others, are plausibly interpreted as espousing this relation:

> *Machresis* – FDR is a non-causal and non-compositional determinative relation that holds "downwardly" from the emergent whole to its parts or between their properties or activities.

The main challenges facing proponents of machresis are, first, to articulate its nature and defend the coherence of this relation and, second, to defend its actual existence, using empirical evidence from concrete scientific cases.

My goal in this chapter is simply to briefly outline and then assess these two accounts of the deeper ontological character of FDR, in downward causation and machresis, having alerted the reader to the other alternatives noted earlier. In this manner, the reader will then have a sense of the terrain concerning a foundational issue confronting contemporary emergentism in the sciences.

I start, in Part 1, by briefly looking at compositional explanations and some features of their compositional relations, since the characteristics of composition constrain any account of FDR because all scientific emergents are composed entities. Part 2 then outlines the positions that take FDR to be "causal". I highlight how different pictures of downward causation and FDR result, depending upon whether we take "causation" to be a "thick" productive relation involving an activity or to be an ontologically "thin" relation of manipulability. I assess each of these views of FDR in Part 3, and I show that there are foundational reasons to believe it is impossible for FDR to be *either* a productive *or* manipulability relation – hence suggesting FDR cannot be a causal relation.

Having found a real need for an alternative to downward causation, in Part 5, I turn to accounts of FDR as a machretic relation. I show that such views avoid the difficulties of downward causation by being compatible with the features of composition, but I highlight the challenges, and extra ontological commitments, involved with machresis. I also note how when we synchronously have machresis acting downwards, alongside compositional relations upwards, then we routinely have a benign form of downward causation existing over time – hence potentially explaining why so many writers think downward causation is involved in scientific emergence even though FDR is not itself a relation of downward causation.

Part 1: compositional explanation in the sciences

To underpin our discussion, we need examples of compositional explanations to get a grip on the nature of such models and the relations they posit, since scientific emergence is found in such cases. I therefore look at a couple of examples of compositional explanation from cell and molecular biology, in the section "Two species of the compositional explanation/model". Then in the section "Some features of compositional relations" I use this work to outline features of composition and how they provide constraints on any adequate account of FDR.

Two species of the compositional explanation/model

Our compositional molecular explanation of cellular protrusion takes the following form. The cell is filled with monomers of globular actin ("G" actin) in the form of unchained actin molecules. One important feature of actin is that it can polymerize swiftly in long filaments ("F" actin). As the model

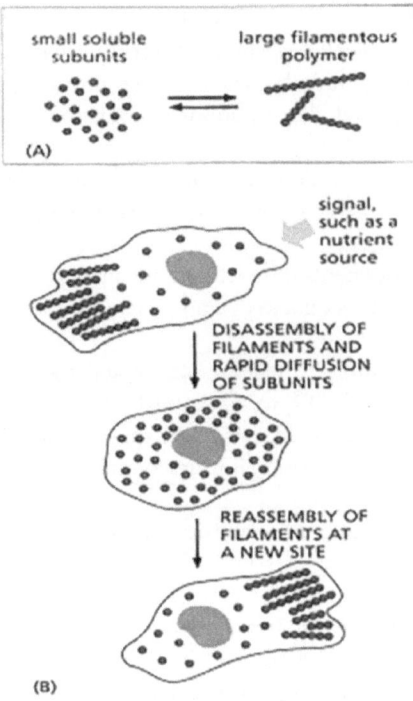

Figure 7.2 Diagram of the molecular basis of cellular protrusion

in Figure 7.2 frames, we consequently explain the protrusion of the cell, an activity of a whole, using directed polymerizations of monomers of G actin into many filaments of actin that press on the lipids in membrane (i.e. using activities of parts). Basically, scientists posit a compositional relation, what I term an "implementation" relation, between the activities of the parts and the activity of the whole. In the assumed compositional context, this compositional relation provides a reason for the existence of the activity of the whole – hence allowing an ontic explanation of it.

It is worth looking at another species of compositional explanation, so consider those that take a property, rather than activity, of a whole as their explanandum. As an example, consider the historically important explanation of the mass-energy of wholes using a compositional relation to the combined mass-energy of their parts at a certain level. For example, the mass-energy of a cell is explained by its being composed, or "realized", by combining the mass-energies of its parts in its proteins, i.e. properties of its parts. There are still further species of compositional explanation, but just using these species, and concrete cases of them, we can highlight some general characteristics of compositional relations.

Some features of compositional relations

Our examples highlight that there are plausibly a number of compositional relations posited in compositional explanations/models – part/whole relations between individuals, as well as realization between the properties of parts and wholes, and implementation between the activities of parts and wholes. Our cases also highlight common characteristics of these compositional relations. Among their shared features, first, we should note that compositional relations are synchronic, since their relata exist at the same time, and that relevant changes to the relata immediately or synchronously accompany each other. For example, increasing the mass-energy of parts synchronously increases the mass-energy of the whole, and vice versa.

Second, we should mark that the relata of compositional relations are in some sense the same, though the relation is weaker than identity. We usually have many components and one composed entity. The cell is in some sense the same as the many proteins, and the mass-energy of the cell is in some sense the same as the combined mass-energy of the many proteins. However, the sameness here cannot be identity – many entities cannot be identical to one entity. Furthermore, qualitatively different entities cannot be identical, and the cell or its mass-energy differs from any protein or its mass-energy.

Fortunately, the sciences do provide more concrete manifestations of this sameness. For example, we have seen how the mass-energy of a whole, like the cell, just is the combined mass-energies of its constituents at some level. Here we see a concretization of the characteristic sameness of the relata of compositional relations in the sciences.

These latter features underlie what I have elsewhere dubbed the "ontologically unifying power" of compositional explanations – when successfully supplied these explanations highlight compositional relations whose existence shows that what we previously thought were independent entities are in some sense the same. There are many other features of compositional relations in the sciences, but these few noted characteristics will suffice for our work here.[4]

It is presently a philosophically contentious issue what the deeper character of such relations of scientific composition actually is, and there are various competing accounts.[5] Some views even take composition to be a causal, or causation-like, relation. However, all these accounts need to accommodate the features just outlined which we find in actual compositional explanations. In these debates, the account I favor takes scientific composition to be a "joint role-filling" relation where a team of component entities together fill the role of, and provide a reason of existence for, some composed entity in a compositional context (Gillett (2016)). However, in my argument below I do not rely on this account, but it will be useful to have a view of composition in hand to highlight claims about the nature of machresis later in the chapter.

Scientific emergentism is committed to emergent entities being composed, so when we have scientific emergence, we always synchronously have the upward compositional relation between components and emergent entity as Figure 7.1 shows. But on top of such composition, the scientific emergentist takes the foundational determinative relation to synchronously hold from the composed entities to their components.

Clarifying some of the features of compositional relations is therefore important because FDR must be compatible with such relations and their singular characteristics. Since scientific emergence involves compositional relations between entities holding upwardly *alongside* FDR holding downwardly between these same entities all at the same time. In the coming sections, I examine how accounts of FDR as downward causation, and then machresis, fare with this important constraint.

Part 2: downward causation as the foundational determinative relation of scientific emergentism

Various scientific emergentists and philosophers have claimed FDR is downward causation.[6] In the section "Two species of downward causation and FDR as direct downward causation", I make this proposal concrete and articulate the specific type of downward causation that FDR must apparently be. I also separate out two distinct theories of causation, since I seek to remain neutral between these accounts and examine what downward causation, and FDR, would be like under each in the sections "FDR as an activity of an emergent whole on its parts" and "FDR as a manipulability relation between an emergent whole and its parts".

Two species of downward causation and FDR as direct downward causation

It is common to talk of "downward causation" in the sciences, since there are techniques that scientists use to illuminate causal relations, and these are often applied across the properties or activities of compositional levels of parts and wholes. On top of this picture, we need to add the scientific emergentist's commitment to FDR. It therefore helps to start by filling out what Langton's diagram looks like under the view that FDR is downward causation in a wider picture of what happens over time as well as at various times.

In Figure 7.3, we now have two times represented and the causal relations over time that result from the emergent whole and its parts in later effects alongside the causal relations now taken to hold at each of these times between parts and wholes.

At the first time, in Figure 7.3, the parts s1-sn compose the whole s*, the properties of P1-P3, of s1-s3, compose the property F of s* and the activities of s1-s3 compose any activities of s*. Furthermore, at this time, through FDR, the emergent whole causally determines its parts and/ or properties have certain powers that we can call "differential powers".

Over time, by time t_2, property F of s* causes effect G in s** at the level of wholes. And as the parts behave as a result of their differential powers, the parts causally bring about certain effects, call them Pz, in other individuals, s'1-s'3, at the level of parts. We therefore have "horizontal" causal relations between wholes and parts, respectively. Although not represented on the diagram, for simplicity, the activity of the whole s* is composed by activities of its parts in s1-s3, In addition, however, we should note that we also appear to have a "diagonal" causal relation of some kind from the emergent whole s*, at t_1, to the effects Pz at the level of parts at t_2, since without the emergent whole and its property F, the parts would not result in Pz – this is represented with the downward diagonal arrow.

The point I want to highlight in this picture of FDR as downward causation is that it is left committed to two different kinds of causal relations, including two kinds of downward causation: a "vertical" downward causal relation at t_1 between the emergent whole and its parts (or their properties or activities) and a "diagonal" downward causal relation from t_1 to t_2 between the emergent whole and later effects at the level of the parts. And it is important to notice that these two types of downward causal relations have importantly different features.

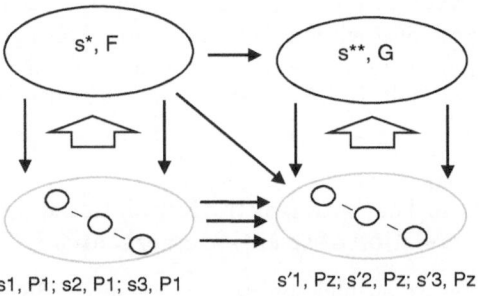

s1, P1; s2, P1; s3, P1   s'1, Pz; s'2, Pz; s'3, Pz

Figure 7.3 Diagram of a scientific emergentism taking FDR to be a causal relation. The enclosed arrows are compositional relations and the other arrows are causal relations.

To bring this out, let us call "Direct" a relation of causation (or manipulability) between property/activity X of one individual and property/activity Y of another individual, where X and Y are in some sense the same, the relation between X and Y is synchronous, and the relation involves synchronous changes in X and Y. In contrast, let us call "Mediated" a relation of causation (or manipulability) between property/activity X of one individual and property/activity Y of another individual, where X and Y are wholly distinct, the relation between X and Y is temporally extended, and involves changes to X and Y that take time.

Applying this rough distinction between Direct and Mediated causation, we see that because FDR is a synchronous relation between compositionally related individuals or their properties/activities, it is a Direct causal relation. In contrast, the "diagonal" relation from the emergent whole to later effects at lower levels is plausibly a Mediated causal relation, since it is diachronic and between entities that are not in some sense the same. Later I return to the differences between Direct and Mediated downward causation, but from this point on I assume causal accounts of FDR are attempts to frame a Direct causal relation.

FDR as an activity of an emergent whole on its parts

Philosophers continue to differ markedly over the nature of causation and endorse a variety of accounts of this relation. I cannot examine all of these proposals, nor do I wish to take sides about what "causation" is either in scientific practice or ultimately. So I look at accounts of FDR framed using the two most prominent families of accounts of "causation" (Hall (2004)) in ontologically "thick" productive and ontologically "thin" manipulability accounts.

Let us start with what FDR as downward causation looks like under accounts treating causation as an ontologically thick relation of what I term "production". As my exemplar of such positions, I focus on views that take "production" to be an activity of an individual, deriving from its powers, that results in and explains certain effects.

This view of FDR takes it to be an activity of the emergent whole that synchronously acts upon its parts to change them. Since we have good empirical evidence that all activities of wholes in nature involve transfer of energy or mediation of force, then there would be a synchronous transfer of energy between the emergent whole and its parts and/or a synchronous exertion of a force between them. We thus have a picture of FDR as an ontologically rich, productive relation between an emergent whole and its parts at a time.

FDR as a manipulability relation between an emergent whole and its parts

On the other side, we have ontologically thin accounts of "causation" like counterfactual and related treatments. Here, given its widespread popularity and connections to scientific practice, I take as my exemplar so-called "interventionist" frameworks that treat "causation" as a relation of manipulability (Woodward 2003, 2015). Within the interventionist framework, causation between X and Y is taken to be a relation of "manipulability" – roughly put, if you can wiggle X and thereby wiggle Y, then we have manipulability between X and Y.

The strength of this approach is that Y is taken to be manipulable by X if we can have an ideal intervention on X with regard to Y where a careful and sophisticated set of conditions for an ideal intervention is then constructed to exclude common causes of X and Y, accidental correlations with Y, and so on. One encouraging sign for those wanting to understand FDR as a manipulability relation is that writers like Craver (2007) have already argued that we standardly have a relation of mutual manipulability between composed and component entities in compositional explanations. So the scientific emergentist would be proposing an account of FDR mirroring this account of composition.[7]

Here Craver (2007)'s adaption of the notion of an "ideal intervention" to apply to such "vertical" relations seems most appropriately used when framing such an account of FDR:

> An *ideal* intervention I on [X] with respect to [Y] is a change in the value of [X] that changes [Y], if at all, *only via* the change in [X]:
>
>> (I1c) the intervention I does not change [Y] directly;
>>
>> (I2c) I does not change the value of some other variable [Z] that changes the value of [Y] except via the change introduced into [X];
>>
>> (I3c) that I is not correlated with some other variable M that is causally independent of I and also a cause of [Y];
>>
>> (I4c) that I fixes the value of [X] in such a way as to screen off the contribution of [X]'s other causes to the value of [X].
>
> *(Craver (2007), p. 154)*[8]

The resulting position claims that FDR is a manipulability relation, and so cases involving FDR will satisfy these conditions on an ideal intervention. We do not have space to go through the various conditions on an ideal intervention, but let me highlight the first condition that X is an ideal intervention on Y only if the intervention on X does not directly change Y. It also is worth emphasizing that the notions of an ideal intervention and manipulability are both technical notions defined by such frameworks.[9]

Part 3: foundational problems for FDR as downward causation

It is striking that compositional explanations have ontologically unifying power but causal explanations do not. There are also apparent mismatches in the features of FDR and causation. However, I put such concerns to one side and instead look carefully at foundational concerns about our two theoretical treatments of FDR. In the section "Assessing FDR as a productive relation", I present reasons to think it is physically impossible for scientific wholes and parts to productively interact – and hence impossible for FDR to be a productive relation. Then in the section "Assessing FDR as a relation of manipulability", I present an argument that it is impossible for a Direct causal relation to satisfy the first condition for an ideal intervention, and hence impossible for FDR to be a manipulability relation.

Assessing FDR as a productive relation: an energetic argument for physical impossibility

A variety of arguments can be given to show that scientific wholes and parts do not productively interact, but let me focus on one centered on mass-energy. In giving this argument, following our successful compositional explanations, I assume the mass-energy of a whole at a time just is the combined mass-energy of its parts at a certain level at that time. Thus, the mass-energy of a cell at a time just is the combined mass-energy of its constituent proteins at this time. And I also assume, again as a result of our empirical findings, that all activities of scientific parts and wholes involve transfer of energy. With these assumptions in hand, let me outline a reductio ad absurdum argument that shows it is physically impossible for a scientific whole to engage in a productive relation with its part, or vice versa.

Consider the cell and a molecule of actin. For the sake of reductio, assume the cell, s* (which is the whole here), productively acts on a molecule of actin, s1 (a part of s*), and changes it – thus framing the picture of FDR as a downward causal relation. Assume the cell transfers energy T to the molecule through this activity. But we know the mass-energy of the cell equals the combined mass-energy of the proteins s1-sn, including our actin molecule s1, and let this equal N. So, the mass-energy of the cell is N. But given the transfer of mass-energy through the productive relation we are assuming FDR to be, from s* to s1, we can conclude the mass-energy of the cell is (N-T). In similar fashion, given this transfer, we can also conclude that the combined mass-energy of the parts, s1-sn, is (N+T). But assuming the cell's mass-energy is the combined mass-energy of the parts, we can thus also conclude that the mass-energy of the cell is (N+T). So we may conclude that the mass-energy of the cell is, and is not, (N+T).

We thus have a contradiction, and the premise that is the most plausibly the candidate for being false is that the scientific whole productively acted upon its own part. So we have a reason to conclude that scientific wholes cannot productively act upon their parts. Consequently, we have reason to conclude that the foundational determinative relation taken by scientific emergentism to hold between scientific wholes, and their parts, cannot be a productive relation.

Assessing FDR as a relation of manipulability: an argument for the impossibility of certain ideal interventions

Let us turn to critically assessing accounts of FDR as an ontologically "thin" relation of manipulability. To see the worry here, let us focus once more on the mass-energy of the emergent whole and its parts, that is, a relation between a property of a whole and properties of its parts, though the same point goes through for compositional relations involving other properties or activities.

Notice that if we intervene to alter a property of the emergent whole that bears FDR to some property of a part, where these properties are compositionally related, then the change in the latter property will be synchronous with the change in the former. For example, if we intervene to change the mass-energy, call it X, of a cell by adding or removing energy to this whole, then this directly changes the energy, call it Y, of at least one part – for the energy of the whole just is the combined energies of the parts. But this consequently prevents satisfaction of the condition for an ideal intervention highlighted earlier, that is, Craver's (I1c), for the intervention on X directly changed Y. But having an ideal intervention is a requirement for having a manipulability relation, so we see that FDR cannot be a manipulability relation.

This type of argument is generally applicable to show that *Direct* causal relations cannot be manipulability relations, for the synchronous character of the changes in their relata precludes satisfaction of the technical requirements for an ideal intervention and hence for the existence of manipulability. We thus see that there is good reason to conclude that FDR cannot be a manipulability relation, since FDR holds synchronously between compositionally related entities.

In contrast, it is worth marking that this argument does not apply to *Mediated* causal relations even when relating properties of wholes and parts. For example, we can press on one side of a cell, changing its shape, X, and this synchronously affects the position, Y*, of a protein that is a part of the cell which composes X. Further, assume the moved protein then productively changes the position Y of a distinct protein where Y does not compose the cell's shape X. Notice that the intervention on X with regard to Y* is not ideal, since it is a Direct causal relation. But the intervention on X with regard to Y is plausibly ideal because, as a Mediated causal relation, a change in Y is not directly produced by the intervention on X – crucially X and Y are not compositionally related properties.

One diagnosis of the appeal of downward causation as an account of FDR is that its proponents have not sufficiently appreciated the features of FDR, or the differences between Direct and Mediated causation. Crucially, the form of causation that has the right features to be FDR cannot involve a relation of manipulability, whilst the species of manipulability often holding between properties/activities of wholes and parts does not have the characteristics of FDR. We can thus see how proponents of FDR as downward causation could easily have fallen into the mistake of thinking it is a viable option.

Part 4: machresis as the foundational determinative relation of scientific emergence?

Although widely popular, we have now found foundational reasons to think that FDR cannot be downward causation. Fortunately, there is another option for scientific emergentism: taking FDR to be a non-causal, and non-compositional, relation of machresis that holds alongside composition. Assuming that productive relations can never hold between parts and wholes, given our earlier argument, we get the type of position outlined in Figure 7.4.

At t_1, we have the upward compositional relations between s1-sn and s*, as well as between their properties and activities, but now we assume we also have the downward machretic relation(s) between some of these compositionally related entities. Over time, we have horizontal productive relations, and a diagonal downward relation of manipulability in a Mediated causal relation – hence highlighting how FDR as machresis results in benign, Mediated downward causation.

This position does not face the type of problems we found with the picture of FDR as downward causation because machresis is like composition in being a synchronous relation between entities that are in some sense the same and it does not involve transfer or energy or mediation of force. However, we know machresis is not a compositional relation because the activities of wholes are such that they cannot fill the roles of their parts – hence stopping them from being joint role-filling relations. Instead, machresis is more plausibly thought of as a role-molding or role-constraining relation which highlights how machresis plausibly involves constraint, reduction in degrees of freedom, or certain kinds of boundary conditions emphasized by the writers noted earlier.

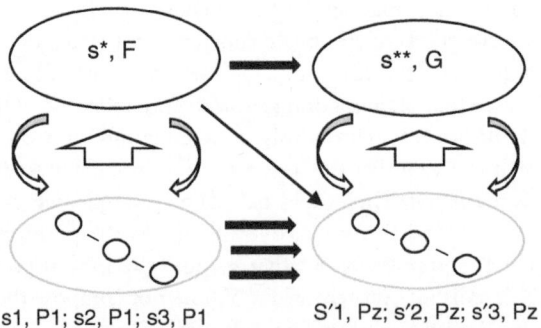

s1, P1; s2, P1; s3, P1   S'1, Pz; s'2, Pz; s'3, Pz

Figure 7.4 Diagram of a scientific emergentism taking FDR to be a machretic relation. The straight enclosed arrows are compositional relations, while the curved enclosed arrows are machretic relations. The thick black arrows are productive relations, and the thin arrow is a manipulability relation.

Under such an account of FDR as machresis, emergent wholes, properties, or activities would be determinative, since the emergent entities machretically determine that their parts have certain differential powers. So "Parts behave differently in wholes" and hence the "Whole would be more than the sum of its parts" – hence securing key claims of scientific emergentism.

As well as such benefits, we need to note some costs of such a view. First, the position accepts both manipulability and production over time along with machresis and composition holding at a time – so embracing a number of relations. Second, the view endorses what I elsewhere term the "Conditioned" view of aggregation (Gillett (2016)) by taking parts in certain aggregations to have differential powers that they would not have if laws in simpler systems were exhaustive. However, both of these commitments are plausibly central to scientific emergentism, rather than deriving from acceptance of machresis.

More importantly, we should mark the two-pronged critique that proponents of machresis need to meet. First, philosophers and scientists argue against the very coherence and possibility of machresis. So a central task of those attracted to machresis is to show it is coherent. However, promising efforts have been made to meet this challenge.[10] Further theoretical work is also needed to simply better understand machresis – for instance, illuminating how, and why, machresis holds in concrete cases.

Second, philosophers and scientists routinely contend that there is no empirical evidence of anything like machresis in nature. So alongside the theoretical work just noted, scientific emergentists need to supply interpretations of compositional explanations defending the existence of machretic relations in these examples. Once again, encouraging efforts have been made in this task. From superconductors to slime mold, working scientists like Anderson, Laughlin, Freeman, and many others have offered a parade of such cases, though more work arguably needs to be done to precisely support their interpretations.

Conclusion

My final conclusion is that scientific emergentists need to give more careful attention to their view of the foundational determinative relation that is at the heart of their positions. Although popular, I have argued that there is *not* (nor could there be) the Direct downward causation required for FDR to be a causal relation. If we take causation to be a "thick" productive relation, we face arguments that scientific wholes and parts cannot bear productive relations. On the other hand, if we take causation to be a "thin" manipulability relation, then we find that it is impossible for the Direct downward causation needed for FDR to satisfy the conditions for an ideal intervention and hence to be a manipulability relation.

Foundational difficulties thus face the common view that FDR is a relation of downward causation. In contrast, taking FDR to be relation of machresis, in a non-causal and non-compositional relation of role shaping or role constraining, avoids such difficulties. As we have seen, when we have FDR as a synchronous machretic relation, then we also have Mediated downward causal relations over time as a result. Substantive work needs to be done to show that machresis is coherent and that empirical evidence supports the existence of machresis in concrete cases. But it bears emphasizing that there is promising work on both of these fronts in contrast to what appear to be intractable foundational problems in understanding FDR as a causal relation.

Notes

1 Anderson (1972, 1995), Freeman (2000), and Laughlin (2005).
2 In Gillett (2016) I term this "Strong" or "S-emergence". In this chapter, wherever I talk of "emergence", unless noted, I mean such scientific emergence.

3 All of the candidates for FDR can be taken to hold between composing individuals, or properties, or activities. For simplicity of exposition, I focus on individuals in parts and wholes, but FDR can also be taken to hold between a composed emergent property or activity and its composing properties and activities.
4 Gillett (2016), chapter 2, gives a fuller, but not complete, list of such features.
5 See Aizawa and Gillett (2016) for an overview of these debates. The existence of treatments of scientific composition as causation-like means that potentially there are views where we have causal relations in both upward and downward directions between the whole and parts. In my discussion, I simply focus on views solely taking FDR to be causal. But in the notes I highlight how the arguments I outline later also undercut positions that take *both* FDR *and* composition to be causal or causation-like.
6 For earlier examples, see Campbell (1974) or Sperry (1986), and for more recent examples see Andersen et al. (2000).
7 The problems I raise later for taking FDR to be a manipulability relation also undermine taking composition to be a manipulability relation.
8 I have changed the variables in the passage to X and Y to be consistent with the rest of my discussion.
9 Woodward (2003, 2015) has a slightly different set of conditions than Craver's (2007) adaption, but the first condition I focus on here is the same.
10 See, for example, Gillett (2016), chapters 7 and 8.

References

Aizawa, K. and Gillett, C. 2016: "Vertical Relations in Science, Philosophy and the World". In K. Aizawa and C. Gillett (eds.) *Scientific Composition and Metaphysical Grounding*. New York: Palgrave MacMillan.
Anderson, P. 1972: "More Is Different: Broken Symmetry and the Nature of the Hierarchical Structure of Science". *Science*, v.177, pp. 393–396.
Anderson, Philip W. (1995). "Physics: The Opening to Complexity", *Proceedings of the National Academy of Sciences*, v.92(15), pp. 6653–6654.
Andersen, P., Christiansen, P., Emmeche, C. and Finnemann, N. (eds.) 2000: *Downward Causation: Minds, Bodies and Matter*. Aarhus: Aarhus University Press.
Campbell, D. 1974: "'Downward Causation' in Hierarchically Organized Biological Systems". In F. J. Ayala and T. Dobzhansky (eds.) *Studies in the Philosophy of Biology*, pp. 179–186. New York: Palgrave MacMillan.
Craver, C. 2007: *Explaining the Brain*. Oxford: Oxford University Press.
Freeman, W. 2000: *How Brains Make Up Their Minds*. New York: Columbia University Press.
Gillett, C. 2016: *Reduction and Emergence in Science and Philosophy*. New York: Cambridge University Press.
Hall, N. 2004: "Two Concepts of Causation". In J. Collins, N. Hall and L. Paul (eds.) *Causation and Counterfactuals*, pp. 225–276. Cambridge, MA: The MIT Press.
Laughlin, R. 2005: *A Different Universe: Reinventing Physics from the Bottom Down*. New York: Basic Books.
Lewin, R. 1992: *Complexity: Life at the Edge of Chaos*. New York: Palgrave MacMillan.
Pattee, H. 1973: "The Physical Basis and Origin of Hierarchical Control". In H. Pattee (ed.) *Hierarchy Theory*. New York: George Braziller.
Polanyi, M. 1968: "Life's Irreducible Structure". *Science*, v.160, pp. 1308–1312.
Sperry, R. 1986: "Macro-Determinism vs. Microdeterminism". *Philosophy of Science*, v.53, pp. 265–270.
Stump, E. 2012: "Emergence, Causal Powers, and Aristotelianism in Metaphysics". In R. Groff and J. Greco (eds.) *Powers and Capacities in Philosophy*. New York: Routledge.
Van Gulick, R. 1993: "Who's in Charge Here? And Who's Doing All the Work?". In J. Heil and A. Mele (eds.) *Mental Causation*, New York: Oxford University Press.
Weinberg, S. 1994: *Dreams of a Final Theory*. New York: Random House.
Woodward, J. 2003: *Making Things Happen: A Theory of Causal Explanation*. New York: Oxford University Press.
Woodward, J. 2015: "Interventionism and Causal Exclusion". *Philosophy and Phenomenological Research*, v.91, pp. 303–347.

8

THE CAUSAL CLOSURE PRINCIPLE

Sophie Gibb

The causal closure argument and emergentism

According to 'strong emergence' – which is the type of emergentism that I shall be concerned with throughout this chapter – certain properties emerge from, they arise out of, more fundamental physical properties, and yet emergent properties are not merely distinct from physical properties, but are novel, 'something over and above' the properties from which they arise. Hence, for example, those who are committed to the existence of emergent *mental* properties hold that mental properties emerge from more fundamental physical properties and yet are something over and above them. How *exactly* to understand 'strong emergence', beyond its negative characteristic of being incompatible with the physicalist claim that all non-physical properties are 'nothing over and above' physical properties is a much-disputed issue, which I do not have the space to explore here. That said, to capture a 'strong emergence' worthy of the name, emergent properties must exist in as robust a sense as the physical properties from which they emerge. Moreover, emergent entities must have full-blooded, independent causal powers that exist over and above the causal powers of the physical entities from which they emerge and which can, given a commitment to downward causation, affect physical entities. In virtue of the combination of its commitment to the claims that emergent properties are distinct from physical properties and that emergent properties are causally relevant in the physical domain, the principle of the causal closure of the physical domain presents a problem for emergence. Indeed, the causal closure argument – one essential premise of which is the causal closure principle – is the central argument against emergence in contemporary discussions of the ontological status of emergent entities.

Taking the existence of emergent mental properties as our example, the causal closure argument can be formulated as follows:

i   *Relevance:* Some mental events are causally relevant to physical effects.
ii *Closure:* All physical effects have sufficient physical causes.
iii *Exclusion:* There is no systematic causal overdetermination.

Therefore, mental events (that are causally relevant to physical effects) are identical with physical events.[1]

Although I have here framed the causal closure argument as one concerning the ontological status of mental entities, the argument could equally have been presented as one concerning the ontological status of chemical, biological or economic entities. If chemical (biological, economic, etc.) events are causally relevant to physical effects, then, given *Closure* and *Exclusion*, the causal closure argument's conclusion is that those chemical (biological, economic, etc.) events must be identical with physical events.

Why is the combination of *Relevance*, *Closure* and *Exclusion* considered to entail the identity of mental events with physical events? Well, in accordance with *Relevance*, say that *M* is a mental event and that it is a sufficient cause of physical event *E*. From *Closure* it follows that *E* must have a sufficient physical cause, call it '*P*'. As *Closure* is consistent with the possibility of physical effects having *both* sufficient physical causes *and* sufficient non-physical causes, the mere combination of *Closure* with *Relevance* does not entail that *M* must be identical with a physical event. It is only the combination of *Closure* with *Exclusion* that entails that the cause of any physical effect must be physical – that there can be no downward causation from *sui generis* mental entities to physical entities. To give a standard example of causal overdetermination, say that two guns are independently fired and the bullets from both guns reach the victim at the same time. If each bullet striking was, on its own, causally sufficient for the victim's death, the death was causally overdetermined by the strikings. *Exclusion* permits *isolated* cases of causal overdetermination but rejects causal overdetermination that is systematic. That is, given *Exclusion*, events cannot be over-determined as a general rule. More precisely, given *Exclusion*, it cannot be the case that *whenever M* causes *E*, *P* also causes *E*, where it is such that if one of the two events *M* and *P* had not existed, the other would have sufficed, in the circumstances, to cause *E*. But the causal overdetermination that the combination of *Relevance* and *Closure* gives rise to *is* precisely of this systematic kind. Whenever *M* causes *E*, given *Closure*, there will be a physical event that is causally sufficient for *E*. Consequently, the combination of *Relevance*, *Closure* and *Exclusion* entail that *M* is identical with *P*. More generally, it entails that mental events are not causally relevant to physical effects, unless they themselves are physical.

The different positions that are held regarding the ontological status of emergent entities can, to a large extent, be distinguished by their response to the causal closure argument. Once again, focusing on the mental causation debate, first take the physicalist stance, which holds that all mental entities are, in some sense, physical. Its proponents can be divided into three main groups. Reductive physicalists identify mental properties with physical properties. Eliminativists reject the existence of mental properties. Non-reductive physicalists hold that mental properties are distinct from, although dependent on (and, hence, 'nothing over and above'), physical properties. Each of these versions of physicalism provides a different response to the causal closure argument. The reductive physicalist accepts its conclusion. According to it, mental properties are identical with physical properties and, hence, mental events are identical with physical events.[2] The eliminativist rejects *Relevance* – according to the eliminativist, there are no mental properties, and, hence, there are no mental causes. The non-reductive physicalist standardly attempts to reject or disambiguate *Exclusion*. Hence, one popular account of non-reductive physicalism argues that although psychophysical causation always involves two distinct, sufficient causes – a mental and a physical one – given non-reductive physicalism, this does not give rise to a worrying systematic causal overdetermination because of the dependence relationship that they posit between the mental and the physical cause.

The response to the causal closure argument that is not available to any physicalist – for this would be to abandon physicalism – is to reject *Closure*. And it is precisely this response that strong emergentism – along with all other forms of dualism that are committed to *Relevance* (that is, all forms of *interactive* dualism) – typically offers to the causal closure argument. In this chapter, I shall consider why this response to the causal closure argument is arguably a plausible one.

Formulating the causal closure principle

In the earlier outline of the causal closure argument, I have formulated the causal closure principle as the principle that 'All physical effects have sufficient physical causes'. This is but one of the many different formulations of the principle presented in the literature on the causal closure argument. Here, I provide examples of some of the different ways in which the principle has been formulated:

1   All physical effects have sufficient physical causes.

(Papineau 1998, p. 375)

2   Every physical event has a physical cause which is enough to bring it about, given the laws of physics.

(Crane 2001, p. 45)

3   All physical effects have complete physical causes ('complete' in the sense that those causes on their own suffice by physical law to fix the chances of those effects).

(Papineau 1993, p. 22)

4   All physical effects are fully determined by law by prior physical events.

(Papineau 2000)

5   If a physical event has a cause at *t*, then it has a physical cause at *t*.

(Kim 2005, p. 15)

6   No physical effect has a non-physical cause.

(Smith and Jones 1986, p. 66)[3]

As I observe elsewhere, clearly these formulations are not all equivalent (Gibb 2015b). Some appeal to the notion of a 'sufficient cause', some appeal to the laws of physics, some are probabilistic in nature. Most importantly, not all of these formulations are of the same strength. Indeed, upon closer scrutiny, not all of these formulations are of the *required* strength. Some are too weak to do their job, whereas others are too strong.

Hence, take formulation (1), the formulation of the causal closure principle that I assumed when outlining the causal closure argument. It is too weak. That is, when (1) is combined with *Relevance* and *Exclusion* it does not entail that mental events (or any other types of emergent events) are identical with physical events. This objection to (1) rests upon the plausible assumption that causation is transitive. If causation is transitive, a physical event *would* have a sufficient physical cause if it had a sufficient mental cause which in turn had a sufficient physical cause. Hence, if physical event *P* caused mental event *M* which in turn caused physical event *E*, this would not be a violation of (1). Consequently, far from entailing that mental events are identical with physical events, given the transitivity of causation, the combination of (1) with *Relevance* and *Exclusion* is compatible with an emergent model of psychophysical causal relevance which holds that neural events cause bodily movement via mental causal intermediaries.

While, on the one hand, a causal closure principle must not be so weak that it renders the causal closure argument invalid, on the other hand, a causal closure principle must not be too strong. Causal closure principles that are so strong that they lack empirical support should clearly be rejected. Equally, causal closure principles that are so strong that they require one to make physicalist assumptions in their defence, and, hence, beg the question against emergentism and other forms of interactive dualism should be rejected. With this in mind, consider formulation (6). Formulations of the causal closure principle such as (1), (2) and (3) rest upon the idea that non-physical causes are never *needed* to account for physical effects. That is, one can tell the complete causal story about any physical effect, purely in terms of physical events. This is not to suggest that non-physical causes could not have physical effects. Rather, the point is that the physical effect would also have a sufficient physical cause, and hence the non-physical cause would be redundant. Unlike these formulations of the causal closure principle, (6) rules out the very possibility of non-physical causes having physical effects. Because of this, given (6), *Exclusion* becomes redundant within the causal closure argument – the following two-premise argument is arguably all that is required to reject interactive dualism:

1   *Relevance:* Some mental events are causally relevant to physical effects.
2   *Closure*: No physical effect has a non-physical cause.

Therefore, mental events (that are causally relevant to physical effects) are identical with physical events.[4]

Lowe (2000, p. 572) argues that any causal closure principle that renders *Exclusion* redundant within the causal closure argument must be too strong. This is because a two-premise argument is being offered, which includes a premise that interactive dualism accepts (*Relevance)* but a conclusion that they reject. Consequently, one can plausibly infer that this formulation of *Closure* is merely an assertion of what the interactive dualist is denying and, hence, that any argument that is presented for *Closure* will inevitably beg the question against the interactive dualist's position.

Ultimately, the goal is to advance a formulation of the causal closure principle which is neither too weak nor too strong. That is, a causal closure principle that is not so weak that it renders the causal closure argument invalid, but not so strong that the causal closure principle is itself untenable. Consequently, when assessing arguments for the causal closure principle – which I will go on to do in the following section – three of the crucial questions to consider are:

1   What strength of causal closure principle does this argument entail?
2   Is this causal closure principle, in combination with *Relevance* and *Exclusion*, strong enough to rule out all plausible forms of interactive dualism, or does it instead render the causal closure argument invalid?
3   Does the argument for the causal closure principle smuggle in assumptions that certain forms of interactive dualism would deny?

In the next two sections I shall briefly consider the two most popular arguments for the causal closure principle and consider why, given certain accounts of emergentism, they fail.

Arguments for the causal closure principle

Despite the central role that the causal closure principle plays in attacks on interactive dualism in contemporary philosophical debates, rigorous arguments for this principle are hard to find. This is presumably because it is assumed that the causal closure principle is an uncontroversial claim that is supported by, and finds its evidence from within, current science. Consequently, it is a principle that requires little further defence from those who make use of it to reject interactive dualism. But how, exactly, does current science provide support for the causal closure principle?

If the principle is a working hypothesis of current science, then as Papineau (a central proponent of the causal closure principle) observes, it is one that is left implicit within current science – it is not written down in any science textbook (Papineau 2000). An argument is therefore needed in defence of the claim that the causal closure principle actually can be inferred from facts of current science.

The no-gap-argument

Within debates about interactive dualism in the philosophy of mind, those that do attempt to defend the causal closure principle most commonly appeal to the 'no-gap argument' to do so (Kim 2010, pp. 112–113; McLaughlin 1998, pp. 278–282; Melnyk 2003, pp. 288–290; Papineau 1993, pp. 31–32). Note, as the no-gap argument is usually specifically targeted against interactive dualist accounts of *psychophysical* causal relevance, in this section my discussion will focus on the mental causation debate and the question of the existence of emergent mental entities. Whether or not the no-gap argument is as persuasive if applied to the existence of other kinds of emergent entities (i.e. chemical, biological, economical, etc.) is a question for another paper.

The no-gap argument begins with the observation that physicists have been incredibly successful in identifying the complete and immediate causes of different kinds of physical events. (A 'complete cause' is the sum of all of the contributory – that is, partial – causes of an event in a particular instance of causation. One event is an 'immediate cause' of another event, if the former does not cause the latter by causing some further event.) To identify the cause of a physical event, physicists have never needed to appeal to *sui generis* mental causes. They have only needed to appeal to physical causes. There is no doubt that there are physical events that are yet to be discovered and, hence, which await a causal explanation. Equally, there are physical events that have been examined but are still to be causally explained. Hence, physics cannot claim to provide the complete and immediate cause of *every* physical event – the causal account that physics provides of physical events contains gaps. But the crucial point is that it is highly implausible that physics will ever need to appeal to *sui generis* mental causes to fill these gaps, or so proponents of the no-gap argument claim. Hence, for example, Kim argues that:

> If a physicist encounters a physical event for which there is no ready physical explanation, or physical cause, she would consider that as indicating a need for further research; perhaps there are as-yet undiscovered physical forces. At no point would she consider the possibility that some nonphysical force outside the space-time world was the cause of this unexplained physical occurrence.
>
> *(Kim 2010, p. 113)*

Furthermore, a more specific no-gap argument can be presented at the level of neurophysiology (Kim 2010, p. 113; McLaughlin 1998, p. 278; Melnyk 2003, p. 187). Interactive dualists in the mental causation debate hold that some bodily events have *sui generis* mental causes. But, the proponent of the no-gap argument reasons, this does not fit with the neurophysiological evidence. In their examination of the causal chains of neural events that bring about bodily movements, neurophysiologists have never needed to appeal to *sui generis* mental causes to provide the complete and immediate cause of any neural event within such a chain. This is not to suggest that for every neural event, current neurophysiology can provide a complete and immediate neural cause. Rather, as with the more general no-gap argument, the crucial point of the neurophysiological no-gap argument is that it is highly implausible that neurophysiology will ever need to appeal to *sui generis* mental causes to fill gaps in the causal chains of neural events that give rise to bodily movement. Quoting from Kim again:

> If a brain scientist finds a neural event that is not explainable by currently known facts in neural science, what is the chance that she would say to herself, 'Maybe this is a case of a Cartesian immaterial mind interfering with neural processes, messing up my experiment. I should look into that possibility!' We can be sure that would never happen.
>
> *(Kim 2010, p. 113)*

I do not wish to assess whether the no-gap argument is plausible here. Instead, my aim is briefly to consider whether, insofar as it is plausible, it generates a successful causal closure argument against emergentism. Does the no-gap argument entail a causal closure principle that is strong enough, when combined with *Relevance* and *Exclusion*, to entail the rejection of emergentism? This question can be broken into two parts: 1) What is the strongest version of the causal closure principle that the no-gap argument entails? and 2) Is this causal closure principle strong enough to defeat emergentism?

What is the strongest version of the causal closure principle that the no-gap argument entails? In the previous section it was observed that if the causal closure principle is taken to be the principle that 'All physical effects have sufficient physical causes', then it is too weak. This principle in combination with *Relevance* and *Exclusion* is consistent with an emergent dualism according to which neural events cause bodily movement via mental causal intermediaries. The no-gap argument, if correct, appears to rule out the existence of such mental causal intermediaries. According to the no-gap argument the discoveries of science and, more specifically, neurophysiology suggest that with regard to any instance of bodily movement, if we trace the causal chain of events leading up to this bodily movement back, we will be faced with a gapless causal chain of purely physical events, with no mental causal intermediaries.

The no-gap argument seems to point to a causal closure principle that is far stronger – something like the following:

> *Closure**: Every physical event that has a cause has an immediate and complete wholly physical cause.

(As noted before, a 'complete cause' is the sum of all of the contributory causes of an event in a particular instance of causation. If each contributory cause of an event is physical, then that event has a complete, wholly physical cause.) *Closure** entails that there will be a seamless causal chain of purely physical events leading up to any bodily movement. And that there will be a seamless causal chain of purely physical events leading up to any bodily movement is precisely the conclusion of the no-gap argument.

Is this causal closure principle, when combined with *Relevance* and *Exclusion*, strong enough to rule out interactive dualism? My view is that it rules out some, but not all, dualist models of psychophysical causal relevance. I shall very briefly summarize the reason why here. (For a detailed defence of these claims, see Gibb 2015a.) The kinds of dualist models of psychophysical causal relevance that are ruled out by this version of the causal closure principle and, hence, which are threatened by the no-gap argument are what I refer to as 'standard' dualist models. According to these models of psychophysical causal relevance, the causal role of a mental event in the physical domain is to cause (either by itself or in conjunction with some other physical event) some neural event or set of neural events, thereby initiating a causal chain of physical events that results in some bodily movement. Descartes' model of psychophysical causal relevance provides an obvious example of a standard dualist model of psychophysical causal relevance. According to Descartes, (simplifying his model greatly) mental events alter the direction of the motion of particles in the brain. These motions initiate a causal chain of physical events which result in bodily movement. This model of psychophysical causal relevance attempts to find a causal role for mental events in the physical domain by suggesting that there is a gap in the causal chain of neural events that result in intentional bodily movement – a gap which is filled by mental events. This claim is in direct opposition to the no-gap argument and *Closure**.

However, not all interactive dualist models of psychophysical causal relevance take the causal role of mental events in the physical domain to be that of causing some neural event which ultimately gives rise to some bodily movement – that is, to fill in gaps in causal chains of neural events. I here have in mind E. J. Lowe's emergent model of psychophysical causal relevance and also my own. (See, for example, Lowe 2000; Gibb 2013.) According to Lowe, the causal role of mental events in the physical domain is not to cause neural *events* but to cause neural *facts*. More specifically, according to Lowe, a mental event is causally responsible for the *fact* that a causal chain of neural events converges upon a particular bodily movement non-coincidentally. According to my own emergent model of psychophysical causal relevance, the causal role of mental events in the physical domain is not to cause neural events, but to *enable* neural events to cause bodily movements. I shall not explore these positions here, but shall simply note that both models depart from standard dualist models of psychophysical causal relevance by denying that the causal role of mental events in the physical domain is to cause some neural event or set of neural events which give rise to some bodily movement. Consequently, both accounts can accept *Closure**, for neither is committed to the claim that any physical event lacks an immediate, complete, wholly physical cause. And as these accounts are not relying on the assumption that there are any gaps in the causal chains of neural events that give rise to bodily movement, neither account would seem to be threatened by the no-gap argument. (For a detailed account of both of these models and why they are not threatened by the no-gap argument, see Gibb 2015a.)

In summary, insofar as the no-gap argument is plausible, it is only a threat to some emergent models of psychophysical causal relevance, not all.

An appeal to the conservation laws

A further argument for the causal closure principle is one that appeals to the conservation laws. Like the no-gap argument, it is most commonly found in the mental causation debate.

According to Harbecke, the causal closure principle is a fact of contemporary physics, because the principle 'draws its main force' from the conservation laws, a cornerstone of contemporary physics (Harbecke 2008, p. 24). Similarly, in Papineau's 'The Rise of Physicalism', which provides one of the most thorough defences of the causal closure principle in the literature, Papineau explains that his original thought was that the principle 'follows from the fact that physics can be formulated in terms of conservation laws' (Papineau 2000, p. 185).

According to these laws:

> *Conservation*: Every physical system is conservative or is part of a larger system that is conservative (where a system is conservative if its total amount of energy and linear momentum can be redistributed, but not altered in amount, by changes that happen within it).[5]

Why is the causal closure principle thought to follow from *Conservation*? Well, as Papineau explains '(i)f the laws of mechanics tell us that important physical quantities are conserved regardless of what happens, then doesn't it follow that the later states of physical systems are always fully determined by their earlier physical states' (Papineau 2000, p. 185)? If, for example, *sui generis* mental events did have neural effects, this would presumably alter the total amount of energy and/or momentum of the brain and, hence, violate the law of the conservation of energy and/or the law of the conservation of momentum.

As Papineau observes, there is an immediate problem with this claim, but it is one which he considers contemporary science to address. This argument for the causal closure principle is threatened if *sui generis* mental energy exists. If *sui generis* mental energy did exist, then provided that it operated conservatively, this would be entirely consistent with the conservation laws – the conservation laws do not regulate what kinds of energy exist, only demanding that all kinds of energy must operate conservatively. If there is *sui generis* mental energy, then the move from *Conservation* to the causal closure principle is brought into question. Indeed, the claim that *sui generis* mental energy does exist and that the occurrence of some physical effects requires the transfer of such energy is precisely the model of psychophysical causal relevance adopted by the interactive dualist, Hart (1988). In light of these considerations, one of Papineau's central concerns in 'The Rise of Physicalism' is to demonstrate that '*sui generis* mental or vital forces should be rejected and physics declared complete' (Papineau 2000, p. 196). Papineau considers that contemporary science gives us very good grounds to conclude that the existence of *sui generis* mental energy is highly improbable. The two arguments that Papineau provides for this conclusion – the 'Argument from Fundamental Forces' and the 'Argument from Physiology' – are not ones that we need to consider here. For the sake of argument, let us assume that Papineau is correct that there is no *sui generis* mental energy. Papineau considers that, from this, it follows that all physical effects are fully determined by law by prior physical events. Hence, we have the following argument for a causal closure principle:

1   Every physical system is conservative or is part of a larger system that is conservative (*Conservation*).
2   There is no non-physical energy (*Energy*).

Therefore:

All physical effects are fully determined by law by prior physical events (*Closure* **).

Closure**, unlike *Closure**, is strong enough to rule out most dualist models of psychophysical causal relevance, including the non-standard dualist models of psychophysical causation that I referred to in the previous section. But does the combination of *Conservation* and *Energy* really entail *Closure*** or even *Closure**? My response is 'No'. Here, I only have the space to briefly indicate why. I consider that to move from the combination of *Conservation* and *Energy*, one must also (at the very least) accept two causal claims without which neither *Conservation* nor *Energy*, nor their combination, could be used to defend *Closure***. First, the causal claim that the redistribution of energy and momentum cannot be brought about without supplying energy or momentum. Second, the causal claim that the only way that something non-physical could contribute to determining an effect in a physical system is by i) affecting the amount of energy or momentum in it or ii) redistributing the energy or momentum in it. Moreover, these causal claims are *denied* by certain dualist models of psychophysical causal relevance. Hence, C. D. Broad's model of psychophysical causal relevance, according to which mental events prompt transfers of energy between physical events without themselves transferring energy, appears to hinge upon the denial of the first causal claim (Broad 1925, p. 109). And the two dualist models of psychophysical causal relevance that I referred to in the discussion of the no-gap argument – that of Lowe's and my own – both reject the second causal claim. (For a detailed discussion and defence of these points see Gibb 2010; Gibb 2015c.)

Hence, while the combination of *Conservation* and *Energy* allows one to rule out certain dualist models of psychophysical causal relevance, such as Hart's model which claims that psychophysical causation consists in the transfer of *sui generis* mental energy, it does not allow one to rule out *all* dualist models of psychophysical causal relevance. More generally, the combination of *Conservation* and *Energy* does not entail *Closure***. To move from *Conservation* and *Energy* to *Closure***, one must make a number of causal assumptions, and whether these causal assumptions are actually correct is the very issue for several dualist models of psychophysical causal relevance.

Concluding remarks

I have suggested that two of the most popular arguments presented in the mental causation debate for the causal closure principle fail. Although they do threaten certain interactive dualist models of psychophysical causal relevance, they certainly do not threaten *all* interactive dualist models of psychophysical causal relevance. Consequently, if these are the best arguments for the causal closure principle, one can conclude that the causal closure argument does not provide a general argument against interactive dualism, or – more specifically – a general argument against emergentism.

And the problems with the causal closure principle do not end here. It has been argued that investigation into what current science itself has to say about the causal structure of the physical domain reveals that the causal closure principle is not a fact of current science – that, far from supporting the principle, current science actually calls it into question. Hence, Hendry (2006) argues that current *chemistry* challenges the causal closure principle. And it has been argued that current *physics* also challenges the principle. It is widely accepted by proponents of the causal closure principle that, given the indeterministic nature of quantum mechanics, causes cannot always be sufficient for their effects. Consequently, formulations of the causal closure principle such as the formulation that 'All physical effects have sufficient physical causes' should be abandoned on the basis of current physics. To attempt to avoid any conflict with quantum mechanics, probabilistic versions of the causal closure principle have been advanced. Formulation (3) of the causal closure principle – 'All physical effects have complete physical causes ('complete' in the sense that those causes on their own suffice by physical law to fix the chances of those effects)' – provides just one example of a probabilistic version of the principle. But the deeper problem that current physics raises for the causal closure principle is that quantum systems are arguably holistic, and the holistic nature of quantum systems appears to conflict with the causal closure principle. For discussions of this particular issue and also other defences of emergence in physics, see Barrett (2006), McGivern and Reuger (2010), Stapp (2005), and Teller (1986).

Consideration of how strong a causal closure principle must be for the causal closure argument to provide a general argument against emergentism and the fact that the arguments presented for the causal closure principle in contemporary philosophical debate fail to support a causal closure principle of this strength, coupled with the questionable status that the causal closure principle has in current science, point to the conclusion that if the causal closure argument is the best argument against emergentism, then emergentism is one of the serious contenders in the debate about the ontological status of certain higher-level entities.

Notes

1 I am understanding causes and effects to be Kimean events. A Kimean event is the exemplification of a property by a substance at a time. Hence, a mental event is the exemplification of a mental property by a substance at a time. A physical event is the exemplification of a physical property by a substance at a time. Given this account of events, two events are identical if and only if they involve the same property, substance and time. It follows that a dualism, not only with regard to mental and physical *substances*, but

also with regard to mental and physical *properties*, entails a dualism with regard to mental and physical events. However, the assumption that the causal relata are Kimean events is not essential to the causal closure argument. (See, for example, Heil and Mele (1993) for further defence of this claim.) Nor is it essential to the arguments that this chapter presents.

2 Note, however, that despite the fact that the causal closure argument entails reductive physicalism, reductive physicalism is an unpopular position in contemporary debate largely as a result of the argument from multiple realizability – the argument that mental properties are multiply realized by, and hence, cannot be identical with, physical properties.

3 For a more extensive list and a thorough examination of the various formulations of the causal closure principle, see (Gibb 2015b).

4 Note, I myself would dispute the validity of this argument, regardless of whether or not *Exclusion* is included as a third premise in it. This is because, according to my own emergent theory of psychophysical causal relevance, a non-physical event does not have to be a cause of a physical event to be causally relevant to it. (See, for example, Gibb 2015b.)

5 See the *Oxford Dictionary of Physics* (Daintith 2005) for a formulation of the conservation laws along these lines.

References

Barrett, J. (2006) 'A Quantum-Mechanical Argument for Mind-Body Dualism', *Erkenntnis*, 65: 97–115.

Broad, C. D. (1925) *The Mind and Its Place in Nature*. London: Routledge & Kegan Paul.

Crane, T. (1995) 'The Mental Causation Debate', *Proceedings of the Aristotelian Society*, Supp. Vol. 69: 211–253.

Crane, T. (2001) *Elements of Mind*. Oxford: Oxford University Press.

Daintith, J. (ed.) (2005) *The Oxford Dictionary of Physics*. Oxford: Oxford University Press.

Gibb, S. C. (2010) 'Closure Principles and the Laws of Conservation of Energy and Momentum', *Dialectica*, 64 (3): 363–384.

Gibb, S. C (2013) 'Mental Causation and Double Prevention', in S. C. Gibb, E. J. Lowe and R. Ingthorsson (eds.), *Mental Causation and Ontology*. Oxford: Oxford University Press, 193–214.

Gibb, S. C (2015a) 'Defending Dualism', *Proceedings of the Aristotelian Society*, 115: 131–146.

Gibb, S. C (2015b) 'The Causal Closure Principle', *Philosophical Quarterly*, 65 (261): 626–647.

Gibb, S. C (2015c) 'Physical Determinability', *Humana Mente: Causation and Mental Causation*, 29: 69–90.

Harbecke, J. (2008) *Mental Causation: Investigating the Mind's Powers in a Natural World*. Frankfurt, Germany: Ontos Verlag.

Hart, W. D. (1988) *The Engines of the Soul*. Cambridge: Cambridge University Press.

Heil, J. and A. Mele (eds.) (1993) *Mental Causation*. Oxford: Clarendon Press.

Hendry, R. (2006) 'Is there downward causation in chemistry?', in D. Baird, E. Scerri and L. McIntyre (eds.), *Philosophy of Chemistry: Synthesis of a New Discipline*. Boston Studies in the Philosophy of Science. Dordrecht: Springer, 242: 173–189.

Kim, J. (2005) *Physicalism, or Something Near Enough*. Princeton: Princeton University Press.

Kim, J. (2010) *Philosophy of Mind*. Boulder, CO: Westview Press.

Lowe, E. J. (2000) 'Causal Closure Principles and Emergentism', *Philosophy*, 57 (294): 571–585.

McGivern, P. and A. Reuger. (2010) 'Emergence in physics', in A. Corradini and T. O'Connor (eds.), *Emergence in Science and Philosophy*. London: Routledge, 213–232.

McLaughlin, B. (1998) 'Epiphenomenalism', in S. Guttenplan (ed.), *A Companion to the Philosophy of Mind*. Oxford: Blackwell.

Melnyk, A. (2003) *A Physicalist Manifesto: Thoroughly Modern Materialism*. Cambridge: Cambridge University Press.

Papineau, D. (1993) *Philosophical Naturalism*. Oxford: Blackwell.

Papineau, D (1998) 'Mind the Gap', *Philosophical Perspectives*, 32 (12): 373–388.

Papineau, D. (2000) 'The Rise of Physicalism', in M. Stone and J. Wolff (eds.), *The Proper Ambition of Science*. New York: Routledge, 174–208.

Smith, P. and O. Jones (1986) *The Philosophy of Mind*. Cambridge: Cambridge University Press.

Stapp, H. (2005) 'Quantum Interactive Dualism: An Alternative to Materialism', *Journal of Consciousness Studies*, 12: 43–58.

Teller, P. (1986) 'Relational Holism and Quantum Mechanics', *British Journal for the Philosophy of Science*, 37: 71–81.

9

COMPUTATIONAL EMERGENCE

Weak and strong

Mark Pexton

Introduction

Imagine you have a piece of square graph paper and a marker pen. You colour in some of the squares in the first column of the paper, leaving others blank. You then move to the second column, and colour in squares in this according to some rule relating to the position of coloured squares in the first column. For example, the rule could be that for every coloured square in column 1, the equivalent square in column 2 is left blank, and every blank in column 1 becomes coloured in column 2. Imagine you now repeat this process for several thousand iterations to see what pattern of coloured squares arises. This setup captures in essence what is known as a cellular automaton in computer science. A cellular automaton is a grid with initial conditions and a set of rules for propagating those initial conditions across the grid. Cellular automata are the basis behind Conway's famous Game of Life (Conway 1970). For the vast majority of initial conditions and rules the Game of Life is uninteresting, either completely random or highly predictable distributions are produced. But for some combinations of rules and initial conditions, striking, almost life-like, distributions arise. These distributions are complex, existing in the hinterland between randomness and regularity. Even though all the rules of the Game of Life, and all the initial conditions, are known perfectly, structures can still appear which defy prediction from that perfect knowledge of the system.

This lack of predictability flies in the face of the reductionist intuition behind Laplace's demon. The standard demon story is that if we had perfect knowledge of the laws of physics and starting conditions, we could predict everything that follows (or objectively specify the probability of everything that follows in a quantum universe). Examples such as the Game of Life, and the growing fields of complexity and chaos science, suggest that even for the demon some systems defy predictability. This realisation has led to a new approach to explicating emergence based on notions derived from computer science. This *computational emergence* has two mutually exclusive forms. The first, developed primarily by Bedau (1997, 2002, 2008) and expanded upon by Huneman (2008, 2012), is called weak emergence. Newman (1996) has also discussed similar ideas (in the context of chaos theory but with a different formalism), as have Boschetti & Gray (2013). Weak emergence is a particular way of classifying a special set of complex resultant phenomena. The second approach is pancomputational emergence, first suggested by Davies (2004, 2007, 2010); Walker, Cisneros & Davies (2012); and Pexton (2015). Pancomputational emergence is a strong form of emergence which identifies higher-level aggregate entities as necessary for the universe to "calculate" the behaviour of complex systems.

Weak emergence

Bedau developed his account of weak emergence inspired by cellular automata, and he makes two claims of them. First, that cellular automata can themselves display weak emergence. Second, that cellular automata are a good model for many real-world systems, and these natural systems also display weak emergence. As stated, in certain scenarios, despite perfect knowledge, cellular automata can produce a confluence of highly context-specific circumstances which produce complex patterns that could not be predicted in advance. This lack of predictability is the hallmark of weak emergence. Imagine we have a macrostate M, with a corresponding microstate m, at time t_1. Then the system evolves at time t_2 to a state (M^*, m^*). If there is no way in principle to predict from m what m^* will be other than running through each intermediate stage between t_1 and t_2, then we have weak emergence. Bedau dubs this running through each stage simulation, since one must use a computer simulation which runs through each step to explain the system's behaviour in microphysical terms.

Weak emergence stands as a subset of resultant behaviour and is perfectly compatible with reduction. It is incompatible with a system displaying strong emergence. This is because all causation in weakly emergent systems is ultimately microphysical causation. All higher-level causal attributions are merely summaries of the aggregate effect of the "micro-causal web". The special category of resultant phenomena that weak emergence captures are those systems that we have to let "crawl the causal web" in order to know what will happen.

> Each concrete physical embodiment of weak emergence is ontologically nothing more than some kind of aggregation of smaller embodied objects . . . each example of macro-level weak emergence is ontologically and causally reducible to micro-level phenomena. However, in practice, typically nobody can understand or follow such a micro-causal reduction unless they simulate the micro-causal web on a computer, because the micro-level causal web is so complex. In a wealth of interesting cases, studied in fields like soft artificial life, computer simulations make it possible to crawl the causal web.
>
> *(Bedau 2008, p. 448)*

Bedau expresses the need to simulate in two very different ways: algorithmic incompressibility and explanatory incompressibility. Explanatory incompressibility is the less precise idea. The basic intuition is that to explain the behaviour of a weakly emergent system, we cannot do better than let it develop and observe its pathway through state space. Explanatory incompressibility is not well defined though, and any definition will no doubt depend on what account of explanation one begins with. Fortunately, Bedau's other way of specifying the necessity to simulate, algorithmic incompressibility, is borrowed from information theory and is more precise.

Algorithmic incompressibility is defined in terms of Kolmogorov complexity. Kolmogorov complexity, or algorithmic complexity, was independently proposed by Solomonoff (1964), Kolmogorov (1965) and Chaitin (1969). Imagine that one wishes to describe an object using a string of binary. The object can have many different possible descriptions, but we are interested in the shortest possible description. The length of the description gives us a measure of the object's intrinsic complexity. We can use computational algorithms to make this more precise: the Kolmogorov complexity of an object X is the length (in bits) of the shortest possible computer program that prints X and then halts. (See Grünwald & Vitányi 2008 for a comprehensive review.)

So in a weakly emergent system the Kolmogorov complexity is at a maximum, and we cannot compress our algorithm at all. There is no shortcut to get from m to m^*. By contrast, in a non-weakly emergent system, there are enough regularities to mean that the Kolmogorov complexity of the system is much reduced. We can therefore compress our algorithm and provide a shortcut to extrapolate $m*$ from m.

The metaphysics of weak emergence

What then of the metaphysical status of weak emergence? As we have seen, weak emergence is a particular form of resultant behaviour; it is not "emergent" at all in the traditional usage of the term. Moreover, it is incompatible with strong emergence. Any strongly emergent system would not be fully determined by microphysical events. Hence, such a strongly emergent system could not in principle also be weakly emergent.

Bedau claims that cases of epistemic emergence merely reflect an epistemic agent's ignorance of microphysical facts. But weak emergence is not like that – even the demon could not compress the algorithms and predict the outcome of a weakly emergent system. In his 1998 paper *Is Weak Emergence Just in the Mind?* Bedau is keen to stress the objectivity of weak emergence. It is not that we happen not to be able to compress the algorithms, it is that no such compression is possible in principle, because of the informational/causal structure of the system.

> It is presumably not an accident that one sort of micro-causal structure is incompressible and another sort is compressible. So, though weak emergence [is a kind of] "epistemological" emergence, weak emergence is not merely epistemological. It is not just in the mind. Instead, weak emergence results from incompressible macro-level structure in the network of micro-level causal connections.
>
> *(Bedau 2008, p. 451)*

And:

> The micro-causal web is real and objective, and the incompressible causal pathways of weak emergent phenomena have a distinctive epistemological consequence. Note that the explanatory incompressibility that defines weak emergence applies to the explanations of any naturalistic epistemic agent, in principle. Just like us, any non-human epistemic agent will have to work through the objective complexity of the local micro-causal interactions. Thus, weak emergence is not merely in the mind, but refers to objective complexity in the objective natural world that is in principle irreducible in practice.
>
> *(Bedau 2008, p. 453)*

As the Game of Life shows, it is not ignorance of laws and starting conditions that prevents predictability, it is just a fact about the complex, context-sensitive, open nature of the system that means one must simulate it. It is the objective micro-causal structure of the system that makes it weakly emergent.

Open questions for weak emergence

The problem of predictability

The first challenge for weak emergentist claims concerns predictability. Predictability is an inherently epistemic concept; hence, weak emergence cannot avoid being a species of epistemic emergence. In order to predict something or not, one must first specify a relevant timescale and level of precision the prediction is to be judged by. Even a chaotic system is predictable on very short timescales (if it weren't, it couldn't be simulated) and on very long timescales. For instance, consider the UK weather system – undoubtedly a chaotic system. Yet in some respects, it is very predictable. If it is sunny one second, it is highly likely it is sunny one second later, and similarly it is highly likely there will be a period of winter at least once a year.

Note that it is impossible to provide a general rule for proving if an algorithm is incompressible in principle but not practice (see Ming & Vitányi 1997). The only way of having a general proof would be to actually compress the string; hence, we can never know for sure whether a string that hasn't been compressed could be compressed or not. However, we can use the fact that attempts to compress an algorithm have failed as empirical evidence that an algorithm is incompressible. So when Bedau claims that no observer in principle could compress an algorithm of a particular system, then this is a provisional statement. We cannot in general say by looking at the micro-causal structure of a system that it is impossible to compress it, only that we have evidence that it may be incompressible. Moreover, such empirically based claims have more weight for artificial systems such as the Game of Life. These are constructed, and we have perfect information concerning every aspect of the system, yet they defy our compression attempts. But for any real system we are not in this position; it is more of an open question whether perfect information concerning starting conditions would allow compression or not. That is not to say that it is not plausible that there are weakly emergent systems in nature, of course.

It is not only timescales that matter but also precision. Bear in mind the distinction between precision and accuracy here. Two people can use the same ruler, where the smallest division marked is 1 mm. One person may be good at using the ruler, the other bad, and so their measurements will have different accuracies. But both measurements will have the same precision: 1 mm. The ruler is not capable of measuring to a greater precision. Of course, one can always use a different ruler, with finer divisions, and this will have a greater precision. There is no mind-independent level of precision; it is an epistemic concept. Now for cellular automata, the level of precision is picked at the point of construction, so all observers agree in advance by design. The "graph paper" is constructed with a certain density of squares. But for real systems, there is no automatic level of precision to view them at or one that all observers must agree to in advance. Now whether a given system is predictable or not depends on the precision one picks to predict it to. So if we are interested in very coarse-grained structural properties, then many chaotic/complex systems will be predictable; that is why there is a science of chaos/complexity in the first place.

Going to the other extreme, if we are fine grained enough, then many ordinary resultant systems cease to be predictable. Consider a very predictable system: a simple pendulum. If we know the pendulum's length, it is a straightforward matter to predict its period to within a second or millisecond, for instance. But say we change our precision and want to predict the period to within a femtosecond or less. For this degree of precision, each actual physical pendulum is completely unpredictable. We might say a pendulum is predictable, but the weather is not, so therefore the weather is weakly emergent. But we can only make this claim because we have implicitly assumed a precision metric. For a different metric both systems would be predictable or both weakly emergent. Which scale we as humans think is relevant is an epistemic concern.

Bedau is correct that whether all epistemic agents would agree on compressibility *once they had agreed on a precision metric* is objective. Compressibility reflects the conjunction of observer-independent facts about the system and the observer-dependent chosen metric. But this objectivity that all *views from somewhere* share is different from weak emergence reflecting a view from nowhere. There is no sense in which an *observerless* universe could be said to have weakly emergent systems.

The problem of randomness

This last point leads to another potential difficulty with weak emergence: Does it adequately distinguish between random states and complex states? As the pendulum example shows, at a fine enough level of precision, nearly all systems contain random elements. This has led McAllister (2013) to contend that all systems are algorithmically incompressible. This is because all systems are made up of structured patterned components and random components. But to compress an algorithm, one must reproduce everything exactly; it is no good replacing a long algorithm with a shorter algorithm that is only nearly the same. Compression requires it to be exactly the same. So even a small random element in a system means that, according to McAllister, it cannot be compressed.

If McAllister is correct, then one cannot use algorithmic incompressibility as the measure of weak emergence. One could still use explanatory incompressibility, but it isn't clear this notion has any rigorous definition. It certainly doesn't have a definition which would apply regardless of which account of explanation one uses. Regardless of this objection, basing compression on explanatory concerns suggests weak emergence is even more epistemically contextualised. As an observer one can simply ignore the aspects of the system that are not compressible because one is uninterested and summarise the rest in some coarse-grained, higher-level rule, but it is difficult to claim that that is not an epistemic activity. Why must all observers agree about which aspects of the system they think are explanatorily salient or not? This will depend on what they wish to explain and what level of detail they wish that explanation to provide.

Even if one rejects McAllister's argument, this only allows weak emergence to distinguish complex systems from regular systems. But weak emergence still faces the challenge of distinguishing complex systems with patterns in randomness from truly random systems. A completely random system is definitely incompressible, yet such systems do not fit into the spirit of what one might think weak emergence should be about. A random data set of pure noise might be incompressible and not predictable, but is it weakly emergent? Emergence implies that there is some pattern – it is defined positively by what the system has, not negatively by what the system lacks (compressibility, predictability, etc.). Complex systems are a balance between order and randomness, where unexpected patterns emerge. Yet characterising this as incompressibility means both complex systems and simply random systems are weakly emergent. Complexity requires randomness, but also patterns and weak emergence, as incompressibility alone says nothing about this structured component of complex systems.

Strong emergence

In contrast to weak emergence there are accounts of strong/ontological emergence based on computational ideas, specifically pancomputationalism. Pancomputationalism is usually accredited to Konrad Zuse, as summed up by "Zuse's Thesis" (ZT):

> ZT: The world is a computer and physical processes are algorithms that are executed on that computer.

The computer could be a cellular automaton as argued by Zuse himself (Zuse 1969), as well as Von Neumann & Burks (1966), Fredkin (2003) and Wolfram (2002). Or it could be a universal Turing machine (Schmidhuber 1997) or a quantum computer (Lloyd 2002).

If the universe is viewed as a computer, then it is natural to consider the physical limitations on that computer. This approach is exemplified by Landauer:

> The laws of physics are essentially algorithms for calculation. These algorithms are significant only to the extent that they are executable in our real physical world. Our usual

laws of physics depend on the mathematician's real number system. With that comes the presumption that given any accuracy requirement, there exists a number of calculational steps, which if executed, will satisfy that accuracy requirement. But the real world is unlikely to supply us with unlimited memory or unlimited Turing machine tapes. Therefore, continuum mathematics is not executable, and physical laws which invoke that cannot really be satisfactory.

(Landauer 1996, p. 192)

Calculations in our world require physical instantiation: computers are real objects bounded by practical limitations. It is the laws of physics that determine the amount of information that a physical system can register (the number of bits) and the number of elementary logic operations that a system can perform (number of operations, or ops).

Information and computation

There are two ways of investigating the limits on computation: 1) make an estimate of the amount of information in a pancomputational universe and 2) make an estimate of how hard the algorithms are that a pancomputational universe executes.

The first of these has been pursued by Lloyd (Lloyd 2002), Davies (2004, 2007, 2010) and Gough (2013) independently. Lloyd has estimated the total number of bits and the total number of operations on those bits since the universe began. Lloyd's calculation is based on looking at some of the physical limitations on computation. These include the constraints on information processing due to entropy, energy and the limit on signalling to be no faster than the speed of light. Lloyd calculates that there are no more than 10^{120} ops on 10^{90} bits (or 10^{120} bits if gravitational degrees of freedom are taken into account) in the universe. A similar number is calculated by Davies from considering the holographic principle and by Gough by considering the current distribution of baryonic matter.

Consideration of these informational limits suggests that reductionism in a pancomputationalist universe may be implausible. The reason is that many higher-level special science systems seem to exceed the limits of microphysical computation. For example, Davies (2004) suggests that the combinatorics of many special science systems exceed Lloyd's information bound. One example is the enzymatic efficacy of a protein (see Luisi 2002). There are 20 varieties of amino acids: a peptide chain of n amino acids can be arranged in 20^n different ways, and each sequence can adopt an enormous number of different conformations. If we assume each amino acid can adopt five different orientations (Fasman 1989) then the total number of conformations is 5^n. Combining 20^n and 5^n means that we have about 10^{2n} different molecular structures. Now n is typically of the order of 100 for a small protein. This means there are 10^{40} times more bits of information needed to simulate the protein step by step at the level of microphysics then there are bits in the accessible universe.

For Davies, it is striking that the upper bound on information processing is exceeded by the complexity of biological systems. Davies argues that this provides good evidence that biological systems are strongly emergent. Davies expresses this form of emergence in causal terms: a pancomputationalist universe cannot be causally closed at the microphysical level, since there is not enough information to compute living systems using informational resources from the microphysical level alone. Davies concludes the informational limits of the universe suggest the causal closure of physics (CCP) is violated.

In a paper co-authored with Walker and Cisneros (Walker, Cisneros & Davies 2012) Davies also suggests that top-down causation can be understood in informational terms. Walker, Cisneros and Davies use major evolutionary transitions in biology as a case study, such as the transition from prokaryotic life to eukaryotic life. They argue that such transitions can be understood as transitions from bottom-up causation to top-down causation, in which higher-organising principles determine the system's behaviour. They argue that such a reversal of the direction of causal power can be identified with a reversal in the flow of information. Bottom-up causation requires information to flow from lower levels to higher levels, whereas top-down causation requires information flow from higher levels to lower levels.

Whilst Davies has identified a new approach to understanding strong emergence, there are challenges to be faced by his account. The most severe is with his identification of information flow with causal determination. There are many different accounts of causation in the philosophical literature. Causation can be based on a set of privileged physical interactions, counterfactual reasoning or powers/capacities. Yet Davies does not make connection to any of these, and it would be desirable to have an explicit account of exactly what is meant by claiming information flow is a form of causal determination.

Algorithmic complexity and computation

The second constraint on computation comes from considering the inherent difficulty of the algorithms themselves (Pexton 2015). In computer science the difficulty of executing an algorithm is evaluated by imagining how difficult such a problem would be to solve for a Turing machine. A Turing machine is a type of thought experiment: an ideal computer invented by Alan Turing (Turing 1937) to probe the limits of calculability. There are two basic constraints on a Turing machine: time and space. Now in computation, each problem, each question, needs to be coded, and some problems will take more code to specify them. Let n be the input size of a problem: we then say a Turing machine can solve this problem in *polynomial time* if the time taken to solve this, $T(n)$, is only a function of n raised to various powers. The problem is dominated by its largest term, so, for instance, if:

$$T(n) = n^2 + 3n + 4$$

then we say that this problem is solvable in polynomial time of order n^2.

Polynomial time contrasts with exponential time. In this it might take on the order of 2^n steps to solve a problem. If we have $n = 100$ and say 10^{12} operations per second, then this exponential time would be on the order of the age of the universe. These limits of ideal computing define different complexity classes as shown in Figure 9.1. It is an open question whether some of these classes are equivalent or not. The aim here is not to rest too much philosophically on the precise boundaries as currently understood. Instead, it is to make the case that some computational problems are difficult in *principle*. The difficulty of these problems is not a limitation of actual technology, but rather an in-principle limitation of what an ideal computer can compute, given that such a computer is a physical object of some kind or another and is therefore not able to use infinite amounts of time or space.

In algorithmic complexity theory, one of the most studied elements of the hierarchy is the relation between the classes P and NP. The class P is the set of all problems that can always be solved by using a polynomial time algorithm. The class NP is the set of problems for which, if we are given a solution, we can verify that solution in polynomial time using a Turing machine. The class P is contained within NP, but it is an open question whether P = NP (mainstream opinion holds that P does not equal NP). The category NP-Hard is the set of problems that are at least as hard as any NP problem. NP-Complete problems are those that are both NP-Hard

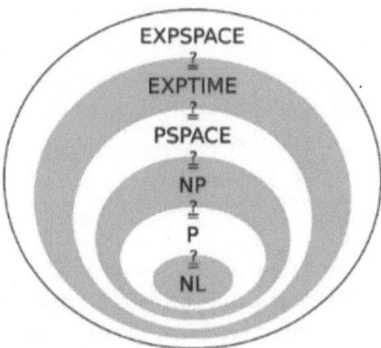

Figure 9.1 The hierarchy of complexity classes, starting with the easiest, NL, P and moving to the most difficult, EXPTIME and EXPSPACE. The question marks reflect the open question of what the relationships between these categories are.

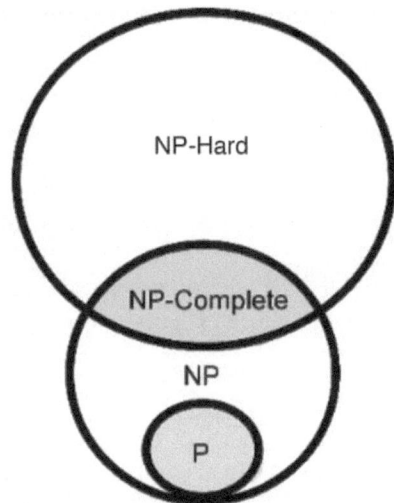

Figure 9.2 Class of NP-Complete for P not equal to NP

and in NP, that is, they are the hardest problems in NP (see Figure 9.2). So although if we are given a solution to an NP-Complete problem we can verify it relatively easily, there is no efficient way to find a solution in the first place. Efficiency is the key concept: for an NP-Complete problem the time required to solve the problem grows quickly as the size of the string inputting the problem increases.

So the lesson from computer science is that some calculations must in principle take longer than the age of our universe in principle (or any physical universe that shared some basic common features such as conservation of energy).

It is possible that some problems in other sciences, when expressed algorithmically, are NP-Complete. For example, Barahona (1982) suggests that spin glass models of phase transitions represent NP-Complete problems, while Welsh (1993) suggests that certain problems in knot theory relating to DNA are potentially NP-Complete or even PSPACE-Complete. One particularly interesting candidate system is protein folding. Fraenkel (1993), Crescenzi et al. (1998) and Berger & Leighton (1998) have each proposed that protein folding is an example of a physical system solving an NP-Complete problem. The status of these examples is controversial though; see Aaronson (2005) for arguments that these cases are not NP-Complete.

So the basic idea is that these problems *expressed in microphysical terms alone could not be calculated given the space, time and informational constraints in a pancomputationalist universe.* But given that these processes definitely occur, this means that a pancomputationalist universe must utilise higher-level terms to reduce both the difficulty of the problems and the amount of information processing needed to solve them.

The Kolmogorov complexity of the algorithms capturing various microphysical processes are reducible if nature can make use of higher-level structural features of those processes. This compression is not merely the epistemic compression of an observer wishing to predict the outcome of a process efficiently. Instead, the compression must be something a pancomputationalist world does itself; otherwise, it does not have the calculational resources to determine the outcome of those microphysical calculations. By using higher-level terms a pancomputationalist world can vastly simplify the informational requirements to perform calculations, thereby allowing it to circumvent the informational limits calculated by Lloyd and to compress algorithms to find solutions to problems that couldn't otherwise be efficiently solved. By applying a pancomputationalist model to the laws of physics, it suggests that the robust higher-level variables of special sciences are not just the epistemic creations of a particular way of viewing the world, but are in fact an essential feature of how nature itself structures the world for calculational purposes.

So we have strong pancomputationalist emergence when:

i   We have a system that cannot be informationally compressed at the microphysical level.

And

ii That system finds a solution in a timescale that exceeds the calculational limits it has at the microphysical level.

Since the system's evolution is "computed", then, given the previously stated constraints, the universe must compress the algorithm (reduce its Kolmogorov complexity) by utilising structural features/terms that are at a higher scale as informational processing resources.

These considerations lead to an information-theoretic definition of strong emergence in terms of the scale-relative compressibility of the Kolmogorov complexity of algorithms. The algorithms can only be compressed to be executable within the salient calculational limits if structural features/kind terms applicable to a higher level are used to reduce the Kolmogorov complexity of the algorithms.

Scale Relative Compressibility-emergence (SRC-emergence)

A system is SRC-emergent if it must use higher-level structural relations and terms to reduce the Kolmogorov complexity of the algorithms that represent that system (such that those algorithms only then become executable given the calculational constraints on the system).

So the basic idea of SRC-emergence is that a pancomputationalist universe uses higher-level kind terms and structural facts as elements in algorithms. By using these higher-level features, those algorithms can be compressed. One way of appreciating this idea is to consider minimum message length encoding, or MML (see Twardy, Gardner & Dowe 2005). In MML, instead of compressing raw data, we look for the most compressed version of a two-part algorithm: the first part states a theory capturing structural features of the data set, and the second part states the data under the assumption that the theory is true. Observational data sets can be thought of as encoding a set of events. Optimal coding theory states that such a sequence will be most efficiently encoded if we use code words of length-log(p) to encode events with probability p. Essentially, the more likely an event is, the shorter the code word we give it; then when we write the whole sequence of events down, the sequence will be compressed. It is knowledge of the structure of a problem that can allow us to allocate the most efficient coding scheme and compress the data.

Let's illustrate this with an example. Andrew is a scientist who wants to test a die with sides A, B, C and D. He rolls it 20,000 times and wishes to send the data set to his friend Jacqui for analysis. The record of the die rolls is an empirical data set with plenty of random noise. Each result can be encoded in binary, say: 00 for A, 01 for B, 10 for C and 11 for D. If McAllister is correct this data set cannot be compressed and the most efficient way of sending the set is just to list the results themselves; therefore, it would take 20,000 bits for Andrew to send Jacqui his results.

But optimal coding theory allows Andrew to do better in certain circumstances. For instance, imagine Andrew suspects the die is biased so that $Pr(A) = 1/2$, $Pr(B) = 1/4$, $Pr(C) = 1/8$ and $Pr(D) = 1/8$. Now using optimal coding theory Andrew can give each outcome a code name of length $-\log(p)$. So outcome A will have code of length 1, 2 for B and 3 for C and D: for example, 0 stands for A, 10 for B, 110 for C and 111 for D. So Andrew now sends Jacqui the set encoded like this and a piece of code relating the theory about the die. If Andrew's theory about the bias of the die is correct, then he has now compressed the data set. Instead of 20,000 bits it will now only take 17,500 bits to send the data to Jacqui.

Remarkably, even if Andrew's theory is only approximately correct, he will still have compressed the data. Let's say the actual probabilities were $Pr(A) = 4/9$, $Pr(B) = 3/9$, $Pr(C) = 1/9$ and $Pr(D) = 1/9$. Now the expected length for sending the data is 17,778 bits; it is not as compressed as if Andrew's theory were exact, but it is still a huge compression in comparison to sending the original data.

How does MML apply to emergence in the context of pancomputationalism? In essence, if we think of laws of nature as algorithms which are executed, then SRC-emergence implies that a pancomputationalist universe must use some special science regularities to allow it to calculate certain physical outcomes. So, for instance, certain higher-level structural facts about the energy landscape of protein folding might be involved in allowing nature to compute which optimal shape a protein should fold to in only 1 sec (rather than the billions of years it would take if it were merely a random microphysical search). Notice that in MML the theory doesn't need to be exactly correct, and hence our special science law doesn't need to be exact and exceptionless — it only has to capture enough facts about probability/counterfactual dependencies to allow a pancomputational universe to compress a calculation. Notice also that MML means that there is no problem with randomness at the microphysical level. The particulars of each concrete microphysical situation (the random "noise" in the data set) do not prevent compression using higher-level algorithmic laws, since these only need to be approximate. (This is contra McAllister 2013.)

Strong and weak pancomputationalism

There are two different strengths of pancomputationalism with different metaphysical commitments:

i   The universe can be adequately modelled as a computer (*weak pancomputationalism*)
ii   The universe literally is a computer (*strong pancomputationalism*)

Davies subscribes to strong pancomputationalism, whereas Pexton (2015) contends that only weak pancomputationalism need be plausible for strong computational emergence to be legitimate. Davies' position is:

> [W]hen it comes to the information bound on the universe, one is forced to confront the status of information: is it ontological or epistemological? If information is simply a description of what we know about the physical world . . . there is no reason why Mother Nature should care about the limit [on the total amount of information]. But if information really does underpin physical reality – if it, so to speak, occupies the ontological basement . . . then the bound on the universe represents a fundamental limitation on all reality, not merely on states of the world that humans perceive.
>
> *(Davies 2007, p. 10)*

Although perfectly defensible, Davies' advocacy of strong pancomputationalism is unnecessary: merely subscribing to weak pancomputationalism still can tell us interesting facts about how the world is constrained. Even though we are only *modelling* the world in computational terms, that modelling step gives us access to physical constraints that would otherwise be obscured. It is not that a weak pancomputationalist universe is constrained by algorithmic features directly; rather, thinking of the world in terms of algorithms replicates certain structural constraints that would be hard to state in non-informational/computational terms.

The Platonic/anti-Platonic debate in mathematics provides a parallel with the debate about whether information theoretic models should be viewed realistically or not. There is little doubt that mathematics is theoretically indispensable to physics – almost all models in physics are expressed mathematically. Certain models could not work without mathematics, and the truths those models reveal to us about the world would be hidden without mathematics.[1] There are two responses to the indispensability of mathematics: the first is Platonism first articulated in modern terms by Frege (1953), in which mathematical objects are real, and, in some versions of Platonism, underpin the physical world and constrain it (see Baker 2009, for an advocate of mathematical constraints).

However, an alternative view is fictionalism about mathematical objects (Balaguer 1996). In this view, mathematical objects do not exist; instead, they are an epistemic construct of mathematicians. However, the fact mathematical objects are fictions does not prevent mathematical models from helping us to express facts about the world. The reason is that there are structural similarities between the constructed mathematical realm and the physical realm. This mapping from mathematics to the real world allows mathematics to capture physical features of the world (Pincock 2004; Leng 2002). Mathematics is applicable in the way a street map is. The map shares structural similarities to the city it corresponds to, allowing a user to identify facts about the city from facts about the map. Note that the map doesn't have to capture every relation in the city, it only has to capture some salient ones. So, for instance, we can model tyres as circles because circles and tyres share some structural similarities, without having to believe that ideal circles exist and constrain the tyre in any way.

Bueno and Colyvan (2011) have extended the mapping idea into an inferentialist account of mathematics. In this, mathematics works not just by a straightforward mapping of features but also by allowing inferences to be made in two stages. The idea is that since certain mathematical objects can share certain structural features of the world, we can then represent the world in abstract mathematical terms. Once we have constructed the mathematical representation, we can use the relation between those directly mapping pieces of mathematics and other (non-directly mapping) pieces of mathematics (dubbed surplus structures) to make mathematical inferences which we can then apply back to the physical world (see Bueno & French 2012; Pexton 2014). The inferences this process allows may otherwise be impossible to make.

Note, however, that for the inferentialist account of mathematics to work, we do not need to believe the world is underpinned by real mathematical objects and laws that constrain physical reality. Mathematical constraints are placeholders for physical constraints that might otherwise have never been revealed without the mathematical modelling step. Information and weak pan-computationalism can be viewed in similar inferentionalist terms. That is, there are structural relations in the physical world which map onto computational relations. By modelling the world in computational terms, we can then make inferences using secondary computational relations and apply them back to the real world. The computational constraints on the universe are then to be viewed as a way of summarising physical constraints on the world. If such an inferential-ist view of pancomputationalism is defensible then, contra Davies, we do not have to defend strong pancomputationalism in order to use information theoretic models to explore issues such as emergence. Although strong pancomputationalism may turn out to be true, weak pancom-putationalism is much easier to establish. In addition, as with the map and city example, weak pancomputationalism does not have to model every aspect of the world, just certain relations that are salient to the question of emergence.

Note

1 Field (1989) has attempted to construct physical theories without mathematics, but his account faces severe difficulties; see Pincock (2007).

References

Aaronson, S. (2005). Guest column: NP-complete problems and physical reality. *ACM Sigact News*, *36*(1), 30–52.

Baker, A. (2009). Mathematical explanation in science. *The British Journal for the Philosophy of Science*, *60*(3), 611–663.

Balaguer, M. (1996). A fictionalist account of the indispensable applications of mathematics. *Philosophical Studies*, *83*(3), 291–314.

Barahona, F. (1982). On the computational complexity of Ising spin glass models. *Journal of Physics A: Mathematical and General*, *15*(10), 3241.

Bedau, M. A. (1997). Weak emergence. *Noûs*, *31*(s11), 375–399.

Bedau, M. A. (2002). Downward causation and the autonomy of weak emergence. *Principia: An International Journal of Epistemology*, *6*(1), 5–50.

Bedau, M. A. (2008). Is weak emergence just in the mind? *Minds and Machines*, *18*(4), 443–459.

Berger, B., & Leighton, T. (1998). Protein folding in the hydrophobic-hydrophilic (HP) model is NP-complete. *Journal of Computational Biology*, *5*(1), 27–40.

Boschetti, F., & Gray, R. (2013). A Turing test for Emergence. In *Advances in Applied Self-Organizing Systems* (pp. 401–416). London: Springer.

Bueno, O., & Colyvan, M. (2011). An inferential conception of the application of mathematics. *Noûs*, *45*(2), 345–374.

Bueno, O., & French, S. (2012). Can mathematics explain physical phenomena? *The British Journal for the Philosophy of Science*, *63*(1), 85–113.

Chaitin, G. J. (1969). On the length of programs for computing finite binary sequences: Statistical considerations. *Journal of the ACM (JACM)*, *16*(1), 145–159.

Conway, J. (1970). The game of life. *Scientific American*, *223*(4), 4.

Crescenzi, P., Goldman, D., Papadimitriou, C., Piccolboni, A., & Yannakakis, M. (1998). On the complexity of protein folding. *Journal of Computational Biology*, *5*(3), 423–465.

Davies, P. C. W. (2004). Emergent biological principles and the computational properties of the universe: Explaining it or explaining it away. *Complexity*, *10*(2), 11–15.

Davies, P. C. W. (2007). The implications of a cosmological information bound for complexity, quantum information and the nature of physical law. *Fluctuation and Noise Letters*, *7*(4), C37–C50.

Davies, P. C. W. (2010). The implications of a holographic universe for quantum information science and the nature of physical law. www.ctnsstars.org/conferences/papers/Holographic%20universe%20and%20infor mation.pdf

Fasman, G. D. (1989) *Prediction of Protein Structure and the Principles of Protein Conformation.* New York: Plenum.

Field, H. H. (1989). *Realism, Mathematics, and Modality.* Oxford: Blackwell.

Fraenkel, A. S. (1993). Complexity of protein folding. *Bulletin of Mathematical Biology, 55*(6), 1199–1210.

Fredkin, E. (2003). An introduction to digital philosophy. *International Journal of Theoretical Physics, 42*(2), 189–247.

Frege, G. (1953). *Foundations of Arithmetic.* Oxford: Blackwell. Transl. by J. L. Austin.

Gough, M. P. (2013). Holographic dark information energy: Predicted dark energy measurement. *Entropy, 15*(3), 1135–1151.

Grünwald, P., & Vitányi, P. (2008). Shannon information and Kolmogorov complexity. https://arxiv.org/abs/cs/0410002

Huneman, P. (2008). Combinatorial vs. computational views of emergence: Emergence made ontological? *Philosophy of Science, 75,* 595–607.

Huneman, P. (2012). Determinism, predictability and open-ended evolution: Lessons from computational emergence. *Synthese, 185*(2), 195–214.

Kolmogorov, A. N. (1965). Three approaches to the quantitative definition of information. *Problems of Information Transmission, 1*(1), 1–7.

Landauer, R. (1996). The physical nature of information. *Physics Letters A, 217*(4), 188–193.

Leng, M. (2002). What's wrong with indispensability? (Or the case for recreational mathematics), *Synthese, 131,* 395–417.

Lloyd, S. (2002). Computational capacity of the universe. *Physical Review Letters, 88*(23), 237901.

Luisi, P. L. (2002). Emergence in chemistry: Chemistry as the embodiment of emergence. *Foundations of Chemistry, 4*(3), 183–200.

McAllister, J. W. (2013). Empirical evidence that the world is not a computer. In *Imagine Math 2* (pp. 127–135). Milan: Springer.

Ming, L., & Vitányi, P. (1997). *An Introduction to Kolmogorov Complexity and Its Applications.* Heidelberg: Springer.

Newman, D. V. (1996). Emergence and strange attractors. *Philosophy of Science,* 245–261.

Pexton, M. (2014). Can asymptotic models be explanatory? *European Journal for Philosophy of Science, 4*(2), 233–252.

Pexton, M. (2015). *An Information Theoretic Critique of the Completeness of Physics.* forthcoming.

Pincock, C. (2004). A revealing flaw in Colyvan's indispensability argument. *Philosophy of Science, 71,* 61–79.

Pincock, C. (2007). A role for mathematics in the physical sciences. *Noûs, 41*(2), 253–275.

Schmidhuber, J. (1997). A computer scientist's view of life, the universe, and everything. In *Foundations of Computer Science* (pp. 201–208). Berlin and Heidelberg: Springer.

Solomonoff, R. J. (1964). A formal theory of inductive inference, parts 1 and 2. *Information and Control, 7,* 1–22, 224–254.

Turing, A. M. (1937). Computability and λ–definability. *The Journal of Symbolic Logic, 2*(4), 153–163.

Twardy, C., Gardner, S., & Dowe, D. L. (2005). Empirical data sets are algorithmically compressible: Reply to McAllister? *Studies in History and Philosophy of Science Part A, 36*(2), 391–402.

Von Neumann, J., & Burks, A. W. (1966). Theory of self-reproducing automata. *IEEE Transactions on Neural Networks, 5*(1), 3–14.

Walker, S. I., Cisneros, L., & Davies, P. C. W. (2012). Evolutionary transitions and top-down causation. *Proceedings of Artificial Life XIII,* 283–290.

Welsh, D. J. (1993). The complexity of knots. *Annals of Discrete Mathematics, 55,* 159–171.

Wolfram, S. (2002). *A New Kind of Science* (Vol. 5). Champaign: Wolfram Media.

Zuse, K. (1969). Rechnender Raum (calculating space). http://philpapers.org/rec/ZUSRR

10

BEING EMERGENCE VS. PATTERN EMERGENCE

Complexity, control and goal-directedness in biological systems

Jason Winning and William Bechtel

1. Introduction

Emergence is much discussed by both philosophers and scientists. But, as noted by Mitchell (2012), there is a significant gulf; philosophers and scientists talk past each other. We contend that this is because philosophers and scientists typically mean different things by *emergence*, leading us to distinguish *being emergence* and *pattern emergence*. While related to distinctions offered by others between, for example, strong/weak emergence or epistemic/ontological emergence (Clayton, 2004, pp. 9–11), we argue that the being vs. pattern distinction better captures what the two groups are addressing. In identifying pattern emergence as the central concern of scientists, however, we do not mean that pattern emergence is of no interest to philosophers. Rather, we argue that philosophers should attend to, and even contribute to, discussions of pattern emergence. But it is important that this discussion be distinguished, not conflated, with discussions of being emergence. In the following section we explicate the notion of being emergence and show how it has been the focus of many philosophical discussions, historical and contemporary. In section 3 we turn to pattern emergence, briefly presenting a few of the ways it figures in the discussions of scientists (and philosophers of science who contribute to these discussions in science). Finally, in sections 4 and 5, we consider the relevance of pattern emergence to several central topics in philosophy of biology: the emergence of complexity, of control, and of goal-directedness in biological systems.

2. Being emergence

Being is a very old subject in philosophy and has been at the center of the branch of philosophy known as metaphysics. Roughly, the more *real* something is, the more *being* it has. Physical objects are thought to be *real* to a degree that imaginary objects are not. Philosophers disagree about whether it makes sense to talk about things being *more* real or *less* real than others, but such debates are nowadays usually carried out not directly in terms of "being" but in terms of *ontology*. An *ontological scheme* defines what types of entities there are (i.e., what *ontological categories* entities can fall into) and how they are related. Instead of talking about entities of one ontological category being "more real" than another, it is now more common for philosophers to say that one ontological category is "more fundamental" or "grounded by" the other, or that the one ontologically "reduces to" the other. However, this represents more of a superficial shift in word choice than a substantial shift in topic.

The idea of levels of being, with the denizens of some levels of being dependent on those at lower levels, is traceable at least to Aristotle, who argued that metaphysics, as the inquiry into being qua being, provided the most fundamental knowledge:

> [H]e who knows best about each genus must be able to state the most certain principles of his subject, so that he whose subject is being qua being must be able to state the most certain principles of all things. This is the philosopher.
>
> *(Metaphysics, 1005b9–11)*

Fundamental to inquiry for Aristotle was determining the true categories of being, with *primary substance* – "that which is neither said of a subject nor in a subject" (*Categories*, 2a14) – providing the foundation because "if the primary substances did not exist it would be impossible for any of the other things to exist" (ibid, 2b6).

Despite his departures from Aristotle and the scholastic tradition on many topics, Descartes also defended the idea of levels of being – "there are various degrees of reality or being: a substance has more reality than an accident or a mode" (CSM 2, p. 117). For Descartes, a single category of being was fundamental: God. Subordinate to God were the categories of mind (thinking substance) and body (corporeal substance). These latter two were "really distinct," that is, ontologically independent: "two substances are said to be really distinct when each of them can exist apart from the other" (CSM 2, p. 114). When characterizing knowledge, Descartes emphasized the importance of knowing what is foundational. He thus gave voice to two ideas that have played an important role in subsequent philosophy: 1) that in order to have true knowledge, we must know what is fundamental in terms of being and 2) what is prior in terms of being cannot come from (i.e., cannot *emerge* from) what is posterior in terms of being.

When Descartes addressed knowledge of the physical world, he defended a mechanistic perspective which derived explanations of compound objects from the properties of their corpuscles. This emphasis on constitution was also developed by Locke, who traced the essence of a thing ("the very being of anything, whereby it is what it is", *An Essay Concerning Human Understanding*, III.iii.15) to its constitution, "which is the foundation of all those properties that are combined in, and are constantly found to co-exist with the nominal essence; that particular constitution which everything has within itself, without any relation to anything without it" (III.iv.6). While acknowledging that we do not know "the internal constitution, whereon their properties depend . . . that texture of parts . . . that makes lead and antimony fusible, wood and stones not" (III.iv.9), he nonetheless assigned them priority both in terms of being and knowledge.

The tradition of Descartes and Locke continues in those contemporaries, exemplified by Kim, who treat the properties at the lowest level of composition as the foundation of the being of all compound entities that provides the explanation for all compound entities. Those who oppose being dependence and being reduction then argue for some version of being emergence.

We can chart these positions by defining the being and being-dependence of any entity:

The **being of X** $=_{def}$: If the ontological category of X is C, then the *being of* X is whatever it is about X that allows it to count as an instance of C.

Y is **being-dependent on X** $=_{def}$: Y's counting as an instance of ontological category C_1 is dependent on X's counting as an instance of ontological category C_2 (for some C_1 and C_2).

With these definitions on the table, we can make more precise the conceptions of reduction and emergence that Kim and many others employ. First, ontological category X *reduces to* ontological category Y if the being of X is *all there is* to the being of Y. An example will clarify. Suppose one type of particle, called an *X particle*, is made up of smaller particles, known as *Y particles*. Something counts as a Y particle if it meets the criteria for a Y particle; in other words, if certain facts obtain. This is also true of an X particle: something only counts as an X particle if the criteria for X particles are met. Now suppose you have a collection of Y particles that form an X particle. If the *facts* that allow the collection of Y particles to count as an X particle are nothing over and above the *facts* that allow the collection of Y particles to count as a collection of Y particles, then X particles *reduce to* Y particles. But if the obtaining of the latter set of facts is only dependent on, not identical to, the obtaining of the former set of facts, then Y is *being-dependent on* X but not reducible to X. Reduction in this sense (we will use the term "being-reduction") entails being-dependence, but being-dependence does not entail being-reduction.

If X is being-dependent on Y but X does not reduce to Y, then X *emerges from* Y. Like reduction, emergence is a relation with two relata: the *emergence base*, from which something is said to emerge, and the *emergent*, that which has emerged. If X particles were emergent in this sense (i.e., *being-emergent*) from Y particles, this would mean that there would be more to the facts that allow the collection of Y particles to count as an X particle than merely the facts that allow the collection of Y particles to count as a collection of Y particles.

Philosophers have used a variety of other terms for being-dependence: ontological dependence, ontological ground, substrate, ontological priority, realization, constitution, truth-maker, componency, noncausal determination, compositional relation, etc. All of these capture the idea that the being of higher-level entities is dependent on those at a lower level. Issues about being-dependence are sometimes raised in terms of "determination," "explanation," or the ability to completely "account for" one ontological category in terms of others. For example, are higher-level entities something more than the components that constitute them or "completely determined by" their constituents? Is a chemical element, such as carbon, completely "accounted for" in terms of the protons, neutrons, and electrons that constitute it? Or are mental states completely "accounted for" in terms of the neurons and other cells that constitute a person's brain? Discussions of *downward causation* are usually centered around being emergence: Do whole entities (e.g., living cells) have properties "over and above" those supplied by their components (genes, proteins, etc.) such that they can have causal effects "independent of" the effects of their constituents? This issue has acquired urgency in the wake of Kim's (e.g., 1999) arguments to the effect that all causation can be "adequately accounted" at the most basic level – assuming that the most basic level is closed so that all effects are determined by causal processes between the occupants at that level – and any causation attributed to wholes built from these constituents is "redundant." Thus, there is nothing to be "explained" in terms of the activities of the wholes. The only way, in Kim's analysis, for minds to exhibit independent causal effects is if dualism is true and minds are neither reducible to nor being-dependent on physical things.

3. Pattern emergence

Having clarified the notion of being emergence, we set it aside. When scientists take up the concepts of emergence and reduction, they are typically not concerned with being and whether the being of one entity can be completely accounted for in terms of its constituents. In part this is due to the focus of scientists on ontology-neutral explanation – on accounting for the phenomena they encounter without taking a metaphysical stance on the underlying ontology. Unlike contemporary philosophers, scientists have to a large extent moved on from the Ancient Greek notion that pursuit of knowledge requires the pursuit for more fundamental levels of being.

Bogen and Woodward (1988) characterize phenomena as repeatable occurrences in the world involving particular types of entities. Although the ontological status of the entities involved may be important for some purposes, phenomena also depend critically on how the entities are *organized*. The same organization can occur among entities regardless of their ontological status. Accordingly, in studying organization, researchers can and often do abstract from considerations about which things are "more real" and focus on patterns exhibited in the phenomena. Explaining a pattern requires an account of how it was generated. Researchers come to treat some patterns as emergent when one cannot account for their generation in the same manner as patterns regarded as more basic. Since patterns are abstract and can be analyzed in disregard of the ontological status of their elements, discussions of pattern emergence are not focused on being.

Studying patterns and their emergence has become important for a wide range of fields over the last several decades. Condensed matter physics deals with "emergent" critical phenomena such as superconductivity, superfluidity, and ferromagnetism. Prigogine pioneered the concept of dissipative structures to understand the emergence of systems that maintain stability far from equilibrium. Chaos theory was developed to understand systems that generate complex and unpredictable, yet determinate, behavior from simple dynamical rules. Mandelbrot developed fractal geometry to understand the emergence of self-similar patterns in nature. Catastrophe theory was developed to understand systems that generate significant or complex effects from simple or minor perturbations; catastrophe theory is part of the more general field of nonlinear systems and complexity theory that examines the mathematics of a wide range of emergent phenomena. Neural network theory deals with systems capable of exhibiting intelligent or adaptive behavior based on nodes that interact in simple ways. Developmental biology is concerned with emergent processes in morphogenesis and tries to understand how from natural selection and environmental constraints complex biological structures and functions can develop. The field of genetic algorithms draws from concepts like evolution and natural selection to develop algorithms that are employed in information processing applications such as optimization and search.

What the areas listed here (sometimes labeled collectively as the "complexity sciences"; Stein, 1989) have in common is that particular patterns *emerge* as entities are configured in particular ways. The emergence of patterns can be addressed independently of questions about ontological fundamentality, or about what is "more real" than what is not. We can recognize the concern with pattern even in some of the statements of philosophers whose primary focus is on being. Sider states:

> Consider questions of ontology, for example. There has been much discussion recently of whether tables and chairs and other composite material objects exist. It is generally common ground in these discussions that there exist subatomic particles that are "arranged tablewise" and "arranged chairwise"; the controversy is over whether there exist in addition tables and chairs that are composed of the particles.
>
> *(2011, p. 7)*

The concern with pattern is captured in the reference to a tablewise *arrangement* of particles. Moreover, we can establish whether or not there is such a pattern independently of addressing questions about whether a level of reality is fundamental and whether any particular arrangement of fundamental stuff will count as an entity. A tablewise arrangement is an example of a pattern, albeit perhaps not a particularly interesting one. Moreover, it is not itself an ontological category; rather, patterns are *candidates* for ontological categories depending on what instantiates them.

Patterns are a central concern for scientists, who investigate how they come about from simpler patterns, what are their properties, etc. Conway's Game of Life (Berlekamp, Conway, & Guy, 2004, chapter 25) illustrates these questions. The Game of Life is laid out on a grid of squares, and simple rules that take into account the state of neighboring squares at a previous instance determine whether a given square is on or off. Given some initial arrangements of on-squares, enduring patterns such as gliders emerge that move as a unit across the grid. Gliders exist, and their emergence and behavior are objects of investigations without raising questions of what type of being they enjoy. Given the gap between the rules that govern squares in the Game of Life and the behavior of gliders, some might view gliders as emergent patterns.

4. Some approaches to characterizing pattern emergence

There are parallels between discussions of being emergence and pattern emergence. With pattern emergence, there is again an emergence base and an emergent: the emergent pattern is in some way dependent on the emergence base pattern. The emergent pattern is also, in some sense, something more than the emergence base pattern. The sense in which the emergent pattern is "something more" differs between contexts, but recently some authors have explored whether there is more to say about what the interesting cases of pattern emergence from the various fields listed earlier have in common, and have taken important steps towards understanding general principles of pattern emergence. With pattern emergence, as with being emergence, the criteria can be ontic, epistemic, semantic, etc. But it is important to keep the criterial dimension (sometimes also referred to as the "weak/strong" dimension) separate from the being/pattern dimension.[1] In the following subsections we review the proposals of several theorists who aspire to develop a general theory of pattern emergence that would enable insights about how patterns that emerge in one field (e.g., theoretical biology) can be applied to pattern emergence in another (e.g., computer science).

4.1. Pattern emergence as bifurcation

Hooker analyzes pattern emergence from the standpoint of dynamical systems theory (DST), a powerful framework for understanding how any kind of system (whether discrete or continuous) changes over time. Central to DST is the concept of a *state space*, "an abstract mathematical space of points where each point is assumed to represent a possible state of the target system" (Bishop, 2012, p. 4). One can then represent the history of the system as a trajectory from an initial state to a final state and employ mathematical tools to analyze the trajectory. What makes DST appropriate is that 1) pattern emergence is usually considered to be something that occurs (or can be modeled as occurring) over time; 2) DST, like the notion of a "pattern," can be applied to any system of elements (as long as they can be described in terms of a state space), regardless of their intrinsic nature; and 3) DST models characterize change in a system using mathematical equations, which assumes that in some sense information about how the system changes over time is compressible (i.e., that it is organized into patterns).

In Hooker's view, we can look at pattern formation as a spectrum running from trivial cases such as the assembly of legs and a top into a table to the emergence of patterns through highly complex processes like biological evolution and creative intelligence. The challenge is to specify when it is useful to appeal to emergence. Hooker rejects epistemic criteria for emergent patterns as problematically subjective, and instead argues in favor of conceptualizing pattern emergence in terms of what is referred to as bifurcation in dynamics, "for then a new behavioural pattern develops, and one whose occurrence is dynamically grounded in a shift in dynamical form"

(2011, p. 209). What Hooker means by a "shift in dynamical form" is that "a differently structured dynamical equation is required to model the behaviours and the pattern of all possible trajectories (the flow) changes" (2013, p. 759). The appeal to equations does not, however, imply that Hooker is invoking a semantic, epistemic, or otherwise mind-dependent criterion. New equations are required as a result of the introduction of constraints, which are objective features of the system itself.

The idea of constraints stems from classical dynamics. Newton's laws fully characterize the behavior of any particle, but they specify each particle's behavior in terms of six variables for the six degrees of freedom it enjoys. Macro-scale objects result from constraints that restrict the degrees of freedom. For example, when two particles are bound together, the particles are constrained to move together. When water molecules are constrained by a pipe, they are restricted to moving in the direction of the pipe. In some cases, one can incorporate the constraints into the equations describing the particles' behavior, but in other cases, one cannot. One cannot, for example, derive the equations Maxwell developed for governors (feedback systems) from basic Newtonian equations.

4.2. *Pattern emergence as nonlinearity or instability*

Hooker's appeal to "differently structured" dynamical equations leaves open a variety of ways to specify the difference other than bifurcations. Bishop (2012), for example, similarly appeals to the DST framework to account for emergence, but appeals to the distinction between linear and nonlinear dynamical equations. A linear equation exhibits superposition: the output of an operation on a variable α is proportional to α. When superposition fails, the equation is nonlinear. In the context of a physical system described by such equations, Bishop states:

> this failure corresponds to a system's output *not* changing proportionally to any change in input. The phenomenon of *sensitive dependence* – the smallest change in the initial conditions can issue forth in a drastic change in a system's behaviour – registers this non-proportional response.
>
> *(2012, p. 4)*

One reason nonlinearity is an interesting point of demarcation is that one can decompose systems described by linear equations into their parts, analyze each independently, and then sum together the results. As Bishop comments, "Reductionist lore tends to work well for such systems." But reductionist strategies are insufficient with nonlinear systems since they respond differently to different inputs given the constraints, which determine the whole. Nonlinear systems often exhibit self-organizing properties: "The interplay between parts and wholes in complex systems and their environments typically leads to the self-organization observed in such systems" (2012, p. 6).[2]

Most traditional scientific analyses have focused on systems that maintain stability under perturbations. Some nonlinear systems, such as two-dimensional planetary systems, are stable, but many are not. Schmidt (2011) argues that nonlinear systems that exhibit instability should be counted as emergent. Stability and instability, though, come in various forms. Schmidt characterizes three kinds of stability that, when violated, give rise to emergence: static, dynamic, and structural. Static instability results from sensitivity to initial conditions at a single point or region of a state space, where "the alternative trajectories from two nearby initial points separate and will never again become neighbors" (2011, p. 228). Dynamic instability begins less localized: "nearly all points in the state space exhibit the property of sensitivity: the trajectories separate exponentially by time evolution" (2011, p. 229). In other words, a dynamical instability is exhibited when the system as a whole is chaotic. A Lorenz system is an example of this: almost all initial conditions lead to chaotic solutions. Finally, Schmidt defines *structural instability* as a kind of higher-order instability: if one were to perturb the structure of a system (i.e., its equations or laws) slightly, then "the overall dynamics changes qualitatively" (2011, p. 230).

4.3. Complementarity

Pattee agrees that dynamical conditions such as bifurcation, nonlinearity, and instability provide for important types of pattern emergence, but contends they do not capture the form of pattern emergence found in living systems. These dynamical conditions exhibit rate-dependent phenomena. But he argued that biology also generates rate-*independent* phenomena. The switching of a light switch provides a simple example of a rate-independent phenomenon. Flipping the switch requires the application of a certain threshold level of energy to the switch, but once it is flipped, the light is turned on or off independently of the energy applied to the switch. Dynamical information about the speed with which it is flipped is filtered out by the system and is irrelevant to the resulting behavior. Only a binary signal is sent to the light from the switch, with no information about the rate at which the switch was flipped. (See Rosen, 1969, for a similar development of the complementarity of multiple descriptions.)

Pattee argued that rate-independence is common in biological systems. For example, molecules act as signals, that is, the molecules consistently have a specific effect, regardless of when they are received by a consumer. Rate-independence is important for many kinds of biological processes: examples include sensor transduction, gene transcription and translation, error correction, enzymatic recognition, any type of memory, and any type of regulation or control. A general theory of pattern emergence needs to be able to account for the emergence of these types of organization.

Pattee (1987) argued that rate-independence can give rise to the emergence of *informational constraints*, in which the information carried by a state, not its dynamics, constrains behavior. Information constraints are distinct from but complementary to dynamical nonlinearity/instability:

> Although it is true that dynamical theory and symbolic information are not associated in our normal way of thinking, they are epistemologically complementary concepts that are nevertheless both essential for a general theory of biological self-organization. Moreover, instabilities are the most favorable condition of a dynamical physical system for the origin of nondynamical informational constraints, and the evolution of self-organizing strategies at all levels of biology require the complementary interplay of dynamical (rate-dependent) regimes with instabilities and nondynamic (rate-independent, nonintegrable) informational constraints.
>
> *(1987, p. 198)*

Although they complement one another, Pattee contended that one cannot describe or model rate-dependent dynamical and rate-independent informational constraints in the same vocabulary. The vocabulary that describes a switch as "closing" or "opening" is different from the vocabulary that refers to the velocity with which the switch was moved. He characterized the complementarity as *semantic closure*, a type of closure in which rate-dependent constraints are dependent on rate-independent constraints, and vice versa, within the same system. The switch must be moved with some velocity in order to close it. Engineering regularly takes advantage of the complementarity of different descriptions of a system. To analyze an electrical circuit involving switches, one ignores rate-dependent features and treats them as rate-independent. Rate-independence is even more fundamental in biology.

5. Pattern emergence applications

Two emergent features of living systems that are challenging to explain are hierarchical control and goal-directedness. We briefly consider how Pattee's framework of treating rate-dependent and rate-independent features as complementary provides a way of understanding these forms of pattern emergence.

5.1. Control

When a system is considered purely in terms of dynamical physical laws, there is no possibility of control. Every detail of what happens is determined by the laws, and no freedom is left open to a controller to make use of:

> [T]he forces that enter the equations of motion determine the change in time of the state of the system as closely as determinism is allowed by physical theory. The whole concept of physical theory is based on the belief that the motions or states of matter are neither free nor chaotic, but governed by universal laws.
>
> *(1973, p. 85)*

If there is no freedom to move in different ways, there seems to be no role for control. Pattee's solution at first seems counterintuitive: the possibility of control only arises though "some selective loss of detail" (1973, p. 80).

The challenge, then, is to explain how "'selective loss of detail' can lead to hierarchical control instead of the usual loss of order in the system" (Pattee, 1973, p. 81). This results from describing the system in a way that leaves out detail. This is what we do when we speak of the degrees of freedom available to a particle. Only in that context can we identify constraints that limit those degrees of freedom. In abstracting from detail, we abandon the lowest level, where everything is determined, and adopt what Pattee speaks of as a higher level of description:

> [T]he physicist's idea of constraint is not a microscopic concept. The forces of constraint to a physicist are unavoidably associated with a new hierarchical level of *description*. Whenever a physicist adds an equation of constraint to the equations of motion, he is really writing in two languages at the same time. The equation of motion language relates the detailed trajectory or state of the system to dynamical time, whereas the constraint language is not about the same type of system at all, but another situation in which *dynamical detail has been purposely ignored*, and in which the equation of motion language would be useless. . . . A constraint requires an *alternative description*.
>
> *(1973, pp. 85–86)*

Why would a scientist ever opt for less detail than is possible? One reason is to characterize macroscopic objects. Macroscopic objects, whether tables or organisms, are not identified in the lower-level dynamical account. They are patterns that can only be identified by recognizing freedom of motion and how this freedom is constrained. They arise in a higher-level language.

Talk of abstraction and languages is usually associated with cognitive activity of observing minds. The talk of constraints and control, however, is intended to refer to something that can operate independently of any mind. The relevant abstracting and imposing of constraints is not done by the scientist, but occurs in the very systems being described. This requires things in that system that can classify "microscopic degrees of freedom of the lower level it controls" (Pattee, 1973, p. 89) – that is, treat different conditions specified at the lower level as the same – and then apply the same rule to all instances. Exercising control, Pattee argues, requires that the system 1) classify situations (Pattee's general term for this is "measurement"), 2) make a *record* (or representation) of what was classified, and 3) respond differentially in light of the record (i.e., the record must be "*read out* inside the system," 1970, p. 132). These together generate what Pattee in the passage earlier from 1987 referred to as *informational constraints*. Together, Pattee refers to this as a *classification-record-control* process (1970, p. 132).

Pattee argues that these conditions are met even in enzymes. An enzyme classifies substrate molecules, changes its conformation when it binds to one with the right shape, and then catalyzes a reaction. The enzyme thereby controls the reaction. The importance of this is even clearer with allosteric enzymes. By binding with one molecule that results from a different reaction, such an enzyme changes how it catalyzes a given reaction. It is thereby sensitive to information about other conditions in the cell than the presence of its substrate. Such an arrangement is also present in the interaction of a neuron with a muscle: the muscle contracts when it recognizes an incoming neurotransmitter, represents this information in a calcium store, and uses that representation to release actin and myosin to slide along each other. In such systems, we can talk about control as existing in an observer-independent, intrinsic way within the system, because the system itself (and its capacity for classifying, recording, and interpreting its own records) defines the necessary complementary mode of description.

5.2. Goal-directedness

The resources required to account for control also provide a basis for explaining the goal-directedness of biological systems, which has long been a point of contention between reductionists and emergentists. In part the controversy reflects ambiguity in what is meant by "goal-directedness." McFarland (1989) usefully distinguishes three senses: goal-achieving, goal-seeking, and goal-directed systems. A goal-achieving system is "one which can recognize the goal once it is arrived at (or at least change its behaviour when it reaches the goal), but the process of arriving at the goal is largely determined by the environmental circumstances" (1989, p. 108). Such a system performs what Pattee calls "measurement," but also progresses towards the goal *by means of* such measurements. A goal-seeking system progresses towards the goal as a result of its own organization or design. In doing so it may rely on what are sometimes called passive control systems (Milsum, 1966) that measure or represent information. Accordingly, neither goal-achieving nor goal-seeking systems exhibit control in the sense Pattee characterized. They pose no challenge for the reductionist, as they do not exhibit interesting pattern emergence.

Interesting pattern emergence arises with what McFarland defines as goal-directed systems since they both represent a goal and produce behavior in response to that representation:

> In the paradigm case of goal-directed behaviour, the difference between the "desired" state of affairs and the actual state of affairs (as monitored in the negative feedback pathway) provides the (error) signal that actuates the behaviour-control mechanism.
>
> *(McFarland, 1989, p. 108)*

What McFarland describes here is a form of negative feedback control: the system that measures some variable, compares it to a set-point value, and based on the comparison responds one way rather than another. This differs from the cases noted earlier in that it is a goal, not a state of the system or the environment, that is represented.

What does it mean for a goal to be explicitly represented in a system such that its behavior is governed by that representation? Dennett offers an ecumenical conception of explicit representation:

> Let us say that information is represented *explicitly* in a system if and only if there actually exists in the functionally relevant place in the system a physically structured object, a *formula* or *string* or *tokening* of some members of a system (or "language") of elements for which there is a semantics or interpretation, and a provision (a mechanism of some sort) for reading or parsing the formula. This definition of explicit representation is exigent, but still leaves room for a wide variety of representation systems. They need not be linear, sequential, sentence-like systems, but might, for instance, be "mapreading systems" or "diagram interpreters."
>
> *(1987, p. 216)*

What is crucial in this account of goal-directedness is the addition to Pattee's account of control the intermediate step of directly comparing measurements of a system or environmental variable with an internal "goal" state variable. Any system that selects behaviors based on such a comparison is a goal-directed system. Such goal-directed behavior is widespread in biology. It is exhibited, for example, in bacteria that "decide" whether or not to sporulate depending on certain variables, such as the concentration of intracellular GTP (guanosine triphosphate) falling in a specified range (Stephens, 1998).

Goal-directedness is a form of pattern emergence that provides for novel forms of behavior in organisms that exhibit it. The goal-directedness of bacteria may seem far removed from the goal-directedness of humans, but what Pattee argues is required in relatively simple biological systems – representations of goals – can be extended by extending the representational machinery. Rosen (1985), for example, explored how representations of anticipated future states of a system and its environment can be used to control the system in the present. More advanced forms of information processing, such as representing alternative futures and selecting between them, require additional representational machinery. The result is the emergence of different patterns of behavior, but the key, in Pattee's analysis, is semantic closure that links the rate-independent informational constraints provided by a representation system to the fully determinate, rate-dependent, dynamical behavior at the lowest level.

6. Conclusions

In the past, scientists and philosophers have run into dead ends by conflating the emergence of phenomena like complexity, control, and goal-directedness with questions about the emergence of being. Increasingly philosophers and scientists who address the emergence of patterns independently of the emergence of being are advancing interesting and useful accounts about when patterns emerge. We have sketched a few of these and then focused on one framework, attributed to Pattee, that offers potential for understanding the emergence of hierarchical control and goal-directedness in biology.

Notes

1 Humphreys (2016, p. 150) uses "pattern emergence" in a different sense from the one used here. In our terminology, Humphreys is referring to a certain type of semantic criterion for being emergence. Humphreys subsumes "pattern emergence" under the larger category of "inferential emergence." Something

like what we are calling "pattern emergence" is also briefly suggested by Clayton (2004, p. 17), but Clayton does not treat this as separate from other distinctions about emergence.

2 Linearity is one of four conditions Wimsatt (2007, pp. 280–281) offers for a system counting as aggregative. The others are intersubstitution of parts, size scaling, and decomposition/reaggregation. When aggregativity fails, Wimsatt counts the behavior of the system as emergent.

References

Berlekamp, E. R., Conway, J. H., & Guy, R. K. (2004). *Winning Ways for Your Mathematical Plays* (Vol. 4, 2nd ed.). Wellesley, MA: A. K. Peters.

Bishop, R. C. (2012). Fluid Convection, Constraint and Causation. *Interface Focus, 2,* 4–12.

Bogen, J., & Woodward, J. (1988). Saving the Phenomena. *Philosophical Review, 97,* 303–352.

Clayton, P. (2004). *Mind and Emergence: From Quantum to Consciousness.* Oxford: Clarendon.

Dennett, D. C. (1987). *The Intentional Stance.* Cambridge, MA: The MIT Press.

Hooker, C. A. (2011). Conceptualising Reduction, Emergence and Self-Organisation in Complex Dynamical Systems. In C. A. Hooker (Ed.), *Philosophy of Complex Systems* (pp. 195–222). Amsterdam: North Holland.

Hooker, C. A. (2013). On the Import of Constraints in Complex Dynamical Systems. *Foundations of Science, 18*(4), 757–780.

Humphreys, P. (2016). *Emergence: A Philosophical Account.* New York: Oxford University Press.

Kim, J. (1999). Making Sense of Emergence. *Philosophical Studies, 95,* 3–36.

McFarland, D. (1989). *Problems of Animal Behaviour.* Harlow, UK: Longman Scientific & Technical.

Milsum, J. (1966). *Biological Control Systems Analysis.* New York: McGraw-Hill.

Mitchell, S. D. (2012). Emergence: Logical, Functional and Dynamical. *Synthese, 185,* 171–186.

Pattee, H. H. (1970). The Problem of Biological Hierarchy. In C. H. Waddington (Ed.), *Towards a Theoretical Biology 3: Drafts* (pp. 117–136). Edinburgh: Edinburgh University Press.

Pattee, H. H. (1973). The Physical Basis and Origin of Hierarchical Control. In H. H. Pattee (Ed.), *Hierarchy Theory: The Challenge of Complex Systems* (pp. 71–108). New York: Braziller.

Pattee, H. H. (1987). Instabilities and Information in Biological Self-Organization. Reprinted in H. H. Pattee & J. Rączaszek-Leonardi (Eds.), *Laws, Language and Life* (pp. 197–210). Dordrecht: Springer, 2012.

Rosen, R. (1969). Hierarchical Organization in Automata Theoretic Models of Biological Systems. In L. L. Whyte, A. G. Wilson & D. Wilson (Eds.), *Hierarchical Structures* (pp. 161–177). New York: Elsevier.

Rosen, R. (1985). *Anticipatory Systems: Philosophical, Mathematical, and Methodological Foundations.* New York: Pergamon Press.

Schmidt, J. C. (2011). Challenged by Instability and Complexity: Questioning Classic Stability Assumptions and Presuppositions in Scientific Methodology. In C. Hooker (Ed.), *Philosophy of Complex Systems* (pp. 223–254). Amsterdam: North Holland.

Sider, T. (2011). *Writing the Book of the World.* Oxford: Clarendon.

Stein, D. L. (Ed.). (1989). *Lectures in the Sciences of Complexity.* Redwood City, CA: Addison-Wesley.

Stephens, C. (1998). Bacterial Sporulation: A Question of Commitment? *Current Biology, 8*(2), R45–R48.

Wimsatt, W. C. (2007). *Re-Engineering Philosophy for Limited Beings: Piecewise Approximations to Reality.* Cambridge, MA: Harvard University Press.

11

COMPLEXITY AND FEEDBACK

Robert Bishop and Michael Silberstein

Introduction

Complexity is a multifarious concept; different researchers focus on different aspects. For instance, a system can be considered complex because it has many interacting parts, yet its behavior can be relatively simple, that is, easy to predict, control and manipulate: car engines and watches would be examples. On the other hand, a system with few interacting parts can exhibit behavior considered complex, that is, hard to predict: three billiard balls on a torus would be an example (Krámli, Simányi, and Szász 1991). Also note that both the terms "complexity" and "emergence" often are used for two seemingly different cases. One case involves simple rules governing uniform elements that generate unpredictable ("complex") behavior such as strikingly stable patterns in finite automata. Another case involves "complex," seemingly random, often nonlinear interactions of uniform elements that generate robust and stable large-scale patterns that are therefore, at least from a particular level of analysis, largely predictable, for example, Rayleigh-Bénard convection (RBC). The idea is that in the first kind of case we have complexity from relative simplicity, while in the second kind of case we have simplicity from relative complexity.

Of course, this all depends on how those terms are defined, so this way of putting things is often misleading. For example, if RBC is just the result of *simple laws* of more fundamental physics, then these two cases are seemingly really not so different after all. That is, given that they both involve macroscopic stable patterns generated from the temporal evolution of more basic microscopic constituents, what, if anything, makes these cases different? Notice that this way of putting the problem immediately ties together the concepts of complexity and emergence, because it defines complex processes as those that *emerge* from more basic processes, and remain stable across time and across changes in the more basic constituents. There are a number of different ways that macroscopic stable patterns apparently can arise in different kinds of systems. This is just to say that such stable patterns per se are, in some sense, multiply realizable.

People often point to a lack of central control as at least a necessary condition for complex behavior (Ladyman, Lambert, and Wiesner 2013). Does that condition hold in both the previous cases? If lack of central control means the absence of a designer or central decision-maker (e.g. a CEO), then both cases meet the condition. Or if a lack of central control means the organization of the system is such that even given different initial conditions or disruptions/deletions of the basic parts, the macroscopic patterns will remain stable, then again, both cases meet that condition. Nevertheless, assuming reality is not actually a finite automaton, then, in the case of RBC and the like, there is no purely computational account for the large-scale stable patterns (i.e., no account in which the program/algorithm itself is a kind of centralized control). Rather, there are various temporal and length scales of physical/chemical processes/elements that actually *interact* to generate the large-scale patterns, where the latter patterns become crucial for how the system components behave.

Ladyman, Lambert, and Wiesner (2013) point out that when the nature of hierarchical physical processes is compositional, one gets a lack of centralized control for free, at least in the sense of no central controller and distributed order. That is true as far as it goes, but doesn't help much with the case of RBC and other actual-world complex systems: How do stable, large-scale patterns arise and maintain themselves, given only the seemingly random interactions of uniform parts governed by nothing but basic physical laws of nature? Certainly, appealing primarily to initial conditions to answer that question in all such cases is inadequate because it ignores the important roles of contingency and context. Moreover, depending on how one thinks of physical laws and physical elements, such as atoms, the mystery only gets more profound. For instance, if laws of nature are not like algorithms that "govern" processes and physical elements are not like digital or information-based processes, then the computational metaphor, the analogy with finite automata, becomes less useful. Put differently, the actual world involves genuine physics and chemistry. It's the physical properties and their physical relations in concrete contexts that make the difference for system behaviors.

We will ultimately argue that while both cases can be said to involve complexity and emergence in *some sense*, the case of RBC involves *ontological* emergence and complexity in an interesting sense. This is in large part because we think the best explanation for the existence of the robust large-scale patterns in RBC involves global or systemic features of such systems that do not reduce to either the basic dynamics of the parts or to the initial conditions of the elements. We call this *ontological contextual emergence*, and we think it is often the best explanatory framework for complexity defined as the emergence of stable and self-maintaining patterns from more basic interactions. Before we get there, we need to say more about the various concepts of complexity and emergence.

Sometimes it is particular properties of systems or dynamical processes that are considered complex: noncomputability such as algorithmic complexity, nonpredictability, nonderivability, inherent statistical complexity (e.g., systems which are neither completely random nor completely ordered), information theoretic limits (e.g., the no-cloning theorem in quantum information theory), stochasticity, multiple realizability, nonlinearity, self-organization, collective effects, feedback loops, hierarchies, interlevel or multiscale relations, phase changes, wholes or global properties constraining and modifying the behavior of their parts, robustness, plasticity and hysteresis would be examples discussed in the literature. One or more of these various aspects have been identified as a signature of complexity in physics, chemistry, biology, geology and computer science, just to name a few disciplines.

However, the fact is neither scientists nor philosophers have made any compelling case for which aspects should be considered necessary or sufficient for a system or dynamics to be classified as complex (e.g., Ladyman, Lambert, and Wiesner 2013). Scientists in many disciplines tend to treat the various aspects of complexity phenomenologically (see later). Furthermore, in the physics literature, where some of the most sustained attention to measures for complexity has been given, there is no agreed-upon measure (e.g., Grassberger 1989; Sporns 2007; Wackerbauer et al. 1994). One possible explanation for the lack of agreement on complexity measures is the alleged observer-relativity of at least some forms of complexity: measures of complexity are not independent of the observer or the observer's choice of measurement apparatus (Crutchfield 1994; Grassberger 1989).

The situation for emergence is very similar to that for complexity in that there is no agreed-upon universal definition of emergence. Both emergence and complexity can be defined in ways that are relatively more epistemic or relatively more ontic, with some definitions straddling both (Silberstein 2012). Likewise, definitions of both can be more or less formal/computational/information theoretic versus physical. There are those who try to define emergence in terms of complexity and vice versa. For example, formal measures of complexity, whether they pertain to computational limits, correlational or causal connections of various sorts that can only be understood statistically or in network terms, say, are often invoked when defining epistemic forms of emergence that focus on failure of predictability and derivability (Silberstein 2012; Ladyman, Lambert, and Wiesner 2013). Clearly, one has to specify exactly what one means by complexity or emergence for any meaningful discussion to take place.

While some might take all of this as reason to be skeptical about either emergence or complexity, we think the pluralism exhibited in the scientific literature about these concepts is perfectly reasonable. Therefore, we accept that emergence and complexity are only contingently related. This means that one has to spell out the particulars of how complexity and emergence relate in some specific case.

If anything passes for a kind of agreement in multidisciplinary "complexity studies," it's that complex systems or dynamics often involve some form of feedback, connecting processes across differing length and time scales. Feedback loops are particular kinds of relations between systems and environment, between system and subsystems or between subsystems. In engineering and design contexts, feedback loops are used as control mechanisms based on goals. For instance, a heating and air conditioning control is set to maintain a particular room temperature. A feedback loop involves a thermometer measuring the temperature and feeding this information back to the control unit, which then turns the heating or air conditioning on or off depending on a comparison of the temperature with the set goal.

Feedback loops relating multiple subsystems with the overall system and/or environment are an example of a more general phenomenon – nonseparability (see later) – that seems to be a characteristic of many systems that are regarded as complex and as possessing emergent properties. For brevity's sake, we will focus on aspects of complexity and feedback that give rise to nonseparable properties and processes in classical physics. It should be noted that nonseparability is often characterized in both formal/mathematical and physical terms. And it should also be noted that complex, nonseparable systems often exhibit both ontic and epistemic features of emergence. In any case, nonseparability often counts as a marker of both complexity and/or emergence of one sort or another.

A relatively simple concrete example of a classical complex system with feedback where nonseparability arises is RBC. Therefore, we will focus on RBC as an example in the third section. In the next section, we will more fully characterize complexity in terms of nonseparability. We are not alone in focusing on RBC as a case study in the literature on complexity. Ladyman, Lambert and Wiesner, for instance, give RBC as an example of "robust order," and they conclude that such "robust order is a further necessary condition for a complex system" (2013, p. 27). While we agree that robust order is a condition for complex systems, we don't think Ladyman, Lambert and Wiesner fully capture what is important about such cases, namely, ontological contextual emergence.

Characterizing complexity

The clearest example in classical systems is when nonlinearity and feedback give rise to nonseparability, as opposed to quantum systems that are linear and also exhibit nonseparability (i.e., quantum entanglement). Here we will focus on such cases of nonlinearity to illustrate the basic concept of nonseparability. Let us be clear that nonlinearity in general is neither necessary nor sufficient for either complexity or emergence in any profound sense. Nonetheless, there are cases of nonlinearity that do exhibit ontological complexity and emergence, such as RBC.

Linear systems obey the principle of linear superposition and can be straightforwardly decomposed into and composed by subsystems. For instance, linear (harmonic) vibrations of a string can be analyzed as a superposition of normal modes, and these normal modes can be treated as uncoupled individual subsystems or parts. Classical linear systems can be thought of as aggregations of parts ("the whole is the sum of its parts"). The linear behavior of such systems is sometimes called *resultant* (as opposed to emergent).[1] Some nonlinear systems, by contrast, cannot be treated even approximately as a collection of uncoupled individual parts.

Some form of global description[2] is required, taking into account that individual constituents cannot be fully characterized without reference to larger-scale structures (e.g., processes, dynamics and constraints that may be system-wide or of even larger/higher scale). Rayleigh-Bénard convection is an example. It exhibits what is called *generalized rigidity*: the individual constituents are so correlated with all other constituents that no constituent of the system can be changed except by applying some change to the system as a whole, via so-called order parameters (Chemero and Silberstein 2008; Cross and Hohenberg 1993). This is an implication of generalized rigidity (e.g., convection cells in RBC systems are modified or disrupted through changing the temperature difference between the bottom and top plates of the confinement system [see Figure 11.1]). These globally constrained behaviors are often referred to as *emergent* (as opposed to resultant).

The intricate coupling between constituents or subsystems in some nonlinear systems is related to or described by the *nonseparability* of the Hamiltonian, a function describing the total energy of the system and characterizing its time evolution. A Hamiltonian is said to be separable if there exists a transformation carrying the Hamiltonian describing a system of N coupled constituents into N equations, each describing the behavior of one of the system's constituents. Otherwise, the Hamiltonian is nonseparable and the interactions within the system cannot be decomposed into interactions among only the individual components of the system (Goldstein 1980). When the constituents of a system are highly coherent, integrated and correlated such that their properties and behaviors are nonlinear functions of one another, the system cannot be treated as just a collection of uncoupled parts. The behavior of the parts is every bit as determined by the state of the whole as the other way around.

Systems in the biological domain often exhibit nonseparability: while there are component parts and processes, their individual behaviors systematically and continuously affect one another in a nonlinear fashion (Silberstein and Chemero 2013; Silberstein 2016). In such cases, mechanisms are not sequential but have a cyclic organization rife with oscillations, feedback loops or recurrent connections among components. In these instances, there is a high degree of interactivity among the components, and the system is nondecomposable, and therefore localization, as expressed by separability, will fail (Bechtel and Richardson 2010, p. 24). Furthermore, if the nonlinearity affecting component operations also affects the behavior of the system as a whole, such that the component states and properties are dependent on a total state-independent characterization of the system (i.e., one sufficient to determine the state and the dynamics of the system as a whole), then the behavior of the system can be called "emergent" (Ibid, 25). Bechtel and Richardson emphasize that when the feedback is system-wide such that almost all "the operations of component parts in the system will depend on the actual behavior and the capacities of its other components" (Ibid, 24), the following features result. First, the behavior of the component parts considered within the system as a whole are not predictable in principle from their behavior in isolation. Second, the behavior of the system as a whole cannot be predicted, even in principle, from the separable Hamiltonians of the component parts (Ibid, 24). All of this applies equally to other systems exhibiting strong forms of nonlinearity.

The Hamiltonian framework provides a useful formal way for understanding the basis for many of the nonseparable phenomena of complexity.[3] Typical definitions of complexity are formalized in terms of probabilities with no explicit reference to physical system variables, though physical variables are required to define the state space over which probability measures are defined (e.g., various forms of algorithmic or statistical complexity – Grassberger 1989; Wackerbauer et al. 1994). Usually, in scientific contexts, complex systems are characterized by phenomenological features that shape our intuitions about complexity. Complex systems are typically many-bodied; they exhibit broken symmetry; there are a number of distinguishable, interdependent scales or nested structures; the hierarchies in complex systems typically are associated with irreversible processes; system constituents are coupled to each other via various relations; constituent properties and dynamics depend upon the structures in which they are embedded as well as the environment of the system as a whole; such systems display an organic unity of function which is absent if one of the constituents or internal structures is absent or if relational coordination among the structures and constituents is lacking; several components are tightly interconnected through feedback loops and other forms of structural/functional relations crucial to maintaining the integrity of the system; system behavior is often situated somewhere between simple order and total disorder; the organizational/relational unity of the system is resilient under small perturbations and adaptive under moderate changes in its environment; and the degree of complexity of systems is relative to how we observe and describe them. Such qualitative features are found in theory and experiments on complex physical systems such as RBC (e.g., Cross and Hohenberg 1993).

The concept of hierarchy is of central importance in complex systems. For noncomplex systems, there is a distinguishable ordering of scales of structure due to the hierarchy of physical forces and dynamical time scales (e.g., elementary particles, molecules, solids).[4] In some of these cases the so-called lower-level constituents may provide both necessary and sufficient conditions for the existence and behavior of the larger-scale structures.

By contrast, in complex systems scales of structure are often only distinguishable in terms of dynamical time scales and are coupled to each other in such a way that at least some of the larger structures are not fully determined by, and even influence and constrain, the behavior of constituents at the smaller scale. That is, these constituents provide some necessary but *no* sufficient conditions for the existence and behavior of some of the larger-scale structures (Bishop and Atmanspacher 2006; Bishop 2012, sec. 3.4). Moreover, the micro-scale constituents may not even provide necessary and sufficient conditions for their own behavior if the larger-scale structures can influence constituent behavior (see later). That is, the laws and properties of the constituents at the smaller scale are not sufficient for their own behavior if large-scale constraints play a role in what behaviors such constituents can exhibit. This latter kind of hierarchy is called a *control hierarchy* to distinguish such cases from systems such as sand grain piles and simple crystals (Pattee 1973, pp. 75–79; Primas 1983, pp. 314–323).

The control exercised in complex systems takes place through constraints and feedback loops. The interesting types of constraints actively change the rate of reactions or other processes of constituents relative to the unconstrained situation (e.g., catalysts). Moreover, these constraints control constituents without removing too many of their configurational degrees of freedom (contrast with simple crystals). Such constraints may be external, due to the environment interacting with the system, and/or internal, arising within the system due to the collective effects of its constituents or large-scale dynamics. Positive feedback loops, for example, push a system further in a particular direction, reinforcing and amplifying the initial change; negative feedback loops push the system in the opposite direction, counteracting the effected change. Obviously in engineering and design contexts there is central control.

An example of a feedback loop without central control would be the Belousov-Zhabotinsky (BZ) reactions (e.g., malonic acid–bromate–cerium catalyst in sulfuric acid medium) in chemical kinetics, where autocatalysis forms a positive feedback loop, while negative feedback loops block unbounded growth and return the system back towards its original state. These complex reactions produce the startling oscillations between two equilibrium states signaled by a series of changes in color from yellow to clear to yellow repeating the pattern. Most self-organizing phenomena, such as these BZ patterns, are due to the interaction of positive and negative feedback loops (Érdi 2008).

Rayleigh-Bénard convection and complexity

As a detailed case study, consider RBC (Figure 11.1). A layer of fluid is sandwiched between two thermally conducting plates, where the lower plate is heated while the upper one is maintained at a fixed temperature. This establishes a temperature gradient, ΔT, in the vertical direction. Being warmer, the fluid near the lower plate undergoes thermal expansion and becomes less dense than that above. In the presence of gravity, this creates an instability resulting in a buoyancy force tending to lift the whole mass of fluid from the lower plate. Whereas in the case of gases, individual molecules move independently of their neighbors except for occasional collisions, in fluids, due to long-range cohesive forces, individual molecules are packed as closely together as quantum repulsion forces will allow. This means that fluid parcels are actually collections of an Avogadro's number of individual molecules acting as a unit (e.g., individual H_2O molecules in a parcel of water). It's the fluid parcels as a whole bulk that have this tendency to start moving vertically.

The upper plate acts as an external constraint against this upward motion. System-wide, global constraints on the motion of parcels due to container boundaries and symmetries, as well as conservation laws, also contribute to this subtle balance in the fluid (Busse 1978). The allowable states of motion accessible to fluid parcels are established by the system as a whole. This includes the so-called body or volume forces acting equally on all matter within a given volume of fluid, such as gravity. In the absence of constraints and body forces, the fluid in the uniform state in principle could flow in an arbitrary number of directions, and an infinite number of convection patterns would be possible. Restrictions on the fluid velocity are imposed by conservation of mass (Batchelor 1967), while fluid flows and allowable pattern formation are strongly influenced by system geometries and symmetries (Cross and Hohenberg 1993). In addition, due to long-range

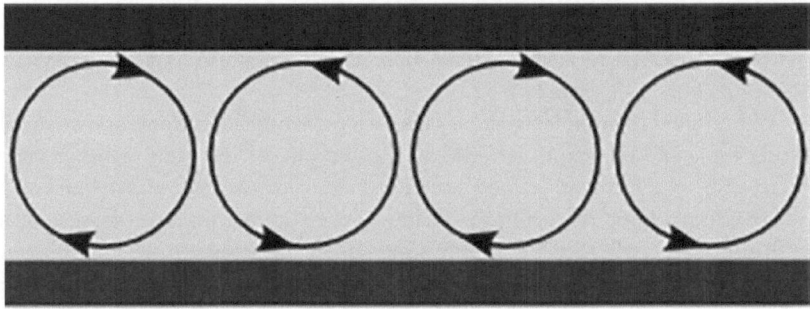

Figure 11.1 Basic Rayleigh-Bénard convection system. The lower plate is held at a warmer temperature than the top plate creating convective cells rotating in the directions indicated by the arrows.

interactions and the symmetry properties of the state vectors describing fluid parcel positions, different values of the relative positions between parcels are correlated. The spatial distribution of parcels is influenced by the presence of such correlations, and this coupling due to correlations leads to collective behavior contributing to the formation of coherent large-scale structures. As the temperature difference surpasses a critical value, the correlation lengths among fluid parcels become the size of the container. It's when all fluid parcels become so highly correlated with each other that the behavior of individual parcels can be modified only by global changes (e.g., by changing ΔT) that the fluid system exhibits the generalized rigidity referred to earlier.

Typically, fluid density variations are relatively small and are rapidly damped out in the initial globally stable conductive state. Nevertheless, when ΔT passes a critical value ΔT_c, the system becomes globally sensitive to small perturbations in fluid density. Eventually, some of the spatial symmetries of the initial stable state of the fluid induced by the container are broken along with the homogeneity of this state of the fluid parcels. The growth rate of perturbations depends only on the wave number, meaning that although the fluctuations in combination with the system geometry and symmetries lead to a selection of one growing mode, its growth rate is determined solely by global system properties (Cross and Hohenberg 1993). The most unstable mode scales are the distance of plate separation. The result is distinguishable large-scale structures: Bénard cells. This new stable pattern is a large-scale, dynamical constraint on the individual motions of fluid parcels due to a balancing among effects due to the dynamics, the structural relations of each fluid parcel to all other fluid parcels, bulk motion of the fluid as a whole, system boundaries and symmetries and conservation laws. In particular, important effects in this balance are the positive feedback in the coupling between the convective flows and the deformation of the isotherms, on the one hand, and the countering of this feedback by the fluid's viscosity tending to slow the convection and the thermal conductivity tending to suppress perturbations to the convective flows due to temperature, on the other hand. This balancing keeps the Rayleigh-Bénard instability in check and produces an organizational unity within the large-scale dynamical and container constraints that is remarkably stable in the face of small perturbations.

Although Bénard cells emerge out of the motion of fluid parcels after ΔT exceeds ΔT_c, these large-scale structures determine modifications of the configurational degrees of freedom of fluid parcels such that some motions possible in the initial uniform state are restricted while other motions not available in the uniform state are now accessible. For instance, while $\Delta T < \Delta T_c$, fluid parcels cannot access rotational states of motion characteristic of RBC. In the new nonequilibrium steady state, the fluid parcels exhibit coherent convective motion, while most of the states of motion characteristic of the uniform state are no longer accessible (e.g., fluid parcels cannot sit motionless).

As $\Delta T > \Delta T_c$, there is a hierarchy distinguished by dynamical time and spatial scales (individual molecules, fluid parcels, Bénard cells) with complex interactions taking place among the different scales. The system as a whole displays integrity as the constituents of various hierarchic scales exhibit highly correlated, cohesive behavior. In particular, Bénard cells act as a control hierarchy, constraining the motion of fluid parcels. These cells not only modify the allowable states of motion for the fluid parcels but also determine which fluid parcels will stay with which cell and which parcels will migrate to a different cell.

The stability of the new patterns of constraint that arise is due to the collective effects and feedback loops of steady, large-scale, shear fluid motion suppressing local deviations within a particular system geometry. Although body forces such as gravity and container boundaries play a role, here, the key constraints and feedback loops arise dynamically to shape constituent behavior within the particular system geometry. The bulk flow of fluids also plays an important role in the formation and dynamics of patterns by contributing nonlocal effects (e.g., acting over many roll widths in RBC; see Paul et al. 2003). So although the fluid parcels and their properties provide some necessary conditions for the existence and dynamics of Bénard cells, they are neither sufficient to determine that dynamics nor sufficient to fully determine their own motions.

Complexity and emergence

Reductionist accounts maintain that properties and behaviors of systems are nothing other than the resultant product of the states and intrinsic properties of their parts (ontological reductionism) or are ultimately explainable in terms of the states and intrinsic properties of their parts (epistemological reductionism). Emergence accounts deny one or both of these claims. The various cases of nonseparability in the form of strong nonlinearity, constraints, feedback loops acting across spatial scales, etc., and the "holism" exhibited by such complex systems have proved fruitful for understanding more about how emergence works in complex phenomena.

For instance, reductive atomism or "mereological supervenience" maintains that "[t]he only law-like regularities needed for the determination of macro features by micro features are those that govern the interactions of those micro features in all contexts, systemic or otherwise" (Van Gulick 2001, p. 18). Control hierarchies with feedback loops operating across length scales are crucial to determining the behavior of system constituents in many complex systems such that underlying constituents in complex systems lack the necessary and sufficient conditions for their own behavior, let alone the system's behavior. Constituent behavior in such systems (e.g., RBC) is *conditioned* by the contexts in which the constituents are situated and is not merely the result of the context-free, law-like regularities envisioned in reductive atomism. When nonlinearities are important, the failure of separability is clearly on display: individual system components are not independent of each other, and the behavior of individual system components is not independent of the wholes and various scales in between wholes and parts. Feedback loops and constraints acting across various length scales, as well as constraints associated with the whole system, enable some possibilities for constituent behavior while restricting other possibilities relative to what would be possible for the constituents if the effects of feedback loops and constraints were absent.

This is a form of *mereological emergence*, wherein the behavior of the whole isn't determined (either synchronically or diachronically) by the intrinsic properties of the most basic parts or even their relations (Silberstein 2002, pp. 91, 96–98, 2012). But there is more to the story than this. The interplay between parts and wholes in complex systems and their environments mediated by positive and negative feedback loops typically leads to the self-organization observed in such systems. Sensitivity to minute changes at the component scale in these systems is partly determined by the relations acting at different scales in such systems. This kind of behavior is indicative of the framework of ontological contextual emergence (Bishop 2005, 2010, 2012; Bishop forthcoming):

> The properties and behaviors of an underlying domain (including its laws) of a system offer some necessary but no sufficient conditions for the properties and behaviors at a larger scales or target system domain.

A helpful way to think of how necessary and sufficient conditions work in contextual emergence is as an INUS condition – an Insufficient though Necessary contributor to a condition that is Unnecessary but Sufficient for the effect. For example, a short-circuit by itself is neither necessary nor sufficient for causing a fire, but can be a necessary contributor to a contingent sufficient condition that started a fire. In contextual emergence, the underlying domain contributes a necessary but insufficient part of a contingent sufficient condition for properties and behaviors in that domain or in other domains or at larger scale. An important set of the sufficient conditions for the contextual emergence of properties and entities are *stability conditions*. Stability conditions represent those conditions guaranteeing the existence and persistence of relevant states and observables for a system, perhaps even for the persisting existence of the system in question. These conditions are never given by "elementary" or underlying domains alone.[5]

To illustrate, think of the domain of H_2O parcels as the putative realizers of the convection cells in RBC. The reference to necessary conditions of the parcels means that properties and behaviors of the convective cells may imply the behaviors of these parcels. Nevertheless, the converse is not true, as the properties and behaviors of the fluid parcels alone do not offer the set of necessary and sufficient conditions for the properties and behaviors of convective cells (Bishop 2012; Bishop forthcoming). Contingent conditions specifying the context for the transition from the fluid parcels domain to the properties and behaviors of fluids and of the convective cells are required to provide such sufficient conditions. In complex systems, such contingent contexts are not given by the laws, properties and behaviors of the underlying domain alone.

In this sense, RBC illustrates how large-scale constraints play an ineliminable role in the determination of the behavior of complex systems. Yet care is needed with typical emergentist intuitions such as that the existence and behavior of wholes cannot be predicted from parts, or that wholes cannot be explained from parts in complex systems. With ontological contextual emergence, relevant information about the properties and behaviors of constituents in the underlying domain *plus* the specification of an appropriate contingent context of the target system not given by the underlying domain allow for the (in principle) prediction or explanation of properties and behaviors of the larger scales in many cases (Bishop forthcoming; Bishop and Atmanspacher 2006; Primas 1998). Stability conditions provided by the large-scale fluid dynamics allow the existence of particular reference states and observables for RBC. States, observables and stability conditions are important ontological features of physical systems, so ontological forms of emergence are the most interesting to explore here.

Rayleigh-Bénard convection and contextual emergence

The Bénard cells emerge out of the local dynamics of fluid parcels as a large-scale, nonlocal, dynamical process that, in turn, constrains or shapes the states of motion accessible to fluid parcels. Bénard cells act to provide functional organization and coherence to fluid parcel behavior. Instant by instant during pattern formation, a corresponding large-scale, evolving, nonlocal constraint modifies the accessible states of fluid parcel motion. Prior to the establishment of this dynamically evolving constraint, fluid parcels' trajectories had the property of accessing various states of motion (e.g., those accessible in the initial uniform state). But they lose this property due to the large-scale, nonlocal, constraining effects of the forming Bénard cells. The large-scale constraints on the motion of the fluid parcels in this case are synchronic: the emergence of the self-regulating large-scale pattern is simultaneous with the instant-by-instant modifications of the accessible states of motion for fluid parcels. The forming patterns exhibit a large-scale constraint on the fluid parcels, making the contribution of the latter at an instant *t* to the large-scale conditional *on the pattern at t*. Large-scale structures arise out of fluid parcels, but they also dynamically condition or constrain the contributions the fluid parcels can make, namely by modifying or selecting which states of motion are accessible to the fluid parcels. Hence, if there was no synchronic relationship between the constraints and the fluid parcels, there would be no pattern.

This is indicative of contextual emergence. There are no "new forces" coming out of nowhere, so no radical emergence. Instead, the fluid is governing itself in a complex set of interactions among parts and wholes, where the collective effects of the action of the Bénard cells on the fluid parcels is the new influence. The jointly necessary and sufficient conditions for the behavior of the convection system and its constituents are provided by the fluid parcels, the dynamics, the large-scale structures and the body forces, the system-wide geometry and the symmetries. An individual fluid parcel can only execute motions allowed to it by *all other fluid parcels and the large-scale dynamics.*

While the fluid parcels and the Bénard cells are not wholly distinct entities – nonseparability – the former can multiply realize the latter, similar to the way gas molecules in a room can multiply realize temperature. The fluid parcels of a homogeneous fluid can be freely rearranged without changing the macroscopic properties of the fluid such as temperature, density and pressure. The different arrangements of these fluid parcels corresponding to the same macroscopic fluid properties form equivalence classes with respect to these fluid properties. Such multiple realizability is typical of emergence in complex systems.

Situating complexity and emergence in context

The literature typically discusses two categories of emergence: 1) Strong or radical emergence, where emergent entities or properties are thought to be the result of irreducible bridge laws or causal powers to produce qualitatively new phenomena. Such strongly emergent phenomena are also said by some to possess novel "downward" causal powers that constrain the behavior of other phenomena at smaller spatial and temporal scales. And 2) weak emergence, where emergent entities or properties fail to be predictable, derivable, explainable or characterizable in terms of the "more basic" entities and properties out of which the emergents arise. Weak emergence has no necessary ontological implications, whereas strong emergence does (Silberstein 2012) – indeed, weak emergence is consistent with ontological reduction and the failure of epistemic reduction. Strong emergence is suspicious to most participants in the reduction/emergence debates, as it's often characterized as inexplicable magic or otherwise impossible (Ibid). As we have seen in the case of RBC, forms of large-scale determination arise that do not involve new brute bridge laws or forces. On the other hand, weak emergence is ubiquitous but does not illuminate the ontological contextual emergence illustrated by RBC. Ontological contextual emergence provides a framework where the emergence of new entities, such as Bénard cells, and their properties, are predictable, derivable or explainable once the jointly necessary and sufficient conditions are understood.

If strong and weak accounts of emergence were the only options, it's easy to see why there is so much skepticism about emergence. Without a credible alternative for ontological (strong) emergence, it can appear that ontological reduction is the only viable option. A diagnosis of this unhappy situation is that the different sides of the reduction-emergence debate seem to share the same metaphysical picture: reality as analogous to an axiomatic system. Just as there are axioms in geometry and derived theorems, so the world has some fundamental features, and all other features are logical or metaphysically necessary consequences of these fundamentals. The finite automata known as the Game of Life can serve as an analogy: There is a small set of rules, and all behaviors of the entities that arise in the game are consequences of these rules plus their initial conditions.

Particular ontological forms of emergence, such as mereological emergence or ontological contextual emergence, are viable alternatives for robust forms of emergence that are not brute, mysterious or merely weak. There is good scientific evidence for these forms of ontological emergence. In the case of classical systems, we have RBC, some types of networks, etc., and in the case of quantum systems, we have molecular structure and entanglement (Bishop 2005, 2010; Silberstein 2002, 2012, 2017).

Unlike the Game of Life, the emergence of large-scale structures, such as Bénard cells and their constraint on fluid parcels' degrees of freedom, is not the logical or necessary consequences of a set of fundamental axioms or laws simpliciter, but rather involve stability conditions not given by the base axioms or laws. In complex systems, context plays as fundamental a role as do physical laws. The relationship between complexity and emergence may be contingent, but ontological contextual emergence is an example of a framework allowing a middle path between weak and strong emergence that illuminates this contingence.

Notes

1 See (McLaughlin 1992) for a discussion of the origin and history of the terms "resultant" and "emergent."
2 Global descriptions in nonlinear dynamics are descriptions that necessarily must refer to irreducible, system-wide and environmental features in addition to local interactions of individual constituents with one another.
3 Any case of quantum entanglement represents an example of a nonseparable Hamiltonian. And although one usually does not write down a Hamiltonian for networks generated by linear matrices, some of these exhibit nonseparability as well (Silberstein 2016).
4 There is no assumption that there are ontologically pre-given levels. We assume that whatever levels there are in the physical world arise over time, contingently as it were.
5 Some contextual information appears in boundary conditions, but note from the description of RBC earlier that concrete contexts and their stability conditions are not captured in boundary conditions.

Bibliography

Batchelor, G. (1967), *An Introduction to Fluid Dynamics* (Cambridge: Cambridge University Press).

Bechtel, W., and Richardson, R. C. (2010), *Discovering Complexity: Decomposition and Localization as Strategies in Scientific Research*, 2nd ed. (Cambridge, MA: MIT Press).

Bishop, R. C. (2005), "Patching Physics and Chemistry Together," *Philosophy of Science* 72:710–722.

Bishop, R. C. (2010), "Whence Chemistry? Reductionism and Neoreductionism," *Studies in History and Philosophy of Modern Physics* 41:171–177.

Bishop, R. C. (2012), "Fluid Convection, Constraint and Causation," *Interface Focus* 2:4–12.

Bishop, R. C., and Atmanspacher, H. (2006), "Contextual Emergence in the Description of Properties," *Foundations of Physics* 36:1753–1777.

Bishop, R. C. (forthcoming), '*The Physics of Emergence*', IOP Concise Physics. (San Rafael, CA: Morgan & Claypool Publishers; Bristol, UK: IOP Publishing).

Busse, F. (1978), "Non-Linear Properties of Thermal Convection," *Reports on Progress in Physics* 41:1929–1967.

Chemero, A., and Silberstein, M. (2008), "After the Philosophy of Mind: Replacing Scholasticism with Science," *Philosophy of Science* 75:1–27.

Cross, M., and Hohenberg, P. (1993), "Pattern Formation Outside of Equilibrium," *Reviews of Modern Physics* 65:851–1112.

Crutchfield, J. (1994), "Observing Complexity and the Complexity of Observation," in H. Atmanspacher and G. Dalenoort (eds.), *Inside Versus Outside* (Berlin: Springer), pp. 235–272.

Érdi, P. (2008), *Complexity Explained* (Berlin: Springer).

Goldstein, H. (1980), *Classical Mechanics*, 2nd ed. (Boston: Addison-Wesley).

Grassberger, P. (1989), "Problems in Quantifying Self-Generated Complexity," *Helvetica Physica Acta* 62:489–511.

Krámli, A., Simányi, N., and Szász, D. (1991), "The K-Property of Three Billiard Balls," *Annals of Mathematics* 133:37–72.

Ladyman, J., Lambert, J., and Wiesner, K. (2013), "What Is a Complex System?," *European Journal for Philosophy of Science* 3:33–67.

McLaughlin, B. (1992), "British Emergentism," in A. Beckermann, H. Flohr and J. Kim (eds.), *Emergence or Reduction? Essays on the Prospects of Nonreductive Physicalism* (Berlin: Walter de Gruyter), pp. 49–93.

Pattee, H. (1973), "The Physical Basis and Origin of Hierarchical Control," in H. Pattee (ed.), *Hierarchy Theory: The Challenge of Complex Systems* (New York: George Braziller), pp. 69–108.

Paul, M., Chiam, K.-H., Cross, M., and Greenside, H. (2003), "Pattern Formation and Dynamics in Rayleigh-Bénard Convection: Numerical Simulations of Experimentally Realistic Geometries," *Physica D* 184:114–126.

Primas, H. (1983), *Chemistry, Quantum Mechanics and Reductionism: Perspectives in Theoretical Chemistry* (Berlin: Springer).

Primas, H. (1998), "Emergence in Exact Natural Sciences," *Acta Polytechnica Scandinavica* 91:83–98.

Silberstein, M. (2002), "Reduction, Emergence and Explanation," in P. Machamer and M. Silberstein (eds.), *The Blackwell Guide to the Philosophy of Science* (Oxford: Blackwell), pp. 80–107.

Silberstein, M. (2012), "Emergence and Reduction in Context: Philosophy of Science and/or Analytic Metaphysics," *Metascience* 21:627–642.

Silberstein, M. (2016), "The Implications of Neural Reuse for the Future of Cognitive Neuroscience and the Future of Folk Psychology," *Brain and Behavioral Sciences* 39:e132.

Silberstein, M. (2017), "Contextual Emergence," *Philosophica* 91:145–192.

Silberstein, M., and Chemero. A. (2013), "Constraints on Localization and Decomposition as Explanatory Strategies in the Biological Sciences," *Philosophy of Science* 80:958–970.

Sporns, O. (2007), "Complexity," *Scholarpedia* 2(10):1623, www.scholarpedia.org/article/Complexity.

Van Gulick, R. (2001), "Reduction, Emergence and Other Recent Options on the Mind/Body Problem: A Philosophic Overview," *Journal of Consciousness Studies* 8:1–34.

Wackerbauer, R., Witt, A., Atmanspacher, A., Kurths, J., and Scheingraber, H. (1994), "A Comparative Classification of Complexity Measures," *Chaos, Solitons, Fractals* 4:133–173.

12

BETWEEN SCIENTISM AND ABSTRACTIONISM IN THE METAPHYSICS OF EMERGENCE

Jessica Wilson*

Both experience and the seeming structure of the sciences suggest that some special scientific phenomena (broadly construed to include artifacts and the like) are synchronically dependent on lower-level physical phenomena, and yet are also distinct from and distinctively efficacious as compared to the lower-level physical phenomena upon which they depend. This characteristic combination of dependence and autonomy serves to motivate the notion of *metaphysical emergence* as important to understanding natural reality. Indeed, and reflecting different specific interpretations of the dependence or autonomy at issue, there is at present a great diversity of contemporary accounts of metaphysical emergence. As I'll here argue, however, many of these accounts – thankfully, not all – fail to satisfy one or both of the following criteria of adequacy:

1 *The criterion of appropriate contrast*: Reflecting that metaphysically emergent phenomena are supposed to be distinct from the lower-level phenomena upon which they synchronically depend, an adequate account of metaphysical emergence must provide a clear (i.e., explicit) basis for ruling out that phenomena it deems emergent can be given an ontologically reductionist treatment – that is, it must provide a clear basis for ruling out that emergent phenomena are *identical* to dependence base phenomena. Relatedly, an adequate account of metaphysical emergence must provide a clear basis for establishing that the phenomena it deems emergent are not just epistemically or representationally emergent. For example, even supposing that seemingly metaphysically emergent phenomena are always surprising to those who first encounter them, then (since being surprising is in itself compatible with being identical to some lower-level phenomenon) an account of metaphysical emergence in terms of such a response would fail to satisfy the criterion of appropriate contrast.

2 *The criterion of illuminating contrast*: An adequate account of metaphysical emergence must not only provide a clear basis for contrasting such emergence with ontological reduction, but moreover do so in a way that provides an illuminating – that is, explanatorily relevant – basis for understanding how such failures of reduction might occur. Relatedly, an adequate account of metaphysical emergence must provide a clear basis for the in-principle resolution of disputes over whether some phenomenon is metaphysically emergent, in a way going beyond appeal to brute intuitions or irrelevant distinctions. For example, even supposing that an oracle exists who can infallibly report whether a given phenomenon is metaphysically emergent, an account pitched in terms of such oracular pronouncements would fail to

illuminate what it is to be metaphysically emergent and so would fail to satisfy the criterion of illuminating contrast. More plausibly, an account according to which metaphysically emergent phenomena are dependent on but ontologically irreducible to lower-level physical phenomena, but which did not provide any explanatory insight into how, exactly, some dependent goings-on might be so irreducible, would lead directly to stalemate between emergentists and reductionists (e.g., with regard to the status as emergent of certain mental states), and so would fail to satisfy the criterion of illuminating contrast.

There are, of course, other important criteria of adequacy on an account of metaphysical emergence. Foremost among these is to provide a basis for accommodating the distinctive efficacy of metaphysically emergent phenomena – and moreover, to do so in a way that is able to address concerns, notably raised by Kim (in, e.g., his 1989 and 1993), according to which certain varieties of metaphysical emergence give rise to problematic causal overdetermination. Treatment of this and other criteria of adequacy is beyond the scope of this chapter, though I believe such treatment would support the methodological morals I will later draw.[1]

I'll start with some preliminaries (§1). I'll then discuss certain representative accounts of metaphysical emergence falling into three broad categories, assessing their prospects for satisfying the earlier criteria; the ensuing dialectic will have a bit of the Goldilocks fable about it. At one end of the spectrum are what I call 'scientistic' accounts, which characterize metaphysical emergence by appeal to one or another specific feature commonly registered in scientific descriptions of seeming cases of emergence; such accounts, I'll argue, typically fail to provide a clear basis for ensuring incompatibility with ontological reduction, and thus fail to guarantee satisfaction of the criterion of appropriate contrast (§2). At the other end of the spectrum are what I call 'abstractionist' accounts, which characterize emergence in terms floating free of scientific notions; such accounts, I'll argue, typically fail to guarantee satisfaction of the criterion of illuminating contrast (§3). I'll then look at several accounts of metaphysical emergence that are what I call 'substantive', in appealing to familiar relations or posits which are properly metaphysical, while being intelligibly connected to scientific relations and posits. As we'll see, the resources of such accounts are up to the tasks of blocking ontological reduction and of enabling investigations into metaphysical emergence to proceed in an illuminating way (§4). I close with some methodological morals (§5).

1. Preliminaries

I start with five preliminaries.

First, the phenomena (or as I'll sometimes put it, the 'goings-on') at issue in discussions of metaphysical emergence might be of any ontological category, including objects, systems, events, processes, properties, and tropes, and might pertain either to types or tokens of such goings-on. That said, accounts of metaphysical emergence are commonly pitched in terms of properties, on the assumption that the emergence of goings-on of diverse categories ultimately comes down, one way or another, to the emergence of properties.

Second, the notion of synchronic dependence at issue in metaphysical emergence is 'broad', in allowing for temporally extended dependence of emergent phenomena on base phenomena (as in the case, e.g., of processes). Synchronic dependence in this sense is compatible with but broader than instantaneous dependence, and is intended to contrast with diachronic dependence of the sort associated, for example, with effects as temporally posterior to causes. The dependence at issue is also typically supposed to have certain modal implications: at a minimum, it is nomologically necessary for the occurrence of a given emergent phenomenon (in worlds with the relevant laws of nature) that some dependence base phenomenon occurs, and the occurrence of the dependence base goings-on is nomologically sufficient (again, in worlds with the relevant laws of nature), in the circumstances, for the occurrence of the emergent goings-on.

Third, discussions of metaphysical emergence typically take as a starting point the supposition that the foundational, or more foundational, goings-on upon which the emergent goings-on synchronically depend are physical. In turn, the typically operative supposition is that the physical goings-on are those that are the target subject matter of physics, modulo the caveat that physical goings-on cannot be fundamentally mental (as, e.g., a panpsychist view would have it).[2] Most participants to the present debate have something like this account in mind, and though there is room for further debate, such a characterization of the physical will suffice for purposes of my discussion.

The common assumption that the dependence base goings-on are physical reflects presumably contingent facts of a broadly scientific variety: as it happens, we have good reason to believe that special science goings-on broadly synchronically depend on lower-level physical goings-on. But it's worth noting that for purposes of characterizing metaphysical emergence, nothing *really* turns on the status as 'physical' of the dependence base goings-on. We can, and should, aim to make sense of what it is, or could be, for some goings-on to metaphysically emerge from some others, irrespective of whether the dependence base goings-on are physical.

Fourth, discussions of metaphysical emergence tend to suppose that the 'level' of physical goings-on includes not just the physical entities and features that are the explicit target of physics but also any and all massively complex pluralities (collections) or relational aggregates of physical goings-on. For example, suppose for the purposes of illustration that physics is atomistic, such that the basic physical entities are individual atoms and the basic physical relations are certain atomic relations holding between small numbers of atoms. The level of physical goings-on would be understood by most participants to the emergence debates to also include any and all pluralities of atoms, as well as any and all relational aggregates of atoms standing in atomic relations. Also standardly taken to be part of the lower-level physical base would be broadly logical combinations of individual atoms and groups of atoms (e.g., conjunctions or disjunctions of atomic aggregates) and linear combinations of the features of individual atoms (e.g., the mass of an aggregate of atoms would be included as among the lower-level physical properties as being just the sum of the individual masses).

Here again there is room to debate the details, not just about the extent of goings-on properly located at the physical level but more generally about how 'levels' of natural reality should be individuated. Again, not much will hinge on exactly how these questions are answered, so long as the level of physical goings-on is sufficiently inclusive that an ontologically reductionist view is not ruled out of court. Consider, for example, a middle-sized rock, and again assume an atomistic physics. If the lower level of physical goings-on fails to include complex aggregates of atoms, then there will be no substantive question of whether a reductive physicalist account of the rock makes sense: such an account will be immediately rendered false. Correspondingly, discussions of metaphysical emergence typically presuppose that we have an independent handle on what it would be for some complex entity or feature (a 'one') to exist and be properly deemed part of the lower-level physical dependence base, and the further question is: What would it come to for some special scientific entity or feature (another 'one') to depend on, yet be appropriately autonomous from, such a complex lower-level physical aggregate or feature?[3]

Fifth, against the operative backdrop assumption of the dependence base goings-on as physical, accounts of metaphysical emergence can be sorted into two broad categories: first, accounts on which the metaphysically emergent goings-on are physically acceptable, as per what is sometimes called 'weak metaphysical emergence'; second, accounts on which the metaphysically emergent goings are physically unacceptable, as per what is sometimes called 'strong metaphysical emergence'. In what follows I'll make clear whether a given account of emergence aims to characterize the weak or rather the strong variety. Whether weak or strong, an account of the metaphysical emergence of some phenomenon should appropriately contrast with a reductive account of the phenomenon, on which the phenomenon is ontologically reducible to – that is, identical with – some (perhaps massively complex, etc.) lower-level physical goings-on.

2. Scientistic approaches to metaphysical emergence

As noted earlier, one of the primary motivations for there being metaphysical emergence comes from the sciences and the seeming combination of dependence and autonomy that frequently characterizes special scientific goings-on. As such, it is highly desirable that philosophers investigating emergence carefully attend to these data. 'Scientistic' accounts have the virtue of doing so, extracting features of emergence from close consideration of scientific case studies.

That said, many scientistic accounts are long on scientific detail and short on substantive metaphysical interpretation or illumination, with the ensuing accounts of emergence often doing little more than recapitulating features read off of scientific descriptions of the case studies in ways which fail to provide a principled basis for establishing the ontological irreducibility of the phenomenon at issue, and which correspondingly fail to shed illuminating light on the nature of metaphysical emergence.

2.1. *Unpredictability/algorithmic incompressibility*

Scientific descriptions of case studies of emergence often emphasize the 'in-principle' unpredictability (i.e., failure of deducibility) of the phenomena at issue, and an appeal to such unpredictability has been a frequent theme in accounts of emergence, including those aiming to characterize emergence as a metaphysical phenomenon. For example, the British Emergentist C. D. Broad offered the following account of emergence:

> The emergent theory asserts that there are certain wholes, composed (say) of constituents A, B, and C in a relation R to each other and that the characteristic properties of the whole $R(A,B,C)$ cannot, even in theory, be deduced from the most complete knowledge of the properties of A, B, and C in isolation or in other wholes which are not of the form $R(A,B,C)$.
>
> *(Broad 1925, 64)*

Broad supposed that such in-principle failure of predictability would falsify 'mechanism', a variant of physicalism; hence he is appropriately seen as aiming to provide an account of strong emergence.

Now, unpredictability is an epistemological, not a metaphysical, notion, so why characterize metaphysical emergence in terms of this notion? Broad's reason for thinking that in-principle unpredictability tracked strong emergence wasn't bad at the time. For one way in which a metaphysically emergent phenomenon might be seen as both dependent on yet 'over and above' lower-level physical goings-on would be if the emergent phenomenon involved the operation of a new fundamental force, interaction, or law, which became operative only at a comparatively high level of complexity. And when Broad wrote, our best science – a relativistic variation on Newtonian mechanics – seemed to presuppose that phenomena that didn't involve any fundamentally novel goings-on could be predicted, in principle, by means of the usual additive force composition laws. Accordingly, cases where some composite phenomena could *not* be so predicted were thought to be indicative of the presence of a new fundamental force or interaction.

As it happens, soon after Broad offered his account of strong emergence, scientists and others came to appreciate and explore the fact that the properties and behavior of some uncontroversially physical phenomena (e.g., gases, hurricanes, and turbulent fluids) evolve via nonlinear rather than linear laws involving their lower-level physical components. Some such systems – so-called 'chaotic' complex systems – are highly sensitive to initial conditions, in ways entailing that even with the resources of the entire universe in hand, one would not be able to use the nonlinear laws governing such systems to predict the evolution after some period. Yet in spite of such 'in-principle' unpredictability, such systems are uncontroversially physically acceptable. And so an 'in-principle failure of predictability' account of strong emergence turned out not to be viable.

But – and here we turn to more recent accounts that are properly deemed 'scientistic' – the advent of complex nonlinear systems did not spell the end of accounts of metaphysical emergence appealing to unpredictability. Rather, this feature became the mainstay of many accounts of emergence of the physically acceptable – that is, weak – variety.[4] Characteristic here is Bedau's (1997) account, on which weak emergence is characterized as involving the absence of antecedent predictability, or 'derivability, but only via simulation'. Bedau starts by flagging that his approach is motivated by scientific accounts of chaotic and nonchaotic nonlinear systems, as seeming cases in point of physically acceptable emergence:

> An innocent form of emergence – what I call 'weak emergence' – is now a commonplace in a thriving interdisciplinary nexus of scientific activity [. . .] that includes connectionist modeling, non-linear dynamics (popularly known as 'chaos' theory), and artificial life.
>
> *(375)*

As Bedau observes, such systems typically fail to admit analytic or 'closed' solutions. The absence of any such means of predicting the evolution of such systems means that the only way to find out what this behavior will be is by 'going through the motions': set up the system, let it roll, and see what happens. This refined version of unpredictability serves as the basis for Bedau's account of weak emergence:

> Where system S is composed of micro-level entities having associated micro-states, and where microdynamic D governs the time evolution of S's microstates: Macrostate P of S with microdynamic D is weakly emergent iff P can be derived from D and S's external conditions but only by simulation.
>
> *(378)*

The broadly equivalent conception in Bedau (2002) takes weak emergence to involve algorithmic or explanatory 'incompressibility', where there is no 'short-cut' means of predicting or explaining certain features of a composite system. In being derivable by simulation from a microphysical dynamic, weakly emergent phenomena are understood to be physically acceptable; as Bedau (1997) says, such systems indicate "that emergence is consistent with reasonable forms of materialism" (376). And as one concrete illustration of the targeted variety of emergence he offers the property of being a glider gun in Conway's well-known Game of Life.

Now, for purposes of characterizing a form of metaphysical emergence, the problem with Bedau's account is that it is unclear why unpredictability in the sense at issue would rule out the ontological irreducibility of the unpredictable phenomena. Indeed, Bedau encourages such a reductive reading when he says that "weakly emergent phenomena are ontologically dependent on and reducible to microphenomena" (2002, 6), and later, that "the macro is ontologically and causally reducible to the micro in principle" (2008, 445). As such, an 'algorithmic incompressibility' account of weak emergence fails to satisfy the criterion of appropriate contrast.

Notwithstanding said compatibility with reductionism, Bedau maintains that emergence on his account is not just epistemological but is also metaphysical – a challenge that might be seen as tacitly aimed at denying that the criterion of appropriate contrast is appropriately applied. He offers two reasons for thinking that emergence on his account is properly metaphysical. The first is that the incompressibility of an algorithm or explanation is an objective fact:

> The modal terms in this definition are metaphysical, not epistemological. For P to be weakly emergent, what matters is that there is a derivation of P from D and S's external conditions and any such derivation is a simulation. [. . .] Underivability without simulation is a purely formal notion concerning the existence and nonexistence of certain kinds of derivations of macrostates from a system's underlying dynamic.
>
> *(1997, 379)*

But such facts about the absence of incompressible derivations or explanations, though objective and hence in some broad sense 'metaphysical', are not suited to ground the metaphysical autonomy of emergent goings-on. It's not clear that anything answering to 'autonomy' is associated with algorithmic incompressibility; in any case, insofar as such compressibility is compatible with ontological reduction, any such autonomy would attach not to the phenomena at issue, but rather to what it would take to predict or explain the phenomena, with knowledge about future states being in some sense epistemically 'autonomous' from knowledge about present states. Again, these are epistemic notions, with no clear bearing on the nature of metaphysical emergence.

Bedau also suggests that emergence on his account is properly metaphysical since the nonlinear phenomena at issue typically instantiate macro-patterns. But first, incompressibility isn't either necessary or sufficient for the occurrence of a macro-pattern, so even if the occurrence of a macro-pattern is in some sense a metaphysical fact, the connection to Bedau's official account of emergence is unclear. Second, and more importantly, the mere presence of a macro-pattern isn't enough to establish ontological and causal autonomy. For ontological reductionists are happy to allow that there are macro-patterns – they just insist that these are ontologically reducible to complex configurations of micro-phenomena. Sure, they will reasonably say, nonlinear complex systems seem to manifest macro-patterns – that's why they are candidates for being emergent. But why think that there is anything metaphysically as opposed to epistemologically real about the phenomenon of macro-patterns? Bedau never addresses or answers this question; hence, an appeal to macro-patterns doesn't clearly establish that his account of emergence is properly metaphysical, either by lights of blocking ontological reduction (as per the criterion of appropriate contrast) or by any other lights.

Relatedly, the presence of seemingly autonomous macro-patterns is the starting point of most investigations into metaphysical emergence. What is primarily at issue in debates over the status of some phenomena as genuinely metaphysically emergent is precisely whether the appearances of seeming macro-patterns can be taken at face value or whether, contrary to appearances, the phenomena are ontologically reducible to some or other lower-level goings-on. Correspondingly, an account based in a bare appeal to the advent of macro-patterns fails to provide any substantive basis for resolving the dispute at issue, and so violates the criterion of illuminating contrast.

2.2. *Universality/stability under perturbation*

Another scientistic approach to emergence adverts to certain other scientifically interesting features of nonlinear complex systems – notably, that such systems exhibit universality, reflected, for example, in the behavior of systems undergoing phase transitions being characterized by a small set of dimensionless 'critical exponents'. As Batterman (1998) motivates the approach:

> What is truly remarkable about these numbers is their universality [. . .] the critical behavior of systems whose components and interactions are radically different is virtually identical. Hence, such behavior must be largely independent of the details of the microstructures of the various systems. This is known in the literature as the 'universality of critical phenomena'.
>
> *(198)*

More generally, in a series of papers, Batterman (2011, 1998, 2000, 2011) has offered an account of emergence as involving universality of certain features, or stability of certain behaviors under perturbation (for short, 'stability'). As Batterman notes, universality and stability are two sides of the same coin; "most broadly construed, universality concerns similarities in the behavior of diverse systems" (Batterman 2000, 120).

Now an account of emergence as involving universality/stability is naturally given a metaphysical interpretation, and some commentators have read Batterman this way (see, e.g., Menon and Callender 2013). On the other hand, Batterman himself disavows concern with questions of ontological reduction or nonreduction, focusing rather on what is required if the critical behaviors of systems exhibiting universality are to be explained, with his key insight being that, for such systems, neither theoretical derivations nor causal-mechanical considerations can do this explanatory work. Reflecting this focus, Morrison (2012, 143) reads Batterman as offering an 'explanatory' account of emergence.

That Batterman's account of emergence is open to both metaphysical and epistemological interpretations is symptomatic of a tendency in scientistic accounts to follow scientists and scientific descriptions of case studies in often failing to properly track epistemological and metaphysical distinctions. Be this conflation as it may, and independent of exegesis of Batterman's work, in any case it is common for both scientists and philosophers of science to appeal to features such as universalizability and stability under perturbation as characteristic of emergence of a weak (physically acceptable) variety (see, e.g., Wimsatt 1996; Klee 1984).

Any such accounts are, however, open to the charge that they fail to provide a clear basis for blocking ontological reduction. The difficulty here can be brought out by observing that universality and stability are close cousins of multiple realizability, which notion has played a large role in the metaphysics of mind; and as debates in the metaphysics of mind concerning the import of multiple realizability reveal, a mere appeal to universal or stable features of seemingly emergent goings-on is in itself insufficient to establish the distinctness of such goings-on. In that debate, non-reductive physicalists initially aim to block type-reduction of mental to physical states by appealing to multiple realizability, with the idea being that if mental state types are multiply realizable, they cannot be identified with any single physical state type. But reductionists respond with various strategies for accommodating multiple realizability. Perhaps the most popular is to grant that such multiple realizability shows that certain mental state types cannot be identified, one to one, with individual physical state types but to maintain that this much is compatible with every multiply realizable mental state type being identical to a disjunction of physical realizer state types. Nonreductionists have a number of responses, which include rejecting disjunctive properties, denying that disjunctive properties track genuine natural kinds, and the sort of response that I will discuss down the line. But in any case insofar as scientistic accounts appealing to universalizability/stability fail to address or respond to reductionist strategies for accommodating these features, they fail to provide a clear basis for blocking ontological reduction, and so fail to satisfy the criterion of appropriate contrast.

Relatedly, insofar as reductionists and emergentists can agree that seemingly higher-level goings-on are multiply realizable/universal/stable under perturbation, compatible with their contrasting views, accounts of metaphysical emergence in terms of these features fail to provide an illuminating basis for resolving debate over the emergent status of a given phenomenon, and so also fail to satisfy the criterion of illuminating contrast.

3. Abstractionist approaches to emergence

At the other end of the spectrum from scientistic approaches are abstractionist approaches, on which metaphysical emergence is characterized in abstract modal or primitivist terms. As I'll now argue, these approaches also fail to guarantee satisfaction of one or both of the operative criteria.

3.1. *Modal accounts of strong emergence*

As noted earlier, core to the notion of metaphysical emergence is that emergent goings-on broadly synchronically depend on base goings-on, in that, at a minimum, the occurrence of the emergent goings-on nomologically requires the occurrence of some dependence base goings-on, and the dependence base goings-on are nomologically sufficient, in the circumstances, for the occurrence of the emergent goings-on. This minimal characterization is pitched in modal correlational terms; in the lingo of supervenience, emergent goings-on supervene with at least nomological necessity on base-level goings-on. Modal accounts of emergence take a page from this sort of minimal specification, aiming to characterize strong emergence in terms of a difference in the strength of modal correlations holding between goings-on that are or are not strongly emergent on lower-level physical goings-on.

More specifically, on modal accounts, strongly emergent goings-on are taken to supervene with nomological, but not metaphysical, necessity on base goings-on. For example, van Cleve (1990) characterizes emergence of the sort intended to contrast with physicalism as follows:

> If P is a property of w, then P is emergent iff P supervenes with nomological necessity, but not with logical necessity, on the properties of the parts of w.
>
> *(222)*

Similarly, Chalmers (2006) characterizes the strong emergence of consciousness as involving nomological, but not metaphysical, supervenience:

> [C]onsciousness still supervenes on the physical domain. But importantly, this supervenience holds only with the strength of laws of nature (in the philosophical jargon, it is natural or nomological supervenience).
>
> *(247)*

Similar positions are endorsed by Witmer (2001), Noordhof (2010), and others.

A modal account of strong emergence is problematic, however. Consider Lewis's (1966) position, according to which mental states are functionally specified states that are identical to the normal lower-level occupiers of those states. Now, according to Lewis, the powers of lower-level states are contingent, varying with laws of nature. Hence, it might be that in worlds with laws of nature similar to ours, every instance of mental state M is identical to an instance of lower-level physical state P, but that in worlds with different laws of nature, physical state P has different powers not suitable unto occupying the functional role associated with M. In other words, on Lewis's view, M might supervene with nomological, but not metaphysical, necessity on lower-level physical goings-on. Nonetheless, as per Lewis's calling his view an 'identity theory', this view is ontologically reductive. Hence, a modal account of strong emergence fails to satisfy the criterion of appropriate contrast.

A modal account also fails to satisfy the criterion of illuminating contrast, not just because mere nomological supervenience is compatible with ontological reduction but also because metaphysical supervenience is compatible with emergence of an 'over and above' (i.e., strong) variety. This last would be the case if, for example, a consistent Malbranchean God brings about certain higher-level features upon the occasion of certain lower-level features in every possible world; or if features are essentially constituted by (all) the laws of nature into which they directly or indirectly enter; or if some strongly emergent features are grounded in nonphysical fundamental interactions and all the fundamental interactions are unified.[5] Correspondingly, disputants might agree on whether some goings-on satisfy the modal conditions, yet disagree about whether the goings-on are strongly emergent. In all such cases, a modal account fails to shed illuminating light on the phenomenon of metaphysical emergence.

The deeper diagnosis of these problems lies in the attempt to use modal correlations as a 'stand-in' for a properly metaphysical distinction. The underlying motivation for a modal account of strong emergence lies in the natural thought that strongly emergent phenomena involve some relation associated with laws of nature, by way of contrast with more intimate relations such as identity or the determinable–determinate relation, whose holding is not thought to hinge on laws of nature. But even if one's independent commitments (e.g., to the contingency of laws of nature) do ensure that there is a modal difference between relations that do, and relations that don't, hinge on laws of nature, it is the substantive distinction between relations, not the distinction between modal correlations, which is doing the work of illuminating the phenomenon of emergence.

3.2. *Primitivist approaches to emergence*

A second abstractionist approach appeals, one way or another, to a primitive conception of metaphysical dependence, as per the notion of 'Grounding' recently advanced by Fine (2001), Schaffer (2009), and Rosen (2010), among others, understood as holding either between worldly items (Schaffer) or facts/propositions (Fine, Rosen).

3.2.1. *A Grounding-based approach to weak emergence*

Proponents of Grounding typically stipulate, first, that Grounding is operative in any and all contexts where the idioms of metaphysical dependence (e.g., 'in virtue of', 'nothing over and above') are operative, and second, that Grounding has the formal features of a partial order: Grounding relations are irreflexive, asymmetric, and transitive. These features together might be thought to motivate a Grounding-based account of weak emergence, according to which any broadly synchronic case of Grounding is one involving weak emergence. Such an account of metaphysical emergence is inadequate, however, in failing to satisfy either of our two operative criteria.

To start, notwithstanding the stipulation that Grounding is a partial order, the account does not clearly block ontological reduction; for as noted earlier, it is also stipulated that Grounding is operative in any and all contexts where the idioms of metaphysical dependence ('nothing over and above') are operative; and in these contexts, one of the most common understandings of nothing-over-and-aboveness is in terms of *identity*, as per reductive physicalism. If Grounding is indeed operative in 'any and all' such contexts, then it is compatible with ontological reduction, in which case a Grounding-based account of weak emergence fails to satisfy the criterion of appropriate contrast.

The proponent of Grounding might respond by revising their ostensive characterization of Grounding to apply only to *nonreductive* contexts where the idioms of dependence are operative. But as the debate between reductive and nonreductive physicalists shows, there is no agreement about which contexts are reductive and which aren't. Hence, there's no ostensive route to a primitive posit of nonreductive Grounding, nor to a Grounding-based account of weak emergence.

The upshot is that a Grounding-based account of weak emergence will satisfy the criterion of appropriate contrast only at the expense of failing to satisfy the criterion of illuminating contrast; for in the absence of any independent handle on Grounding *qua* primitive nonreductive metaphysical dependence relation, a Grounding-based account will foreseeably lead to immediate stalemate between those weak emergentists and their ontologically reductive opponents.

3.2.2. Grounding-based approaches to strong emergence

Might primitive Grounding do better as a basis for characterizing strong emergence? There are a couple of strategies by which one might aim to develop such an account, but as I'll now argue, none succeeds.

One strategy would be to characterize strong emergence as involving a failure of Grounding. But a failure of Grounding is – at least going by what its original proponents have said – compatible with either an antirealist eliminativist stance concerning the non-Grounded goings-on or with a realist strong emergentist stance with regard to the non-Grounded goings-on. Before one can settle on a metaphysical interpretation, one must look to the specific reasons for the supposed failure of Grounding, but *qua* thinly described primitive, Grounding is silent on these reasons. Hence, such an account will fail to guarantee satisfaction of the criterion of appropriate contrast (see Wilson 2014 for further discussion).

Yet more problematically, if Grounding corresponds to metaphysical dependence, then a failure of Grounding corresponds to a failure of metaphysical dependence. But it is a live position with regard to strongly emergent phenomena that these are, while over and above lower-level physical phenomena, nonetheless *partially*, though not completely, metaphysically dependent on lower-level physical phenomena (see Wilson 2002). In response, the proponent of a Grounding-based characterization of strong emergence might introduce a notion of partial Grounding and move to an account of strong emergence as involving a failure of full *or partial* Grounding. But it is unclear how to add partial Grounding to the mix without introducing additional primitives, thus rendering the account incapable of satisfying the criterion of illuminating contrast.

To see this, first note that one cannot define full Grounding in terms of primitive partial Grounding, by analogy with accounts taking proper parthood as primitive and defining improper parthood in terms of identity (such that P is a part of Q iff P is a proper part of Q or P is identical with Q); for fully Grounded goings-on may be distinct from Grounding goings-on.[6]

Nor (as noted by Leuenberger, 2015) can a proponent of Grounding define partial Grounding in terms of primitive full Grounding, with X being a partial Ground of Z iff there is some Y such that

X and *Y* fully Ground *Z*. For this move problematically imports a conjunctive or weak supplementation structure to dependence. As Fine (2012) notes, "a partial ground need not always be part of a full ground" (53). More specifically: perhaps truth is an unsupplementable component of knowledge (Williamson 1995); the soul is an unsupplementable constituent of a person (Brentano 1874), open interiors are unsupplementable proper parts of closed regions (Whitehead 1929), determinables are unsupplementable constituents of determinates (see Wilson 2017 for discussion), and physical goings-on are unsupplementable dependence bases for strongly emergent features (see Wilson 2016a).

The upshot is that the 'partial dependence strategy' for accommodating strong emergence on a Grounding-based view will require a second 'partial Grounding' primitive – and likely a third, corresponding to the unexplained relation between full and partial Grounding. Such a plethora of primitives is ontologically costly. More importantly for present purposes, a conception of metaphysical emergence in terms of multiple primitives is incapable of shedding any light on or providing a basis for resolving disputes over metaphysical emergence, and hence fails to satisfy the criterion of illuminating contrast.

3.2.3. Primitive fundamentality and primitive dependence

One last abstractionist approach is worth considering – namely, Barnes's (2012) 'meta-ontological' account of emergence, according to which emergent goings-on are those which are both fundamental and dependent, and where the notions of fundamentality and dependence are each primitive. Barnes does not distinguish between strong and weak forms of metaphysical emergence in her discussion, but in appealing to fundamentality her account plausibly aims to characterize emergence of the strong, 'over and above' variety. In characterizing strongly emergent goings-on as fundamental, this account is a step in the right direction. In particular, such a characterization, if it makes sense, clearly blocks ontological reduction, thus satisfying the criterion of appropriate contrast: if some goings-on are fundamentally novel vis-à-vis the goings-on upon which they depend, then the former are clearly not identical to the latter.

The problem is that such an abstract characterization doesn't clearly make sense of strong emergence. An initial concern is that such a characterization is at odds with a common understanding of independence as core to our understanding of fundamentality:

> [I]ndependence is a – the – central aspect of our notion of fundamentality. Things that are [fundamental] do not depend on anything else.
>
> *(Bennett 2017)*

Even if one denies, as I do, that independence is core to our understanding of fundamentality on grounds that there are cases of fundamental structure on which the fundamental goings-on are self-dependent or mutually dependent (see, e.g., Wilson 2014; Bliss 2014; Tahko forthcoming),[7] these cases don't involve positing fundamentality both at low levels and at higher levels of compositional complexity. Hence it is that an understanding of strong emergence as combining dependence with fundamentality is the traditional starting point of discussions of strong emergence (to be discussed in more detail shortly), with the bulk of effort devoted to filling in the operative notions of dependence and fundamentality so as to illuminate how these characteristics could be jointly instantiated.[8]

An account of strong emergence on which the operative notions of dependence and fundamentality are each taken to be primitive goes no distance towards illuminating how this could be. Even if what makes it the case that some goings-on are fundamental at a given world is a primitive matter (in that, following Wilson 2014, the fundamental goings-on are analogous to metaphysical axioms or postulates), in combination with a primitivist account of dependence, the resulting conception of strong emergence is simply too abstract to provide a clear basis for establishing that such emergence is metaphysically coherent (a necessary condition of satisfying the criterion of appropriate contrast), much less for illuminating the nature of metaphysical emergence or for allowing debate to proceed in substantive fashion (as per the criterion of illuminating contrast).[9]

4. Substantive accounts of metaphysical emergence

I turn now to considering several cases in point of what I call 'substantive' accounts of metaphysical emergence. As I'll argue in what follows, in each of these accounts there is sufficient substance in the operative characterizations of emergence to provide a clear basis for satisfaction of the criteria of appropriate and illuminating contrast.

4.1. Substantive accounts of weak emergence

Substantive accounts of weak emergence are frequently pitched as accounts of 'realization' of one form or another and are typically advanced in service of establishing the viability of ontologically nonreductive physicalism, either as a general thesis or with regard to a specific phenomenon. In what follows, I'll use 'nonreductive realization' and 'weak emergence' as synonyms.

4.1.1. Functional realization accounts

The original account of nonreductive realization was one according to which higher-level features are associated with certain functional roles, which roles are played or implemented on any given occasion by some lower-level physical feature, and where a functional role consists in a characteristic syndrome, so to speak, of typical causes and effects. Higher-level features, so understood, are functionally realized by lower-level features.

An early version of the view was proposed in Putnam (1967):

> We shall discuss 'Is pain a brain state?' [. . .] I shall, in short, argue that pain is not a brain state, in the sense of a physical-chemical state of the brain (or even the whole nervous system), but another *kind* of state entirely. I propose the hypothesis that pain, or the state of being in pain, is a functional state of a whole organism.
>
> *(53–54)*

Putnam suggested that the functional role of pain was along the following lines:

> [T]he functional state we have in mind is the state of receiving sensory inputs which play a certain role in the Functional Organization of the organism. This role is characterized, at least partially, by the fact that the sense organs responsible for the inputs in question are organs whose function is to detect damage to the body, or dangerous extremes of temperature, pressure, etc., and by the fact that the 'inputs' themselves, whatever their physical realization, represent a condition that the organism assigns a high disvalue to. [. . .] [T]his does not mean that the Machine will always *avoid* being in the condition in question ('pain'); it only means that the condition will be avoided unless not avoiding it is necessary to the attainment of some more highly valued goal.
>
> *(57)*

Putnam's talk of 'the Machine' reflects the analogy motivating a functionalist treatment, of a software program which can be run or implemented on any number of hardware systems. Relevantly similar accounts of functional realization have been advanced by Fodor (1974), Papineau (1993), Antony and Levine (1997), Melnyk (2003), among many others.

In picking up on, developing, and applying the idea that multiple hardware systems may implement the instructions associated with a given piece of software, there is a sense in which functional realization accounts appeal to the same phenomenon underlying scientistic appeals to universality and stability under perturbation – that is, multiple realizability. However, in highlighting that the basis for the universality/stability/multiple realizability reflects the association of the higher-level feature with a specific functional role, functional realization accounts introduce certain metaphysically substantive posits which, in turn, provide a principled basis for blocking ontologically reductive strategies for accommodating multiple realizability.

Here's how. To start, if a feature (e.g., pain) is associated with a functional role, then it is associated with certain powers – in particular, powers to give rise to certain effects (e.g., grimacing) when in certain circumstances.[10] Philosophers disagree on how best to understand powers; they disagree, for example, about whether properties and powers are associated with nomological or rather metaphysical necessity. But as we will see, nothing in the weak emergentist strategy to come hinges on this or other controversial issues in the metaphysics of properties.

So, a higher-level functionally realized feature F will be associated with a distinctive set of powers, reflecting the 'forward-facing' aspects of its functional role. Now on the assumption that F is physically acceptable (as the weak emergentist assumes), every such power of F on any given occasion of F's instantiation will be numerically identical with a token power of whatever lower-level physical feature P realizes F on that occasion. But importantly, F will not inherit *all* of P's powers. Again, recall the motivating software/hardware analogy. Here the realizing systems are similar in each having whatever powers are needed to implement the software, but are different in having other powers associated with their distinctive hardware bases. More generally, in cases where a type of functionally characterized higher-level feature is multiply realized, it is plausible that each of its realizing types will have all of the powers associated with its functional role, and more besides. Correspondingly, a proper subset relation will hold between the powers of the realized type and those of any of its realizing types (for discussion, see Wilson 1999).

Also importantly, this subset relation between powers will hold not just between higher-level and lower-level types but will also hold on any occasion of realization involving tokens of the types (see Wilson 2015). Again, attention to the powers of the features involved lights the way. To start, a type of functionally realized feature has a distinctive set of powers – namely, those tracking the effects characteristic of its functional role; and as just argued, any lower-level realizer of such a functionally realized feature will have these powers, and more. Now if a token of this functionally realized type were to be identical with a token of its realizing type on a given occasion, then the token of the functionally realized type would have token powers not associated with its type. But that would be a reason for denying that the token was an instance of the functionally realized type! Correspondingly, an account of weak emergence in terms of functional realization blocks ontological reduction at both the level of types and the level of tokens of the features involved, and hence satisfies the criterion of appropriate contrast.

A functional realization account of weak emergence also illuminates the nature of (this form of) weak emergence in a way that allows debate to proceed in conformity to the criterion of illuminating contrast. Is a given seemingly emergent feature associated with a distinctive functional role? Relatedly, can a functional role capture all the distinctive characteristics of the seemingly emergent feature? If so, can the role be implemented by multiple lower-level physical systems in such a way as to support taking a proper subset relation between powers to be in place? Even if the higher-level feature is not multiply realizable, are there other ways (e.g., by appealing to distinct systems of laws) to establish that a proper subset relation between powers is in place? Here it is worth noting that in answering these questions, a functional realization account can appeal to special science laws in ways reflecting that, notwithstanding that functional realization is a properly metaphysical relation in being generally applicable to a wide range of phenomena (including broadly mathematical phenomena of the sort associated with software programs and the like), this relation is nonetheless intelligibly connected to the more specific causal roles encoded in special science laws. An account of weak metaphysical emergence as based in functional realization thus does not float free of, but rather encodes and generalizes, the special scientific considerations that motivate attention to metaphysical emergence in the first place.

4.1.2. *Determinable-based accounts*

Next, consider accounts of nonreductive realization appealing to the determinable–determinate relation, the relation of increased specificity paradigmatically holding between properties such as *being red* (as determinable) and *being scarlet* (as determinate) or *being shaped* (as determinable) and *being rectangular* (as determinate).[11] On determinable-based accounts of the sort advanced by Mac-Donald and MacDonald 1986, Yablo 1992, and Wilson 1999, the suggestion is that higher-level features are nonreductively realized by lower-level features specifically in being determinables of lower-level physical determinates. Here again the appeal to a substantive metaphysical posit provides the basis for satisfaction of the criteria of appropriate and illuminating contrast.

To start, as with functionally realized features, there is a case to be made that the causal powers associated with determinable features are a proper subset of those of their realizing determinate features at both the type and token levels (see Wilson 1999). Like other broadly scientific features, determinable features are plausibly taken to be associated with a distinctive set of powers. For example, if a patch is red, it has the power to get a pigeon, Sophie, who is trained to peck at any red patch, to peck at it. Now, determinable properties, like functionally realized properties, can be multiply 'determined': a red patch might be scarlet, or burgundy, or crimson. Each of the distinct determinates of a determinable will share the powers of the determinable – a scarlet patch has the power to get Sophie to peck at it, as does a crimson patch, and so on. But these determinate features will have powers associated with their more determinate nature that the determinable feature does not have. For example, a scarlet patch has the power to get Alice, a pigeon trained to peck only at scarlet patches, to peck at it, whereas a red patch does not have this power (since a red patch might be crimson, not scarlet). So, determinable features plausibly have a proper subset of the powers of their associated determinates. And here again, this proper subset relation will be preserved when determinable and determinate are instantiated, on pain of undercutting the supposition that the determinable is instantiated. Consequently, an account of weak emergence as involving the determinable–determinate relation satisfies the criterion of appropriate contrast.

A determinable-based account also satisfies the criterion of illuminating contrast in characterizing weak emergence in terms of a metaphysical relation with which we are experientially and theoretically familiar. This is not to say, of course, that a determinable-based account is correct. Indeed, debate over the status of determinable-based realization has been vigorous and substantive, proceeding largely via attention to the question of whether, given the characteristic features of the determinable–determinate relation, it really makes sense to see mental or other higher-level features as determinables of lower-level physical determinates (see, e.g., Ehring 1996; Walter 2006; Worley 1997; Funkhouser 2006; see Wilson 2009 for replies). The jury is still out, but what is crucial for present purposes is that such debate has been able to proceed precisely because we do have an independent and substantive handle on the relation between determinables and determinates.

4.2. Substantive accounts of strong emergence

4.2.1. *Fundamental powers, forces, interactions, and laws*

Though, as previously, the British Emergentists sometimes characterized strong emergence in terms of 'in-principle unpredictability', they did so not because they thought emergence was at bottom an epistemological phenomenon, but because they thought that in-principle epistemic gaps were a reliable criterion of fundamental novelty at the level of the emergent goings-on. Though their epistemological criterion turned out to be unsuited for this purpose, still intact is the conception of British Emergentism as "the doctrine that there are fundamental powers to influence motion associated with types of structures of particles that compose certain chemical, biological, and psychological kinds" (McLaughlin 1992, 52), where the powers at issue might be thought to "generate fundamental forces" (71), or more generally, fundamental interactions. A related concept is one according to which emergent features are governed by fundamental laws (tracking or otherwise associated with having new powers to produce fundamental forces, etc.), as in Broad's (1925) remarks:

> [T]he law connecting the properties of silver-chloride with those of silver and of chlorine and with the structure of the compound is, so far as we know, an unique and ultimate law.
>
> *(64–65)*

Several contemporary accounts of strong emergence similarly appeal to fundamentally new powers, forces, or laws. So, for example, Silberstein and McGeever (1999) understand emergent features as having irreducible causal capacities (that is, fundamentally new powers) of the sort that would undermine physicalism:

> Ontologically emergent features are neither reducible to nor determined by more basic features. Ontologically emergent features are features of systems or wholes that possess causal capacities not reducible to any of the intrinsic causal capacities of the parts nor to any of the (reducible) relations between the parts.
>
> *(186)*

And O'Connor and Wong (2005) characterize strongly emergent features as conferring fundamentally novel powers:

> [A]s a fundamentally new kind of feature, [an emergent feature] will confer causal capacities on the object that go beyond the summation of capacities directly conferred by the object's microstructure.
>
> *(665)*

Finally, Wilson (2002) suggests that strong emergence involves the coming into play of a new fundamental interaction.

As with substantive accounts of weak emergence, substantive accounts of strong emergence introduce resources that provide a basis for satisfying the criteria of appropriate and illuminating contrast by explicating how it can or could be that some goings-on are both fundamental and dependent.

First, in many and indeed perhaps all of these accounts, the explanatory basis at issue will plausibly ultimately advert to the notion of a power, comparatively neutrally understood (as noted earlier). We have previously seen that both functionalist and determinable-based accounts of weak emergence block reduction by taking the higher-level features at issue to have a proper subset of the powers of their realizers on any given occasion. In being associated with only a proper subset of powers of their realizers, weakly emergent features are nonfundamentally novel. By way of contrast, substantive accounts of strong emergence take such emergence to involve posits that plausibly entail that the higher-level features at issue have novel powers – powers that their lower-level physical realizers do not have – which in turn provide a basis for the usual supposition (again, tracing back at least to the British Emergentists) that strongly emergent features are fundamentally novel. For example, insofar as laws of nature pertaining to some goings-on register what these goings-on do when in certain circumstances, such laws are plausibly seen as registering what powers are associated with those goings-on; hence, the supposition that strong emergence involves fundamentally novel laws is appropriately seen as entailing that strongly emergent features have new powers – powers that their lower-level physical goings-on do not have. Similarly for views on which strong emergence involves fundamentally novel features, with powers to produce new forces, and the like.

To be sure, there has been debate over whether it really makes sense to attribute novel powers to dependent goings-on, with the charge being that any such powers will 'collapse', one way or another, so as to be possessed by the dependence base goings-on (see, e.g., van Cleve 1990; Howell 2009; Taylor 2015). However, here again there is sufficient metaphysical content to the operative account(s) of emergence for debate to proceed, and as discussed in Baysan and Wilson (2017), there are several available means of responding to the collapse objection (one of which is to relativize strong metaphysical emergence to sets of interactions and powers associated with these interactions, as I will later discuss).

Moreover, this feature – the having of at least one new power – is compatible with strongly emergent features' being dependent on lower-level physical features, whether this dependence is cashed in terms of minimal nomological necessitation or in some stronger terms (e.g., a view on which strongly emergent goings-on share some of the token powers of their physical dependence bases). Such a powers-based explication, it seems to me, provides a reasonable basis for taking strong emergence as coupling dependence with fundamentality to make sense, and moreover to do so in a way that clearly blocks ontological reduction. Thus, a 'new power' conception of strong emergence plausibly satisfies the criterion of appropriate contrast. Moreover, since we have in hand working means of exploring whether fundamentally novel laws or features and associated novel powers are in place, substantive debate can proceed, as it in fact has (as in, e.g., O'Connor 2002; Chalmers 2003), over whether a given phenomenon is strongly emergent.

Second, an approach to strong emergence as involving a novel fundamental interaction (as per Wilson 2002) provides a yet clearer means of explicating the joint compatibility of fundamentality and broadly synchronic dependence. Consider, by way of illustration, the weak nuclear interaction. This interaction was originally posited as a novel fundamental interaction whose operation broadly synchronically depends on certain other phenomena (i.e., certain complex nuclear arrangements and associated fundamental interactions). Correspondingly, the weak nuclear interaction is, or in any case clearly could have been, an actual case of a fundamental feature of natural reality that broadly synchronically depends on other fundamental goings-on. The advent of this sort of novel fundamental interaction provides a clear sense in which fundamental novelty can make an appearance only in comparatively complex situations; moreover, the case of the weak nuclear interaction also makes sense of how such higher-level fundamental goings-on might also be broadly synchronically dependent on lower-level goings-on – since the operation of the weak nuclear interaction in nuclear circumstances clearly synchronically depends on other fundamental interactions (e.g., the strong nuclear interaction) being in operation. Now, such an instance of strong emergence would not be such as to falsify physicalism, for reasons having to do with the usual ways of characterizing the physical goings-on (see Wilson 2006 for discussion). But the more general point remains that this scientific case study provides support for the coherence of an interaction-based account of strong emergence, and moreover illuminates the nature of such emergence in ways that (since we have some working handle on the criterion for positing a novel fundamental interaction; see Wilson 2002 for details) provide an in-principle basis for resolving debates over the status of a given phenomenon as metaphysically emergent.

5. Methodological morals

I close with some methodological morals concerning the metaphysics of metaphysical emergence.

First, regarding scientistic accounts. As philosophers interested in the metaphysics of emergence, we should of course attend to scientific case studies of the sort inspiring attention to metaphysical emergence in the first place. But these case studies should constitute input into our theorizing as opposed to the theorizing itself. This is not to say, of course, that features discussed in scientific descriptions of scientific case studies might not serve as an apt basis for an adequate account of metaphysical emergence – indeed, my own work has often taken advantage, for metaphysical purposes, of certain previously underexplored scientific notions, including the notion of a fundamental interaction (as fruitfully entering into an adequate account of strong emergence; see Wilson 2002) and the notion of a degree of freedom (as fruitfully entering into an adequate account of weak emergence; see Wilson 2010). Still, philosophers appealing to such features need to do more than simply highlight the fact that certain scientists have taken these features to be characteristic of emergence. This is especially true since, as illustrated earlier, the features making an appearance in scientific case studies of emergence are often either explicitly epistemological, in ways not having any clear or relevant metaphysical import, or else are – at least for all proponents of scientistic accounts of emergence established – subject to ontologically reductive interpretation, contra the criterion of appropriate contrast.

Second, regarding abstractionist accounts. Though abstractionist accounts of emergence appeal to posits (modal correlations, primitive Grounding, and the like) that are properly metaphysical, the overly abstract characterizations of these posits renders them too devoid of content to do any useful theoretical work. As with attempts to characterize emergence in epistemological terms, modal patterns are, at best, a sign of emergence or its absence; they do not constitute emergence. And accounts in terms of primitive Grounding, as well as other accounts characterizing emergence in primitive terms, provide no illuminating or working handle via which debate with regard to the status of some phenomenon as metaphysically emergent might proceed. If these primitivist gestures were the best we could do, then metaphysical emergence, weak or strong, would have turned out to be simply beyond our ken.

Third, luckily, there are ways of theorizing about metaphysical emergence that lie between the extremes of scientistic reportage and abstractionist gestures. The substantive accounts of weak and strong metaphysical emergence discussed earlier, while sensitive to the proffered features of scientific case studies of seeming emergence, bring to the theoretical table a range of substantive metaphysical resources for theorizing about emergence, which provide a clear basis for ensuring the contrast between emergent and ontologically reducible phenomena, and for illuminating the nature of metaphysical emergence in a way that allows substantive debate to proceed.[12]

This comparative result is a specific case in point of the virtues of what I have called (in Wilson 2016b) the 'embedded' conception of metaphysics. On this conception, the posits and claims of metaphysics do not float free of other disciplines treating its target subject matter, as abstractionist accounts are prone to do; nor does metaphysics simply take on board the posits and claims of these other disciplines, as scientistic accounts are prone to do. Rather, on this conception, metaphysical posits and claims are embedded in the posits and claims of other disciplines; though typically both more abstract and more metaphysically substantive, the components of metaphysical accounts grow from, rather than float free of, these other notions in ways that are, dare I say, just right.

Notes

* Department of Philosophy, University of Toronto; jessica.m.wilson@utoronto.ca

1 See Wilson (forthcoming) for expanded discussion.

2 See Wilson (2006) for motivation and defense of a 'no fundamental mentality' account of the physical.

3 That said, one might also consider an account of emergence as a 'many-one', broadly compositional phenomenon, taking individual physical goings-on (e.g., atoms and atomic features and relations) as input and composed aggregates with distinctive features as output (see, e.g., Gillett 2002). Here the further question is: What would it be for an entity or feature (the 'one') to emerge from a complex plurality of entities or features (the 'many')? As I've discussed elsewhere (in, e.g., Wilson 2015), the 'many-one' and 'one-one' approaches to emergence need not be seen as competitors. Rather, phenomena that are output of many-one accounts are in many cases plausibly seen as the dependence base goings-on input into one-one accounts.

4 For detailed discussion of nonlinearity and its historical and contemporary bearing on accounts of emergence, see Wilson (2013).

5 See Wilson (2005) for discussion.

6 Indeed, as previously noted, proponents typically stipulate that Grounded and Grounding goings-on are distinct.

7 Cases involve, among others, a self-sustaining God, a world of mutually dependent objects (as per a certain read on Leibnizian monads or Huayan buddhism), and mutually constituting quarks.

8 Problematically, Barnes does not cite or engage with any of these historical precursors.

9 Barnes does offer a 'criterion' for dependence of the sort at issue in her account of emergence:

> *Ontological Dependence* (OD): An entity x is dependent iff for all possible worlds w and times t at which a duplicate of x exists, that duplicate is accompanied by other concrete, contingent entities in w at t.

But this criterion is subject to immediate fatal counterexamples. Suppose that a form of nonreductive physicalism (weak emergence) is true about mental states and that according to this account some presumed fundamental lower-level physical state P metaphysically necessitates some mental state M, distinct from P, in any world where P exists and where both P and M are contingent – that is, do not exist at some worlds. Then physical state P turns out to be dependent (on M, in particular) by lights of OD, since for all possible worlds w and times t at which a duplicate of P exists, that duplicate will be accompanied by a distinct contingent entity (namely, M). That's wrong: by assumption, M depends on P, not vice versa. Moreover, since the physicalist takes P to be fundamental, then given OD, P turns out to be strongly emergent. That's wrong. More generally, when coupled with OD, Barnes's account of strong emergence renders all fundamental physical goings-on strongly emergent from any contingent goings-on that completely metaphysically depend upon them. For example, according to Barnes's account any fundamental lower-level physical determinate feature (say, being a specific charge) will be strongly emergent from any (e.g., special scientific) determinable of that determinate. See Pearson (2018) for a number of other reasons to think that Barnes's account of emergence is either unilluminating (if appealing to primitive notions of dependence and fundamentality) or extensionally incorrect (if the notion of dependence is filled in via OD).

10 The feature might also be thought to be associated with what Shoemaker (1980) calls 'backwards-facing' powers. Such talk is misleading, however. Strictly speaking, features do not have 'powers' to be caused in this or that way – they simply are caused to be instantiated, one way or another. Once instantiated, features can then contribute to the causing of certain effects – that is, be associated with powers.

11 See Wilson (2017) for further discussion.

12 It will not be lost on the reader that the strategies via which the substantive accounts of weak and strong emergence noted earlier manage to satisfy the operative criteria proceed by attention to one or another relation between the powers associated with emergent and dependence base features, respectively. In

particular, substantive accounts of weak emergence provide resources for accommodating the criteria by appealing to metaphysical relations (functional realization, the determinable-determinate relation) which plausibly entail that weakly emergent features have a proper subset of the powers of their dependence base features, whereas substantive accounts of strong emergence provide resources for accommodating the criteria by appealing to metaphysical posits (fundamental features, forces/interactions, laws) that ensure that strongly emergent features have one or more powers that their dependence base features don't have. Such patterns are suggestive; and indeed, in previous and forthcoming work (see especially Wilson 2015, forthcoming), I argue in detail that existing substantive accounts of metaphysical emergence, of both weak and strong varieties, are plausibly and profitably seen as instantiating one or another pattern, or schema, for emergence.

References

Antony, Louise M. and Joseph M. Levine, 1997. "Reduction with Autonomy". *Philosophical Perspectives*, 11:83–105.

Barnes, Elizabeth, 2012. "Emergence and Fundamentality". *Mind*, 121:873–901.

Batterman, Robert, 1998. "Why Equilibrium Statistical Mechanics Works: Universality and the Renormalization Group". *Philosophy of Science*, 65:183–208.

Batterman, Robert W., 2000. "Multiple Realizability and Universality". *British Journal for the Philosophy of Science*, 51:115–145.

Batterman, Robert W., 2011. "Emergence, Singularities, and Symmetry Breaking". *Foundations of Physics*, 41:1031–1050.

Baysan, Umut and Jessica Wilson, 2017. "Must Strong Emergence Collapse?". *Philosophica*, 91: 49–104.

Bedau, Mark A., 1997. "Weak Emergence". *Philosophical Perspectives 11: Mind, Causation and World*, 11:375–399.

Bedau, Mark A., 2002. "Downward Causation and the Autonomy of Weak Emergence". *Principia*, 6:5–50.

Bedau, Mark A., 2008. "Is Weak Emergence Just in the Mind?". *Minds and Machines*, 18:443–459.

Bennett, Karen, 2017. *Making Things Up*. Oxford: Oxford University Press.

Bliss, Ricki, 2014. "Viciousness and Circles of Ground". *Metaphilosophy*, 45:245–256.

Brentano, Franz, 1874. *Psychology from an Empirical Standpoint*. Abingdon, UK: Routledge.

Broad, C. D., 1925. *Mind and Its Place in Nature*. Cambridge: Kegan Paul. From the 1923 Tanner Lectures at Cambridge.

Chalmers, David J., 2003. "Consciousness and Its Place in Nature". In Stephen P. Stich and Ted A. Warfield, editors, *Blackwell Guide to the Philosophy of Mind*, 102–142. Oxford: Blackwell.

Chalmers, David J., 2006. "Strong and Weak Emergence". In *The Re-Emergence of Emergence*. Oxford: Oxford University Press.

Ehring, Douglas, 1996. "Mental Causation, Determinables, and Property Instances". *Nous*, 30:461–480.

Fine, Kit, 2001. "The Question of Realism". *Philosophers' Imprint*, 1:1–30.

Fine, Kit, 2012. "Guide to Ground". In Fabrice Correia and Benjamin Schnieder, editors, *Metaphysical Grounding*, 37–80. Cambridge: Cambridge University Press.

Fodor, Jerry, 1974. "Special Sciences (Or, the Disunity of Science as a Working Hypothesis)". *Synthese*, 28:77–115.

Funkhouser, Eric, 2006. "The Determinable-Determinate Relation". *Nous*, 40:548–569.

Gillett, Carl, 2002. "The Dimensions of Realization". *Analysis*, 4: 316–323.

Howell, Robert J., 2009. "Emergentism and Supervenience Physicalism". *Australasian Journal of Philosophy*, 87:83–98.

Kim, Jaegwon, 1989. "The Myth of Nonreductive Materialism". *Proceedings and Addresses of the American Philosophical Association*, 63:31–47. Reprinted in Kim 1993b.

Kim, Jaegwon, 1993. "The Non-Reductivist's Troubles with Mental Causation". In John Heil and Alfred Mele, editors, *Mental Causation*, 189–210. Oxford: Oxford University Press. Reprinted in Kim 1993b.

Klee, Robert, 1984. "Micro-Determinism and Concepts of Emergence". *Philosophy of Science*, 51:44–63.

Leuenberger, Stephan, 2015. "Emergence and Failures of Supplementation". Talk at the Emergence and Grounding Workshop at the University of Glasgow.

Lewis, David, 1966. "An Argument for the Identity Theory". *The Journal of Philosophy*, 63:17–25. Reprinted in Lewis 1983.

MacDonald, Cynthia and Graham MacDonald, 1986. "Mental Causes and Explanation of Action". *Philosophical Quarterly*, 36:145–158.

McLaughlin, Brian, 1992. "The Rise and Fall of British Emergentism". In Ansgar Beckerman, Hans Flohr and Jaegwon Kim, editors, *Emergence or Reduction? Essays on the Prospects of Non-Reductive Physicalism*, 49–93. Berlin: De Gruyter.

Melnyk, Andrew, 2003. *A Physicalist Manifesto: Thoroughly Modern Materialism.* New York: Cambridge University Press.

Menon, Tarun and Craig Callender, 2013. "Ch-Ch-Changes Philosophical Questions Raised by Phase Transitions". In Robert Batterman, editor, *The Oxford Handbook of Philosophy of Physics*, 189–212. Oxford, USA: Oxford University Press.

Morrison, Margaret, 2012. "Emergent Physics and Micro-Ontology". *Philosophy of Science*, 79:141–166.

Noordhof, Paul, 2010. "Emergent Causation and Property Causation". In Cynthia Macdonald and Graham Macdonald, editors, *Emergence in Mind.* Oxford: Oxford University Press.

O'Connor, Timothy, 2002. "Free Will". *The Stanford Encyclopedia of Philosophy (Summer 2016 Edition).*

O'Connor, Timothy and Hong Yu Wong, 2005. "The Metaphysics of Emergence". *Nous,* 39:658–678.

Papineau, David, 1993. *Philosophical Naturalism.* Oxford: Basil Blackwell.

Pearson, Olley, 2018. "Emergence, Dependence, and Fundamentality". Erkenntnis, 3:391–402.

Putnam, Hilary, 1967. "The Nature of Mental States". In W.H. Capitan and D.D. Merrill, editors, *Art, Mind, and Religion,* 1–223. Pittsburgh, PA: Pittsburgh University Press.

Rosen, Gideon, 2010. "Metaphysical Dependence: Grounding and Reduction". In B. Hale and A. Hoffmann, editors, *Modality: Metaphysics, Logic, and Epistemology,* 109–136. Oxford: Oxford University Press.

Schaffer, Jonathan, 2009. "On What Grounds What". In D. Manley, D. Chalmers and R. Wasserman, editors, *Metametaphysics: New Essays on the Foundations of Ontology,* 347–383. Oxford: Oxford University Press.

Shoemaker, Sydney, 1980. "Causality and Properties". In Peter van Inwagen, editor, *Time and Cause,* 109–135. Dordrecht: D. Reidel.

Silberstein, Michael and John McGeever, 1999. "The Search for Ontological Emergence". *Philosophical Quarterly,* 50:182–200.

Tahko, Tuomas E., forthcoming. "Fundamentality and Ontological Minimality". In Ricki Bliss and Graham Priest, editors, *Reality and its Structure.* Oxford: Oxford University Press.

Taylor, Elanor, 2015. "Collapsing Emergence". *Philosophical Quarterly,* 65:732–753.

van Cleve, James, 1990. "Mind-Dust or Magic? Panpsychism versus Emergence". *Philosophical Perspectives,* 4:215–226.

Walter, Sven, 2006. "Determinates, Determinables, and Causal Relevance". *The Canadian Journal of Philosophy,* 37:217–243.

Whitehead, Alfred North, 1929. *Process and Reality: An Essay in Cosmology.* New York: Free Press.

Williamson, Timothy, 1995. "Is Knowing a State of Mind?". *Mind,* 104:533–565.

Wilson, Jessica M., 1999. "How Superduper Does a Physicalist Supervenience Need to Be?". *The Philosophical Quarterly,* 49:33–52.

Wilson, Jessica M., 2002. "Causal Powers, Forces, and Superdupervenience". *Grazer Philosophische-Studien,* 63:53–78.

Wilson, Jessica M., 2005. "Supervenience-Based Formulations of Physicalism". *Nous,* 39:426–459.

Wilson, Jessica M., 2006. "On Characterizing the Physical". *Philosophical Studies,* 131:61–99.

Wilson, Jessica M., 2009. "Determination, Realization, and Mental Causation". *Philosophical Studies,* 145:149–169.

Wilson, Jessica M., 2010. "Non-Reductive Physicalism and Degrees of Freedom". *British Journal for the Philosophy of Science,* 61:279–311.

Wilson, Jessica M., 2013. "Nonlinearity and Metaphysical Emergence". In Stephen Mumford and Matthew Tugby, editors, *Metaphysics and Science.* Oxford: Oxford University Press.

Wilson, Jessica M., 2014. "No Work for a Theory of Grounding". *Inquiry,* 57:1–45.

Wilson, Jessica M., 2015. "Metaphysical Emergence: Weak and Strong". In Tomasz Bigaj and Christian Wüthrich, editors, *Metaphysical Emergence in Contemporary Physics: Poznan Studies in the Philosophy of the Sciences and the Humanities.* Leiden: Brill, 251–306.

Wilson, Jessica M., 2016a. "Grounding-Based Formulations of Physicalism". *Topoi,* 37(3):495–512.

Wilson, Jessica M., 2016b. "The Question of Metaphysics". *The Philosophers' Magazine,* 74:90–96.

Wilson, Jessica M., 2017. "Determinables and Determinates". *Stanford Encyclopedia of Philosophy.*

Wilson, Jessica M., forthcoming. *Metaphysical Emergence.* Oxford: Oxford University Press.

Wimsatt, William, 1996. "Aggregativity: Reductive Heuristics for Finding Emergence". *Philosophy of Science,* 64:372–384.

Witmer, Gene, 2001. "Sufficiency Claims and Physicalism". In Carl Gillett and Barry Loewer, editors, *Physicalism and Its Discontents,* 57–73. Cambridge: Cambridge University Press.

Worley, Sara, 1997. "Determination and Mental Causation". *Erkenntnis,* 46:281–304.

Yablo, Stephen, 1992. "Mental Causation". *Philosophical Review,* 101:245–280.

PART 2

Emergence and mind

13

EMERGENT DUALISM IN THE PHILOSOPHY OF MIND

Hong Yu Wong

Motivations for emergent dualism

Emergence has always been seen as a third way between reductive materialism and substance dualism. The concept of emergence is built on three basic commitments: dependence, irreducibility, and novelty. The basic idea of emergence is that novel, irreducible phenomena arise when a system goes beyond some threshold. Often the threshold is some required level of complexity of dynamics or structure. A further characteristic of emergence is that the novel behaviour of the system is not predictable (or at least unexpected) given the behaviour of the system under conditions below the threshold of complexity. The appeal of emergence is clear. On the one hand, because emergent phenomena inhere in and arise from material systems, they do not introduce anything external to material systems. Novel phenomena conceived of as emergent can be explained without alluding to alien, immaterial substances. Thus, substance dualism can be rejected and naturalistic scruples can be upheld. On the other hand, because emergent phenomena are novel and, in some way, unpredictable, they are irreducible to the underlying material phenomena from which they emerged. Thus, reductive materialism can be rejected. Emergence allows for the attractive prospect of a naturalist, materialist, but also anti-reductionist account of the world.

The basic doctrine of emergence has been around since antiquity. Galen is said to have been an emergentist (Caston 1997), and emergentist ideas have also been found in classical Indian philosophy (Ganeri 2011). However, emergence really rose to prominence only in the nineteenth century with the intensifying focus on chemical and biological processes. The golden period of emergentism was between the mid-nineteenth century to the first three decades of the twentieth century (McLaughlin 1992). The doctrine re-emerged in the final decade of the twentieth century with the growing interest in complexity and self-organisation (Bedau 1997; Kauffman 1993), in non-reductive physicalism (Beckermann et al. 1992; Kim 1993), and in the 'hard' problem of consciousness (Chalmers 1996).

The examples of choice for what is emergent have often been the phenomena which most stubbornly resisted the pursuits of science – the favoured example being the frontier of science of that epoch. Examples of emergent phenomena over the years have included chemical properties, biological properties – prominently that of life – and mental properties, such as consciousness, intentionality, or agency. In the heyday of British Emergentism, spanning Mill (1843) up to Broad (1925), the critical case was that of chemical properties and chemical bonding. Once that was solved with the advent of quantum mechanics and the development of molecular biology, a mechanistic account of life could be provided, and the last frontier was that of the mind (McLaughlin 1992; Papineau 2000). This is the explanatory target of most accounts of emergence today (e.g. Beckermann et al. 1992; Corradini and O'Connor 2010; Macdonald and Macdonald 2010).

We can distinguish between two forms of emergence: epistemological and ontological emergence. Epistemological emergence is supposed to be the weaker form of emergence. As the label suggests, it is a form of emergence that relates to what we can know. The key notion here is predictability. Are we able to predict the behaviour of some complex system based on the behaviour of its constituents acting individually? The thought is that knowing what the constituents do when acting individually may not entail that one knows how the complex system they constitute will behave. This unpredictability, however, is not supposed to introduce any new ontology into the world; it is merely epistemic. Ontological emergence goes beyond this. It does not only concern what we are required to posit due to our cognitive and epistemic limitations, but rather marks out new entities that are part of the basic furniture of the world. Elizabeth Barnes (2012) has proposed an elegant way to characterise ontological emergents as entities which are both fundamental (as opposed to derived) and dependent (as opposed to independent). Since the article is on emergent dualism, I shall set aside the epistemological variety of emergence because it is not opposed to physicalism.

Having said this, outside of metaphysics, the distinction between epistemological and ontological emergence isn't as clear as one might want or expect, because a criterion for emergence requires higher-level phenomena that resist reduction to lower-level phenomena. In actual practice, the feasibility of reduction will have to be evaluated through examining specific theories of the higher-level and lower-level phenomena relative to a set of standards for inter-theoretic reduction. In case of a failure of reduction, whether this is to be interpreted as epistemic or ontological emergence is a complex issue. Even taking a set of stringent standards for inter-theoretic reduction, a failure of reduction could be interpreted as indicating autonomous patterns of explanation, but without any corresponding fundamental ontology (Alexander 1920; Fodor 1974).

Today's dialectic sees emergentism pitched against physicalism, a descendent of the previous materialism. The key transformation from materialism to physicalism is the shift from a claim about all individuals being material (McLaughlin 1992) to a stronger modal claim, roughly, that physical facts fix all the other facts (Jackson 1998; Wilson 2005). From a theoretical point of view, the kinds of emergent entities posited could be as diverse as the kinds of metaphysical categories one accepts: emergent events, processes, facts, properties, individuals, substances, what have you. But there have been two major options canvassed in the philosophy of mind: emergent properties and emergent substances. Emergent substances (Hasker 1999; O'Connor and Jacobs 2003) are problematic for similar reasons as substance dualism. It is unclear what evidence could warrant us recognising anything other than material substances (Strawson 1974; Shoemaker 1976; Sosa 1987). Furthermore, the consequences of having such substances in one's ontology are dire (Wong 2007). Having one substance emerge out of another does little to mitigate the objections to substance dualism. That leaves us with emergent properties, which I will focus on. Emergent property dualism is the dominant view amongst emergentists today. Typically, proponents opt for this because of arguments having to do with some special feature of mind that appears to be irreducible and intractable, for example, consciousness (Chalmers 1996, 2006), intentionality (Crane 2001), or agency (O'Connor 2000).

Two forms of emergent dualism

We find two forms of emergent dualism in the literature: supervenience emergentism and causal emergentism. On both views, the emergence relation between basal properties and emergent properties is usually understood to hold as a matter of nomological necessity. On supervenience emergentism, this is because there are fundamental emergent laws that guarantee supervenience. On causal emergentism, this is because basal properties cause the emergent properties.

Supervenience emergentism

Supervenience emergentism (SE) has been the most influential emergentist view in the last few decades (Kim 1992, 1993, 1999; McLaughlin 1992; Van Cleve 1990; O'Connor 1994; Crane 2010). C. D. Broad first proposed SE in his *The Mind and Its Place in Nature* (1925), where he also provided its canonical formulation and defence (including a version of the knowledge argument). On SE, in certain complex systems, irreducible higher-level properties emerge from lower-level properties due to fundamental emergent laws. SE provides an attractive package of a materialist (but not physicalist) view that combined irreducibility, novelty, and dependence. The dependence between emergent and basal properties is unusually tight because, on SE, emergence holds as a matter of fundamental emergent laws, which govern modal-dependent variation between the basal properties and the emergent properties, ensuring that emergent properties supervene with nomological necessity on basal properties. The emergent laws are fundamental because they are not entailed by any other laws, even together with initial conditions (McLaughlin 1997; Wong 2006: Appendix).

Despite its attractions, SE faces three major objections. First, how does SE differ from non-reductive physicalism (NRP)? Other than the emergent laws being fundamental laws, SE bears a striking resemblance to NRP (Kim 1992). If this is so, in what sense is SE an anti-physicalist view? This is the converse of the question that Kim has posed to NRP. Second, emergent phenomena are explained in terms of fundamental emergent laws that can receive no further elucidation. This is striking because we are dealing with high-level properties of complex systems here. Samuel Alexander (1920) counselled that we should accept the brute connection between emergent and basal properties with 'natural piety'. Can we really swallow this bitter pill? Third, there are worries from causal exclusion. These worries come in two varieties, both originating from Jaegwon Kim. The standard version of the argument relies on the premise of causal closure of the physical. It observes that for any downward causal effect of an emergent property, due to the causal closure of the physical, its basal property will also be sufficient for that effect. Thus, if the effect is not to be overdetermined, the emergent property must be causally impotent (Kim 1993, 1998). Another variant of the argument tries to show that SE is condemned to epiphenomenalism without the causal closure premise by arguing that the downward causal effect of an emergent property will be overdetermined (Kim 1999). (I consider responses to these arguments in the next section.) Whatever one's favoured response to these causal exclusion worries is, Kim's arguments bring out what might be perceived as a tension in SE: How might SE allow for novel, downward causal powers for emergent properties which are supervenient (Wong 2010)? Partly driven by these considerations, some emergentists have proposed another form of ontological emergence, where the emergence relation is understood to be causal.

Causal emergentism

A second form of ontological emergence is causal emergentism (CE). We can trace CE back at least to Mill's (1843) discussion of heteropathic (= emergent) as opposed to homopathic laws and effects. Roughly the distinction was between two classes of causes and the laws governing these causes: those causes whose results could be understood additively ('homopathic') and those which could not ('heteropathic'). Mill's favoured cases were biological and chemical interactions. For example, mixing an acid and an alkali produces salt and water, but the latter is not the additive sum of the effects of the causes. The basic idea of CE, unlike SE's 'vertical' synchronic relation, is a 'horizontal' diachronic relation. The emergence relation can be seen as a distinctive form of causal relation that generates an emergent property. Depending on the conception of emergence in question, the emergent property can then be bound to co-vary with basal properties in ways that are constrained by law or can lead a causal life of its own, independent from basal properties.

There are two major contemporary versions of such a view. One is Humphreys's (1996, 1997a, 1997b, 2016) notion of fusion emergence – which he sees as a subset of cases of what he now calls transformational emergence (Humphreys 2016). Not all instances of transformational emergence are causal – for example, the case of quantum entanglement is not one where the entangled molecule is caused by the constituent atoms – but the key idea is that the emergence relation here involves a non-supervenient relation of 'interaction'. On Humphreys's distinctive view, when basal properties fuse to become the emergent property, the basal properties are destroyed in the fusion operation. Because there are no basal properties when there are emergent properties, they do not supervene on the basal properties. (But see Wong 2006.) Thus, the exclusion problem does not arise.

Another is O'Connor's dynamic view of emergence (O'Connor 2000; O'Connor and Wong 2005). In this picture, emergent properties are caused and sustained by basal properties, but have novel causal powers that can influence the dynamics of both emergent and basal properties. Emergent properties do not supervene on basal properties even though they are caused and sustained by basal properties, because there can be differences in emergent properties without a change in basal properties due to the novel causal powers of the emergent properties. In this picture, because emergent properties have effects which basal properties could not themselves cause, emergent properties supplement the causal dynamics of systems. Once again, the exclusion problem does not arise.

Emergence and non-reductive physicalism

Non-reductive physicalism (NRP) is the dominant form of physicalism (Fodor 1974; Loewer 2001). It combines adherence to supervenience physicalism (SP) – the claim that any minimal physical duplicate of the world is a duplicate simpliciter (Jackson 1998) – with a rejection that higher-level properties reduce to physical ones. This anti-reductionism usually derives from considerations about multiple realisability – the observation that higher-level kinds can be realised by different physical kinds and hence cannot be identical with them – and externalism about content.

NRP seemingly allows one to have one's cake (anti-reductionism) and eat it too (physicalism). Despite its apparent attractions, Kim has argued that NRP is an unstable view. NRP and SE seem to share the same basic commitments: SP and anti-reductionism. Yet SE is explicitly an anti-physicalist view. If NRP has the same basic commitments as an anti-physicalist view, it cannot be an acceptable form of physicalism. Kim argues that NRP's anti-reductionism undermines its physicalist credentials. Hence, emergence has come to play an important role as a foil in the debates surrounding how physicalism ought to be formulated (Horgan 1993; Wilson 2005).

Horgan has argued that what is required for NRP beyond adherence to SP is 'superdupervenience' – that the supervenience of higher-level properties must be explained in a 'physicalistically acceptable way' (Wilson 2005). It is an open question what counts as 'physicalistically acceptable'; possible ways include, *a priori* deducibility from physical truths or *a posteriori* functional reduction. Ultimately the idea is that the high-level properties must be made sense of in a broadly physical way. I concur that any form of physicalism needs to meet this requirement and that this has powerful ramifications for how one must respond to the explanatory gap (Crane 2010). However, I disagree that there is no difference between SE and NRP. Partly this comes down to whether SE holds SP. And this depends on what a minimal physical duplicate comes down to. If emergent laws are not part of the physical base, fixing the physical facts will not be enough to fix the emergent facts. These will require fixing the physical facts and the emergent laws. However, I agree that there is a major issue about the epistemic status of higher-level facts. Even if one rejects *a priori* deducibility of higher-level truths from physical truths, NRP cannot rest content with taking the explanatory gap as a brute necessity, unlike SE, but must discharge the supervenience of high-level facts in a 'physicalistically acceptable way'.

Prospects for emergent dualism

What are the prospects for emergent dualism? In closing, I discuss three major challenges for it.

Causal closure of the physical

The first challenge comes from the apparent causal closure of the physical (CCP) (Kim 1998; Papineau 2000), which is a key premise both in causal exclusion arguments against emergence and in arguments for physicalism. Roughly, this principle states that if a physical effect has a sufficient cause, then it has a physical cause. Though there is some controversy about how best to state the principle (Montero 2003; Gibb 2015), it is widely accepted that some such principle is correct. How can the emergentist respond?

Any specific response would require settling on a formulation of the principle and then assessing the situation for emergent properties, but let me consider four responses. First, the emergentist can reject causal closure. Are there principled grounds for rejecting CCP other than the emergent dualist's conviction that emergent properties make a causal difference? The most straightforward way to reject CCP is because the emergent dualist believes in new, configurational forces. But is there evidence for such causes? The status of this rejection will be considered in the next two subsections. Second, the emergent dualist can attempt to show that any formulation of CCP would be question begging against the emergentist. The status of this move depends on whether there are independent grounds to think that causal closure is correct. Third, the emergent dualist can accept CCP but see emergent properties as overdetermining causes (e.g. Sturgeon 1998). However, he will claim that this kind of overdetermination is unlike the vicious cases of causal overdetermination involving more than one independent, sufficient cause. Fourth, the emergent dualist can accept CCP and resign himself to emergent properties being epiphenomenal (Chalmers 1996).

Empirical evidence for emergent dualism?

The second challenge comes from the paucity of positive empirical evidence for emergent dualism. Even proponents of the view admit that positive empirical evidence for emergent dualism has been scant. It is wedged between empirical facts that cannot be explained in physical terms and an acknowledgement that such brute facts are to be taken with 'natural piety'. The flirtation of philosophers attracted to emergence with epiphenomenalism is evidence of the attempt to insulate themselves from empirical commitments and consequent counterexamples.

In this vein we find physicalists going on the offensive. McLaughlin (1992) boldly asserted that there is 'not a scintilla of evidence' in favour of ontological emergence. Papineau (2000) mounts an inductive argument where there is increasing evidence from the sciences, including physiology, that all accelerations can be accounted for by a small stock of basic forces and that there is no need for the configurational forces that the emergentists posited.

If there is little by way of positive empirical evidence for emergent dualism, then it seems that, epistemically, the view is always a 'last resort' kind of view. One has to show that no materialist treatment is in principle possible, and thus one has to accept emergent dualism with 'natural piety'. If that is indeed the case, then two issues arise. First, how can we ever be sure that no materialist treatment is in principle possible? If the argument is based a priori, then this might work, but not if it's based on induction and inference to the best explanation. Here our credence that physical theory will never show it to be the case must be so much weaker than our credence that current physical theory does not succeed. So it leaves us in the agnostic state of wait and see. Second, if other ontological pictures which can accommodate the same range of observations are possible without taking on the heavy epistemic and ontological commitments of emergent dualism, then it would seem that these are preferable. It is to such pluralist ontologies that we finally turn.

Other pluralist pictures

The evidential situation for new forces or causes from emergent properties would suggest that another response is to develop a more ecumenical vision of physicalism. Are there such alternative pluralist pictures that would undercut the motivation for emergentist dualism? If a pluralist picture could be developed that would combine the following elements, it might satisfy some of the demands which engendered emergentism in the first place: supervenience physicalism, anti-reductionism, an interventionist account of causation, and, consequent on this, a rejection of CCP. Adherence to supervenience physicalism is a necessary condition of physicalism, whilst the rejection of reductionism is key to the autonomy of higher-level phenomena.

Very roughly, an interventionist account of causation claims that X causes Y if and only if intervening on the values of variable X is a way of intervening on the values of variable Y (Woodward 2008). In this case, we can say that X is a control variable for Y. Note that since the notion of an intervention used to explicate causation is itself a causal notion, this is a non-reductive account of causation. Armed with an interventionist account of causation, we can consider causation in the special sciences (Woodward 2008). Two things follow from the interventionist account. First, it is not the case that whenever we have an emergent or higher-level property with downward causal effects, then the basal property that the emergent property supervenes on will also be a cause of the basal effect (*pace* Kim). This is because it does not follow from the emergent property being a control variable for a basal effect that its supervenience base must also be a control variable for that effect. There are independent conditions that a control variable must satisfy that the basal property may not meet (Woodward 2008). Second, in this picture, causation is a macroscopic phenomenon rather than a microscopic phenomenon and is not found at the level of fundamental physics. If causation is seen as a macroscopic phenomenon and not a ground-level physical phenomenon, then it will not be true that there is always a sufficient physical cause, whenever there is a cause of a physical effect. So the thesis that physics is causally closed is not true – but not because there is evidence of any alien configurational forces. Interventionism allows us to capture the sense in which higher-level properties may be genuine causes, because they provide for genuine control variables.

Through an interventionist lens, emergence is more commonplace and less exotic than one might expect. Whenever we can identify control variables which are higher level, we have a case of emergence. The interventionist framework allows us to make sense of these higher-level causal claims and to see how their status as causes is not undermined from below. Thus, this way of characterising emergence gives us much of what we want from emergence – novelty, anti-reduction, and causal relevance. Seen in this light, emergence is less metaphysically dubious and more consistent with scientific practice. An interventionist account makes it clear why the question of reduction does not arise, tying the issue nicely to the existence and explanatory power of higher-level control variables. This allows for high-level causal claims to be vindicated, without any need for undue metaphysics. In this picture, we can have novel control variables without having either to introduce alien forces or reject physicalism.

An interventionist account of emergence thus allows theorists to recognise both that we have new entities which are causally relevant, and, hence, explanatorily useful, and that there is nothing mysterious about this. It allows us to navigate between the Scylla and Charybdis of theories of emergence: typical accounts of ontological emergence build in unacceptably strong assumptions, whilst epistemological accounts are unsatisfactorily weak.

References

Alexander, Samuel. *Space, time, and deity: The Gifford lectures at Glasgow, 1916–1918* (London: Macmillan, 1920).

Barnes, Elizabeth. "Emergence and fundamentality." *Mind* vol. 121 no. 484 (2012): pp. 873–901.

Beckermann, Ansgar, Hans Flohr, and Jaegwon Kim, eds. *Emergence or reduction?* (Berlin: Walter de Gruyter, 1992).

Bedau, Mark A. "Weak emergence." *Noûs* vol. 31 no. 11 (1997): pp. 375–399.

Broad, C. D. *The mind and its place in nature* (London: Kegan Paul, Trench and Trubner, 1925).

Caston, Victor "Epiphenomenalisms, Ancient and Modern." *Philosophical Review* vol. 106 (1997): 309–363.

Chalmers, David J., *The Conscious Mind: In Search of a Fundamental Theory* (Oxford: Oxford University Press, 1996).

Corradini, Antonella, and Timothy O'Connor, eds. *Emergence in science and philosophy* (Abingdon, UK: Routledge, 2010).

Crane, Tim. *Elements of mind* (Oxford: Oxford University Press, 2001).

Crane, Tim. "Cosmic hermeneutics vs. emergence: The challenge of the explanatory gap." *Emergence in Mind*, edited by Cynthia Macdonald and Graham Macdonald (Oxford: Oxford University Press, 2010): pp. 22–34.

Fodor, Jerry A. "Special sciences (or: The disunity of science as a working hypothesis)." *Synthese* vol. 28 no. 2 (1974): pp. 97–115.

Ganeri, Jonardon. "Emergentisms, ancient and modern." *Mind* vol. 120 no. 479 (2011): pp. 671–703.

Gibb, Sophie. "The causal closure principle." *The Philosophical Quarterly* vol. 65 no. 261 (2015): pp. 626–647.

Hasker, W. *The Emergent Self* (Ithaca: Cornell University Press, 1999).

Horgan, Terence. "From supervenience to superdupervenience: Meeting the demands of a material world." *Mind* vol. 102 no. 408 (1993): pp. 555–586.

Humphreys, Paul. "Aspects of emergence." *Philosophical Topics* vol. 24. no. 1 (1996): pp. 53–70.

Humphreys, Paul. "How properties emerge." *Philosophy of Science* vol. 64 no. 1 (1997a): pp. 1–17.

Humphreys, Paul. "Emergence, not supervenience." *Philosophy of Science* vol. 64 (1997b): pp. 337–345.

Humphreys, Paul. *Emergence* (Oxford: Oxford University Press, 2016).

Jackson, Frank. *From metaphysics to ethics: A defence of conceptual analysis* (Oxford: Oxford University Press, 1998).

Kauffman, Stuart. *At home in the universe: The search for the laws of self-organization and complexity* (Oxford: Oxford University Press, 1996).

Kim, Jaegwon. "'Downward causation'in emergentism and nonreductive physicalism." *Emergence or Reduction: Prospects for Nonreductive Physicalism*, edited by Ansgar Beckermann, Hans Flohr and Jaegwon Kim (Berlin; De Gruyter, 1992): pp. 119–138.

Kim, Jaegwon. *Supervenience and mind: Selected philosophical essays* (Cambridge: Cambridge University Press, 1993).

Kim, Jaegwon. *Mind in a Physical World: An Essay on the Mind-Body Problem and Mental Causation* (Cambridge, MA: MIT Press, 1998).

Kim, Jaegwon. "Making sense of emergence." *Philosophical Studies* vol. 95 no. 1 (1999): pp. 3–36.

Loewer, Barry. "From physics to physicalism." *Physicalism and its discontents*, edited by Barry Loewer and Carl Gillet (Cambridge: Cambridge University Press, 2001): pp. 37–56.

Macdonald, Graham, and Cynthia Macdonald, eds. *Emergence in mind* (Oxford: Oxford University Press, 2010).

McLaughlin, Brian. "The rise and fall of British emergentism." *Emergence or Reduction? Prospects for Non-reductive Physicalism*, edited by Ansgar Beckermann, Hans Flohr & Jaegwon Kim (Berlin: De Gruyter, 1992): pp. 49–93.

McLaughlin, Brian. "Emergence and supervenience." *Intellectica* vol. 25 no. 1 (1997): pp. 25–43.

Mill, John Stuart. *A System of Logic: Ratiocinative and Inductive* (8th ed., 1872) (New York: Harper and Brothers, Publishers, 1843).

Montero, Barbara. "Varieties of causal closure." *Physicalism and Mental Causation: The Metaphysics of Mind and Action*, edited by H.D. Heckmann and S. Walter (Exeter: Imprint Academic, 2003): pp. 173–187.

O'Connor, Timothy. "Emergent properties." *American Philosophical Quarterly* vol. 31 no. 2 (1994): pp. 91–104.

O'Connor, Timothy. "Causality, mind, and free will." *Noûs* vol. 34 no. 14 (2000): pp. 105–117.

O'Connor, Timothy, and Hong Yu Wong. "The metaphysics of emergence." *Noûs* vol. 39 no. 4 (2005): pp. 658–678.

O'Connor, Timothy, and J.D. Jacobs. "Emergent individuals." *The Philosophical Quarterly* vol. 53 no. 213 (2003): pp. 540–555.

Papineau, David. "The rise of physicalism." *The Proper Ambition of Science* vol. 2 (2000): pp. 174.

Shoemaker, Sydney. "Embodiement and behavior." *The identities of persons*, edited by Amelie Rortey (Berkeley: University of California Press, 1976): pp. 109–137.

Sosa, Ernest. "Subjects among other things." *Philosophical Perspectives* vol. 1 (1987): pp. 155–187.

Strawson, P.F. 1974. 'Self, Mind & Body', in *Freedom & Resentment and other essays*. London: Methuen. (Reprinted by Routledge 2008): pp. 186–195.

Sturgeon, Scott. "Physicalism and overdetermination." *Mind* vol. 107 no. 426 (1998): pp. 411–432.

Van Cleve, James. 'Mind-Dust or Magic? Panpsychism Versus Emergence.' *Philosophical Perspectives* 4 (1990): 215–226.

Wilson, Jessica. "Supervenience-based formulations of physicalism." *Nous* vol. 39 no. 3 (2005): pp. 426–459.

Wong, Hong Yu. "Emergents from fusion." *Philosophy of Science* vol. 73 no. 3 (2006): pp. 345–367.

Wong, Hong Yu. "Cartesian Psychophysics", in P. van Inwagen and D. Zimmerman (eds.), *Persons: Human and Divine* (Oxford: Oxford University Press, 2007): pp. 169–195.

Wong, Hong Yu. "The secret lives of emergents." *Emergence in Science and Philosophy*, edited by A. Corradini and T. O'Connor (London: Routledge, 2010): pp. 7–24.

Woodward, James. *Making things happen: A theory of causal explanation* (Oxford: Oxford University Press, 2005).

Woodward, James. "Mental causation and neural mechanisms." *Being Reduced: New Essays on Reduction, Explanation, and Causation*, edited by Jakob Hohwy and Jesper Kallestrup (Oxford: Oxford University Press, 2008): pp. 218–262.

14

EMERGENT MENTAL CAUSATION

David Robb

1. Introduction

The twentieth-century neuroscientist Roger Sperry characterized emergent mental causation like this:

> [C]onscious phenomena as emergent functional properties of brain processing exert an active control role as causal determinants in shaping the flow patterns of cerebral excitation. Once generated from neural events, the higher order mental patterns and programs have their own subjective qualities and progress, operate and interact by their own causal laws and principles which are different from and cannot be reduced to those of neurophysiology. . . . The mental entities transcend the physiological just as the physiological transcends the molecular, the molecular, the atomic and subatomic, etc. The mental forces do not violate, disturb or intervene in neuronal activity but they do supervene.
>
> *(Sperry 1980, 201)*

As described in this passage, mental causation is novel, dependent, harmonious, and unexceptional. (1) Novel: Mental phenomena "transcend the physiological" and manifest distinctive "causal laws and principles" at the psychological level. This results in an "active control role" over neural processes, a point I'll often put – even if Sperry would not – by saying emergent mental phenomena have novel *causal powers*, powers that cannot be found among (or reduced to) powers at these lower levels. (2) Dependent: Mental phenomena, while novel, nevertheless depend or "supervene" on the lower-level processes from which they emerge. (3) Harmonious: Emergent mental causation does not disturb or intervene in lower-level causal processes. (4) Unexceptional: Emergent mental causation is one instance of a phenomenon found at many levels of nature, perhaps all levels except (if there is such) the most basic.

One immediate question is whether these four features are consistent. For example, it's difficult to see how the mind's new causal powers could exist harmoniously with those at the lower levels. How could there be mental powers with "active control" over neural processes but that do not "violate, disturb or intervene in" those same processes? We might also ask whether novelty fits with dependence: mental phenomena as described here causally influence the lower-level processes from which they emerge, an apparently vicious causal circle (Kim 1999). Moreover, at least on some conceptions of dependence, any supervening mental powers must be found in the lower levels, contradicting novelty (Wilson 2011). Not surprisingly, then, much of the literature on emergent mental causation – as well as on emergence generally – has focused on whether a "strong" form of emergence is even coherent (e.g., O'Connor and Wong 2005). I will discuss some of these internal issues later, but given the narrow aims of this chapter, it will be useful to focus not so much on the viability of emergence generally, but on the prospects for emergent *mental* causation. Throughout, Sperry's characterization will provide a useful, if defeasible, framework for discussion.

2.   The autonomy of psychology

One argument for emergent mental causation starts with the frequently cited claim that psychology is an autonomous discipline. Autonomy has been a standard assumption in the study of the mind since the rise of cognitive science in the second half of the last century (Bermúdez 2014, ch. 4). Psychology is autonomous in the sense that it has distinctive kinds, laws, and styles of explanation. While mental kinds – such as belief, memory, and intention – are "realized" or "implemented" in us by biochemical states of the central nervous system, psychology abstracts away from these details to find broader, higher-level patterns (laws) that would not be visible to natural scientists working at lower levels of detail. Such patterns have turned out to be remarkably successful in explaining and predicting human behavior (Dennett 1987; Baker 1987; Antony 2007).

The autonomy of psychology has been challenged from a number of directions, but here we are more interested in how it might secure emergent mental causation. And indeed, all of Sperry's requirements seem to follow straight away.

First, autonomous psychological kinds, and their causal powers, are novel. By abstracting away from details of implementation, psychologists pick up on similarities – couched in terms of meaning, goals, and the like – that would not be considered salient or natural at the lower levels. Consider, for example, all those who believe in a liberal conception of justice. Psychologically, they have something in common, a similarity that would, in specified circumstances, cause them to act in much the same ways. Psychologists could formulate laws to capture these similarities. But a neuroscientist or physicist examining our subjects is unlikely to find such patterns, claiming that what the psychologist calls a "shared belief" is a unnatural, gerrymandered state in each person, one that moreover shows no interesting common effects across the population. In this sense, autonomous psychological states, and their causal powers, appear novel.

Second, autonomy is compatible with psychophysical dependence. Indeed, proponents of autonomy usually insist on it, often putting the point in terms of supervenience: whenever a subject is in a psychological state Ψ, that person is in some physical state Φ such that necessarily, anything in Φ is in Ψ. Novelty is preserved, as there is no hope of identifying Ψ with any candidate Φ. Any such physical state in an individual merely realizes (implements) the psychological state, and physical realizations can vary dramatically among individuals – this is the "multiple realizability" of the mental (Putnam 1980).

Third, while the psychologist picks up on novel causal patterns, these patterns are, like psychological states themselves, always implemented in lower-level causal mechanisms (Fodor 1974, 1989), resulting in the harmony required by Sperry's emergentism. Earlier I suggested that novelty may be incompatible with harmony, but the two appear compatible so long as the *way* psychological states shape neural processes (and thus produce behavior) is via their own neural realizers. That is, the causal powers of the mental work in harmony with the powers of the neural, because the former are implemented in the latter. Thus, if I volunteer for a political party because I hold a liberal conception of justice, my belief causes my behavior by virtue of my belief's realizing physical state causing that behavior. (The physical state may achieve this indirectly by causing the series of movements that realize the behavior.)

Fourth and finally, we should expect that the kind of autonomy claimed for psychology should be found at multiple levels, both above and below the psychological. There are many "special sciences" which abstract from basic physics; they include, among others, neuroscience, economics, and geology. While emergence may not be as ubiquitous as Sperry imagines – it's not clear, for example, that chemistry is emergent – it is not confined to the psychological and is found in many disciplines, resulting in an important sense in which emergent mental causation is not exceptional.

Here then is a model of emergent mental causation, one grounded in the widely accepted thesis of psychological autonomy and exhibiting all of Sperry's core features. However, there is a problem here, one that comes from the attempt to reconcile novelty and harmony. Far from saving the causal powers of mentality, our harmonious picture may deprive it of such powers, at least if these are to be new powers to downwardly shape neural processes and thereby produce behavior. The problem is this: I noted earlier that when an emergent mental state causes some lower-level effect, it does so only via the causal powers of its implementing neural state. Any other way, it seems, would involve a disruption of the sort prohibited by Sperry's harmony thesis. But this means that the powers of the mental state are found in its neural base, contrary to the thesis of novelty. We can put this in the form of a dilemma: either emergent mental causation involves powers not found at the lower level, violating harmony, or it involves powers already found there, violating novelty (cp. Kim 1993, ch. 17).

One option at this point is to give up novelty, at least in the strong form requiring downward causal powers. This can preserve much of what's in Sperry's emergentism. But without downward causation, the view loses much of its distinctive theoretical interest. Indeed, this sort of "emergentism" seems hardly distinguishable from its mainstream rival, nonreductive physicalism (Antony 2007). So perhaps the best option for an emergentist is to give up harmony: emergent mental causation alters the otherwise unbroken causal processes at the neural and other lower levels. So characterized, emergentism is clearly an empirical thesis, predicting that the biochemical processes in the central nervous system will show systematic causal disruptions not predicted by the lower-level laws and powers of the natural sciences. Some opponents of emergentism have argued that we already have enough scientific evidence to rule this out (McLaughlin 1992; Bedau 1997; Papineau 2001), but this remains controversial, and in any case, this broader evidence is evaluated by other chapters in this volume. I will thus continue my more limited strategy of looking at features of our mental lives that are encouraging signs of (i.e., prima facie evidence for) emergent mental causation. The next topic is the causal role of reasons, and following that, the causal role of consciousness. I conclude in the final section by looking more closely at whether novelty and harmony might be reconciled after all.

3. The space of reasons

We move in the "space of reasons" (Sellars 1956). We think and act for reasons, we justify what we do in terms of reasons, and we can be evaluated as rational or irrational, sensible or silly. Three days a week, a student walks the same winding route around campus, always visiting the same buildings in the same order; moreover, most of the time during this ritual, she walks backwards. What explains this pattern of behavior? She is a tour guide for the admissions department. Her being a tour guide is a reason, a good one, for her to walk this way. No doubt she would provide such a reason if asked to justify her behavior. And what might have initially seemed strange now makes sense.

For our purposes, the most important feature of the space of reasons is its normativity. Reasons rationally motivate, justify, and make sense of how we act. The student walks backwards because that's what she should do, given the task at hand. Similar points apply to mental actions such as the forming of justified beliefs (cp. Moore 2014). My reasoning that it will rain tomorrow is not just a sequence of thoughts: I conclude that it will rain because I know this is what's sensible to think in light of my evidence. If I were to conclude otherwise, I would be considered unreasonable, and the resulting belief unjustified. We do not always act or think according to the norms of rationality, but we often do, and the assumption that we move in the space of reasons is indispensable to explaining and predicting our behavior.

Why should these commonplace observations push us towards emergentism? First, it appears as if normativity plays a causal role: my student walks backwards *because* this is the sensible thing to do; I conclude that it will rain tomorrow *because* I should infer this in light of my evidence. These seem to be causal claims. But second, there is no such normativity to be found with the biochemical processes operating in our bodies: rationality, justification, and the like "have no echo in physical theory" (Davidson 1980, 231). Neural and other physical processes are mechanical, not in the old-fashioned clockwork sense, but in the sense that they are governed by brute, impersonal laws of nature. When neurons fire, for example, this is not because it's the rational or sensible thing to do. Granted, a scientist may say that some microphysical event is what "ought" to happen or what "makes sense", but such talk reflects the scientist's expectations, not the physical processes producing the event, which are normatively blind.

If normativity plays a causal role in our lives, such powers must come from a source other than the biochemical level, and this is where emergentism starts to look promising. The desired novelty is built in from the start, as the power of mental phenomena to respond to (and act on) normative constraints cannot be found in the lower levels. Furthermore, we can respect dependence: while the normative cannot be reduced to the lower levels, it is usually assumed – even by philosophers with otherwise different approaches – that the normative supervenes on the descriptive (Nuccetelli and Seay 2012). But in the interests of saving the distinctive causal powers of the mental, let us abandon Sperry's harmony: swayed by reasons, mental phenomena "break into" the otherwise smooth and mechanistic operations of neural processes so that sometimes, affected by these new powers, neurons really do fire because they should, because it's rational that they do so. (Strictly, it may be the resulting behavior, not the processes leading to it, that is the primary object of evaluation. The preceding neural processes would then be "rational" by virtue of their role in producing the behavior.) Our emergentist picture might also require that we reject or modify Sperry's claim that emergent mental causation is unexceptional, for we have so far not permitted any normative properties at the lower levels that could support a thesis of emergence for biological or chemical causation. If there is emergence at these levels, it will have to come from another source.

Our ability to think and act in the space of reasons seems to be what motivates some emergentists about mental causation (e.g., Searle 1992; Hasker 1999; Lowe 2008). But an opposing camp rejects any such picture as entailing an "intolerable intrusion" of the normative into the causally closed physical world (Kim 1993, 208). We should reject emergentism and instead "naturalize" the role of reasons in our lives. But how? Here is one influential line of thought: suppose the mechanical processes in our brains are structured by our evolutionary and learning histories to *match* the appropriate normative relations so that the space of mechanical causes is (approximately) isomorphic to the space of reasons. When it's good or rational that the tour guide walk backwards around campus, there is, if all is working properly, an internal, mechanical stand-in for this normative fact causing her to do so. On such a picture, the relevant causal powers are at the lower levels, but we nevertheless move within the space of reasons because these powers are arranged to match the normative relations in this space. This is, in broad outline, what deserves to be called the classical naturalist account of mental causation (Crane 2003; Bermúdez 2014, ch. 6); it might also be extended to nonhuman animals (Gallistel 1989) and to artificial minds (Haugeland 1985).

We should expect emergentists to object that in this picture, we merely appear rational, mechanically mimicking a trajectory through the space of reasons but without acting or think-ing *for* those reasons. I can't adjudicate this dispute here. As I suggested earlier, it is an empirical matter – to be settled by the natural sciences – whether, when we act for reasons, higher-level events causally disrupt the biochemical processes in our brains and bodies. If no such disruptions are discovered – if the physical world, or at least our part of it, is causally closed – then it looks as if downward mental causation will need to be abandoned. In that case, some form of natural-ism, classical or otherwise, will look more attractive, and we will try to find normativity within the world of natural causes (Dretske 1988; Neander 1995; Taylor 2000). Alternatively, and more drastically, we will abandon talk of moving in the "space of reasons" as prescientific and obsolete (Quine 1960; Stich 1983).

4. Consciousness

So far I have not discussed the feature of mentality that plays a starring role in the earlier quota-tion from Sperry: consciousness. Sperry writes of "conscious phenomena" and the "subjective qualities" emerging from, then shaping, the lower levels. Consciousness seems to have been what motivated many of the early emergentists, such as Broad (1925), and it's at the center of contemporary discussions of emergent mentality (in, e.g., Clayton and Davies 2006). Perhaps McLaughlin – himself no emergentist – is right that consciousness is "the last refuge of an Emergentist" (1992, 91).

Three features of consciousness make it especially attractive as a basis for emergent mental causation. First, its existence is difficult to doubt. We might be skeptical of the earlier evidence presented for emergentism – perhaps psychology is not autonomous, or reasons are not causes – but we cannot doubt that we are conscious. While consciousness may generate some mysteries, its existence is, as Descartes argued, impossible to rationally deny. Rational doubt is generated by a potential gap between how things seem and how they are, but in the case of consciousness, the reality *is* the seeming: consciousness just is the way things seem to us.

Second, there are good reasons, also originating from Descartes, for thinking that conscious-ness is novel in the sense required by Sperry's strong form of emergentism. Antireductionist arguments about the conscious mind have swayed even philosophers who are otherwise sympa-thetic to reductionism about other aspects of our mental lives (Chalmers 1996; Kim 2005). These arguments cannot be summarized here, but many of them are based on thought experiments apparently showing that the physical facts do not suffice to fix the phenomenal facts so that the latter are in this sense additions to the physical world (Gertler 2006).

Third, consciousness seems clearly to affect our behavior. I stop working and walk to lunch because I consciously experience hunger. Your conscious "a-ha" moment during a period of deep thought causes you to begin writing the correct solution to an exam problem (Koriat 2000). By consciously imagining how a large chair could be rotated, I can orient it so that it will fit through a door (Arp 2005). The causal role of consciousness may not be as extensive as we ordinarily think (Wegner 2003), but there is plenty of evidence that it affects a wide range of behavior.

These three features of consciousness jointly form powerful evidence for a strong kind of emergent mental causation. While we should, as before, let go of Sperry's harmony thesis, novelty is secured in from the start. And dependence is present as well: while consciousness cannot be reduced to its physical base, it supervenes – via the appropriate psychophysical laws – on goings-on at lower levels. What about Sperry's fourth feature, exceptionality? On the face of it, consciousness is unique to the psychological level, so that if there is emergence at lower levels, such as those of neuroscience or chemistry, or at higher levels, such as those of sociology, politics, or economics, it will have to be grounded in something other than consciousness. That said, there are views – speculative, to be sure, but intriguing nonetheless – on which consciousness is more widespread in nature than common sense would suggest. There are, for example, varieties of panpsychism on which consciousness, or something like it, is found at lower levels (Seager 2016, ch. 14). And some philosophers have explored the possibility of conscious group minds (Schwitzgebel 2015). Evaluating such views would take us far beyond the scope of this chapter, but here it is enough to note that if consciousness brings with it a strong form of emergentism, then downward mental causation could be even more widespread than imagined by Sperry's already ambitious account.

If there is mental causation of the form just sketched, this should, like emergent causation by reasons, be empirically detectable. There should be disruptions in the physiological processes occurring in the human body when, say, a conscious decision is made. If such disruptions are considered unlikely, emergentists have a venerable fall back: deny that consciousness has novel causal powers, or indeed, that it has causal powers at all. This is the epiphenomenalist's position (Jackson 1982; Kim 2005; Seager 2006). Here we keep Sperry's dependence as well as harmony, but unfortunately the resulting "vindication" of emergentism is pyrrhic, as the view loses its most theoretically distinctive feature: the downward causal powers of the mental.

5.   Can novelty and harmony be reconciled?

It appears so far that even if there is evidence for a strong form of emergent mental causation, Sperry's criteria cannot simultaneously be fulfilled. While dependence is available on all versions and unexceptionality on some, it is difficult to see how we might secure both novelty and harmony on any of them. Indeed, it seems almost trivial that mental causal powers that are novel in the relevant sense will fail to work harmoniously with the physical powers from which they emerge. An emergentist about mental causation must, it seems, choose one or the other. Choose harmony over novelty, and emergentism is drained of much of its theoretical interest. Choose novelty over harmony, and emergentism makes a bold empirical prediction about the long-term results of human physiology. But this dilemma might not be genuine. I'll conclude this chapter by looking at some ways novelty and harmony might be reconciled after all.

One way to reconcile them is by first distinguishing types of causal powers from tokens (Wilson 2011). I've assumed so far that if the mind's emergent causal powers are novel, they are of novel types, that is, powers of a sort not found in the lower levels. It's for this reason that we would expect disruptions at those levels, thereby violating harmony. But suppose instead that mental causal powers are "novel" only in the sense that they are new tokens of the same types of powers found at lower levels. That is, mental powers duplicate what's in the neurophysiological base, bringing more (token) powers on the scene, but powers of the same kind that were already present. What results is a kind of overdetermination: mentality causes behavior, but in a way that simply matches what the physical base was already doing. Mental interventions in this picture thus appear nondisruptive in accordance with harmony.

However, this attempt to reconcile novelty and harmony faces some problems. First, we should ask whether this brand of emergentism will in fact respect harmony as desired. If mental causation always involves two sets of duplicate powers producing behavior (or the processes leading to behavior), this may alter the lower levels after all. Here is an empirical conjecture: overdetermining powers, even when type-identical, always leave distinctive marks on their effects, that is, marks not left by just one set of powers. Our present emergentist – wanting to reconcile novelty and harmony – is betting against this principle. Second, it's not clear in any case that mere token novelty will deliver what the emergentist requires. Consider our earlier discussion of the causal role of reasons. There, the point of postulating emergent powers was that there's no trace of normativity at the lower levels. But if mentality merely duplicates the mechanistic powers at these lower levels, mental powers will also fail to be normative. We might say the same about the causal powers emergentists want to attach to consciousness.

An alternative set of strategies comes from E. J. Lowe, an emergentist who has in a number of works tried to reconcile novelty and harmony (see his 1992, 1996, 2003, 2008). Lowe has proposed several models of downward mental causation to achieve the desired harmony; here I consider just one. Let us avoid overdetermination and say that the token powers of the emergent mind are also of new types – that is, types not duplicating those found at the lower, implementing levels. According to Lowe, these mental powers could constrain causal processes at the neuro-physiological level without disrupting them. A central analogy here is the spider and its web. As a spider crawls along its web, the structure of the web constrains the spider's movements, yet with-out breaking into the spider's internal physiology. Similarly, emergent minds could, like the web, constrain neural pathways without violating their internal integrity. When decisions are made in the brain, multiple independent neural pathways converge in a way that would look entirely coincidental were it not for some more global causal influence guiding them to a particular end. Such an influence comes from the mind's novel powers, but someone looking at the internal workings of any given neural pathway would find it unremarkable and fully in accordance with physical law. As a bonus, we also find Sperry's dependence in this picture: just as the spider pro-duces (i.e., spins) the web that constrains its movements, so neurophysiological processes produce mentality, and with it, emergent causal powers.

The spider web analogy is suggestive, but it cannot fully illuminate the metaphysics of mental causation, at least not if such causation is to be novel and harmonious. While the web's causal powers are distinct from those of the spider, and this in sense are new, the web could not con-strain the spider's movement without exerting forces (for example, reaction forces) on the spider, thereby affecting the shape and orientation of the spider's legs, not to mention the distribution of energy within the spider's body. This kind of causal influence isn't harmonious in the required sense. (See Lowe 1996, 82–84 for more discussion web analogy and its limitations.)

Whether there are other kinds of constraining, downward causes more friendly to harmony is an open question (Gibb 2010; Paoletti and Orilia 2017). As mentioned earlier, Lowe has other models of emergent mental causation on offer, and it's also worth considering Sperry's own homespun example of downward causation: the way a rolling wheel's macro-features constrain the movement of the wheel's micro-parts (Sperry 1980, 201). In any case, how this issue is resolved is no mere technical difficulty: the nature of emergent mental causation, as well as its empirical viability, rests on it.

References

Antony, L. 2007. Everybody Has Got It: A Defense of Non-Reductive Materialism. In B. P. McLaughlin and J. Cohen, eds., *Contemporary Debates in Philosophy of Mind*. Oxford: Blackwell.

Arp, R. 2005. Scenario Visualization: One Explanation of Creative Problem Solving. *Journal of Consciousness Studies* 12: 31–60.

Baker, L. R. 1987. *Saving Belief*. Princeton: Princeton University Press.

Bedau, M. A. 1997. Weak Emergence. *Philosophical Perspectives* 11: 375–399.

Bermúdez, J. L. 2014. *Cognitive Science*. Second Edition. Cambridge: Cambridge University Press.

Broad, C. D. 1925. *The Mind and Its Place in Nature*. London: Kegan Paul.

Chalmers, D. A. 1996. *The Conscious Mind*. Oxford: Oxford University Press.

Clayton, P. and P. Davies, eds. 2006. *The Re-Emergence of Emergence*. Oxford: Oxford University Press.

Crane, T. 2003. *The Mechanical Mind*. Second Edition. London: Routledge.

Davidson, D. 1980. *Essays on Actions and Events*. Oxford: Clarendon Press.

Dennett, D. C. 1987. *The Intentional Stance*. Cambridge, MA: MIT Press.

Dretske, F. 1988. *Explaining Behavior*. Cambridge, MA: MIT Press.

Fodor, J. A. 1974. Special Sciences (Or: The Disunity of Science as a Working Hypothesis). *Synthese* 28: 97–115.

Fodor, J. A. 1989. Making Mind Matter More. *Philosophical Topics* 17: 59–79.

Gallistel, C. R. 1989. Animal Cognition: The Representation of Space, Time and Number. *Annual Review of Psychology* 40: 155–189.

Gertler, B. 2006. Consciousness and Qualia Cannot Be Reduced. In R. J. Stainton, ed., *Contemporary Debates in Cognitive Science*. Oxford: Blackwell.

Gibb, S. 2010. Closure Principles and the Laws of Conservation of Energy and Momentum. *Dialectica* 64: 363–384.

Hasker, W. 1999. *The Emergent Self*. Ithaca: Cornell University Press.

Haugeland, J. 1985. *Artificial Intelligence: The Very Idea*. Cambridge, MA: MIT Press.

Jackson, F. 1982. Epiphenomenal Qualia. *Philosophical Quarterly* 32: 127–136.

Kim, J. 1993. *Supervenience and Mind*. Cambridge: Cambridge University Press.

Kim, J. 1999. Making Sense of Emergence. *Philosophical Studies* 95: 3–36.

Kim, J. 2005. *Physicalism, or Something Near Enough*. Princeton: Princeton University Press.

Koriat, A. 2000. The Feeling of Knowing: Some Metatheoretical Implications for Consciousness and Control. *Consciousness and Cognition* 9: 149–171.

Lowe, E. J. 1992. The Problem of Psychophysical Causation. *Australasian Journal of Philosophy* 70: 263–276.

Lowe, E. J. 1996. *Subjects of Experience*. Cambridge: Cambridge University Press.

Lowe, E. J. 2003. Physical Causal Closure and the Invisibility of Mental Causation. In S. Walter and H. Heckmann, eds., *Physicalism and Mental Causation*. Charlottesville, VA: Imprint Academic.

Lowe, E. J. 2008. *Personal Agency*. Oxford: Oxford University Press.

McLaughlin, B. P. 1992. The Rise and Fall of British Emergentism. In A. Beckermann, H. Flohr and J. Kim, eds., *Emergence or Reduction?* New York: de Gruyter.

Moore, D. 2014. The Epistemic Argument for Mental Causation. *Philosophical Forum* 45: 149–168.

Neander, K. 1995. Misrepresenting & Malfunctioning. *Philosophical Studies* 79: 109–141.

Nuccetelli, S. and G. Seay, eds. 2012. *Ethical Naturalism*. Cambridge: Cambridge University Press.

O'Connor, T. and H. Y. Wong. 2005. The Metaphysics of Emergence. *Noûs* 39: 658–678.

Paoletti, M. P. and Orilia, F. 2017. *Philosophical and Scientific Perspectives on Downward Causation*. London: Routledge.

Papineau, D. 2001. The Rise of Physicalism. In C. Gillett and B. Loewer, eds., *Physicalism and Its Discontents*. Cambridge: Cambridge University Press.

Putnam, H. 1980. The Nature of Mental States. In N. Block, ed., *Readings in Philosophy of Psychology*, Vol. 1. Cambridge, MA: Harvard University Press.

Quine, W. V. O. 1960. *Word and Object*. Cambridge, MA: MIT Press.

Schwitzgebel, E. 2015. If Materialism is True, the United States Is Probably Conscious. *Philosophical Studies* 172: 1697–1721.

Seager, W. 2006. Emergence, Epiphenomenalism and Consciousness. *Journal of Consciousness Studies* 13: 21–38.

Seager, W. 2016. *Theories of Consciousness*. Second Edition. London: Routledge.

Searle, J. R. 1992. *The Rediscovery of the Mind*. Cambridge, MA: MIT Press.

Sellars, W. 1956. Empiricism and the Philosophy of Mind. In H. Feigl and M. Scriven, eds., *Minnesota Studies in the Philosophy of Science*, Vol. 1. Minneapolis: University of Minnesota Press.

Sperry, R. W. 1980. Mind-Brain Interaction: Mentalism, Yes: Dualism, No. *Neuroscience* 5: 195–206.

Stich, S. 1983. *From Folk-Psychology to Cognitive Science*. Cambridge, MA: MIT Press.

Taylor, K. A. 2000. What in Nature Is the Compulsion of Reason? *Synthese* 122: 209–244.

Wegner, D. M. 2003. The Mind's Best Trick: How We Experience Conscious Will. *Trends in Cognitive Sciences* 7: 65–69.

Wilson, J. 2011. Non-Reductive Realization and the Powers-Based Subset Strategy. *Monist* 94: 121–154.

15

EMERGENCE AND NON-REDUCTIVE PHYSICALISM

Cynthia Macdonald and Graham Macdonald

Non-reductive physicalism is the view that although all empirical entities and phenomena are physical, mental properties or kinds are irreducibly distinct from physical ones. Early contemporary versions of physicalism (such as those endorsed by brain state theorists J.J.C. Smart (1959) and Herbert Feigl (1958)) were committed to the identity of mental properties or kinds with physical ones. Such type–type identity was taken to be the earmark of reductive physicalism. The phenomenon of multiple realizability of mental properties or kinds by physical ones – the thesis, made famous by Hilary Putnam (1967), that one and the same mental property or kind may be realized in the same or in different organisms by not one but rather many distinct physical properties or kinds – was one of the early motivations for non-reductive physicalism (Putnam (1967); J. Fodor (1974)). Other proponents of physicalism were struck not only by the multiple realizability of mental properties or types but, importantly, also by the intentional nature of mental phenomena, which seemed to them to have no place in physical theory, but who, for this reason, were inclined to be eliminativist about the mental (cf. W.V.O. Quine (1960); P. Feyerabend (1963)). For present purposes this entry will focus on the version of non-reductive physicalism that has its source in the seminal work of Donald Davidson (1970) and Jerry Fodor (1974). Davidson focused attention on the phenomenon of intentionality, which he considered to be constitutive of the mental domain. He argued for a position he termed 'anomalous monism' – 'anomalous' because according to it mental events do not fall under universal, exceptionless psychophysical or psychological laws, and 'monism' because such events are nevertheless held to be identical with physical ones. The monism argued for is a physical monism because mental/physical events *do* fall under universal, exceptionless physical laws governing their causal interactions with physical events. Fodor's emphasis was on the taxonomic divergence between upper-level sciences and physics: money, for example, has no corresponding physical kind, given that monetary tokens are constituted from diverse physical items (e.g., gold, silver, paper, etc.). Both variants of the view combine token physicalism (the view that each individual mental event or phenomenon is identical with a physical one) with type or kind distinctness. Subsequent discussions of the position typically formulate the type distinctness claim in terms of properties, although Davidson himself was never comfortable with property talk.

Because non-reductive physicalism (hereafter, NRP) is a nonreductive position in the philosophy of mind, questions have arisen as to how it can legitimately claim to be a form of physicalism and how it can preserve genuine causal efficacy of the mental in the light of several theses to which physicalism seems committed. Specifically, physicalism is deemed to be committed to the theses that (1) if physical effects have causes, they have sufficient physical causes (sometimes referred to as the closure principle, or *Closure*) and (2) there is no systematic overdetermination of physical events by events distinct from and independent of them (sometimes known as *Exclusion*). Some have claimed that the irreducible distinctness of mental properties or kinds from physical ones is incompatible with a genuine physicalist monism, since the latter is committed to the thesis that everything – mental properties included – is, at root, physical. Others have claimed that NRP only awards physical causal powers to mental events, those that are sanctioned by the causal laws that govern the physical events with which mental events are identical, and that, as a result, the causal powers that make mental events *mental* can make no causal difference to the world. (This charge has taken two forms: one in the charge that mental events are causally inefficacious, and another, more challenging one in the charge that mental events are not causally potent *in virtue of* their mental properties – or *qua* mental – often referred to as the *qua* problem or the problem of the causal relevance of the mental.) Attempts to resolve the tension inherent in the position between its commitment to irreducibility, physicalism, and genuine causal efficacy of the mental has led to recent interest in doctrines of emergence in the philosophy of mind.

This entry begins by setting out the main objections to NRP formulated in terms of the combination of a token identity and a type distinctness claim, and some attempts that have been made by proponents of the position to refute these objections by developing the metaphysics of the view. Dissatisfaction with two well-known attempts to develop the metaphysics in a way that satisfactorily addresses these objections has motivated weaker forms of NRP which do not commit to a token identity thesis but rely instead on a notion of physical realization. Although the weaker forms of NRP do not suffer from problems stemming from commitment to token identity, they are no less vulnerable than are the stronger versions of NRP to the charge that they either lead to reduction or to the causal impotence of the mental. Because all forms of NRP are committed to reconciling non-reducibility of the mental with a broadly physicalist world view, abandoning non-reducibility is not an option for its proponents. It is this refusal to abandon non-reducibility in the face of the 'reduction or causal impotence' charge that has shaped recent interest in emergentism.

1.   NRP, token identity, and metaphysics

Davidson's original position had two notable features: it presumed, first, that events are 'unstructured particulars', individuals with no essential intrinsic properties, whose identity conditions consist in their causal relations to other individuals of the same kind and nothing more, and second, that the position could be supplemented by a doctrine of supervenience, roughly specified as 'there can be no mental difference without a physical difference', or 'sameness with respect to the physical guarantees sameness with respect to the mental', as a way of attempting to capture a suitable relation between mental and physical types. Subsequent defences of NRP focussed on both of these features, attempting to resolve the tension inherent in the position by supplying a metaphysics of events and an appropriate doctrine of supervenience strong enough to ward off the charge that mental properties 'float free' from physical ones but weak enough to ensure non-reducibility. Development of a metaphysics of events was designed to help de-fuse the objection that the only causal powers that mental events could exercise in the physical world are the physical ones that they have by virtue of being identical with physical events governed by physical laws. And development of an appropriate doctrine of supervenience was designed to help de-fuse the objection that a world in which irreducible mental properties 'float free' from all physical ones is not a world in which physicalism could be true.

Two main metaphysical theories of events emerged as ones favoured for resolving the causal tension inherent in NRP: one was anchored in what is known as the Property Exemplification Account (or PEA), a position pioneered by Jaegwon Kim (1976), and the other was anchored in a trope view of events whose origins stem from the work of D.C. Williams (1966). According to the former, events are exemplifications of act-or event-properties at (or during intervals) of times in objects, where properties are understood as universals, entities capable of multiple exemplification in many entities at the same and at different times. They can be represented schematically in terms of structures of the form '$[x, P, t]$' (known as their canonical descriptions), where x, P, and t are variables ranging over objects, properties, and times, respectively, and two conditions govern them, an existence condition and an identity condition, specified thus:

> Existence Condition: Event $[x, P, t]$ exists if and only the object x has property P at time t.
> Identity Condition: Event $[x, P, t]$ is identical with event $[y, Q, t']$ if and only if $x = y$, $P = Q$, and $t = t'$.

According to Kim (1976), mental properties are constitutive properties of mental events. As a result, in Kim's view the property identity requirement on mental and physical events conflicts with the irreducibility thesis of NRP and so with NRP itself.[1]

Those who have made use of the PEA developed it differently from Kim but consistently with the two noted conditions. Specifically, Kim made two important but contestable assumptions that are not part of the core commitments of the PEA. The first is that (monadic) events have just one constitutive property, and the second is that the mental properties of mental events are constitutive properties of them. The first is rejected by, for example, Lawrence Lombard's (1986) theory of events, which takes events to be the exemplifying of first one, than another, of a pair of incompatible properties in the same 'quality space' at or during intervals of times. And both the first and the second are rejected by Cynthia and Graham Macdonald (1989, 1995, 2006), who maintain that mental properties are not constitutive properties of mental events, but rather supervene on physical properties that are constitutive properties of those events and are co-instantiated with them (i.e., there is a single instancing or exemplifying of two properties, one lower level and the other – mental – one higher level). It is this departure, they maintain, that makes possible the combination of the PEA with a token identity thesis.

The second departure from Kim's way of developing the account is one that can be appealed to not only by those who rely on the PEA metaphysics of events but also, in a slightly different way, by those who subscribe to a trope metaphysics of events in order to resolve the causal tension inherent in NRP and, more specifically, to resolve the threat of epiphenomenalism. According to one well-known version of trope metaphysics, known as the Classic Account, tropes, or 'particularized' properties or qualities, such as this particular whiteness of this white cup, or this particular weight of this particular apple, are the foundations of all things – ordinary particular things and phenomena as well as properties understood as types that may be possessed by many different particulars. Individual objects such as the white cup sitting on my desk are bundles of compresent tropes. And events, in this account, are either tropes or bundles of compresent tropes. Property types are derivative entities whose natures are both dependent on and individuated as the types of properties they are by the tropes that fall under them. In this account, property types are classes of exactly resembling tropes. Physical ones are classes of exactly resembling physical tropes. Mental ones are higher-level properties: they are classes whose members are physical tropes that fall into them in virtue of falling into distinct, lower-level classes of exactly resembling physical tropes. Given that NRP is committed to the multiple realizability of mental types (in this context, the thesis that different members of a given class of mental tropes may fall into many wholly distinct classes of exactly resembling physical tropes), the class of mental tropes has as members physical tropes that are not exactly resembling.

Two charges have repeatedly been voiced against both ways of developing the metaphysics of the version of NRP committed to token identity. One is that NRP leads to causal impotence of the mental. The other is that the position is inconsistent.

Consider the first charge. According to both metaphysical accounts, there is a distinction to be drawn between mental tokens and mental types or properties. And, according to both, mental tokens are the primary bearers of causal efficacy. One, weaker, version of the causal impotence charge is that mental events are not causally efficacious. This version of the charge can be addressed by both types of metaphysics. Epiphenomenalism is avoided by the adoption of an additional commitment, and this is to a version of what is known as the Co-instantiation Thesis (Macdonald and Macdonald 1986, 1995; Whittle 2007). The claim by the proponent of the PEA is that two properties, one mental and one physical, can be co-instantiated in a *single* instance, that is, that there can be just one instancing of two distinct properties. Examples where this seems most plausible are cases of determinates and their associated determinables. A scarlet cardinal bird is both an instance of the determinate colour scarlet and an instance of the determinable red, but it would be very implausible to suppose that these are distinct instances in the cardinal. In instancing scarlet, the cardinal just does instance red. On the version of the PEA developed by those such as the Macdonalds (1995, 2006), mental properties are higher-level properties whose instancings just are (i.e. are identical with) instancings of lower-level, physical properties that are constitutive of events. It is in virtue of the Co-instantiation Thesis that the token identity claim is true. In a similar vein, the proponent of the trope account will claim that the token identity claim just is the claim that mental tropes and physical tropes are identical (irrespective of whether events are tropes or whether events are bundles of tropes). Since tropes themselves are particularized properties, the Co-instantiation Thesis amounts to the claim that one and the same trope is both a mental trope and a physical trope (i.e., there is just one trope), and that one trope falls into two classes, a class of physical tropes, and a higher-level class, a class of mental tropes (cf. Whittle 2007).

This may satisfactorily address the causal efficacy charge at the level of events. However, a stronger version of the causal impotence objection remains, one which questions whether either metaphysics can award mental events causal efficacy *in virtue of* being mental (the *qua* problem). On both accounts, mental events are deemed capable of bringing about physical effects because they are identical with physical events that cause those effects. This is essential to the NRP position under discussion because it is a form of physicalism, and physicalism is deemed to be committed to (1) *Closure* and (2) *Exclusion*. The charge is that neither the PEA nor the trope view can account for a mental event's being causally effective in virtue of its mental properties. On the PEA, it is claimed that the identity of property instancings combined with (1) and (2) has the consequence that it is in virtue of being instancings of physical properties that mental events are causally effective. With the trope account as developed earlier, the claim is that it is only in virtue of falling into physical classes that mental tropes are causally effective given (1) and (2). If this criticism succeeds, it pushes the proponent of NRP toward one of two positions: either mental properties are, *qua* mental, causally impotent (risking sacrificing non-reduction), or the causal powers of mental properties are more independent of the causal powers of physical ones than supervenience allows (suggesting an emergentist position similar to that of C.D. Broad (1925) and J.S. Mill (1843)).

Both charges have been responded to by proponents of these accounts. Those who favour the PEA argue that causal potency of mental properties is secured by virtue of a combination of Co-instantiation and two further theses concerning property relations: one a supervenience thesis that relates higher-level mental properties to the physical properties that they supervene on and with which mental property instances are co-instantiated, and a dependency thesis according to which a higher-level property can be instanced just by instancing a lower-level property on which it supervenes and which realizes it (this is the intuition underlying the determinate-determinable relation, but in more general form) (Macdonald and Macdonald 2006). Given that supervenience is not a reductive relation between properties but permits multiple realizability of the higher-level by lower-level subvening ones, mental explanations are distinctive causal explanations because they invoke distinctive causal patterns of relations between mental properties that mis-match the patterns of relations of physical properties on which they supervene and depend. As a result, although it is true that mental events are causally effective in virtue of their physical properties, they are not causally effective *only* in virtue of their physical properties. This is not to say that mental properties contribute causal powers to the events that have them that no physical property has, for that would be a breach of *Closure*. It is to say that mental/physical events fall into at least *two* patterns of causal relations, one of which does not reduce to the other, because the causal powers of mental properties do not match the causal powers of any lower-level, physical property (some, e.g., Macdonald and Macdonald (2010), claim that this is a form of 'weak' emergence).

Those who favour the trope account argue that mental and physical tropes are causally effective in virtue of being the tropes they are, not in virtue of falling into the classes they do. Because classes are dependent on tropes, no trope is causally effective in virtue of falling into any class. However, even if this is correct, it doesn't follow that there isn't a *qua* problem that arises at the level of tropes. For any given mental/physical trope, we can ask, in virtue of what does that trope have the effects it does? Suppose that mental event trope M1, which is identical with physical event trope P1, causes physical effect trope B1. We can ask, is it in virtue of event M1/P1's being an M1-ing that it causes B1, or is it in virtue of event M1/P1's being a P1-ing that it causes B1? And that just is the *qua* problem analogue for tropes. If, in order to respect *Closure* and *Exclusion*, we must answer that it is in virtue of M1/P1's being a P1-ing, then it seems we must conclude that that trope is not causally effective *qua* mental.

Some trope theorists (e.g., Robb 2013; Heil and Robb 2003) respond by saying that tropes are not entities that *have* causal powers; a fortiori they are not entities that cause in virtue of being entities that have mental powers or physical powers. So it is a mistake to think that being an M1-ing (/being a P1-ing) is a property in virtue of which event trope M1/P1 causes effect trope B1. Rather, tropes *are* (identical with) powers: event trope M1/P1 just is an M1/P1 power, and it causes what it does in virtue of being *that* power. *Qua* questions are illegitimate here because powers do not have higher-level properties. However, if this resolves any residual *qua* problem, it seems to do so by identifying mental with physical powers and ruling out as illegitimate any features or respects in virtue of which they are capable of discharging their powers, thereby eliminating whatever difference mental powers can make to the world that physical ones do not.

The second primary objection that has been voiced against the two main ways of developing the metaphysics of NRP committed to token identity is that they are inconsistent. Specifically, it has been claimed that the commitment to a version of the Co-instantiation Thesis that lies at the heart of both the PEA and the trope metaphysics is incompatible with the non-reductive commitment of NRP. If this objection succeeds, non-reductive physicalists of this kind are faced with a dilemma; in order to be consistent, they must either abandon irreducible distinctness of mental and physical types, or they must abandon token identity. Either spells the death of the position.

Consider first the PEA. Given that NRP's need to avoid the consequence that mental properties 'float free' from physical ones and so to at least some kind of supervenience thesis relating the physical, realizing, properties to the mental, realized ones, commitment to token identity seems to carry with it commitment to something like the Causal Inheritance Principle (Kim 1993a), according to which each instance of a mental property that supervenes on and is realized by some physical property or properties has the same causal powers as the physical instance with which that mental instance is identical. That is to say, instances of realized mental properties inherit all of the causal powers of instances of physical properties that realize them and with which they are identical. Kim's version of this principle originally understood it as stating that instances of mental properties inherit all of the causal powers of *all* of the physical instances that might realize them (call this 'full causal inheritance'). And he took this to lead to a problematic kind of overdetermination of causal powers by mental properties that threatens and undermines the non-reductive part of NRP.

Proponents of the version of NRP who commit to the PEA and Co-instantiation Thesis might attempt to rescue the position from the threat of reduction by insisting that mental properties are multiply realizable, mis-matching with all of their actual and possible physical realizers precisely because mental properties themselves bear distinctive causal potentialities to one another – ones that are not and cannot be echoed in the patterns of causal potentialities their physical realizers bear to one another. One such pattern of causal potentialities is what one might call the 'rational' one; events that are instantiations of mental properties such as intentional ones (e.g., believes that p, desires that q, for some propositional contents p, q) do not just cause other instantiations of mental properties and intentional behaviour. They make their effects intelligible in ways other than in terms of mere instances of regular, causal-cum-statistical nomological patterns. But if this is correct, mental and physical properties are irreducible not simply because they are *different* types of properties; they are irreducible because they are *incompatible* properties. And now the threat to the position is that if such properties are indeed incompatible with one another, the Co-instantiation Thesis cannot be true; mental and physical properties could not be co-instantiated in a *single* instance (Schneider 2012). (A different but related objection argues that although the Causal Inheritance Principle is true, it can be respected by what is known as the powers-based subset account of the relation between mental properties and their physical realizers, a view to which we return in Section 2). According to this, mental properties do indeed have causal powers that are different and distinctive, since each mental property has causal powers that are a proper subset of the causal powers of each and every one of the physical properties that might realize it. So mental properties are irreducible to their physical realizers. But precisely because of this, co-instantiation is ruled out; mental properties and their physical realizers cannot be co-instantiated in a single instance because the physical realizers have more causal powers than any mental realized property, powers which the mental instance would also need to have if it were to be co-instantiated with its physical realizer (Shoemaker 2001; Wilson 2011).

The trope account, inasmuch as it is a version of NRP, suffers from a similar charge, since according to it mental and physical types are held to be irreducibly distinct compatibly with trope identity and so with a version of the Co-instantiation Thesis. According to the Classic account, types are classes of exactly resembling tropes. So a given mental type, say, excitement, if it is to be a genuine type, must be a class of exactly resembling physical tropes (for recall that mental types are higher-level classes whose members fall into them in virtue of falling into lower-lever classes of physical tropes). But then, the argument goes, given that mental types are classes of exactly resembling physical tropes, mental types must be physical types (Gibb 2004). The only way to avoid this consequence would be to insist that because of multiple realizability, the physical tropes that fall into a given mental class are not exactly resembling. However, this now looks incompatible with trope identity. For it seems to have the consequence that one and the same trope, in being both fully mental and fully physical, both *does* exactly resemble certain other physical tropes and *does not* exactly resemble those same physical tropes.

2. From realization physicalism to emergence

Many proponents of a weaker form of NRP focus on the realization relation between mental and physical properties and their instances. Sydney Shoemaker tells us that to be realized is to be made real 'in a sense that is constitutive rather than causal' (2007: 10). Realization physicalism is the view that every property either is a physical property or is realized by a physical property; in the case of NRP the claim is that mental properties are realized by physical properties. Proponents of the position are motivated by a desire to formulate a version of NRP that (1) respects multiple realizability, (2) ensures non-reducibility of mental to physical properties, and (3) awards genuine causal potency to mental (and more generally, to higher-level) properties. Although there are a number of versions of realization physicalism (Poland 1994; Kim 1998;[2] Levine 2001; Melnyk 2006), our focus will be on a very influential and well-known one: the powers-based subset view of the relation between higher-level properties and their lower-level realizers.

In a number of seminal works, Shoemaker (2001, 2007) and Jessica Wilson (1999, 2011) (see also Clapp 2001; Watkins 2002) have argued that the most effective way for NRP to view the relation between mental and physical properties is to view mental properties as realized by physical ones and to understand realization in terms of an inclusion relation between sets of causal powers of the properties that have them. Mental properties are thus construed as having causal powers that are proper subsets of the sets of causal powers of the physical properties that realize them. Both Shoemaker and Wilson treat the determinable/determinate relation as a case of realization more generally. If, they argue, the conditional causal powers (hereafter, simply causal powers) of any determinable property are a proper subset of the causal powers of any of its determinates, the determinable will bestow *different* causal powers from those bestowed by any such determinate, and this distinctive causal profile is thought to be sufficient to ward off threats of overdetermination.

Wilson and Shoemaker hold that the subset view is incompatible with the Co-instantiation Thesis and so with token identity, remarking that where properties have different causal powers, so, too, do their instances. The subset relation thus holds not just between the causal powers of higher-level properties and those of the lower-level ones that realize them, but it also holds between the causal powers of instances of higher-level properties and the causal powers of instances of their lower-level realizers. This is one way in which subset realization is viewed as differing from the stronger version of NRP. Shoemaker suggests that instances of mental properties are parts of instances of physical ones, which is a kind of mereological view of the relation between them (see also Gillett 2006; Pettit 1993, 2007). Wilson prefers instead to view the relation between the mental and physical instances as non-mereological; mental property instances are realized by physical property instances.[3]

Because the subset strategy takes the causal powers of mental properties to be proper subsets of the causal powers of physical ones, proponents do not subscribe to Kim's version of the causal inheritance principle (noted in Section 1), according to which instances of mental properties inherit *all* of the causal powers of the instances of physical properties that realize them and with which they are identical. This we earlier called 'full causal inheritance'. They do, however, subscribe to a weaker version of that principle which we might call 'partial causal inheritance'. According to this, instances of mental properties inherit some, but not all, of the causal powers of the instances of physical properties that realize them.

Subset realization physicalism is intended to meet all of (1)–(3) above in the following way. (1) Since the causal powers of any higher-level property form a proper subset of the powers of properties that realize them, all realizers have those powers. But they also have additional powers, and each such realizer has powers that individuate it from other realizers sharing the same subset of powers that individuate the higher-level, realized property. For example, every determinate of the colour property red has the causal powers of red. But scarlet has further, additional powers, ones that distinguish it from red and from other determinates of red, such as burgundy. Multiple realizability is a matter of lower-level properties having all the causal powers of their higher-level realized one plus additional ones not possessed by the higher-level, realized property. (2) Since the causal powers of realized properties form proper *subsets* of the sets of causal powers individuating any realizer property, no realized property (mental properties included) can be identified with any realizer property, and so, it is argued, the threat of reducibility is avoided. (3) Genuine causal potency is awarded to higher-level, realized mental properties, since the causal profile of such a property is distinct from that of any of its realizing properties. Moreover, proponents argue that there are cases where we can see that the higher-level property is implicated as cause of a given effect where no lower-level realizing property is sufficient. This can be illustrated with the following, well-worn example.

Consider two pigeons, Sophie and Alice, trained to peck at certain coloured patches. Sophie is trained to peck at red patches but not at patches of any colour other than red. Alice, on the other hand, is trained to peck at scarlet patches, but not at patches of any other shade of red. Suppose now, on a particular occasion, Sophie is presented with a scarlet patch and she pecks. Which of the two properties, red or scarlet, is the causally relevant one, the one whose instancing is causally responsible for Sophie's pecking? Shoemaker and other proponents of the subset view argue that it is the determinable, red, that is the causally relevant one in this case. Indeed, Shoemaker (2001: 31) takes difference-making considerations to show not only that red is the causally relevant property but that the *instance* of red, *rather than* the *instance* of scarlet, is the cause of Sophie's pecking:

> Red is the difference-maker here because, had Sophie been presented with patch of any other shade of red, say, burgundy, she still would have pecked. Alice, on the other hand, who pecks at the scarlet patch, would not have pecked at any patch presented to her of any other shade of red, so in her case scarlet, not red, is the difference-maker.[4]

The view that instances of realized properties are distinct from, but are realized by or are parts of, instances of realizing ones is difficult to sustain even in the relatively benign case of determinables and their determinates: having *distinct* instances of scarlet and red when scarlet is instantiated is metaphysically puzzling. More to the point, to attribute causal potency to the higher-level (determinable) property instance and not to the lower-level realizer instance (as Shoemaker and Yablo do), while looking plausible in the case of Sophie's pecking at all and only red instances, proves to be problematic in the context of NRP (cf. Wilson 2011). As we have noted, the proponents of the subset view share with other proponents of NRP a commitment to both the non-reducibility and the causal potency (autonomy) of mental properties. But recall that there are two further conditions imposed by physicalism: *Closure* and *Exclusion*. And these cause trouble for the subset view in the case of mental realization.

The problem can be put in the form of a dilemma. If the mental part (in Shoemaker's terms) of a physical instance is, by virtue of being an instance of a mental property, a *non*-physical part of the physical instance that realizes it, denying co-instantiation in order to rescue distinctive causal relevance of the mental properties has the consequence that it is the non-physical part of the physical property instance that is causally effective in bringing about the effect it does. Where that effect is physical (where there is 'downward causation'), there is a breach of *Closure*, given no systematic overdetermination (*Exclusion*). And this is a breach of the physicalist commitments of NRP. With *Exclusion* and *Closure* in place, the non-physical part of the physical property instance cannot be the causally effective one. The mental property instance is thus causally inefficacious. There is also the additional problem of seeing how an instance of a physical property can be physical if it has a non-physical part, and correspondingly, how a physical property can have causal powers that include as subsets the causal powers of a non-physical one without compromising its physicality.

Of course, nothing in the subset view prohibits construing the proper subsets of the causal powers of physical properties as themselves physical (likewise for causal powers of instances), leaving it open to the proponent of the subset view to maintain that instances of realized mental properties are themselves physical parts of instances of their realizers. But – and this is the other horn of the dilemma – it is hard to see how treating the causal powers of mental properties as themselves physical – and correspondingly treating instances of mental properties as physical parts of physical property instances – can offer any solution to the problem of mental causation for subset theorists. To do is simply to retreat to reductionism of a different form than that originally envisaged by type–type identity theorists.

Further consideration signals a deeper alignment with traditional type–type physicalist theories. On the subset view the causal powers of the higher-level, multiply realized property are included in of every one of its realizing properties' causal powers. On the present assumption we are treating these subsets of the causal powers of realizer properties as physical, which means that amongst the diverse realizers there are some causal powers of realized and realizer that are numerically identical with each other – those realizing the causal powers of the higher-level property. Given this identity, it is difficult to see why what the realized and realizer properties have in common is not itself physical. On the subset view, the mental property cannot be identical with any of the realizing physical properties, but that does not show that it cannot be identical to some other, non-realizing, physical property. The identity of (some of the) causal powers shared by the realizers with all of those of the realized property suggests that there is such a physical property, having just those causal powers, with which to identify the mental property. (See Morris (2011) for an in-depth discussion of this difficulty for subset theory.) This undermines the point of appealing to NRP as a metaphysics of mind that reconciles the causal autonomy of the mental and with it psychology as a special science with a unified, naturalistic view of mental causation in a physical world (see Kim (2010) for similar remarks). Effectively, it sacrifices non-reduction.

The difficulties encountered in formulating and defending a coherent version of NRP has prompted some to opt for different strategies, either choosing reduction (cf. Kim 1998) or sacrificing one or more of the commitments that have been the source of NRP's problems. The major tensions in both of the versions of NRP discussed in this entry are induced by the desire to maintain causal autonomy for the mental while respecting key elements of physicalism, especially the commitment to *Closure* and *Exclusion*, and with this the commitment to the causal inheritance principle. Proponents of NRP have claimed that their position is a form of weak emergence, so called because it is committed to either partial or full causal inheritance while still permitting a higher-level (e.g. mental) property to have a distinctive causal profile (Macdonald and Macdonald (2010); Wilson (2015)). The higher-level property inherits its causal powers from those of its realizing (physical) properties while having a distinctive causal profile (i.e., not having exactly the same powers as any particular physical property). Those who think this to be an unstable position and who are averse to reduction opt for a stronger form of emergence of the sort favoured by Broad (1925) and Mill (1843), one which rejects any form of causal inheritance and which gives to the higher-level property causal powers not possessed by any physical property. This rejection of causal inheritance also leads to the rejection of *Closure*.[5]

Notes

1 Note that Kim subsequently abandoned this commitment (1993b, 1998).
2 Note that Kim's version does not reject token identity.
3 Although Yablo (1992) does not explicitly endorse the subset view, he takes the realization relation to be the same as the determination relation, and he also maintains that where properties related as determinable to determinate are instantiated, their instances are not identical (hence that the Co-instantiation Thesis is false). Since he takes mental and physical properties to be related as determinables to determinates, he rejects token identity for mental and physical events. For someone who rejects the view that determination is the same relation as that of realization and who endorses co-instantiation for determination but not for realization, see Funkhouser (2014). Effectively, he sides with Yablo, Shoemaker, and Wilson on whether the Co-instantiation Thesis is true in the psychophysical case.
4 The main idea in a difference-making account of causation is that if we manipulate the values of the one variable and find that this is accompanied by systematic changes in the values of the second variable, then we can take it that there is a causal relation between the values of the variables. See List and Menzies (2009) for further details.
5 See Jessica Wilson (2015) for a thorough survey of the various options.

References

Broad, C.D. 1925: *Mind and Its Place in Nature*. New York: Harcourt, Brace & Company.

Clapp, L. 2001: 'Disjunctive Properties; Multiple Realizations'. *Journal of Philosophy* 98: 111–136.

Davidson, D. 1970: 'Mental Events'. In L. Foster and J. Swanson (eds.), *Experience and Theory*. Amherst: Massachusetts University Press. Reprinted in Davidson 1980/2001: *Essays on Actions and Events*. Oxford: Oxford University Press, pp. 207–227.

Feigl, H. 1958: 'The "Mental" and the "Physical"'. In H. Feigl, M. Scriven, and G. Maxwell (eds.), *Concepts, Theories and the Mind-Body Problem* (*Minnesota Studies in the Philosophy of Science* 2). Minneapolis: University of Minnesota Press, pp. 370–497.

Feyerabend, P. 1963: 'Mental Events and the Brain'. *Journal of Philosophy* 40: 295–296.

Fodor, J. 1974: 'Special Sciences: Or the Disunity of Science as a Working Hypothesis'. *Synthese* 28: 77–115.

Funkhouser, E. 2014: *The Logical Structure of Kinds*. Oxford: Oxford University Press.

Gibb, S. 2004: 'The Problem of Mental Causation and the Nature of Properties'. *Australasian Journal of Philosophy* 82: 464–476.

Gillett, C. 2006: 'The Hidden Battles Over Emergence'. In P. Clayton and Z. Simpson (eds.), *Oxford Handbook of Religion and Science*. Oxford: Oxford University Press, pp. 291–296.

Heil, J., and Robb, D. 2003: 'Mental Properties'. *American Philosophical Quarterly* 40: 175–196.

Kim, J. 1976: 'Events as Property Exemplifications'. In M. Brand and D. Walton (eds.), *Action Theory*. Dordrecht: D. Reidel, pp. 310–326.

Kim, J. 1993a: 'The Non-Reductivist's Troubles with Mental Causation'. In J. Heil and A. Mele (eds.), *Mental Causation*. Oxford: Oxford University Press, pp. 189–210.

Kim, J. 1993b: *Supervenience and Mind*. Cambridge, MA: Cambridge University Press.

Kim, J. 1998: *Mind in a Physical World*. Cambridge, MA: Cambridge University Press.

Kim, J. 2010: 'Thoughts on Sydney Shoemaker's *Physical Realization*'. *Philosophical Studies* 148: 101–112.

Levine, J. 2001: *Purple Haze*. Oxford: Oxford University Press.

List, C., and Menzies, P. 2009: 'Nonreductive Physicalism and the Limits of the Exclusion Principle'. *Journal of Philosophy* 106: 475–502.

Lombard, L. 1986: *Events: A Metaphysical Study*. London: Routledge & Kegan Paul.

Macdonald, C. 1989: *Mind-Body Identity Theories*. London: Routledge & Kegan Paul.

Macdonald, C., and Macdonald, G. 1986: 'Mental Causes and Explanation of Action'. *Philosophical Quarterly* 36: 145–158.

Macdonald, C., and Macdonald, G. 1995: 'How to Be Psychologically Relevant'. In C. Macdonald and G. Macdonald (eds.), *Philosophy of Psychology: Debates on Psychological Explanation*. Oxford: Basil Blackwell, pp. 4–28.

Macdonald, C., and Macdonald, G. 2006: 'The Metaphysics of Mental Causation'. *Journal of Philosophy* 103: 539–576.

Macdonald, C., and Macdonald, G. 2010: 'Emergence and Downward Causation'. In C. Macdonald and G. Macdonald (eds.), *Emergence in Mind*. Oxford: Oxford University Press, pp. 139–168.

Melnyk, A. 2006: 'Realization and the Formulation of Physicalism'. *Philosophical Studies* 131: 127–155.

Mill, J.S. 1843: *A System of Logic.* London: Longman, Green, Reader, and Dyer.

Morris, K. 2011: 'Subset Realization and Physical Identification'. *Canadian Journal of Philosophy* 41(2): 317–336.

Pettit, P. 1993: *The Common Mind: An Essay on Psychology, Society, and Politics.* New York: Oxford University Press.

Pettit, P. 2007: 'Joining the Dots'. In G. Brennan, R. Goodin, and M. Smith (eds.), *Common Minds: Essays in Honour of Philip Pettit.* Oxford: Oxford University Press, pp. 215–338.

Poland, J. 1994: *Physicalism: The Philosophical Foundations.* Oxford: Oxford University Press.

Putnam, H. 1967: 'Psychological Predicates'. (Later reprinted as 'The Nature of Mental States'). In W.H. Capitan and D.D. Merrill (eds.), *Art, Mind, and Religion.* Pittsburgh: University of Pittsburgh Press, pp. 37–48.

Quine, W.V.O. 1960: *Word and Object.* Cambridge, MA: MIT Press.

Robb, D. 2013: 'The Identity Theory as a Solution to the Exclusion Problem'. In S.C. Gibb, E.J. Lowe, and R.D. Ingthorsson (eds.), *Mental Causation and Ontology.* Oxford: Oxford University Press, pp. 215–232.

Schneider, S. 2012: 'Non-Reductive Physicalism Cannot Appeal to Token Identity'. *Philosophy and Phenomenological Research* 85: 719–728.

Shoemaker, S. 2001: 'Realization and Mental Causation'. In C. Gillett and B. Lower (eds.), *Proceedings of the Twentieth World Congress of Philosophy*, Volume 9. Cambridge, MA: Cambridge University Press, pp. 23–33.

Shoemaker, Sydney 2007: *Physical Realization.* Oxford: Oxford University Press.

Smart, J.J.C. 1959: 'Sensations and Brain Processes'. *Philosophical Review* 68: 141–156.

Watkins, M. 2002: *Rediscovering Color.* Dordecht: Kluwer Academic Publishers.

Williams, D.C. 1966: 'The Elements of Being'. In *Principles of Empirical Realism.* Springfield, IL: Charles Thomas.

Wilson, J. 1999: 'How Superduper Does a Physicalist Supervenience Need to Be?'. *Philosophical Quarterly* 50: 33–52.

Wilson, J. 2011: 'Non-Reductive Realization and the Powers-Based Subset Strategy'. *Monist* 94(1): 121–154.

Wilson, J. 2015: 'Metaphysical Emergence: Weak and Strong'. In T. Bigaj and C. Wüthrich (eds.), *Metaphysics in Contemporary Physics.* Amsterdam/New York: Rodopi/Brill, pp. 345–402.

Whittle, A. 2007: 'The Co-Instantiation Thesis'. *Australasian Journal of Philosophy* 85: 61–79.

Yablo, S. 1992: 'Mental Causation'. *Philosophical Review* 101: 245–280.

16

INTENTIONALITY AND EMERGENCE

Lynne Rudder Baker

Intentionality and emergence are both huge and contested topics in philosophy. Rather than surveying the enormous literature that each topic has generated, I'll begin with some general remarks about the idea of intentionality, followed by a brief sketch of a mainstream construal of emergence (Jaegwon Kim's). Then I shall consider four possible ways that intentionality and emergence may be related and criticize three of them. The remaining view – the one that I support – is strong emergence based on my constitution view of the material world.

Intentionality

Intentionality is the generic term for an important mental capacity and for properties that presuppose exercise of that capacity: the capacity is one that is manifested in thoughts, sentences, and actions; it is the property *of being about*, or representing, or being directed toward something. For example, my promising to call you tomorrow is about calling you tomorrow. My remembering that Vienna is in Austria is about Vienna's being in Austria. Your shooting at the target is directed toward the target. So intentional phenomena include not only explicitly mental properties – such as having and expressing propositional attitudes like hoping, remembering, deciding, believing, wishing, feeling, experiencing, pretending, and so on – but also thoughts, sentences, and actions.[1]

Human affairs bristle with intentional properties. Someone *runs for political office*; someone *votes*; someone *is under the illusion* that she hears a ghost; the team *practices* on Friday; the Senate *confirms* a nominee; someone *pursues* a PhD.

Indeed, the category of intentional properties is even broader still. Intentional properties also comprise all manner of representational properties – for example, symbols (a crucifix *represents* the death of Christ), symphonies (Beethoven's Emperor Concerto was written *in honor of* Napoleon), novels (*Crime and Punishment* is about a person's *moral transformation*), and paintings (Goya's *The Third of May* represents the *horrors of war*). Not only do these examples illustrate the extent of intentional properties, but also the very production of symbols, symphonies, novels, and paintings requires intentionality. No artifact or artwork or institution could exist in a world without intentionality.

In light of the vast range of intentional properties, I propose that we characterize as *intentional* any entity, property, state, activity, or event that could not exist, be exemplified, or occur in a world without minded people (or animals) who can represent and misrepresent things.

Let me mention some distinctive features of intentionality. Three stand out: (1) Famously, thoughts can be about things that do not exist (unicorns, Zoroaster, phlogiston). Our amazing ability to think of things that do not exist is called "intentional inexistence." There is an enormous literature on the status of nonexisting items and how we can think about them. (See, e.g., Reicher 2015). (2) One cannot freely substitute co-referring terms *salva veritate* in sentences that contain intentional terms. For example, you might be on the lookout for the bank robber without being on the lookout for the mayor, even though (unbeknownst to you) the bank robber is the mayor. (3) Intentional thoughts or sentences cannot be existentially generalized. For example, you may want to buy a car, but there may be no particular car that you want to buy. Although these are fascinating features of intentionality, we cannot address them here. The question at issue is this: Are intentional properties emergent properties? So let us turn to the idea of emergence.

Emergence

There are two major concepts of emergence. Both appeal to an idea of "levels," but each conceives of levels differently. One conception (called "weak emergence") takes levels to be determined linguistically by predicates, descriptions or concepts. The other conception (called "strong emergence") takes levels to be determined ontologically by different kinds of causally efficacious properties (Kim 2002, 16). When we say that electrical conductivity makes its first appearance at a lower level than having a metabolism, if we are talking about the descriptions "electrical conductivity" and "having a metabolism," we are assuming weak emergence, but if we are talking about the causally efficacious properties *electrical conductivity* and *having a metabolism*, we are assuming strong emergence.

In "Making Sense of Emergence," Jaegwon Kim takes the core idea of emergence to be this:

> [A]s systems acquire increasingly higher degrees of organizational complexity, they begin to exhibit novel properties that in some sense transcend the properties of their constituent parts, and behave in ways that cannot be predicted on the basis of the laws governing simpler systems.
>
> *(Kim 1999, 3)*

Kim's response to this "core idea" is two-fold. He agrees that there are properties that may legitimately be called "emergent" and that they cannot be predicted, but he also holds that these emergent properties fail to be genuine properties with novel causal powers. The failure of emergent properties to be genuine (or causally efficacious) follows from two of Kim's basic assumptions.[2]

First, Kim endorses mereological supervenience: "the doctrine that properties of wholes are fixed by the properties and relations that characterize their parts" (Kim 2002, 18). Since Kim, along with classical mereologists, takes wholes to be aggregated out of parts, with their properties and relations, he endorses a closure principle for wholes (entities): "Any entity aggregated out of physical entities is physical" (Kim 2000, 114).

Second, Kim endorses the claim that causally efficacious properties are micro-based, where a micro-based property is a property of being completely decomposable into properties of non-overlapping proper micro-parts of a whole, W, and the relations between the properties of the micro-parts (Kim 2000, 84). Kim adds to this a closure principle on the physical: "Any property that is formed as micro-based properties in terms of entities and properties in the physical domain is physical" (Kim 2000, 114–115).

Since micro-properties are uncontroversially physical properties, combining the claims about micro-properties with mereological supervenience yields a strong result: Although micro-based properties are properties of wholes, they are not on a higher level than the properties of the micro-parts into which they decompose. So an emergent property, E, is just a new way "of indifferently picking out, or grouping, first-order properties, in terms of causal specifications that are of interest to us" (Kim 1999, 17).

Indeed, Kim suggests that we "define 'higher' and 'lower' not for properties but instead for predicates and concepts" (Kim 2002, 19). In that case, there is no difficulty with supposing that two predicates, one on a higher level than the other, both designate the same property. So the hierarchy of "levels" turns out to be a hierarchy of predicates, not a hierarchy of properties at all (Kim 2002, 19).

Nevertheless, Kim does not reject emergence altogether. Rather, he refigures the levels that support emergence as linguistic, rather than ontological; he construes emergence to be a matter of predicates or concepts, not of genuine properties. So in Kim's view – "weak emergence" – emergent properties are aggregates of physical (nay, microphysical) properties described in a higher-level vocabulary.[3] In this view, emergent properties are epiphenomenal. All the causal work goes on at a microphysical level.

In sum, in a theory of weak emergence, the appearance of an ontological hierarchy of levels is dispelled as, ontologically, the levels all collapse into the basic level of microphysics. So Kim makes sense of emergence as a feature that concerns predicates or descriptions; emergence is not ontological at all.

The alternative conception of emergence is ontological. In my view – "strong emergence" – emergence is an ontological idea that is not just a matter of description. If an entity or property is emergent, it comes into being or comes to be exemplified in time, and it has causal powers. An emergent property is ontologically distinct from whatever properties it emerges from (the base properties): it is on a higher ontological level than its base properties, and is unpredictable and unexplainable from the base properties alone. So emergent properties do not reduce to microphysical properties. Thus, in my view – unlike Kim's (Kim 1999) – emergence is an engine of genuine ontological novelty.

Is intentionality emergent?

Logically, there is a spectrum of possibilities for whether intentional properties are emergent. I'll mention four possibilities (and criticize one and defend another):

1   Nonexistence: Intentional properties have never been exemplified; we are just deluded in thinking otherwise.
2   Weak Emergence: Intentional properties are emergent only in the sense that they lead to new ways to describe physical phenomena; they introduce no new causal powers (Kim's view).
3   Strong Emergence: Intentional properties add to the basic ontology and introduce new causal powers but do not introduce any immateriality into the world (my view).
4   Ontological Dualism: Intentional properties not only add to the basic ontology but also add properties whose bearers are immaterial entities (e.g., souls). Let's consider each of these construals in turn.

Eliminativism

The first alternative is that intentionality is eliminable. This alternative is known as eliminativism, a view that has a forceful proponent in Alex Rosenberg (2011b, 2011a, 2009). If, as I have urged, intentional phenomena presuppose that there are minds, then one way to express the view that intentionality is eliminable is to say that there are no intentional phenomena and never have been.

According to Rosenberg, "the foresightless play of fermions and bosons produc[e], in us conspiracy theorists, the *illusion* of purpose" (Rosenberg 2011b; emphasis his). But there really is no purpose or intentionality in the world at all: "There is no *aboutness* in reality"(Rosenberg 2011a, 191; emphasis his).

Rosenberg says: "When you consciously think about your own plans, purposes, motives, all you are really doing is stringing together silent 'sounds' . . . in your head" (Rosenberg 2011a, 208). But this claim contradicts his denial of intentionality. If there is anything that "you are really doing," you are exhibiting intentionality. If you are really "stringing together silent 'sounds,'" then you are doing something. You may be mistaken about "what you are really doing," but what is going on is not just events occurring in your brain. Stringing together silent "sounds" is itself an intentional activity; it's not the same sort of phenomenon as neurons firing.

Eliminativists like Rosenberg owe us an error theory of how we could be so comprehensively mistaken about aboutness. He should give us a clue in terms of bosons and fermions and the like as to how the illusion of intentionality is even possible; what is it about bosons and fermions, etc., that can give rise to such illusion?

If eliminativism about intentionality were correct, then it would be false that there is something that Dostoevsky's *Crime and Punishment* is about. Certainly, we cannot explain what Dostoevsky was trying to do in wholly physical terms. To try to give a complete explanation, or even an accurate description, of what is going on in *Crime and Punishment* completely in nonintentional terms would only succeed, at best, in changing the subject.

Without intentionality, we would just be mute. We may emit noises, but they would not mean anything. I shall leave it to someone else to give a successful defense of eliminativism.

Weak emergence

Consider the second alternative: intentional properties are just descriptive; they have no ontological significance. This view is dominant today in Anglophone philosophy. As we have seen, Jaegwon Kim is a good example of the descriptive view. Kim holds that every material object has a microphysical description, in terms of which it can be completely described (Kim 2000, 6). Call the property referred to by the complete microstructural description of the macro-object its "total microstructural property." "Systems with an identical total microstructural property have all other properties in common. Equivalently, all properties of a physical system supervene on, or are determined by, its total microstructural property" (Kim 2000, 7).

Kim characterizes mereological supervenience as "the doctrine that properties of wholes are fixed by the properties and relations that characterize their parts" (Kim 2000, 18). Levels are "structured by the mereological relation 'is a part of.': Entities at level n are mereologically composed of (i.e., are aggregates of) entities belonging to lower levels" viz., n-1 (Kim 1993, 337).

In the mereological view, emergent properties are properties of mereological wholes, the parts of which are the emergent base. But a mereological whole is just an aggregate of entities that are its parts appropriately related. Emergent properties are had by "aggregates of basic entities standing in an appropriate 'relatedness.' Thus, no new concrete entities emerge . . . it's only that some of these [basic] entities come to be characterized by novel characteristics not had by their constituents" (Kim 1992, 123). As I mentioned earlier, Kim takes emergence to be a matter of redescription.

So the levels generated by mereological supervenience are not ontological levels. Indeed, ontologically, they collapse all back into the lowest ("most fundamental") level. A mereological conception of levels (à la Kim) will not yield any ontological differences between emergent properties and others. Emergent properties are just differently characterized from their parts.

For example, "certain aggregates of water molecules have such emergent properties as transparency . . . properties not had by individual water molecules." What is mereologically higher is not ontologically higher. This is so because according to mereological supervenience, what is mereologically higher is just an aggregate of items at a lower level.

With this non-ontological construal of emergence, we can use intentional descriptions without adding anything to the basic ontology. All that exists in an ontologically significant way are basic (microphysical) entities. So intentional properties that (weakly) emerge must be reducible. Unreduced, they are simply at a higher *descriptive* level than their base. However, all attempts to make intentionality naturalistically acceptable by reducing intentional properties that I know of have failed (Baker 1991b, 1991a, 1989).

In any case, Kim's whole edifice rests on the pair of assumptions that I mentioned earlier – mereological supervenience and the claim that all causally efficacious properties are micro-based. I think that these assumptions are dubious, because they restrict the domain under discussion in a question-begging way.[4] Indeed, I think that they both fall to counterexamples.

For example, many of the properties of wholes that have causal efficacy do not depend on anything about parts, *pace* mereological supervenience. Consider legal properties (e.g., the property of being a suspect) or political properties (e.g., the property of being a registered voter) or financial properties (e.g., the property of being in debt). All of these properties obviously have effects in the world: the property of being a suspect has the effect that your telephone records can be subpoenaed; the property of being a registered voter has the effect of being handed a ballot; the property of being in debt has the effect of being turned down for loans. So mereological supervenience seems inadequate as a basis of causally efficacious properties. Likewise, the same properties – being a suspect, being a registered voter, being in debt – are not micro-based and hence are also counterexamples to the second dubious assumption. "Being a suspect" is not just an alternative description of a micro-based property, nor is it a description that lacks causal consequences.

Unless these non-micro-based properties are ruled out by stipulation, they are counterexamples of Kim's view: some macroproperties are causally efficacious but are neither micro-based nor supervenient on parts of wholes. This fact provides evidence that we need an ontological construal of emergence and not just a language-based construal.

Strong emergence

The third alternative gives emergent properties, including intentional properties, ontological weight. If intentionality emerges and introduces new causal powers but is not reducible to something else (*pace* (Fodor 1987, 97)), it is ontologically basic. Here is my (not uncontroversial) understanding of strong emergence:

> (SEP) Properties of kind K are strongly emergent if and only if: (1) they issue from already-exemplified base properties K' (K' $\neq$ K); (2) they begin to be exemplified during some time interval; (3) they confer causal powers on their bearers; and (4) they are not reducible to more fundamental – ultimately physical – properties.

Here is some background for (SEP). At the Big Bang, there were no exemplifications of intentional properties. Intentional properties did not appear until animals (perhaps birds or fish) began to respond to their environments in ways more flexible than rigid instinct. Contrast a bird's seemingly choosing materials with which to build a nest and the *Sphex* wasp that displays only mindlessly mechanical behavior (Dennett 1984). The bird displays a flexibility in responding to items that it could pick up; the wasp has a behavioral routine that is unalterable. The difference between the insect and the bird is each of their places on the evolutionary scale. Evolution occurs by natural selection. Since there is a time at which there is no exemplification of intentional properties and a later time at which intentional properties (e.g., the bird's) are exemplified, on the assumption that there is no outside intervention, intentional properties begin to be exemplified in the course of evolution.

Intentionality in the first instance resides at the "common-sense" level – the level at which we fall in love, children learn how to behave, young adults pursue careers, and so on. Although I believe that the common sense level is prescientific, parts of it can be made accessible to science by selecting – from the hodgepodge of reality that we share – certain properties that recur in different situations and can be measured. The sciences look for correlations (among other things), and correlations appear between items that recur. One-off properties or idiosyncratic properties are ignored. For example, sociologists do not study correlations between wearing green on Thursday and eating carrots; rather, they study correlations between wealth and philanthropy. When the correlations get robust enough, the scientists suspect that the relation in question is causal.

The top level of reality is the level at which we live our lives – the level of the midnight snack or traffic jams or comfortable offices. Although this level is not per se in any science's domain, it can partially be studied by many sciences. For example, midnight snacks may be studied by sociologists (to determine eating habits) or by biologists (to determine ingredients); although there is much to be discovered about junk food, there is also much that is not scientifically accessible or relevant (e.g., the distribution of places where it is consumed).

There is causation at every level of reality, even at the (lawless) top. Although there is no science that measures any correlations between speakers' remarks and hearers' irritation, your remark at dinner may irritate me (cause me to be irritated). So causation is not proprietary to science: we all, scientists and nonscientists alike, have causal powers (when you itch, sometimes you scratch and sometimes you don't) and make causal judgments (when you see the empty bowl and the peaceful dog, you judge that the dog ate the food).

Different levels are "home" for different kinds of causes. The relation between levels is not a matter of mereology, but of what I call "constitution." With a nod to Aristotle, I think that each entity is of some primary kind, and the entity has its primary-kind property essentially. (X's primary kind is the answer to the question: What most fundamentally is X?) Each entity known to us is constituted by something of a different primary kind at a lower level. For example, a US dollar bill is constituted by a piece of specially treated paper. There may be multiple kinds of constituters for a single constituted kind: for example, a boat may be constituted by pieces of fiberglass, by pieces of wood, or by something else. What makes the item a boat is not what constitutes it, but rather appropriate items (constituters) that are in what I call "boat-favorable circumstances," where boat-favorable circumstances include things like intended for transport on water. Not only concrete entities like us are constituted, but exemplifications of properties are constituted as well.

Since, in my view, levels are produced by the relation of constitution and constituted items are not reducible to their constituters, I take levels to be irreducible pieces of reality. Different kinds of properties first appear at different levels, for example, *spin* and *charge* appear on the microphysical, along with electrons; *having a brain* appears at biological levels; *having a good reputation* appears at social levels. New causally efficacious properties get exemplified at every level.

Here is what distinguishes my view from Kim's: in my view, levels are related by constitution and not, as Kim would have it, by mereological entities.[5] (See Baker 2007, 181–198.) If x constitutes y at t, then y is on an ontologically higher level than x, and y's primary-kind property is on a higher level than x's. Property-constitution is analogous to entity-constitution (Baker 2000,

2007, 2008, 2013). With the addition of one more term, "nonderivatively," we have a sufficient condition for one property's being at a higher level than another. Say that x has F nonderivatively if and only if x's having F is independent of x's constitution relations (e.g., my being a philosopher is independent of any facts about my body or its relation to me). Now here's the sufficient condition: property Q is a higher-level property than property P if:

There are x, y such that:

(i) y nonderivatively has Q, and
(ii) for any x, if x is a lower-level entity than y, then it is not the case that x nonderivatively has Q, and
(iii) there is some z such that: z is a lower-level entity than y and z nonderivatively has P.

Now we can make more precise the idea of constitution of property-exemplifications (x's having F at t).

(P-C) x's having F at t constitutes x's having G at $t =_{df}$

a   G is a higher-level property than F; and
b x has F at t and x has G at t; and
c x is in G-favorable circumstances at t; and
d   It is necessary that: anything that has F in G-favorable circumstances at t has G at t; and
e   it is possible that: x has F at t and x lacks G at t.

We can generalize: constitution is a contingent relation between entities and property exemplifications at different levels. No constituter entails what it constitutes. Hence, no constituted object is reducible to its constituter. What is constituted is at a higher ontological level than what constitutes it. It remains to show that intentional properties are strongly emergent. Here is a superficially simple argument for the emergence of intentionality:

1   Intentional properties begin to be exemplified in the course of evolution.
2   Intentional properties confer causal powers on their bearers.
3   Intentional properties are not reducible to more fundamental – ultimately physical – properties.
4   If 1–3, then ∴5.
5   Intentional properties are strongly emergent properties.

The argument is valid. Is it sound? Here are justifications for the premises: Line 1 follows from the discussion of strong emergence earlier. Line 2 follows from what intentional properties enable us to do and the extent to which they enable us to be agents at all. Indeed, intentional properties shape all recognizable human life and hence are causally efficacious. Line 3 does not lead to any kind of dualism, but rather to (extravagant?) pluralism. It is a bad misreading of my constitution view to take it to be dualistic. My characterization of constitution itself implies nonreduction of constituted to constituting items (Baker 2013). Line 4 just follows from (SEP) earlier. So it seems that intentional properties are strongly emergent properties that belong in the ontology – as we would suspect from the counterexamples to Kim's view.

Why is strong emergentism so unpopular? I think that there are two reasons, both stemming from the fact that strong emergence implies downward causation: (1) Downward causation has been deemed to be incoherent (Kim 1992, 136). Downward causation conflicts with certain of

Kim's (unargued for) principles, for example, "the only way to cause a [strongly] emergent property to be instantiated is by causing its emergent base property to be instantiated" (Kim 1992, 136). However, the constitution view recognizes no such principle. (2) Downward causation seems to conflict with the completeness of physics. But why, other than blind ideology, endorse the completeness of physics? We may well reject the modus ponens, "If physics is complete, then there are no strongly emergent properties," in favor of the related modus tollens, "If there are strongly emergent properties, then physics is incomplete." (Indeed, my position in Baker (2015) is tantamount to holding that the first-person perspective is strongly emergent.) We cannot make sense of human life without appealing to intentional properties, which seem not to be reducible to physics, but all we have as evidence for the completeness of physics is a (weak, I think) inductive argument. So strong emergence seems to me a better bet than the completeness of physics.

Ontological dualism

The final alternative is ontological dualism. Perhaps intentional properties have existed since the Big Bang (and hence did not emerge) and confer on their bearers irreducible causal powers. (This seems to be the view of ontological dualists, e.g., Lavassa and Robinson 2014.) Or perhaps minds, "endowed with novel causal powers" emerge from brains of biological individuals (Hasker 1999, 188), and both intentionality and teleology are "basic-level phenomena" (Hasker 1999). The emergent mind is to be part of nature, something naturally generated by natural processes; nevertheless, the mind, with its intentionality, is an "entity actively influencing the brain but distinct from it" (Hasker 1999, 193).

Although a minority view, emergent dualism is an important new form of substance dualism. Speaking for myself, however, I'll stick with the seamlessness of nature afforded by the constitution view with its commitment, not to dualism, but to ontological pluralism.

Conclusion

I conclude that intentionality is strongly emergent. Although I argued for that conclusion from my view of constitution, an argument by elimination is another path by which the same conclusion could be reached. Assuming that there are only four possible answers – eliminativism, weak emergence, strong emergence, and ontological dualism – to the question, "How is intentionality related to emergence?" there are powerful reasons against three of the possibilities: eliminativism makes illusion a mystery, weak emergence (at least Kim's version) has dubious presuppositions (for a list of six such principles, see Baker (2009, 113–114)), and ontological dualism fractures the natural world. The only remaining possibility is strong emergence.

Notes

1 In ordinary language, "intentional" refers to what is done deliberately or on purpose. As used in philosophy, "intentional" refers to many other kinds of phenomena.
2 Kim has a raft of unargued-for principles. For a detailed argument concerning Kim's assumptions, see (Baker 2007).
3 This feature of Kim's seems reminiscent of Davidson's view of mental events as physical events under nonphysical descriptions.
4 For a discussion of mereology and constitution, see (Baker 2007, 181–198, 2008, 16–22).
5 To see how the constitution view handles downward causation, see (Baker 2007, 115–119).

References

Baker, Lynne Rudder. 1989. "Instrumental Intentionality." *Philosophy of Science* 56: 303–316.

Baker, Lynne Rudder. 1991a. "Dretske on the Explanatory Role of Belief." *Philosophical Studies* 63: 99–111.

Baker, Lynne Rudder. 1991b. "Has Content Been Naturalized?" In *Meaning in Mind: Fodor and His Critics*, edited by Barry Loewer and Georges Rey, 17–32. Oxford: Blackwell Publishing.

Baker, Lynne Rudder. 2000. *Persons and Bodies: A Constitution View.* Cambridge: Cambridge University Press.

Baker, Lynne Rudder. 2007. *The Metaphysics of Everyday Life: An Essay in Practical Realism.* Cambridge: Cambridge University Press.

Baker, Lynne Rudder. 2008. "A Metaphysics of Ordinary Things and Why We Need It." *Philosophy* 83(1): 5–24.

Baker, Lynne Rudder. 2009. "Nonreductive Materialism." In *The Oxford Handbook of Philosophy of Mind*, edited by Brian P. McLaughlin, Ansgar Beckermann, and Sven Walter. Oxford: Clarendon Press.

Baker, Lynne Rudder. 2013. *Naturalism and the First-Person Perspective.* Oxford: Oxford University Press.

Baker, Lynne Rudder. 2015. "Ontology, Down-to-Earth." *The Monist* 98: 145–155.

Dennett, Daniel C. 1984. *Elbow Room: The Varieties of Free Will Worth Wanting.* Cambridge, MA: The MIT Press.

Fodor, Jerry A. 1987. *Psychosemantics: The Problem of Meaning in the Philosophy of Mind.* Cambridge, MA: MIT Press.

Hasker, William. 1999. *The Emergent Self.* Ithaca, NY: Cornell University Press.

Kim, Jaegwon. 1992. "'Downward Causation' in Emergentism and Nonreductive Physicalism." In *Emergence or Reduction? Essays on the Prospects of Nonreductive Physicalism*, 119–138. Berlin: Walter de Gruyter.

Kim, Jaegwon. 1993. "The Nonreductivist's Trouble with Mental Causation." In *Supervenience and Mind: Selected Philosophical Essays*, 336–357. Cambridge: Cambridge University Press.

Kim, Jaegwon. 1999. "Making Sense of Emergence." *Philosophical Studies* 95: 3–36.

Kim, Jaegwon. 2000. *Mind in a Physical World.* Cambridge, MA: MIT Press.

Kim, Jaegwon. 2002. "The Layered Model: Metaphysical Considerations." *Philosophical Explorations* 5: 2–20.

Lavassa, Andrea, and Howard Robinson, eds. 2014. *Contemporary Dualism: A Defense.* Routledge Studies in Contemporary Philosophy. New York: Routledge.

Reicher, Maria. 2015. "Nonexistent Objects". *The Stanford Encyclopedia of Philosophy.* Edward n. Zalta, ed.

Rosenberg, Alex. 2009. "The Disenchanted Naturalist's Guide to Reality." *On the Human.* http://onthehuman.org/2009/11/the-disenchanted-naturalists-guide-to-reality/.

Rosenberg, Alex. 2011a. *The Atheist's Guide to Reality: Enjoying Life without Illusions.* New York: W.W. Norton & Company.

Rosenberg, Alex. 2011b. "Disenchanted Naturalism." In *The Metaphysics of Evolutionary Naturalism*, 1–24. Beirut. https://nationalhumanitiescenter.org/on-the-human/2009/11/the-disenchanted-naturalists-guide-to-reality/

17

EMERGENCE AND CONSCIOUSNESS

Robert Van Gulick

Consciousness is at once the most familiar aspect of mind but also the most mysterious. It is a ubiquitous feature of our daily lives, and yet we lack a clear understanding of how it fits in to the larger world. As Thomas Nagel (1974) noted, "Consciousness is what makes the mind-body problem really intractable", and thus it is the focus of many mind–body debates. Critics of physicalism argue that it is unable to adequately explain consciousness, while physicalists have offered many specific theories that aim to do so. Whatever their degrees of success, an air or mystery remains about how the "trick might be done". Both sides of the debate agree that explaining consciousness as part of the physical world seems to involve special difficulties, and thus a wide variety of ideas and concepts have been invoked to explain the relation of consciousness to the physical, including identity, reduction, realization, panpsychism and emergence. Can the notion of emergence help us understand the nature and basis of consciousness and explain how it fits into the larger world? And would viewing it as emergent confirm or refute physicalism? The answer depends in part on how key concepts are defined. "Consciousness", "emergence" and "physicalism" can each be defined in a variety of ways, and thus the question of whether or not consciousness is emergent is ambiguous, as are its implications for physicalism.

Consciousness

The terms "conscious" and "consciousness" are used in a variety of ways that differ both in their meaning and in the sorts of items to which they are applied. Philosophers often distinguish two families of relevant concepts: those concerning *creature consciousness* and those that concern *state consciousness*. Concepts of the former sort aim to explain the distinction between those creatures or systems that are conscious and those that are not. For example, are dogs conscious? And what of fish, honeybees, trees or robots? In what sense of "conscious" are such organisms or systems conscious or not? By contrast, concepts of state consciousness aim to demarcate a boundary within our mental life between those states and processes that are conscious and those that are not. Which of our memories, perceptions or desires are conscious? And what does a mental state's being conscious consist in? What is the difference between those mental states or processes that are conscious and those that are not? Both families encompass a diversity of more specific concepts, specific ways of understanding creature consciousness or state consciousness.

Concepts of creature consciousness aim to distinguish between those organisms or systems that are conscious and those that are not. There are many such concepts, of which the following four provide a good sample.

> *Sentience.* A creature counts as conscious insofar as it is sentient. Being sentient is essential and necessary to being conscious, but whether it is sufficient is less clear insofar as sentience admits of many degrees and varieties. A creature might count as sentient insofar it as has any sensors that allow it to gather and respond to information about its environment. In this basic sense, not only would dogs and crows count as sentient, but so too would ants, clams, earthworms and perhaps even plants, bacteria and off-the-shelf drones. It seems implausible to count them all as conscious. Thus, interpreting creature consciousness in terms of sentience would seem to require some additional constraints on the quality and nature of the relevant sentience, but it is not clear just what those conditions should be.
>
> *Wakefulness.* A creature might have the capacity for sentience but still be judged unconscious if it is not actively exercising that capacity. We typically regard someone in deep sleep as unconscious and as returning to consciousness only upon awakening. Thus, consciousness is sometimes defined in terms of wakefulness or alert awareness, though questions about the exact conditions remain unclear. Wakefulness and alertness admit of degrees, and what of dreams? Should the dreamer count as unconscious because she is asleep and not aware of her current surroundings, or does dreaming itself suffice to count as being conscious?
>
> *Self-Awareness.* A creature counts as conscious insofar as it is not only aware but also aware that is aware (i.e. self-aware of its own awareness). Animals that count as sentient because they are aware of their environments might still not count as conscious insofar as they lack any reflective awareness of their own minds. Such a creature might seem to be just an unconscious information processor that lacks any reflective self-awareness of its own awareness. If we interpret this in terms of explicit conceptual self-awareness, we may wonder whether most nonhuman animals or even young children are conscious (Carruthers 2000). However, the required self-awareness might be interpreted more loosely to involve a wider range of less explicit and conceptual forms of reflective understanding (Rosenthal 1997; Van Gulick 2004). Dogs may be not able to report or comment on their mental states, but it does not mean that they lack any self-awareness of their mental life; a dog in pain is aware that it is in pain.
>
> *Phenomenal/Qualia/What It Is Like.* A creature is said to be phenomenally conscious (p-conscious) in so far as it has mental states that involve phenomenal or experiential qualities (i.e. properties involved in how things appear to us in experience). Of special note are so-called "qualia", simple qualities of which we are supposedly directly aware in experience, such as the phenomenal look of red, the smell of coffee, or the hurt of pain. However, the phenomenal can also be interpreted to include not just simple sensory qualities but also more structural or relational features of our experience such as our subjective experience of time or causality. A phenomenally conscious creature is one that has a subjective experiential life. As Thomas Nagel (1974) put it in his seminal paper, "What Is It Like to Be a Bat?", a creature is conscious just if there is *something that it-is-like-to-be* that creature, some subjective way in which it *seems from the inside.*

State consciousness also gets defined in a diversity of ways, of which the following three specific concepts provide a good range:

State one is aware of. Emphasizing the link between consciousness and self-awareness, conscious states are sometimes defined as simply those of one's own mental states *of which one is aware.* The difference between a conscious desire or memory and an unconscious one is simply that one is aware of having that desire or memory. We may be in unconscious mental states that affect our behavior, but we are not aware that we are in those states. Defining state consciousness in terms of self-awareness is especially common in so-called *higher-order theories* of consciousness according to which a mental state is made conscious by the presence of a simultaneous meta-state whose content concerns the fact that one is in the relevant first-order state (Lycan 1996; Rosenthal 1997; Carruthers 2000). My desire for coffee is a conscious desire because it is accompanied by a higher-order thought or perception about my having that desire.

Phenomenal or qualitative state. Conscious mental states might be defined as those that have phenomenal or qualitative properties, such as the experienced color of red or the subjective taste of chocolate. Other mental states might have content or carry information, but if they lack phenomenal or qualitative properties, then they would not count as conscious.

Access consciousness. Conscious mental states can be defined in terms of their supposed distinctive functional roles, in particular, the degree to which multiple personal and subpersonal systems are able to access and utilize the information or content associated with a given mental state. Ned Block (1995) thus contrasts phenomenal state consciousness with access consciousness. A mental state is access conscious just if its content is available for report, inference and widespread use by a diverse range of behavior-controlling systems. By contrast, the content or information associated with a state that is not access conscious is encapsulated and restricted to a specific and limited range of applications within the mind or brain.

Emergence

The notion of emergence can be defined in a variety of ways, but all the relevant concepts involve two key elements. On one hand, emergent entities or properties are said to *emerge from* some underlying *base* upon which they *depend,* and yet they are also said to *go beyond* what is present in the base introducing an element of genuine *novelty.* The relation between base and emergent is thus one of *dependence yet novelty.* Both conditions can be unpacked in a variety of ways, generating a range of concepts of emergence.

Concepts of emergence divide into two basic categories in terms of whether emergence is understood as an epistemological claim or an ontological claim (Van Gulick 2001). Claims of *epistemological emergence* concern what we can or cannot know or understand about the relation between the base and emergent, while claims of *ontological emergence* concern the metaphysical or logical status of the relation itself, the nature of the existential relation between base and emergent independent of what we can know or understand about it.

Both ontological and epistemological concepts of emergence occur in multiple forms that can vary in strength. Emergence concepts are often classed as *weak* or *strong* in terms of how radically they interpret the novelty aspect of emergence. In general, the strength of the emergence relation varies with the degree to which the novelty aspect is independent; the more independent the novelty, the stronger or more radical the emergence.

The nature of the novelty, as well as its degree, varies across different concepts of emergence. At a minimum, the relevant complex or system must have a specific property not possessed by its parts or components, nor by the mere sum of its components. In a case of emergence, the whole is supposed to be *more than the sum of its parts*, and in that sense *emergents* are sometimes distinguished from mere *resultants* (McLaughlin 1992). The mass of a complex or system will have a specific value different from the mass of any of its proper parts, but that greater mass is merely the additive sum of the masses of its parts and thus simply a resultant. By contrast, in cases of emergence the properties of the system or whole differ from those of their parts not merely in their specific value but also in their basic type. For example, a complex (e.g. a structured collection of atoms) may have a color even if none of its basic atomic parts has any such property. Similarly, a large collection of H_2O molecules at room temperature will have the property of being liquid, but it makes no sense to attribute liquidity to individual H_2O molecules. Liquidity may be a *micro-based* property, but it is a *macro-property* that applies only to large collections and structured wholes.

The macro level of the system may also show novel *patterns*, *regularities* and *dynamics* that conform to special science laws or models that have no general application at the underlying base level. For example, living organisms show many patterns of behavior that biologists aim to describe and explain in distinctively biological terms (e.g. population genetics) that appeal to macro biological and environmental properties and relations that do not appear in the laws of chemistry, even though the properties of living organisms ultimately depend on the base facts of their chemistry.

As a matter of actual scientific practice, the special science laws and models that describe the macro-behavior of the complex systems may be explanatorily autonomous from those of the underlying base level (Fodor 1974, 1997). Explanations at the macro and micro levels may use very different concepts and formal structures; biology, psychology and economics do not frame their explanations and laws in the language and concepts of physics. Indeed, given our human limitations, we may be unable in practice, or even in principle, to explain or derive those macro laws or regularities from the laws and facts at the base level (Van Gulick 1992, 2004). However, that need imply only a form of epistemological emergence; the higher-level laws appear emergent from our cognitive perspective.

By contrast, with respect to ontological emergence the issue is not whether we can *explain* the dependence relation, but whether the higher-level laws in fact *logically follow* from the base-level laws and facts. Do the base-level facts and laws by themselves logically entail or metaphysically necessitate the higher-level properties and laws? In cases like that of the liquidity of water, the answer is apparently yes. The macro liquidity of the water *logically supervenes* on the micro-interactions among the H_2O molecules that compose it, and those interactions are fully explained by the laws of the micro-level base. It supervenes logically because liquidity can be functionally defined at the macro level, and the micro-interactions determined by the micro-laws guarantee that the collection of H_2O molecules will satisfy those functional conditions. Thus, the liquidity of water is typically regarded as, at most, a case of weak ontological emergence. It might be regarded as emergent because the higher-level property is novel in type, but only as weakly emergent insofar as it can be fully explained in terms of base-level facts and laws that logically entail it.

The move from weak to strong ontological emergence typically involves the addition of new special emergence laws, fundamental laws that link properties at different levels as a brute nomic fact (McLaughlin 1992). These emergence laws are fundamental, in the sense that they do not derive from the base-level laws or facts. If there were such special emergence laws, then they would be an essential part of the way in which the emergent property depends upon its base. The emergent property would not follow logically from the base-level laws and facts alone, but only if the emergence laws were also added in. In that sense, the emergent property is independent of the base-level laws and facts. They in themselves do not necessitate the emergent property; more is needed, new special emergence laws. The philosophical debate about ontological emergence turns largely on whether or not any properties are emergent in that strong, more independent sense. Most physicalists deny that there are any such cases of strong ontological emergence, but other philosophers argue that cases do exist (Hasker 1999) and that consciousness provides a prime example.

Concepts of epistemological emergence can also vary in strength, though the dimension of strength is somewhat different than in the case of ontological emergence. Properties or regularities are epistemologically emergent when we are unable to predict or explain how they arise from their dependence base. Epistemological emergence is a claim about our lack of knowledge or inability to comprehend the link between base and emergent property. Thus, strength with respect to epistemological emergence can be interpreted in terms of the severity of the epistemic limitations. A claim of weak epistemic emergence might merely concern our actual current lack of understanding how the higher-level property arises from its base. *In practice*, we do not see at present how the facts and laws of the base level could give rise to the higher-level property, but perhaps we will be able to do so in the future with better theories. A stronger claim of epistemic emergence might deny that we humans will *ever be able to understand* the relevant link, perhaps because of some limitations in our cognitive structure (McGinn 1989). Going even further, one might assert that it is *in principle* impossible to explain or understand the emergent property solely in terms of base-level facts and laws. No cognitive agent, no matter its cognitive capacities, could do so.

The various forms of epistemological and ontological emergence are distinct, and one should avoid conflating them. Though some forms are no doubt linked, specific supporting arguments are needed to infer one sort of emergence from another.

Physicalism

Physicalism also occurs in many versions. In essence, physicalism is the metaphysical thesis that everything real is physical. But just what counts as the physical and in what sense everything must be constituted by the physical is less clear. Most contemporary versions of physicalism appeal to the entities and properties that fall within physics proper, that is, to whatever items are posited in our best physical theories. However, recognizing that our current physical theories may not be complete or fully correct, the claim is usually relativized to future ideal physical theories. In what sense then are other real entities that are not mentioned in physics supposed to be physical? Physics proper does not mention mountains, birds or economies, yet they all seem real. How must they be related to the strictly physical items in order to also count as physical? There are many proposals appealing to notions such as identity, composition and realization. On realization accounts, for example, the physical includes the strictly physical plus everything that is ultimately realized by the strictly physical, which may involve many stages of realization. If atoms are strictly physical, then molecules realized by atoms would also be physical. Membranes and proteins realized by molecules would count as physical, as would cells realized by membranes and proteins and organisms realized by collections of cells. Whether the physicalist appeals to realization, composition or some other relation, the bottom line is that the existence of everything real can ultimately be accounted for solely in terms of the entities, properties and laws of physics.

Consciousness and emergence

Having distinguished among varieties of emergence and types of consciousness, we can now ask which forms of consciousness might be emergent in one or another relevant sense. Several forms of consciousness appear to be at least weakly ontologically emergent, though it is far less clear if any are strongly ontologically emergent. As to epistemological emergence, again weak emergence seems plausible for several forms of consciousness, and there may also be reasons for regarding at least some types of consciousness as strongly epistemologically emergent, though the matter remains controversial. However, even if some forms of consciousness are strongly epistemically emergent, it is unclear that this would entail any conclusions about strong ontological emergence or pose any problems for physicalism.

Creature consciousness in the sense of sentience and wakefulness both appear to be at least weakly ontologically emergent. Both are novel properties of whole systems or organisms that are not properties of their parts or components. Neither eyes nor cortical neurons are sentient, though they may contribute to the sentience of a sighted organism. Moreover, the novel system-level properties give rise to distinctive patterns of activity and new dynamics that are not present in nonsentient or nonwakeful systems, and thus such properties play a key role in our practical ability to explain, predict and understand the behavior of such systems. However, neither sentience nor wakefulness per se appears to be strongly ontologically emergent. There is a good deal that we do not presently understand about both, but in so far as we think of them in terms of information processing systems, there is little reason to believe that they cannot in principle be explained in terms of the underlying structure and organization of the relevant systems and the lower-level interactions among the components that compose them.

If we interpret creature consciousness as self-awareness, a similar result seems to follow: self-awareness seems to involve weak but not strong ontological emergence. Endowing a system with reflexive awareness, the ability to monitor and regulate its own states, can change its operation and dynamics in major ways that are not present in its underlying components nor in systems without it. New models, theories and perhaps even logical structures may be needed to adequately describe and understand systems that exhibit reflective self-awareness. Nonetheless, it seems the relevant functional capacities and operations of self-aware systems are realized by the organization of their underlying non-self-aware components. In so far as reflexive self-awareness can be realized in complex systems through such functional organization, there seems little reason to regard it as strongly ontologically emergent.

Thus, at least as in so far as they are understood as information processes, self-awareness, sentience and wakefulness would all seem to fall within what David Chalmers (1995, 1996) has called the "easy problems of consciousness", which are to be distinguished from the so-called "hard problem" of explaining how phenomenal consciousness might arise from purely physical, neural or neurochemical processes. Indeed, it is our supposed inability to analyze phenomenal consciousness in solely functional terms that may seem to prevent us from explaining its existence by appealing to the organization and interaction of components lacking any phenomenal properties. Thus, phenomenal, qualitative or what-it-is-like consciousness is the form of creature consciousness most likely to lead toward stronger claims of ontological emergence. The relevant system-level property (i.e. phenomenal qualitative experience) is not only of a radically different type than those present in its physical, neural and neurochemical components, but it is difficult to see how its existence could result from those components acting solely according to lower-level laws. There is said to be an *explanatory gap* (Levine 1983, 1993), one that perhaps cannot be bridged. At least at present, we have no satisfactory story to tell about how the trick might be done, no way to make intelligible how qualitative phenomenal consciousness might be realized by the organized activity of non-phenomenal components. Indeed, some have argued that *in principle* no such explanation is possible, and thus more radical metaphysical options must be considered, including dualism and strong ontological emergence (Chalmers 1996; Strawson 2006; Nagel 2012). If it were impossible *even in principle* to explain how phenomenal consciousness is produced by the interactions of micro-level components obeying micro-level laws, then strong emergence with its appeal to special inter-level emergence laws might seem a plausible alternative. However, the possibility or impossibility of giving such an explanation remains a matter of dispute.

With respect to the various forms of state consciousness, the result is similar. Most seem to involve weak but not strong ontological emergence, though again it is the phenomenal qualitative forms of state consciousness that most often inspire claims of more radical emergence. For example, if we interpret conscious states as *mental states we are aware of*, we get a result like that we saw earlier with creature self-awareness. Self-awareness can transform systems and produce states that behave in novel ways. Nonetheless, it seems possible at least in principle to explain how it might be realized by the interactions of suitably organized lower-level mechanisms governed solely by lower-level laws. The same would seem to be true of *access consciousness*, which is defined in terms of functional features, such as broadcasting information for widespread use, that are of just the sort that seem open to explanation in terms of functional realization by lower-level mechanisms.

It is only the phenomenal and qualitative concepts of state consciousness that seem metaphysically problematic in ways that have led some to consider the option of strong ontological emergence (Hasker 1999; Silberstein and McGeever 1999). As noted earlier, there is an air of mystery about how a state with phenomenal properties might result purely from the interactions of nonphenomenal states governed solely by lower-level laws. We cannot, at least at present, see "how the trick is done", and that raises the question of whether special emergent laws might be required. The attractiveness or plausibility of strong ontological emergence thus depends in part on what one takes the status of the hard problem to be. If one believes that phenomenally conscious creatures and states cannot even in principle be explained in terms of lower-level processes and mechanisms governed solely by lower-level laws, then strong ontological emergence may constitute an attractive metaphysical option, especially if one is not inclined to full-blown mind–body dualism. Emergence may seem to offer at least the possibility of still making consciousness in some way depend on the physical even if special emergence laws are required.

Regarding epistemic emergence, weak emergence again seems likely for many forms of consciousness, and even strong epistemic emergence may apply with respect to phenomenal, qualitative and "what-it-is-like" forms of creature and state consciousness. With regard to the functionally definable forms of consciousness associated with Chalmers' so-called "easy problems" – such as sentience, wakefulness, self-knowledge, and access consciousness – there is still much we do not understand about the details of how they are produced by physical or neural substrates. Thus, at least at present, there is some degree of practical epistemic emergence with respect to some aspects of those forms of consciousness. However, despite those current practical limits, there is little reason to believe there are insurmountable *in principle* barriers to our eventually knowing or understanding how such forms of consciousness are produced or realized. It is a difficult research problem, but one that seems solvable. Thus, consciousness of those sorts seems at most weakly epistemically emergent and not strongly so. With respect to phenomenal, qualitative, "what-it-is-like" consciousness, the gap in our current understanding is far greater. It is not merely that we do not have a complete theory, but at a deeper level we do not really know what such a theory might look like. There appears to be a problematic "explanatory gap" (Levine 1983, 1993). We have a sense of staring at a blank wall with little idea of how the gulf might be bridged. How could phenomenal qualitative consciousness – like my present experience of smelling and tasting dark roast coffee as I type these words – be produced or realized by the physical activity of neurons? Though we have many promising neural and psychological theories of the functional aspects of consciousness (e.g., Koch 2012; Prinz 2012; Dehaene 2014), the phenomenal link remains mysterious, and we have as yet no intelligible story to tell about how to bridge the gap.

However, the depth of our current perplexity does not itself entail strong epistemic emergence. Perhaps our puzzlement can be relieved by future explanatory theories. The limits on our understanding may be merely practical and simply the result of our early state of research rather than an insurmountable epistemic obstacle. To the contrary, supporters of strong epistemic emergence argue that the two sides of the gap are so unlike each other that there is no possibility of making the link intelligible. Thus, they claim that if the phenomenal in fact depends upon the physical or neural, it must do so in a way that cannot be intelligibly explained, that is, it cannot be explained in a way that lets us see in more general terms *why* and *how* the link holds, rather than holding as a matter of brute fact or as the result of special primitive psycho-physical emergence laws. There is no consensus on these issues, and thus the question of whether phenomenal consciousness is strongly epistemically emergent remains open.

Moreover, even if the answer were yes, it is not clear what the implications would be for ontological emergence or physicalism. Some have appealed to the supposed strong epistemic emergence of p-consciousness to argue for its strong ontological emergence (Jackson 1982, 1986; Chalmers 1996). Though the former does not strictly entail the latter, it might give a reason in support of it, perhaps as a form of inference to the best explanation. If p-consciousness were strongly ontologically emergent, that would explain why it is strongly epistemically emergent. Strong ontological emergence would entail and thus explain strong epistemic emergence.

However, whether it would be the best hypothesis to explain strong epistemic emergence would depend on the alternatives. Are there better options for explaining strong epistemic emergence that do not appeal to strong ontological emergence? One could, of course, be an ontological dualist about the phenomenal and the physical. Dualism would also explain our epistemic limits: we cannot explain the phenomenal in terms of the physical because they are separate. However, most contemporary philosophers and scientists would prefer to avoid dualism and its extra ontological commitments if possible.

An attractive and less radical option might be some version of nonreductive physicalism (NRP). According to NRP, the mental, including phenomenal consciousness, is ontologically physical in that it is *fully realized by* underlying physical processes and mechanisms. Realization in the relevant sense is a *constitutive* relation in that the mental features and properties are not merely *caused by* but also fully *constituted by* the underlying physical structures and are thus in a strong ontological sense *nothing over and above* the physical.

However, the nonreductive aspect of NRP denies that the concepts and resources of physical or neural theories suffice to describe or understand all that we want to about the mental. Indeed, the supporters of NRP argue that in many scientific cases the representational and conceptual resources of lower-level theories do not suffice to allow one to fully describe and understand all the higher-level phenomena that are constituted or realized by lower-level structures and processes. We should not in general expect that higher-level theories can be conceptually reduced to lower-level theories or that higher-level concepts or contents can always be translated into lower-level ones (Fodor 1974, 1997; Van Gulick 1992).

Moreover, some versions of NRP explain why we should not expect such conceptual or representational reductions even when the higher-level phenomena are fully realized by lower-level structures (Van Gulick 2011). The key move in such accounts is an appeal to the pragmatic and contextual nature of understanding and to the view that knowledge is not simply a matter of correctly picturing reality but of being able to successfully engage it as situated epistemic agents. On such a view, theories and models are better thought of as cognitive tools rather than as mere pictures of reality. Given the affordances available to us as the embedded cognitive agents that we are, we cannot use the "tools" available from physics or neuroscience to engage p-consciousness in the ways that duplicate all those provided by our first-person mental and phenomenal concepts. Thus, there are forms of understanding that cannot be achieved even in principle through the use of physical science theories and models. Such versions of NRP thus explain the strong epistemic emergence of p-consciousness as a predictable and expected result of the essentially pragmatic and contextual nature of knowledge and understanding. Rather than conflicting with strong epistemic emergence as other versions of physicalism do, NRP and strong epistemic emergence are mutually supporting views. Each predicts and explains the other.

To sum up, it seems that most forms of consciousness are weakly ontologically emergent. Most forms do not seem strongly ontologically emergent, though some philosophers argue that phenomenal consciousness is, especially if they are inclined to regard the so-called "hard problem" as unsolvable. With respect to epistemic emergence, again most forms of consciousness are to some degree weakly epistemically emergent, and some forms, especially p-consciousness, may also be strongly epistemically emergent. However, even if they are strongly epistemically so, that in itself does not support any claims about strong ontological emergence. Though the latter entails the former, the reverse does not hold. Nor does the latter obviously provide the best explanation of the former. In particular, nonreductive physicalism provides a plausible means of accommodating strong epistemic emergence while remaining ontologically physicalist in a robust sense.

References

Block, N. 1995. "On a confusion about the function of consciousness." *Behavioral and Brain Sciences*, 18: 227–247.

Carruthers, P. 2000. *Phenomenal Consciousness*. Cambridge: Cambridge University Press.

Chalmers, David. 1995. "Facing up to the problem of consciousness." *Journal of Consciousness Studies*, 2: 200–219.

Chalmers, David. 1996. *The Conscious Mind*. Oxford: Oxford University Press.

Dehaene, S. 2014. *Consciousness and the Brain*. New York: Viking.

Fodor, Jerry. 1974. "Special sciences." *Synthese*, 28: 77–115.

Fodor, Jerry. 1997. "Special sciences: Still autonomous after all these years." *Philosophical Perspectives*, 11: 149–163.

Hasker, William. 1999. *The Emergent Self*. Ithaca, NY: Cornell University Press.

Jackson, Frank. 1982. "Epiphenomenal qualia." *Philosophical Quarterly*, 32: 127–136.

Jackson, Frank. 1986. "What Mary didn't know." *Journal of Philosophy*, 83: 291–295.

Koch, Chiristof. 2012. *Consciousness: Confessions of a Romantic Reductionist*. Cambridge, MA: MIT Press.

Levine, Joseph. 1983. "Materialism and qualia: The explanatory gap." *Pacific Philosophical Quarterly*, 64: 354–361.

Levine, Joseph. 1993. "On leaving out what it's like." In M. Davies and G. Humphreys, eds. *Consciousness: Psychological and Philosophical Essays*. Oxford: Blackwell.

Lycan, William. 1996. *Consciousness and Experience*. Cambridge, MA: MIT Press.

McGinn, Colin. 1989. "Can we solve the mind: Body problem?" *Mind*, 98: 391, 349–366.

McLaughlin, Brian. 1992. "The rise and fall of British emergentism." In A. Beckermann, H. Flohr, and J. Kim, eds. *Emergence and Reduction?: Prospects for Nonreductive Physicalism*. Berlin and New York: De Gruyter, pp. 49–93.

Nagel, Thomas. 1974. "What is it like to be a bat?" *Philosophical Review*, 83: 435–456.

Nagel, Thomas. 2012. *Mind and Cosmos: Why the Materialist Neo-Darwininan Conception of Nature Is Almost Certainly False*. Oxford and New York: Oxford University Press.

Prinz, Jesse. 2012. *The Conscious Brain*. Oxford: Oxford University Press.

Rosenthal, David. M. 1997. "A theory of consciousness." In N. Block, O. Flanagan, and G. Guzeldere, eds. *The Nature of Consciousness*. Cambridge, MA: MIT Press.

Silberstein, Michael and John McGeever. 1999. "The search for ontological emergence." *The Philosophical Quarterly*, 49: 182–200.

Strawson, Galen. 2006. *Consciousness and Its Place in Nature: Does Physicalism Entail Panpsychism?* Exeter: Imprint Academic.

Van Gulick, Robert. 1992. "Nonreductive materialism and intertheoretical constraint." In A. Beckermann, H. Flohr, and J. Kim, eds. *Emergence or Reduction?: Prospects for Nonreductive Physicalism.* Berlin and New York: De Gruyter, pp. 157–179.

Van Gulick, Robert. 2001. "Reduction, emergence and other recent options on the mind-body problem." *Journal of Consciousness Studies*, 8: 9, 1–34.

Van Gulick, Robert. 2004. "HOGS (Higher-Order Global States): An alternative higher-order model of consciousness." In R. Gennaro, ed. *Higher-Order Theories of Consciousness.* Amsterdam & Philadelphia: John Benjamins, pp. 67–92.

Van Gulick, Robert. 2011. "Non-reductive physicalism and the teleo-pragmatic theory of mind." *Philosophia Naturalis*, 47/48: 1–2, 103–123.

18

EMERGENCE AND PANPSYCHISM[1]

John Heil

> On the one hand I have a clear and distinct idea of myself, in so far as I am simply a thinking, non-extended thing; and on the other hand I have a distinct idea of body, in so far as this is simply an extended, non-thinking thing.
>
> *(Descartes,* Meditation *VI)*

> I therefore define [matter] to be a substance possessed of the property of *extension* and of *powers of attraction or repulsion.* And since it has never yet been asserted, that the powers of *sensation* and *thought* are incompatible with these (*solidity*, or *impenetrability* only, having been thought to be repugnant to them) I therefore maintain that we have no reason to suppose that there are in man two substances so distinct from each other, as have been represented.
>
> *(Priestley, Disquisitions Relating to Matter and Spirit, Introduction, ii)*

Introduction

How might someone who wants to take both physics and conscious experiences seriously reconcile the two? Contemporary defenders of panpsychism argue that there are two options. First, consciousness might be an emergent phenomenon, something that 'arises' when you put enough of the right kinds of particles together in the right way. This would mean that consciousness is an addition to the physical universe, an apparently nonphysical add-on. Second, consciousness, far from being an add-on, might be present at the root level of the physical realm: panpsychism.

Panpsychists contend that the first option, in addition to rendering consciousness epiphenomenal, involves an unacceptably mysterious form of emergence. Under the circumstances, anyone who hopes to reconcile consciousness and physics should opt for panpsychism. This line of argument is the subject of my remarks in what follows.

Emergence

Emergence was much discussed in the late 19th to early 20th centuries when influential philosophers and scientists came to doubt that classical physics had the resources to explain a range of complex chemical and biological phenomena. The popularity of emergence waned with the advent of quantum physics and its applications to chemistry (see McLaughlin 1992). Nowadays much of the focus on emergence – among philosophers, at least – concerns consciousness. The problem is thought to be not merely the apparent inability of the physical sciences as they stand to accommodate consciousness, but a profound skepticism concerning the prospects for *any* explanation of consciousness in purely physical terms.

Ironically, the original emergentists were sanguine about consciousness, believing that questions about conscious phenomena would be resolved with advances in what we now call neuroscience. Today, however, consciousness is at the epicenter of discussions of emergence. We confront the 'mystery' or 'puzzle' of consciousness, which Colin McGinn (1989: 349) expresses rhetorically: '[H]ow can Technicolor phenomenology arise from soggy grey matter?'

Presumably, consciousness is a product of the brain. When you recall a vivid conscious experience and describe its qualitative character, however, properties that figure in your description seem nothing at all like the properties of brains. Conscious experiences are qualitatively imbued. Physical properties, in contrast, are qualitatively bereft. Particles that make up the brain have mass and charge, for instance, but particles are colorless, odorless, tasteless. How are you supposed to get something qualitatively rich from something that isn't? The puzzle of consciousness.

Descartes prescribed ground rules for subsequent discussions of consciousness when he argued for what his medieval predecessors had called a 'real distinction' between mental and physical entities. Some distinctions are merely distinctions of conception or reason. The distinction between Phosphorus and Hesperus, the Morning Star and the Evening Star, is one of conception only. The terms 'Morning Star' and 'Evening Star' differ in meaning but are simply alternative ways of characterizing one and the same heavenly body, Venus. Something is 'really distinct' from something else, however, when the somethings differ *in reality*. Venus and Mars are, in this sense, really distinct, each is a *res*, a being, in its own right; either could exist without the other.

Descartes regarded the mental and the physical as kinds of substance, kinds of entity that were, as are Venus and Mars, distinct in reality, not merely in conception. This thought survives today in the idea that mental and physical *properties* are incommensurable species of property. Mental properties are nonphysical; physical properties are nonmental. Think of some familiar mental quality – the taste of peppermint, for instance, or a feeling of dizziness – and you will be hard-pressed to imagine how any physical occurrence, and in particular any soggy gray neurological occurrence, could *be* either of these.

So although few participants in the debate accept that there are distinctively mental and physical *substances*, it is widely agreed that there are distinctively mental and physical *properties*. Most hold that mental properties depend in some way on physical properties: take away parts of my brain, and you take away my conscious experiences. The question is: What brings distinctively mental properties on the scene? Where do you locate such properties in the universe comprehended by physics? The history of attempts to answer this question is a history of failures and excursions up blind alleys. Here is a brief recap.

Historical interlude

Materialism, a perennial favorite, takes the bull by the horns and simply denies that there are any mental properties. The only properties are those sanctioned by fundamental physics, or perhaps those together with properties figuring in laws belonging to the special sciences. Belief in mental properties is akin to belief in spirits. After all, whatever happens is governed by physical law. Mental – that is, nonphysical – properties could have no role in the physical world; they would bestow no evolutionary benefit. Appeals to mental properties explain nothing that could not be explained better by appeals to purely physical characteristics of brains. Ockham's Razor tells us to dispense with what we do not need, so we should dispense with mental properties.

Materialists accept Descartes's mental–physical distinction as real. They simply deny that the mental exists. This is sometimes put in terms of *reduction*: the mental is reducible to the physical; mental properties are 'nothing but' physical properties inadequately conceived.

One difficulty facing reductive forms of materialism is that it is hard to conceive of plausible candidates for reduction. Take pain. Most of us find it natural to ascribe pains to a wide range of creatures. Yet it is difficult to find a single physical property shared by every creature in pain. Of course, even if there were such a property, it need not follow that this property, or a creature's possession of it, *is* pain. Perhaps it is only *correlated* with pain, a possibility Descartes could have happily accepted.

Difficulties of this kind led philosophers with materialist leanings to embrace functionalism. Functionalists argue that it is a mistake to try to *reduce* the mental to the physical, to identify mental properties with physical properties. States of mind are not to be identified with *physical* states; states of mind are *functional* states of complex creatures. A functional state is a state defined by relations – typically causal relations – the state bears to other states. Very different kinds of physical state could play the same causal role in systems to which they belong, and hence count as functionally the same despite intrinsic physical differences.

If mental states are functional states, then the same mental state could be 'realized' in very different physical configurations. The inspiration here is the computing machine. Computing machines can be made of very different materials yet run the same programs and be in the same computational states. Think of Charles Babbage's Analytical Engine, made of steel gears and rods; early computing machines comprising rooms full of vacuum tubes; and today's transistor-based iPhones and laptops.

Return to pain. Functionalists hold that what is definitive of pain is pain's 'causal profile', pain's 'functional role'. Your being in pain is a matter of your being in a kind of state that 'plays the pain role', a state that exhibits certain typical causes and effects. You are in pain by virtue of being in a state caused by tissue damage, pressure or heat, a state that gives rise to aversive behavior, to the belief that you are in pain and to the desire to take ameliorative action. The mind is akin to a computer program representable by nodes standing in complex relations to one another. What matters is not what is *at* the nodes – that could be almost anything, really – what matters are relations nodes bear to other nodes.

One long-standing worry about functionalism is that it ignores *qualities* of conscious experiences. Suppose your being in pain is your being in a state with characteristic causes and effects. This seems to omit an essential feature of pains, their qualitative character: pains *hurt*! Whatever else a pain is, it is *painful*. And this brings us back to those pesky mental properties.

Functionalists have adopted a variety of strategies to respond to the charge that they leave out what distinguishes conscious experiences from everything else: their 'phenomenology', their singular qualitative character. Some have tried to 'reduce' the qualitative aspect of conscious states to functional relations. Really, your feeling of pain *is* just your being in a state with the right causal profile. Such attempts have won few converts. Worse, they have generated an unfortunate literature on the possibility of 'zombies', creatures exactly like us functionally but altogether lacking in conscious experiences (see Kirk 1974; Chalmers 1996).

Other functionalists have embraced *epiphenomenalism*, the idea that when you are in a state that plays the pain role, the physical occupant of this role 'gives rise to' the feeling of pain. This means that feelings of pain are going to be perfectly correlated with physical states that 'realize' pain. To you it seems that it is the pain you feel when you drop a large lump of coal on your foot that leads you to whimper and limp to the first aid station in search of an ice pack, but really the feeling is just a side effect of your going into a particular physical state, a state that causes you to whimper and seek an ice pack.

Epiphenomenalism addresses the puzzle of consciousness by allowing that conscious mental properties are real, but merely add-ons to the physical world, add-ons that themselves have no physical effects: ontological lagniappe. In this regard consciousness resembles the heat given off by a computing machine. The heat is a by-product of the activities of a complex physical system. Of course, although the heat plays no role in a computing machine's computational activities, it does have effects – on the surrounding air and on the machine itself: it activates a heat-dissipating fan. Unlike heat, consciousness is taken to have *no* physical – and for that matter, no mental – effects.

One way to think about this is to imagine that the physical laws form a kind of axiom system governing the behavior of every physical object, including the behavior of complex physical objects that happen to be conscious. Accounting for consciousness requires new fundamental laws governing the emergence of conscious experiences when the physical conditions are right. These laws are independent of the physical laws in the way the parallel axiom is independent of Euclid's other axioms. This provides an accounting of consciousness that makes consciousness fully real, but not in a way that poses a threat to the autonomy of the physical sciences.

This is sometimes put in terms of 'closure': the physical world is 'closed' under the physical laws. Every physical effect has a complete physical explanation, every physical occurrence is a product of purely physical causes governed by purely physical laws. Emergent laws are required, not to explain goings-on in the universe, but simply to account for appearances, simply to account for the occurrence and nature of qualitatively imbued conscious experiences, the existence of which could scarcely be denied.

Panpsychism

Despite its appeal to scientists who dismiss consciousness as an illusion cultivated by mystics and philosophers, many philosophers, including this philosopher, find epiphenomenalism deeply unsatisfying. This has led in some cases to a renewed interest in *panpsychism*.

Before discussing panpsychism, I should warn you that the panpsychism abroad today is not a traditional romantic, Emersonian panpsychism. Contemporary panpsychists do not see all things as full of mind, or subscribe to Preacher Casey's conviction that we're all 'just little pieces of a great big soul'. The idea, rather, is that conscious qualities are not emergent, not epiphenomenal, not add-ons. Conscious qualities 'go all the way down'. Electrons might not have minds, but, just as they possess charge and mass, electrons possess conscious, experiential qualities. There is something it is like to be an electron. The brain might appear to be soggy gray matter, but its component parts exhibit flickers of consciousness. The brain has a Technicolor phenomenology because its parts do.

One route to this kind of panpsychism begins with a consideration of ways you might account for characteristics of complex objects, objects made up of parts that are themselves objects. Suppose you have a complex object, o, and o is F. What is the explanation for o's being F? One possibility is that o is F because o's individual parts, or some of o's individual parts, are F. This seems to be how it is in the case of mass or extension: an object has mass because the object is made up of parts that have mass; an object is extended because the object's parts are extended. F-ness – mass, for instance – 'goes all the way down'.

In other cases you might explain o's being F because o's parts are G and you can see how assemblages of G's could result in something that is F. F-ness 'emerges' from G-ness. Think of water, a liquid. As Galen Strawson puts it,

> Liquidity is often proposed as a translucently clear example of an emergent phenom-
> enon, and the facts seem straightforward. Liquidity is not a characteristic of individual
> H_2O molecules. Nor is it a characteristic of the ultimates of which H_2O molecules are
> composed. And yet when you put many H_2O molecules together they constitute a liq-
> uid (at certain temperatures, at least). So liquidity is a truly emergent property of certain
> groups of H_2O molecules. It is not there at the bottom of things, and then it is there.
>
> *(Strawson 2008: 61)*

Strawson characterizes emergence in terms of 'intelligibility' thereby rejecting as 'mysterious' the
kind of emergence appealed to by epiphenomenalists, which is based on the introduction of new
fundamental laws. It is, he thinks, intelligible that nonliquid particles could make up a liquid. You
can see how this would be so. No magic, no mystery.

Now consider mass or, Strawson's favorite attribute, extension. Could an object's having
mass or its being extended result from collections of parts lacking either? If an object has
mass, this is because the object's parts, or some of them, have mass. Were an object's parts
unextended, it would, Strawson thinks, be utterly mysterious how they could make up something extended. Extension and mass can be accounted for only by supposing that they 'go
all the way down'. Wholes with mass must have parts with mass; extended wholes must have
extended parts.[2]

What of consciousness and qualities of conscious experiences? Suppose that the brains of
conscious agents are responsible for those agents' being conscious, responsible for their undergoing qualitatively distinctive conscious experiences. Brains are complex physical entities, the
constituent parts of which are no different from the constituent parts of tables, trees, lumps of
coal and planets. How could consciousness result from the arrangements of such constituents?

Two possibilities come to mind.

First, consciousness might resemble liquidity. The constituent parts of brains might possess
nonconscious characteristics that, given the right organization, unmysteriously result in something with experiential qualities.

The difficulty here is that it is hard to see what the pertinent nonconscious characteristics
could be. The situation resembles, not that of liquidity, but that of mass or extension. Just as it
is mysterious how extended, massy wholes could result from assemblages of unextended parts
lacking mass, it is hard to see how you could get consciousness by putting together nonconscious
components.

At this point a second option suggests itself: conscious agents are made up of parts that *themselves* possess some measure of consciousness – if a complex whole has experiential qualities, this
is because its parts have experiential qualities. But then, given that the kinds of particles that
make up brains are no different from those that make up everything else, experiential qualities
must be present in the fundamental things. This is panpsychism – or, if not quite *pan*psychism,
something close to panpsychism.

This qualification is needed because the argument shows at most that experiential qualities
must be present in *some* of the fundamental things. Perhaps a body with mass must be made up
of parts that themselves have mass, but not *all* of its parts need have mass. So if panpsychism is
the doctrine that *everything* has experiential qualities, the line of reasoning under consideration
does not establish panpsychism, only something weaker: experiential qualities must be present
somewhere at the fundamental physical level – just as extension and mass must be present somewhere at the fundamental level. Call this *micropsychism*. (I shall, however, continue to speak of
panpsychism to spare you a gratuitous proliferation of jargon.)

You might object to this line of reasoning in the following way. This tomato is red. This must be so either because the parts of the tomato are red or because the parts of the tomato are G, and it is intelligible how assemblages of G's could amount to something that is red.

It seems clear that the first option is false. The tomato's parts – the leptons and quarks (or whatever) – that make up the tomato are not red; indeed, they are colorless. But the second option seems false as well. How could you get something red from something colorless? That would seem impossible – except that we know it is *not* impossible. So, by parity of reasoning, from the fact that we cannot make intelligible how you could get experiential qualities by combining items altogether lacking in such qualities, it does not follow that conscious agents must be made up of parts that are themselves conscious.

Permit me to pause for a moment to reflect on this thought.

We *know* the tomato's qualities result from combinations of quarks and leptons (or whatever the fundamental things are). No scientist would dream of taking the trouble to work out an intelligible route from the quarks and leptons (or fields, or superstrings) to qualities of the tomato. How exactly are experiential qualities special in this regard? Why is it any *more* mysterious how conscious experiences could result from combinations of nonconscious quarks and leptons than it is mysterious how qualities of tomatoes result from combinations of particles that, individually, lack those qualities?

Ah, but is the tomato example apt? You might doubt that it is. Color science reveals that a tomato's color depends on the configuration of particles that make it up. One configuration reflects ambient light in a way that makes the tomato look red to ordinary human perceivers. Change this configuration and the tomato will turn green. Change the light and the tomato will look brown. If objects' colors are just particle configurations, there is no special problem in *these* being made up of colorless parts.

This response, however, merely relocates the problem without solving it. Why should one configuration of particles constitute something red, another, something green? We know that it does without being in possession of an intelligible story as to *why* it does. (That red things can fail to look red under certain conditions is neither here nor there.) So, again, the fact that we lack an account of how parts that lack experiential qualities could result in a whole that has such qualities does not appear to require that the parts have experiential qualities.

At this point, a panpsychist might make what could be called the Galilean move, banishing 'secondary qualities' such as colors to the mind. The distinctive qualitative redness you experience when looking at the tomato in bright sunlight is not in or on the tomato; the redness is in *you*, in your mind. Your seeing the tomato as red is a matter of your *projecting* redness onto the tomato.

Why think that? Well, it is mysterious how a configuration of colorless particles could make up something *actually* red, something qualitatively akin to your experience of red.

But wait! The tomato's color looked like a counterexample to the claim that qualities of wholes must either be present in their parts or emerge 'intelligibly'. To appeal to the absence of an intelligible connection between colorless parts and colored wholes as a reason for denying that colors could be characteristics of ordinary objects is to reassert precisely what is in question.

Even if you thought there were good independent reasons to locate colors in the mind, however, a difficulty remains. Suppose your experience, not the tomato, is what is qualitatively suffused with color – *Technicolor experiences*! Then, either your brain's constituent particles are colored, or it is mysterious how they could result in colorful experiences. If they *are* colored, however, then the tomato's parts – which are, after all, no different in kind from parts of your brain – must be colored, contrary to hypothesis. If they are not colored, then it looks as though we can accept that wholes can have experiential qualities lacked by their parts. Your colorful experiences result from combinations of colorless parts.[3]

It is hard to avoid the impression that it is simply being assumed that a mental–physical gap is in principle unbridgeable, but that any physical–physical gap is, *solely by virtue of being physical–physical*, intelligibly bridgeable, at least in principle. Given the nature of considerations advanced to establish the mental–physical chasm, however – chiefly appeals to intelligibility – the assumption is ungrounded. If there is an intelligibility gap, there is a gap in both cases.

So? Well, no one thinks that qualities of tomatoes must be present in their parts, so why should anyone think that qualities of experiences must be present in the parts that make up conscious agents? No one regards qualities of tomatoes as deeply mysterious, their existence obliging us to ascribe such qualities to the particles. Conscious experiences would seem to be in the same boat. Sauce for the goose is sauce for the gander.

Fly and fly bottle

Panpsychism represents a desperate response to the apparent impossibility of understanding how you could get qualitatively rich conscious experiences from 'soggy grey matter'. But the utter hopelessness of the situation intimates that the problem itself is almost certainly ill conceived, that the puzzle of consciousness is of our own making. The puzzle stems from a bevy of largely unexamined, and hence invisible, background assumptions. These assumptions determine a logical space of possible answers, each of which proves unsatisfactory in its own way.

This would be grounds for acute anxiety among the friends of consciousness were it not the case that the assumptions are entirely optional. In each instance, attractive alternatives exist, alternatives that reveal the way out of the fly bottle.

Chief among the assumptions I have in mind is the Cartesian idea that the mental–physical distinction is a distinction, not merely in conception, but a distinction in reality. This yields a sharp division between the mental and the physical. As a matter of historical fact, however, many philosophers *have* denied that the mental–physical distinction is a distinction between two dramatically different kinds of things or properties. I am not thinking of committed materialists. Materialists do not deny a distinction between the mental and the material; they accept the distinction and deny the mental. I am thinking of philosophers – Priestly, Spinoza, and perhaps Locke come to mind – who saw the mental–physical distinction as one of conception only.

In the 20th century, this idea was appropriated by Donald Davidson, who held that physical things are those things describable using a physical vocabulary, a vocabulary borrowed from fundamental physics. Mental things are whatever answers to mental descriptions. In Davidson's view, anything that can be given a mental description could be given a physical description as well. One reality, two ways of describing it.

Is there any reason to prefer Davidson to Descartes?

Start with the idea that physical theories designate properties by reference to the contributions those properties make to what their bearers do or would do. This turns physical properties into *powers*. In contrast, mental properties, or some mental properties, seem to have an intrinsic *qualitative* nature. Does this show that mental and physical properties are dramatically different species of property?

Why should it? If physics tells us that a property is a power, this in no way implies that the property is *not* a quality. This is an elementary logical point. Accepting that *a* is *F* is not to accept that *a* isn't *G*. It is at least consistent with a property's being a power that it be a quality: properties could turn out to be *powerful qualities*.

You might be willing to grant that this *could* be so, yet doubt that it *is* so. But consider. The identity of a power is determined by what it is a power *for*: a power is a power to manifest itself in particular ways with particular kinds of reciprocal power.

Reflect on what it is for a china bowl to be fragile. The bowl's being fragile is a matter of (among many other things) the bowl's being of a kind that would shatter when dropped on a hard surface or struck by a solid object. The shattering is a *mutual manifestation* of powers of the bowl and reciprocal powers possessed by the surface of the solid object. But notice that this manifestation can be understood only when you invoke assorted qualities of the bowl and qualities of whatever leads to its shattering. You might insist that these qualities are themselves really powers. I would agree, but I would want to note that this does not by itself imply that they are not qualities, and powerful qualities at that.

Such considerations do not amount to a proof that properties appealed to in causal explanations are not purely powers, but perhaps they give you a feel for why this might not be so, why it is not entirely crazy to regard properties as powerful qualities (see Heil 2012: chapter 4 for extensive discussion).

You can come at the same point from the opposite direction. Consider paradigmatic qualities – sphericity, for instance. A tomato is spherical. It would seem that, due to its sphericity, the tomato would roll (rather than tumble) down an inclined plane, would make a circular (rather than square) concave impression in the carpet, and would pass smoothly through a round hole of the same diameter (but not through a square hole of equal area). The tomato's sphericity, a quality of the tomato, would seem to endow the tomato with a power.

One reaction to such examples is to accept that sphericity is a power and deny that it is a quality. You would make this move, however, only if you were antecedently convinced that the power–quality distinction is a real distinction, only if you assumed that qualities could not themselves be powers. But this is just what is being called into question.

The suggestion on the table is that the power–quality distinction is one of conception only. How might this help?

Think again of functionalism as described earlier. Functionalism is taken by many philosophers and nonphilosophers to provide a decent account of mental states and processes. You could think of a functional system as a network, the nodes of which are powers. Functionalists hold that mental states owe their identities, not to their 'intrinsic natures' (if they have any), but to their place in the network. You could replace the occupants of nodes in such a network with intrinsically different occupants. So long as this did not affect its causal contingencies, the network's mentality would be unaffected. (Recall thought experiments in which neurons are gradually replaced by transistors.) As Wittgenstein says of the beetle in the box, the intrinsic character of what is in the box is irrelevant (the box could even be empty). Whatever plays the beetle role *is* a beetle.

Functionalists contend that something is a pain because it plays the pain role. Philosophers fond of mental qualities – qualia – think that this could not possibly capture the nature of what it is to be in pain. Surely the qualitative character of pain is all-important. One possibility is that these qualities 'arise' or 'emerge' when you have the right kind of functional organization (Chalmers 1996). You might think this if you thought that qualities were of necessity causally inert, impotent.

But, again, *why think that?* Suppose qualities were *themselves* powerful. Once you admit this possibility, you are in a position to reverse the functionalist order of explanation. According to functionalism, something is a pain by virtue of playing the pain role (or by virtue of being apt to play the pain role, or some such). But why not turn this around and say that something is apt for the pain role if it is qualitatively painful? Not just *anything* could play the pain role, for the same reason that not just *anything* could play the role played by a tomato's sphericity. Try replacing the tomato's sphericity with a transistor.

This leads to the thought that the relations functionalist nodes bear to one another are determined by the nature of items at the nodes. If properties turned out to be powerful qualities, then there would be no separating the qualitative nature of a pain, say, from its aptness for playing the pain role, from pains occupying the nodes they occupy.

The deeper picture

Earlier I contended that a central argument for panpsychism is, or should be, unpersuasive. The argument begins with the assertion that it is unintelligible how a conscious creature could result from combinations of parts, none of which itself had conscious qualities: if complex systems are conscious, then their parts must be conscious. My response was to note that tomatoes are red, but not because their parts are red. The connection between properties of the parts and qualities of the whole, qualities of the tomato, is not regarded as a deep mystery, and is certainly not one scientists or the rest of us care about. How is the connection between the qualities of conscious experiences and the properties of the parts that make up those creatures any different? If you think the case of consciousness requires conscious parts, why not require that red objects have red parts?

As it happens, features of exceedingly complex objects answering to everyday predicates can result from highly organized interactive combinations of fundamental things. We know as much. Only a philosopher would expect the connection to be 'transparent'. This is not to say that it is impossible to work out how it is that particular features of complex objects result from combinations of the fundamental parts of those objects. In most cases, however, we have better things to do.

This is not the end of the story, of course. Objects' colors are explained by features of their surfaces, for instance, and perhaps ultimately by colorless features of objects' constituents. So what business have I appealing to colors in engaging in serious ontology?

This misses the point. First, I have been trying to defuse the idea that qualities could not be powerful, so everyday examples are fair game. Aristotle puts it nicely: 'we must follow this method and advance from what is more obscure by nature, but clearer to us, towards what is more clear and more knowable by nature' (*Physics* 184a 19–21). Second, the idea that lots of qualities of ordinary things could, on closer examination, dissolve into qualities of their constituents interactively arranged in particular ways, does not imply that qualities 'go away' once we move to particles and fields. To think so is to go beyond physics. If physics tells us about the powers, it does not thereby tell us that those powers are nonqualities. This is a gratuitous metaphysical addition that appears wholly unmotivated.

What of consciousness? Well, if you start with the assumption that the deep story about the universe as described by physics is qualitatively barren, you are bound to have difficulty understanding how conscious experiences could be qualitatively rich. Qualities would have to be nonphysical, 'emergent' add-ons perhaps. At the same time, you will hear the suggestion that properties quite generally are powerful qualities as implying panpsychism. If qualities are nonphysical and if electrons have qualities, then electrons must possess some spark of consciousness: consciousness must be spread on the world.

When it comes to consciousness, are the only options emergence or some form of panpsychism? Suppose you take qualities seriously. And suppose you thought as I do that properties are powerful qualities. If powers go 'all the way down', so do the qualities. Assuming that features of ordinary things stem from features of the fundamental things, it is quite possible that the resources required for the qualitative character of conscious experiences are to be found in the qualitative natures of the fundamental things. And this no more requires that the fundamental things be conscious than the redness or roundness of a tomato requires that the fundamental things be red or round.

A final observation. I have remarked that the friends of consciousness have a way of being their own worst enemies. One respect in which this is so stems from the casual way in which we identify conscious qualities. Often we describe conscious experiences by reference, not to their intrinsic character, but by reference to what the experiences are experiences of. This in itself is harmless, but when you start thinking about experiences themselves, there is a risk of confusing features of the experience and features of what is being experienced, committing what U.T. Place called the *phenomenological fallacy* (Place 1958). Your experience of a red, spherical tomato is not itself red and spherical. It is, I believe, a nontrivial matter to identify the intrinsic features of experiences.

But how does it work? How do you get conscious experiences from collections of constituents that make up brains? Even assuming qualities at the fundamental level, how do we get qualities of conscious experiences – whatever these might be?

I have no idea. For that matter I have no idea how ordinary objects such as tomatoes get their powers and qualities from the fundamental things duly organized. I am not offering a solution to the mind–body problem that makes it obvious how dynamic, interactive collections of quarks and leptons could constitute fully conscious agents undergoing Technicolor conscious experiences, however; I am only suggesting that, by bifurcating qualities and powers, we deny ourselves resources that could be useful in solving – or dissolving – the problem: we make a hard problem impossible.

Notes

1 Versions of this chapter were presented at the University of Durham's Institute of Advanced Study, January 27, 2015, and at Glasgow University, February 16, 2015. The chapter benefited substantially from comments and questions raised on those occasions.
2 In contrast to Strawson, I take this to be a plausible but empirically defeasible assumption. It could turn out that extended bodies are extended in virtue of their parts being some way other than extended, and it was intelligible how their being this way resulted in the whole's being extended.
3 Hiding experiential colors 'inside' conscious states of the particles would, again, shift the problem, not solve it.

References

Chalmers, D. J. (1996) *The Conscious Mind: In Search of a Fundamental Theory.* New York: Oxford University Press.
Heil, J. (2012) *The Universe as We Find It.* Oxford: Clarendon Press.
Kirk, R. (1974) 'Zombies versus Materialists'. *Aristotelian Society Proceedings* (suppl.) 48: 135–152.
McGinn, C. (1989) 'Can We Solve the Mind: Body Problem?'. *Mind* 98: 349–366.
McLaughlin, B. P. (1992) 'The Rise and Fall of British Emergentism'. In A. Beckermann, H. Flohr, and J. Kim, eds. *Emergence or Reduction?* Berlin: de Gruyter, pp. 49–93.
Place, U.T. (1956), 'Is Consciousness a Brain Process?'. *British Journal of Psychology* 47: 44–50.
Strawson, G. (2008) *Real Materialism and Other Essays.* Oxford: Clarendon Press.

PART 3

Emergence and physics

19

PHASE TRANSITIONS, BROKEN SYMMETRY AND THE RENORMALIZATION GROUP

Stephen J. Blundell

Introduction

A piston is pushed into a cylinder and compresses a gas. The pressure goes up as the volume goes down, in accordance with Boyle's law. In such a simple thermodynamic description, it is not necessary to worry about what is going on at the atomic scale. Nor do we need to concern ourselves with the fact that this simple lab experiment is being carried out on the surface of planet Earth, a 12,800-km-diameter rotating sphere filled with iron and coated with a silicate crust. Nor is our Boyle's law analysis using the fact that such a rotating sphere glides along its orbit through largely empty space. Nor indeed do we need to keep in mind that the galaxy in which the solar system resides is an unusually high-density region of the universe. Just as our thermodynamics experiment can be performed in a lab without fussing over such cosmological niceties, a cosmologist in the office next door who is studying the Friedmann model of the expansion of the universe will take the average mass density of the universe (equivalent to around a few protons per cubic metre) as a simple parameter. She will blissfully ignore the "clumpiness" of matter at smaller scales, and of course will neglect the presence of the experiment with the piston and the cylinder of gas sitting on planet Earth.

Thus we take it as a given that when a portion of the universe is selected for study, be it a gas or a galaxy, we are allowed to blissfully ignore what is going on at scales that are much larger, or indeed much smaller, than the one we are considering. We might perhaps use those different scales to provide appropriate matching conditions. For example, the fact that the experiment with the piston is located on the surface of planet Earth would inevitably introduce a gravitational field into our experiment, which might perhaps be relevant. By considering the atomic-level motion of the gases we would be able to match a kinetic-theory description containing atomic masses and velocities on to our thermodynamic macroscopic description, but we could also ignore the existence of atoms entirely (as most scientists did in the nineteenth century) and still perform perfectly acceptable thermodynamics. These other scales, the atomic or the planetary, only provide at most the parameters that determine the boundary conditions, and the details of physics arising at other scales are neglected.

However, it turns out that there is a class of very familiar problems where this simple neglect of what is going on at other length scales fails. In these problems it is not only important, but also unavoidable, to consider the physics on multiple length scales that can vary over many orders of magnitude. A very good example of such a problem is a phase transition. In fact, you may become aware of this issue whenever you boil the water in a kettle to make yourself a cup of tea, if you pay attention and listen carefully. As the water heats up, its temperature increases. There is a simple relation between the amount of energy you put into the water per second (the power of the kettle, probably written on a manufacturer's label attached to the base) and the temperature rise per second. Initially this process of warming proceeds steadily and quietly (save a brief hissing noise you hear shortly after switching the kettle on, as dissolved gases are driven out of the water). However, as the heating process nears completion, the kettle starts to become rather noisy due to the hot water bubbling and frothing, and the sound gets progressively louder. When the temperature hits 100 degrees Celsius, the liquid water does something remarkable. It begins to change into a gas and steam starts to rise upwards from the spout. This is an example of a phase transition, a change of state of water. It is utterly discontinuous and strongly contrasts with the smooth continuous changes that have occurred previously (42-degree water is similar, but just a bit hotter, than 41-degree water, which is similar, but a bit hotter still, than 40-degree water, etc.; conversely, steam at 100 degrees Celsius is radically different from water at 100 degrees Celsius). It also involves a finite slug of energy supplied from the kettle to make it happen; this is known as latent heat and indicates that the two phases have different entropy.

But not only is the transition remarkable in its discontinuity, the region close to the transition is remarkable too. Fluctuations occur near a phase transition and are produced on all length scales. Thus, not only does water close to the boil contain lots of little bubbles of steam, but there are big bubbles, and occasionally very big bubbles. This gives rise to the noisy, boisterous behaviour of hot water close to the boil. Bringing water to the boil in a pan on a stove demonstrates this even more vividly, but the same effect is apparent already in the kettle.

We can obtain a visual representation of this behaviour near a phase transition by examining snapshots of configurations of a magnetic system (a two-dimensional Ising model, which I will explain later) that also exhibits a change of state at a critical temperature T_c at which it changes from a ferromagnetic state $(T < T_c)$ to a paramagnetic state $(T > T_c)$. The snapshots are shown in Figure 19.1 and illustrate the magnetic configurations after the system has had time to settle down at different temperatures. Each picture has regions that are either black or white, corresponding to atomic magnetic moments (spins) that are only allowed to be up or down. At low temperature all the spins are aligned and the region is uniformly coloured black, representing identical spin states pointing up. As the temperature increases, a few white blobs begin to appear as small fluctuations start to occur more readily. Close to the transition, the fluctuating regions are very large and the system starts to lose its net magnetic moment. Above the transition, fluctuations occur more readily, but the size of correlated regions of spins (known as the correlation length) begins to decrease and, at the highest temperature shown, the state of each individual spin is only weakly correlated with its neighbour. At temperatures close to T_c, the fluctuations become very slow, due to the correlation length of fluctuating spins becoming very large. This is because to cause a fluctuation, it is necessary to flip a large, correlated block of spins, which is a slow process. Both the correlation length and the fluctuation time diverge at T_c (the latter phenomenon is known as "critical slowing down"). This slowing down affects the Monte Carlo simulations that were used to generate Figure 19.1 just as much as in the real system (in both cases, interactions between more distant spins are becoming important; both the real system and the simulated system become more correlated and complicated near T_c), and this increases the convergence time of calculations near T_c. In general, what we call critical behaviour (a class of phenomena observed at temperatures close to phase transitions) is characterized by the onset of long-range correlations. A diverging correlation length leads to singularities in response functions, and at T_c the system becomes scale-invariant. Techniques such as the renormalization group, which I will describe

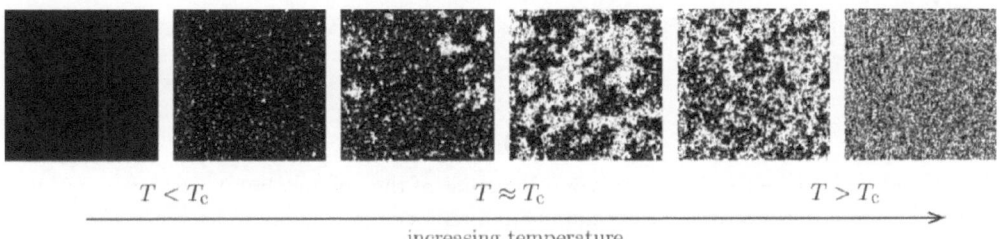

$T < T_c$ $T \approx T_c$ $T > T_c$

increasing temperature

Figure 19.1   A Monte Carlo simulation of spins in the two-dimensional Ising model at various temperatures. Spins are coloured black or white according to whether they are in the up or down state, respectively.

[From Blundell and Blundell (2010)]

later, exploit the scale invariance. It is found that length scaling (Wilson and Fisher 1972; Wilson 1972) accounts for the universality of behaviour found in systems with very different underlying physics because many of the system parameters turn out not to be important at the transition (in the language of the renormalization group, they are said to be "scaled away") and only a very few relevant parameters (such as the system and order parameter dimensionality) remain.

Phase transitions are thus remarkable phenomena in which it is not possible to work at one scale and ignore all the others. You have to think about all scales of length. Counterintuitive behaviour can be expected in such scale-invariant systems, and there are important analogies with self-similar geometric structures, such as fractals (Stinchcombe 1989), that also possess scale invariance.

It is worth recalling that the "usual" scaling behaviour is at the heart of much familiar physics. This use of calculus in physics, due to Newton and Leibniz, relies on the notion that scaling is simple and limits behave sensibly. Thus, quantities such as velocity v can be evaluated by considering a small displacement Δx in a small time interval Δt and writing $v = \Delta x/\Delta t$. This formula is taken in the limit when Δt tends to zero and allows differential equations to be constructed. For the usual $x(t)$ that describes planetary orbits and the motion of cannonballs, "taking the limit" works because, on a sufficiently short time scale, $x(t)$ becomes linear. Thus, halving Δt means halving Δx. The trouble occurs in systems for which these limits do not behave so straightforwardly in the limit of the very smallest distance and timescales. Consider an electron with charge Q. If you model it as a little ball of radius R, with the charge Q smeared uniformly over its surface, then the electric field energy stored in the charge is given by $Q^2/(8\pi\varepsilon_0 R)$. This stored energy is equivalent to a mass, so the particle mass m is then equal to $m = m_{bare} + Q^2/(8\pi\varepsilon_0 R c^2)$, where c is the speed of light and m_{bare} is the "bare mass", the mass the particle would have without the effect of the electric field energy. This means that the mass m that we measure in the experiment is not the intrinsic mass of the particle but contains this additional contribution. But worse, if $R = 0$ (and we think the electron "is" a point-like particle), then this electric field energy contribution is infinite! A similar effect complicates our understanding of the electric charge of the electron, since during very short time intervals vacuum fluctuations result in the creation (and very quickly afterwards, the annihilation) of particle–antiparticle pairs and these pairs will screen the electric charge. This means that the "bare charge" of the particle will appear to be reduced when viewed from a distance R, and in fact the apparent charge Q will diverge as R is sent to zero. This worrying state of affairs has been addressed in quantum field theory by renormalization techniques (Delamotte 2004; Weinberg 1995). I will not review these ideas here insofar as they have been applied to the problem of the mass and charge of elementary particles, but rather concentrate on how they can illuminate the problem of phase transitions. First, it is necessary to consider the effect of symmetry breaking.

Symmetry breaking

As a liquid cools there can be a very slight lowering of the volume, but it retains a very high degree of symmetry. However, below the melting temperature, the liquid becomes a solid and that symmetry is broken. This may at first sight seem surprising because a picture of a solid with all the atoms lined up in a regular lattice "looks" more symmetrical than that of the liquid in which the atoms are floating around all over the place. The crucial observation is that any point in a liquid is, on average, exactly the same as any other. If you average the system over time, atoms visit each position in space with uniform probability. There are no unique directions or axes along which atoms line up. In short, the system possesses complete translational and rotational symmetry. In the solid, however, this high degree of symmetry is nearly entirely lost. Rather than being invariant under arbitrary rotations, the lattice of atoms might be invariant under, say, four-fold rotations; rather than being invariant under arbitrary translations, it is now invariant under a translation of an integer combination of lattice basis vectors. Therefore, not all symmetry has been lost but the high symmetry of the liquid state has been, to use the technical term, "broken". It is impossible to change symmetry gradually. Either a particular symmetry is present or it is not. Hence, phase transitions are sharp and there is a clear delineation between the ordered and disordered states.

Not all phase transitions involve a change of symmetry, and one such example is the liquid–gas transition, illustrated in the phase diagram in Figure 19.2 that shows the three phases: solid,

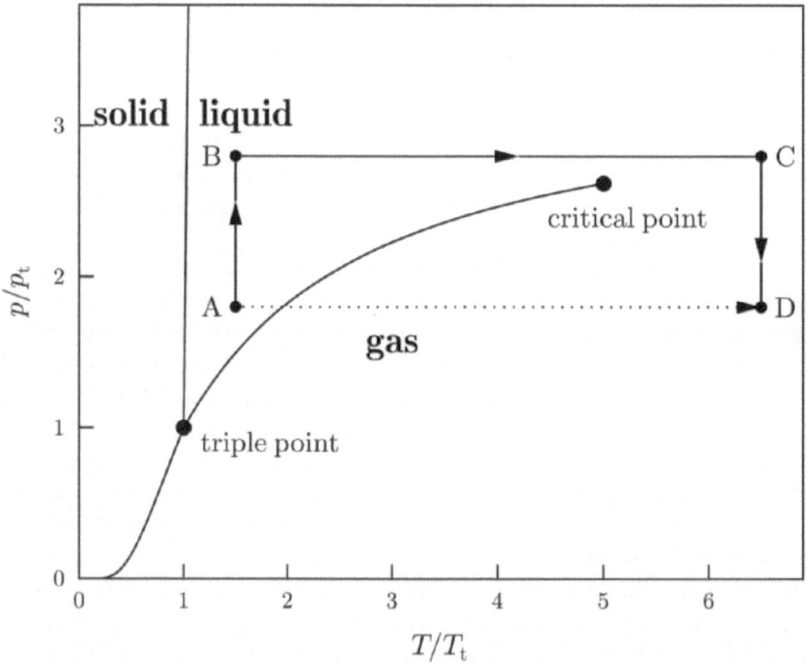

Figure 19.2 The phase diagram of a substance, showing how one can avoid the phase transition by going around the critical endpoint (route ABCD) rather than straight across the phase boundary (route AD)

liquid and gas. There really is no fundamental difference between a very dense gas and a liquid. Thus, the boundary line between the liquid and gas regions is terminated by a critical point. This means that it is possible to "cheat" the sharp phase transition by taking a circuitous path (ABCD) through the phase diagram that avoids a discontinuous change because you never cross the phase boundary; you go around it. For example, you can start with a liquid at low pressure and temperature (configuration A), pressurize it (B), then warm (C) and finally release the pressure (D). Configuration A is firmly in the liquid region, while configuration D is firmly in the gaseous region, but the route from A to D has avoided the phase transition. Of course, if you take the system straight from A to D by warming at constant pressure (exactly as you do when you boil the water in a kettle), then you observe a sharp discontinuity in properties as you cross the phase boundary. The fact that you can, if you want, *avoid* the transition by a judicious choice of route through the phase diagram is a consequence of the two phases having the same symmetry.

The same trick cannot be played when you transform a solid into a liquid because the two phases have different symmetries, and symmetry is not a property that you can have in anything other than full measure. To restate, either the system possesses a particular symmetry or it does not, and so symmetry-breaking transitions cannot be smooth crossovers. Consequently, there is no critical point for the melting curve and no way to cheat. The set of symmetry-breaking phase transitions includes as members those between the ferromagnetic and paramagnetic states (in which the low-temperature state does not possess the rotational symmetries of the high-temperature state) and those between the superconducting and normal metal states of certain materials (in which the low-temperature state does not possess the same symmetry in the phase of the wavefunction as the high-temperature state).

The role of system size

To understand some of the conceptual issues underlying symmetry breaking, it is helpful to consider a very simple model. Take N^2 spins (magnetic moments) arranged on a lattice of $N \times N$ sites. We will start with four spins on a 2×2 lattice. Now allow each spin to take one of only two possible configurations: up and down. Now add an interaction between the spins, operating only between nearest neighbours, and which saves energy J if the spins are aligned and costs energy J if the spins are antialigned. The 16 possible configurations are shown in Figure 19.3. This is known as the Ising model (which is usually studied on a much bigger lattice). If we take $J > 0$, then the configurations with the aligned spins (the two ones on the left) are the lowest in energy.

We might expect to find magnetic order in our simple model, but as we shall see, it does not emerge as simply as one might at first think. In statistical mechanics, we would usually

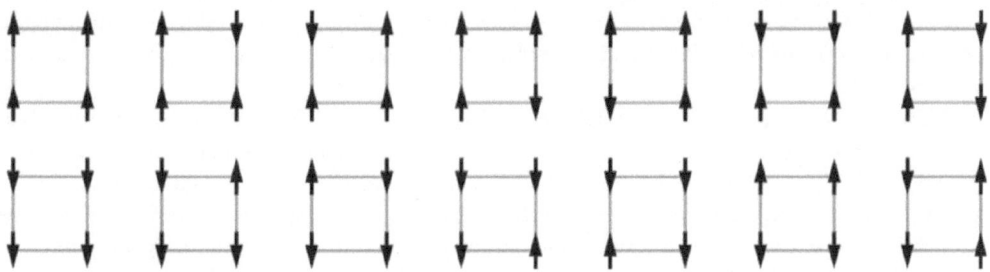

Figure 19.3 The 16 configurations of the Ising model on a 2×2 grid. Each configuration is shown with their symmetry-reversed partner above or below.

assume that if our system were connected to a thermal reservoir (a heat bath) at temperature T, then on average it would adopt a configuration given by a thermal average. The two easy limiting cases to consider are $T = 0$ (when there is no energy to be borrowed from the thermal reservoir and only the lowest energy configuration can be occupied) and $T = \infty$ (when energy can easily be exchanged with the thermal reservoir and every configuration is occupied with equal likelihood). Note that the problem is insensitive to which direction we call up. Thus, at $T = 0$, for example, the two configurations with (i) all spins up and (ii) all spins down (the two ones on the left of Figure 19.3) both minimize the energy and are therefore occupied with probability ½. The expected total number of up-spins minus the down-spins (the net magnetization) is therefore zero (because we weight the two configurations the same and they cancel out). The magnetization is also zero at $T = \infty$ because all configurations are now occupied and there is no net preference for up over down.

In fact, this is true at all temperatures as we can easily show. At a general temperature, the probabilities will be proportional to the Boltzmann factors $\exp(-E/k_{B}T)$, where E is the energy of a particular configuration. The expected magnetization can be written as $<M> = \sum_i M_i \exp(-E_i/k_{B}T)/\sum_i \exp(-E_i/k_{B}T)$, where E_i and M_i are the energies and moments in the ith configuration. $<M>$ will still be zero at any temperature because each term in the sum, which corresponds to a particular configuration, has a corresponding term with opposite magnetization, corresponding to a configuration in which all the spins are reversed (the configurations in Figure 19.3 have been arranged so that the symmetry-reversed configurations are grouped together, one drawn above the other). Thus, statistical mechanics, on its own, does not give us the non-zero order that we are looking for. This should not be surprising because the underlying model (in this case our Ising model) is entirely symmetrical between the up-spins and the down-spins; therefore, the solutions to this model possess this symmetrical property and there is nothing to select the up-spins over the down-spins. What is needed is some kind of symmetry *breaking*, where the system chooses one particular configuration and ignores its symmetrically related partner.

Experiment shows magnetic order does occur, so where does it come from? We can try studying the Ising model on a computer to learn more about it. A magnetic system modelled in a Monte Carlo simulation adopts a particular configuration and, as it fluctuates randomly, explores the configuration space defined by the 16 states. It periodically gets "stuck" in particular configurations or evolves between certain arrangements, and its speed of fluctuation will depend on the temperature we have set in the simulation. But symmetry-breaking behaviour does not become really apparent until we increase the system size.

Practically, analysing the Ising model becomes increasingly difficult as N increases. When $N = 2$, we have 16 configurations. In general, the number of configurations is 2^{N^2}, and this grows very fast with N. Thus, when $N = 10$, the number of configurations is around 10^{30}, well beyond the reach of humanity's computing resources to store all of them (though trivial to write down just one of those configurations; see Figure 19.4). New methods are needed to gain meaningful insight into this sort of problem and, as explained later, the renormalization group is one such method.

Phase transitions exhibit an infinitely sharp discontinuity as a function of temperature. This arises from the supposed infinite size of the system. In finite systems the partition function $Z = \sum \exp(-E_i/k_{B}T)$ (on which an analysis of $<M(T)>$ can be based) exhibits no singularities, and we expect only a smooth crossover between an aligned state at zero temperature and a disordered one at infinite temperature. Phase transitions in nature are, of course, exhibited in finite systems, so can they be said to be really phase transitions? Strong arguments can be given to answer "yes" to this question. A typical macroscopic lump of matter contains a large number N of atoms, perhaps around 10^{23}, and so the discontinuity in properties will occur over such a narrow temperature

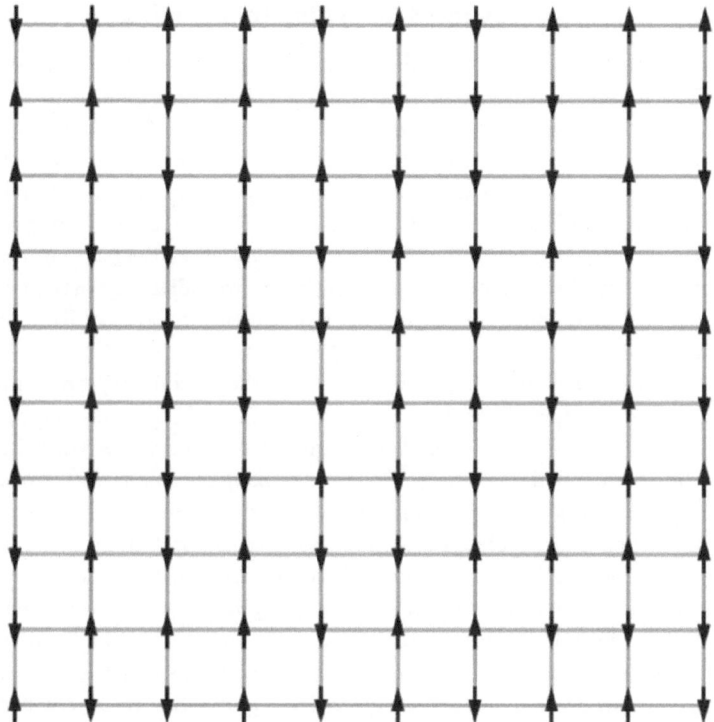

Figure 19.4   One of the approximately 10^{30} configurations for a 10×10 grid of Ising spins

interval that its width is effectively impossible to measure. Moreover, when simulations of finite systems are performed, it is possible to study how the sharpness of the phase transition scales with system size. Such studies show that the temperature range of the crossover falls dramatically as N increases, and so we can describe the phenomenon as a limiting process as N tends towards infinity. The mathematics is most straightforward in the infinite limit, but the finite-size corrections to this model are usually negligible for practical purposes (except in very small systems), but more importantly, they are well understood.

However, we don't need the infinite N limit to be fully made to discern the extraordinary properties of broken symmetry states. In a typical macroscopic lump of matter $N \approx 10^{23}$, and that is big enough. A key insight (Anderson 1984) is the realization that once in a broken symmetry state, it is very difficult to reach other broken symmetry states via fluctuations when the number of particles becomes large. In fact, the probability of all N particles fluctuating at once is proportional to $\exp(-\sqrt{N})$ and becomes negligible as N becomes large. The broken symmetry state is not an eigenstate of the system Hamiltonian H (it breaks one of the symmetries of H) and so it is not a stationary state. Yet it is a stable state of the system (stable, at least, on timescales greater than the age of the universe), an emergent feature of the large-number nature of matter. How do these states form? In an infinite system, an infinitesimal perturbation is sufficient to cause the system to choose one of the symmetry-broken states. Such states can then be modelled by introducing small symmetry-breaking fields (to break the symmetry), and then taking the limit such that N tends to infinity before the symmetry-breaking field is removed; taking the limits in the reverse order leads to a situation in which fluctuations diverge (van Wezel and van den Brink 2006).

Consequences for symmetry breaking

When symmetry breaking occurs, we will expect a sharp change of behaviour at the critical temperature. The region near the phase transition is called the *critical region*. Having broken the symmetry, the system will have a strong energetic preference for staying in that particular broken symmetry state, and attempts by the experimenter to change the choice the system has made in breaking the symmetry will meet with resistance. Thus crystals (which have broken translation symmetry) don't bend or stretch very easily. Ferromagnets show permanent magnetism, and the spins are said to be "stiff". These *rigidity phenomena* are common to broken symmetry states. If symmetry is broken differently in two adjacent parts of a macroscopic sample, the boundary will contain a defect (e.g. a dislocation in a crystal or a domain wall in a ferromagnet).

There are further consequences for the excitations of these symmetry-broken phases. At $T = 0$, the system is perfectly ordered. At finite temperature this order is weakened by excitations in the order parameter. In crystals these excitations are called lattice waves, while in ferromagnets the analogous modes are called spin waves. These wave-like phenomena are normal modes of the system and are therefore susceptible to similar analysis to that applied to the harmonic oscillator, so that quantum mechanically these modes can only acquire energy in discrete lumps. Thus, the energy in these wave systems behaves like a collection of particles. The quanta (i.e. the quantized particles) of lattice waves are termed phonons; those in spin waves are termed magnons. To this menagerie, we can add rotons (quantized rotational waves in liquid helium), ripplons (quantized surface waves), plasmons (quantized plasma waves) and polaritons (particles resulting from coupled light and lattice waves). The mathematical structure underlying these new particles forming in condensed matter systems is entirely analogous to that used to describe the quantization of the electromagnetic waves that results in photons.

So now we come to the key question: Are these emergent particles real? From the perspective of quantum field theory, the answer is a resounding yes. Each of these particles emerges from a wave-like description in a manner that is entirely analogous to that of photons. These emergent particles behave like particles: you can scatter other particles off them. Electrons will scatter off phonons, an interaction that is involved in superconductivity. Neutrons can be used to study the dispersion relation of both phonons and magnons using inelastic scattering techniques. Yes, they are the result of a collective excitation of an underlying substrate. But so are "ordinary" electrons and photons, which are excitations of quantum field modes.

Thus, all of these excitations that arise as a consequence of broken symmetry possess the features you expect of an emergent quantity: they are made up of the combination of another apparently different type of thing but have the properties of something new. Thus, for example, the phonon has a particle-like nature that can propagate through a crystal, but it is entirely composed of atoms. In the same way, a Mexican wave in a crowd has a life of its own, even though it is entirely composed of individuals with their hands in the air, none of whom could do a Mexican wave on his or her own. These emergent particles fall into the class of what one can describe as "real patterns" (Dennett 1991), thereby deserving their metaphysical status from the intellectual coherence of their description of physical phenomena. Just as the "life-forms" in Conway's Game of Life are not simply mirages imagined in a sea of flickering pixels (see Blundell 2017), so these condensed matter particles which emerge from broken symmetry states can be legitimately claimed to be *real* and not just *apparent*. Just as real, in fact, as the electrons that emerge from the vacuum as excitations in the electron quantum field.

The renormalization group and universality

The renormalization group breaks big problems down into small ones. This is easy to understand in a magnetic system of $N \times N = N^2$ spins by considering the Kadanoff block-spin transformation, in which na $m \times m$ cluster of m^2 sites is considered together and the majority spin in the cluster measured (Kadanoff 1966). The old $N \times N$ lattice is then replaced by a new $(N/m) \times (N/m)$ lattice that is now composed of these "block spins" that each represent the average behaviour of the cluster, thinking of it as a unit in itself. This transformation has therefore acted like a blurring process, reducing the information in the system and only preserving the large-scale structure. One might guess that the couplings in this reduced system are identical to the original system, but usually this is not the case. Kadanoff's block-spin transformation can then be applied repeatedly in a series of successive blurrings, and one can see how the couplings develop as a consequence.

This is the key idea behind the so-called renormalization group procedure, which is to look at how a model behaves when the scale changes (and naturally generalizes Kadanoff's original argument). Typically, the rescaling procedure results in identical physics but with altered coupling constants. Then with repeated rescaling we can study how the coupling constants change (or to use the lingo of the subject: how they "flow"). The coupling constants g_i describe a point (g_1, $g_2, g_3, \ldots$) in a multidimensional space, and repeated rescalings result in a trajectory through this space, known as the renormalization group flow. In some cases, the rescalings cause the point to be blasted out to infinity, in others to converge on to a fixed point. At a fixed point, the coupling constants are invariant to further rescalings and describe scale-invariant physics.

The renormalization group is not actually a group in the mathematical sense. The transformations consisting of rescaling and integrating up to a cut-off have no inverse because the integration process results in blurring and a loss of detail. The renormalization group should probably be called the renormalization semigroup, but sometimes, contradictory terminology sticks (as Groucho Marx quipped, think of "military intelligence"). Nevertheless, the renormalization group procedure provides important insights because it shows quantitatively how fine-scale structure is progressively ignored and the physics of critical phenomena depend on these larger-scale, what you might call "structural", features of the theory. It suddenly becomes clear why certain measurable features of the physics of many systems are seemingly independent of the microscopic details, why phase transitions in certain magnets can share common features with the liquid–solid transition or with transitions that can occur in helium films or in traffic flow.

These common features, which include scaling relations of the order parameter and correlation lengths, give rise to the concept of "universality". This is the notion that some of the key parameters of a phase transition depend only on some broad characterizations of the system but not on certain fiddly details. Sometimes those fiddly details are not so obviously ignorable. For example, it might be important to know whether you are dealing with a two-dimensional lattice or a three-dimensional one, but not whether that two-dimensional lattice contains points on a square or a triangular grid. Whether you are studying a square or a triangular lattice is a detail that is important on the scale of the lattice, but its importance diminishes as your perspective widens and you examine the properties of the system on a larger and larger scale. However, whether the lattice itself is two- or three-dimensional is important on all scales.

The features that determine which "university class" will apply to the problem in hand are (i) the dimensionality of its order parameter field, (ii) the dimensionality of the system and (iii) simply whether the interactions are short or long range. They do not depend on the details of the microscopic interactions between constituent parts or whether the system is made up of polymers in a beaker or particles in the early universe. Thus, for example, the properties of a fluid near its critical point and a three-dimensional Ising ferromagnet are both in the same universality class. Both have one-dimensional order parameter fields (the difference between the liquid and vapour densities, $\rho_v - \rho_l$, in the case of the fluid; the magnetization M, in the case of the Ising ferromagnet) that are defined in a three-dimensional space, and both vanish as the temperature is increased above the critical temperature. Both have a corresponding response function (the compressibility κ or the susceptibility χ, respectively) that diverge in the same way (i.e. with the same critical exponents) as the critical point is reached. The "multiple realizability" of phenomena produced at critical points (Batterman 2001) is a feature of the oft-neglected role of singularities in physics and the importance of considering the asymptotic behaviour close to the critical point (Berry 1994), exactly what the renormalization group aims to do. Here I want to underline the way in which critical behaviour *emerges* from a substrate of microscopic interactions and that the predominant narrative for understanding such behaviour operates almost exclusively at that higher, emergent level. In many physical situations, the emergent description simplifies a more complex lower-level account; here, the emergent description makes the lower-level account irrelevant.

Conclusion

Broken symmetry phases are all around us and include crystals, magnets, liquid crystals and superconductors. By breaking symmetry, these phases forfeit the status of being a "stationary state" of the sort beloved of elementary quantum mechanics treatments. Nevertheless, because they are composed of a macroscopic number of particles, these phases can be stable on timescales longer than the universe and give rise to a range of emergent ordered properties. The excitations of such phases give rise to a new set of emergent particles, such as phonons and magnons. Theoretical approaches, such as the renormalization group method, allow one to study in detail how certain microscopic parameters become irrelevant in the critical regime, while other features remain crucially relevant. Broken symmetry can therefore be seen as a well-studied paradigm of emergent behaviour in the physical world.

References

Altland, A. and Simons, B. (2006) *Condensed Matter Field Theory*, Cambridge: Cambridge University Press.

Anderson, P. W. (1984) *Basic Notions of Condensed Matter Physics*, Menlo Park: Benjamin-Cummings.

Batterman, R. W. (2001) *The Devil in the Details: Asymptotic Reasoning in Explanation, Reduction and Emergence*, Oxford: Oxford University Press.

Berry, M. (1994) "Asymptotics, singularities and the reduction of theories", in *Proc. 9th Int. Cong. Logic, Method., and Phil. of Sci.* 9: 597–607, eds. D. Prawitz, B. Skyrms, and D. Westerståhl, Amsterdam: Elsevier.

Binney, J. J., Dowrick, N. J., Fisher, A. J., and Newman, M. E. J. (1992) *The Theory of Critical Phenomena*, Oxford: Oxford University Press.

Blundell, S. J. (2001) *Magnetism in Condensed Matter* (2nd edition), Oxford: Oxford University Press.

Blundell, S. J. (2017) "Emergence, causation and storytelling: Condensed matter physics and the limitations of the human mind", *Philosophica* 92: 139–164.

Blundell, S. J. and Blundell, K. M. (2010) *Concepts in Thermal Physics*, Oxford: Oxford University Press.

Delamotte, B. (2004) "A hint of renormalization", *Am. J. Phys.* 72: 170–184.

Dennett, D. C. (1991) "Real patterns", *J. Philosophy* 88: 27–51.

Fisher, M. E. (1974) "The renormalization group in the theory of critical behavior", *Rev. Mod. Phys.* 46: 597–616.

Goldenfeld, N. (1992) *Lectures on Phase Transitions and the Renormalization Group*, Boulder, CO: Westview Press.

Kadanoff, L. P. (1966) "Scaling laws for Ising models near T_c", *Physics* 2: 263–272.

Kadanoff, L. P. (2013) "Theories of matter: Infinities and renormalization", in R. Batterman (ed.) *The Oxford Handbook of Physics and Philosophy*, Oxford: Oxford University Press, pp. 141–188.

Lancaster, T. and Blundell, S. J. (2014) *Quantum Field Theory for the Gifted Amateur*, Oxford: Oxford University Press.

Le Bellac, M., Mortessagne, F., and Batrouni, G. G. (2004) *Equilibrium and Non-Equilibrium Statistical Thermodynamics*, Cambridge: Cambridge University Press.

Ma, S.-K. (1973) "Introduction to the renormalization group", *Rev. Mod. Phys.* 45: 589–614.

McComb, W. D. (2004) *Renormalization methods*, Oxford: Oxford University Press.

Nishimori, H. and Ortiz, G. (2011) *Elements of Phase Transitions and Critical Phenomena*, Oxford: Oxford University Press.

Stinchcombe, R. B. (1989) "Fractals, phase transitions and criticality", *Proc. Roy. Soc. London* 423: 17–33.

Tolédano, J.-C. and Tolédano, P. (1987) *The Landau Theory of Phase Transitions*, Singapore: World Scientific.

van Wezel, J. and van den Brink, J. (2006) "Spontaneous symmetry breaking in quantum mechanics", *Am. J. Phys.* 75: 635–638.

Weinberg, S. (1995) *The Quantum Theory of Fields*, Cambridge: Cambridge University Press.

Wilson, K. G. (1972) "Feynman-graph expansion for critical exponents", *Phys. Rev. Lett.* 28: 548–551.

Wilson, K. G. (1983) "The renormalization group and critical phenomena", *Rev. Mod. Phys.* 78: 583–600.

Wilson, K. G. and Fisher, M. E. (1972) "Critical Exponents in 3.99 Dimensions", *Phys. Rev. Lett.* 28: 240–243.

Wilson, K. G. and Kogut, J. (1973) "The renormalization group and the ε expansion", *Phys. Rep.* 12(2): 75–200.

Zinn-Justin, J. (2007) *Phase Transitions and Renormalization Group*, Oxford: Oxford University Press.

Further reading

A very good introduction to the renormalization group may be found in Nishimori and Ortiz (2011) and McComb (2004). In the context of quantum field theory, a non-technical account may be found in Lancaster and Blundell (2014) and Altland and Simons (2006), with more details in Weinberg (1995). Good reviews of the renormalization group in critical phenomena may be found in chapter 4 of Le Bellac et al. (2004), and also in Binney et al. (1992), Ma (1973), Fisher (1974), Wilson (1983), Wilson and Kogut (1973) and Zinn-Justin (2007). A particularly clear exposition is given in Goldenfeld (1992), and historical background is covered in Kadanoff (2013). Philosophical implications of asymptotic limits are described in Batterman (2011), and Berry (1994) provides the scientific background. The implications of broken symmetry to condensed matter systems are described in Anderson (1984), though see Blundell (2001) for applications to magnetism. For a technical account of the relationship between symmetry groups and phase transitions, see Tolédano and Tolédano (1987).

20

SOFT MATTER – AN EMERGENT INTERDISCIPLINARY SCIENCE OF EMERGENT ENTITIES

Tom McLeish

From the decade of the 1990s, terms such as 'soft matter' and 'soft matter physics' began to appear with increasing regularity in conference announcements, review articles and books. Previously a much more fragmented disciplinary landscape had been evidenced by repeated reference to the individual research topics that 'soft matter' later subsumed: colloid physics, polymer physics and chemistry, liquid crystal science and more.[1] The language strongly indicates not simply a summation of research programmes, but the recognised emergence of a new field of science. The coherence of soft matter principally draws on both conceptual and experimental foundations, as we shall see, but the superficial consequence is that materials falling under the labels of colloids (Lekkerkerker and Tuinier 2011), polymers (Doi and Edwards 1986), liquid crystals (de Gennes and Prost 1995), self-assembly (Witten 2004), membranes (Safran 2003), foams (Weare and Hutzler 1999), granular materials (Duran 2000), biological materials (Nelson 2004), glasses (Jones 2002) and gels (Rubinstein and Colby 2003) now find a common scientific home within soft matter (Jones 2002). A more recent extension into biological science benefits from the ubiquity of soft matter material structures at the nanoscale level within biological cells (McLeish 2011). Conversations between these subfields, as well as within them are, as a consequence, now commonplace.

Furthermore, the nature of experimental and theoretical methodologies, epistemology and ontology in soft matter means that the field furnishes a rich conceptual hunting ground of emergence, top-down causation (Ellis 2012) and potential anomalies in supervenience (Batterman 2002) and multiple realisability (Aizawa and Gillett 2009) for the philosopher of emergent phenomena. This is not surprising given that one of the sources of the new subfield was the wider field of 'condensed matter physics' employed by Anderson (1972) in a landmark articulation of (weak) emergence as *More Is Different*. Soft matter provides many illustrations of Anderson's claim that the notion of the 'fundamental' in physics should not be tied to any one scale of length or energy.

The scientific history of soft matter science was propelled by a combination of communication within the scientific community (through conferences, research departments and journals that attracted scientists from more than one subfield), intrinsic conceptual overlap and commonality and visionary leadership from a small number of pioneering scientists. Pre-eminent among those were two theoretical physicists, Pierre-Gilles de Gennes in France (and later Nobel Laureate in physics for 1993) and Sam Edwards in the UK. Both of these leaders realised that broad conceptual frameworks and powerful theoretical techniques from other areas in physics could be applied to soft matter systems and that, as a result, simple and deep structure appeared beneath what had previously appeared as a disparate collection of very complicated materials. The conceptual leaps were considerable: ideas from fields dominated by the structures of quantum mechanics and many-body physics required translation into systems dominated by thermal physics, and even from those traditionally thought of as branches of chemistry (polymers, liquid crystals). In the case of de Gennes (a former student of Friedel), this background was superconductivity; for Edwards (a former student of Schwinger), quantum field theory. We examine the deep mathematical structures that generate this surprising connectivity in the next section, but first identify the physical characteristics of soft condensed matter systems. This chapter continues with a discussion of two relevant examples of philosophical interest and closes with an assessment of the application of soft matter physics to biological systems.

Characteristics of soft matter

First, the energy scale of internal interactions in soft matter is comparable to the quantum of thermal energy, $k_B T$. So in contrast to systems whose physics are dominated by quantum mechanics, thermal transitions between microscopic energy levels are frequent, and quantum coherence is (usually) negligible. Classical statistical mechanics furnishes, as a consequence, the appropriate set of tools to model and calculate with. Fluctuations in structure are large, and local equilibrium the dominant paradigm.

Second, mesoscopic structure or order at the length scale of several nanometers is almost ubiquitous in soft matter systems. For example, colloids are suspensions of particles at this scale suspended in a solvent – stabilised because the thermal energy is large compared to their typical gravitational potential energy in the bulk fluid. To give some other examples: polymers are long-chain molecules of very high molecular weight whose internal flexibility results in configurations in solutions or in melts such that the random macromolecular coils are typically several tens of nanometers in scale. Liquid crystals are solutions of molecules with strong directional order that support lamellar and defect structures dominated by this length scale (much larger in this case than the scale of the molecules themselves). The dominance of this structural, 'mesoscopic' length-scale (neither macroscopic – mm to m, nor truly microscopic at the atomic dimension of sub-nanometer) is also responsible for the epithet 'soft'. The existence of a typical structural length l, together with the condition of strong thermal dominance, leads to a natural estimate of the elastic modulus $G \sim (k_B T/l^3) \sim 10^6$ Pa, a thousand times smaller than the modulus of metals or ceramics, and in the region of that of rubbers (of course another soft matter exemplar – a class of gels) (McLeish, Cates, Higgins and Olmsted 2003).

Third, the dynamics of soft matter systems are often very rich and contain one or more 'slow variables' – coordinates that due to constraints or internal energy barriers, return to equilibrium on much longer relaxation timescales than the typical intermolecular ballistic trajectory time (at 300 K) of ~10ps (Larson 1999). An example is the set of very slow viscoelastic relaxations in solutions and melts of polymers (Doi and Edwards 1986). These are generated from the multiple topological interactions between polymer chains, which typically (for chemically formed polymers) cannot cross each other. Another example is given by the long, and sometimes extremely long, structural rearrangement times in foams and glassy materials in the approach to the glass transition. In these cases dynamic arrest can arise from the requirement that any dynamical move implies the simultaneous rearrangement of increasingly large coherent volumes of the material. This in turn demands a large energy fluctuation ΔE, suppressed by the Boltzmann weight $e^{-\Delta E/kT}$.

A fourth characteristic of soft matter is the thermodynamic (and emergent – Ellis 2012) property of multiple realisations at lower levels than the operative structure. For example, even when states of a colloidal or polymeric system are characterised by a specific configuration of colloidal particles, or entire polymer chains (at suitable small-scale resolution), there are many configurations of solvent molecules and/or of subchain states that correspond to the same 'meso-state'. For many purposes (e.g. the analysis of scattering experiments, or the measurement of osmotic pressure) even coarser variables, such as local mean density averaged over a mesoscopic volume, are sufficient and hyper-exponentially more multiply realised in microstates. The techniques of statistical mechanics can therefore be applied even to these mesoscopic structural volumes.

Fifth, and consequent on the other four, the variety of soft materials and their phases exhibits a high degree of 'universality'. Essentially the same emergent material arises from different underlying chemistries. With a rescaling of a few coarse-grained parameters, the mapping may be essentially exact. So, for example, the linear elastic modulus of a polymeric gel is dependent on the density and distribution of the cross-links between its constituent polymer chains, not on the chemistry of the chains themselves. The similarity may persist into even non-linear response.

The interdisciplinary nature of the (multiple) sub-discipline of soft matter is remarkably broad. The nature of the materials required in each of the exemplars listed earlier frequently implies just as significant challenges to synthetic chemistry in their fabrication as it does to theoretical and experimental physics in explanation and characterisation. In nearly every case soft matter has also found application in structural and functional materials (personal care products, plastic materials, displays and more) so engineering disciplines are frequently involved, often at a fundamental level, while the scientific community conversation over soft matter spans industry and academia. The more recent application of soft matter science to the analysis of biological and bio-inspired phenomena (Nelson 2004) increases the interdisciplinary palette even further. The consequences are as yet hard to predict, but already two promising directions for research have been generated by the confluence of biology and the statistical physics of soft matter. The first sheds new light on the physical basis biological phenomena; the second draws inspiration from biology to define new research programmes in physics. An example is the rapidly growing field of 'active matter'. The pickings for historians and philosophers of science interested in the social functioning of scientific communities will, I suspect, prove very rich when a little more dust (some of it 'active') has settled (McLeish 2011).

Theoretical and experimental provenance of soft matter

The selection of de Gennes and Edwards as the two leading pioneers of soft matter as a recognised sub-discipline is not a controversial one. But in each case their contribution illustrates the power of theoretical and explanatory concepts to unify and propel research programmes, especially when accompanied by ready-made mathematical machinery. In this case, the techniques of path (and functional) integration from quantum field theory and symmetry breaking (especially within the Landau-Ginzburg theory of superconductivity) proved particularly powerful tools to the new field, witnessing at the same time the deep commonality enjoyed by the underlying structures of physics.

Polymer physics and path integrals

The concept of a polymer within the mathematical physics of statistical mechanics illustrates the point. Feynman's early work on quantum theory provided a third mathematical representation of quantum mechanics, in addition to the two previous instantiations arising from the wave mechanics of Schrödinger and the matrix mechanics of Heisenberg. Feynman's contribution was the framework of *path integrals*. In particular, he found a new way of conceiving of the propagator $G(r,t)$ for the wave function of a quantum particle. This central and generative object is the ('Greens-') function that gives the amplitude that the particle is at position r at time t given that it was at the origin $r = 0$ at time $t = 0$. After some formal care in definitions, this function can be calculated as a formal sum over all possible classical trajectories from $(0, 0)$ to (r, t), weighted by a phase S, itself a functional of the path, denoted as follows:

$$G(r,t) = \int D[r(t)] e^{\frac{i}{\hbar} S([r(t)])} \tag{1}$$

Here, the physical contribution to the phase from each path $S([r(t)])$ is the classical action for the single path $[r(t)]$, and the integral is the *functional* or *path* integral − summing not over the measure of a single variable, but over all possible functions (in this case paths) connecting the origin to the space-time point (r,t). This extension of the notion of an integral to a functional integral is denoted by the special measure $D[r(t)]$. The action is, in turn, the (standard) time integral of the classical Lagrangian for that path, so that, for a single spinless particle of mass m moving in a potential $V(r)$, the expression for the propagator becomes (Feynman and Hibbs 1965):

$$G(r,t) = \int D[r(t)] e^{\frac{i}{\hbar} \int_0^t \left[\frac{m}{2}\left(\frac{\partial r}{\partial t}\right)^2 - V(r(t)) \right] dt} \tag{2}$$

The path-integral formulation illustrates very powerfully the connection between classical and quantum physics − the former provides phases for the state-addition of the latter. It can also be shown that, in the classical limit, where these phases change very sensitively in response to perturbations of the paths, that the propagator is dominated by a quantum mechanical superposition of paths that lie very (exponentially) close to the classical trajectory. Far from the classical path, phase contributions from close trajectories cancel each other. Techniques for the computation of such path integrals in various potentials, and under perturbation theory, were developed in the 1950s and 1960s.

In statistical mechanics, the generative object within a theory for any physical system is the *partition function*, $Z(\{X_i\}, T)$, where the $\{X_i\}$ are the external constraints on the system and T the temperature. Z is calculated as a sum over all accessible microstates of the system, weighted by a 'Boltzmann Factor':

$$Z(\{X_i\}, T) = \sum_{states\ i} e^{-E_i/kT} \tag{3}$$

All thermodynamic properties can be calculated from this object − for example, the free energy $F = kT\ln Z$. In the case of a polymer, a single microstate is just one of the spatial paths accessible to it. Providing that the molecular chain is long enough, even the single steps of the path can be large compared to the scale of atoms, so that an appropriate mathematical coarse-graining becomes a trajectory $[r(s)]$ in terms of a contour variable s that runs from 0 at one end of the chain to the total number of steps ('monomers') N at the other end. A small subchain of Δs steps of mean length b is a small random walk and so has a Gaussian distribution of end-to-end displacements Δr, $\exp(-3(\Delta r)^2/2b^2\Delta s)$. Adding in this local weight for each small section of the polymer and allowing the chain to pass through a general potential energy for its monomers $V(r)$ generates an expression for the partition function of a polymer chain that takes the form of a path integral, since the sum over states is simply a sum over paths constrained to the same end-to-end displacement $\boldsymbol{R}$:

$$Z(\boldsymbol{R},T) = \int D[r(s)] e^{\frac{1}{kT}\int_0^N \left\{\frac{-3}{2b^2}\left(\frac{\partial r}{\partial s}\right)^2 - V(r(s))\right\}ds} \qquad (4)$$

Comparing the expressions (4) and (2) demonstrates the remarkable analogy between the quantum mechanics of single particles and the statistical mechanics of polymers. Both 'sum over paths' follow the visual recipe of Figure 20.1.

The contour length s takes the form of an imaginary time variable in the Feynman propagator. The (square) step length maps onto an (inverse) particle mass, and the temperature (suitably converted to the units of energy) plays the role of Plank's constant. Varying the temperature of the polymer system is the correspondence of varying the scale of dimensionless action of the quantum system between quantum and classical limits. So, for example, extreme tension applied to a polymer chain stretches it so that all fluctuations in its typical path lie close to a 'classical' trajectory of a straight line between its endpoints. Such a situation is equivalent to the large-mass limit in the quantum case, where this is the dominant classical trajectory of the particle.

For all that these considerations might appear arcane, they provide the physical insight, as well as computational strategies, into phenomena such as rubber elasticity and nanoscale structuring of reinforced plastics. When a rubber band is stretched, the force with which it responds comes not from any physics at the atomistic level, such as the stretching of molecular bonds. Rather, it is an emergent effect arising from the increased restriction on the multiplicity of realisations of all the polymer chains belonging to the rubber network into which they are chemically linked. Similarly, the local separation of nanoscale volumes of rubbery polymer within a glassy matrix of more brittle polymer – an example is the self-composite styrene-butadiene (SBS) – can be conceived of as the interplay between the two terms in the 'action' of the polymer partition function (4). A repulsive effective potential $V(s)$ between the two different chemistries of polymer chain, which would on its own drive them into two bulk phases, is moderated by the 'kinetic energy' term penalising polymer chain stretch. Together the two effects create the local length scale and structure of the all-important separated structure. This in turn leads to the remarkable emergent property of toughness.

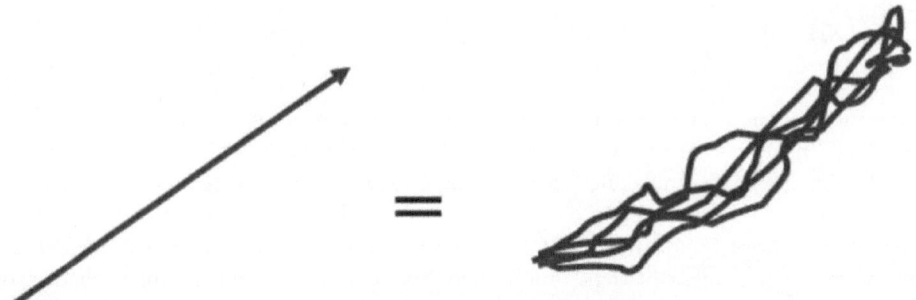

Figure 20.1 The propagator (arrow on left) can be calculated as a weighted sum over all possible paths from starting to finishing points. A situation close to the classical limit (for the quantum particle) or high-tension limit (for the polymer) is illustrated, where the highly weighted paths cluster around a single trajectory.

Liquid crystals and symmetry breaking

A similar deep structure from condensed matter physics (superconductivity) and field theory (the so-called 'Higgs mechanism') provided a conceptual framework for the physics of liquid crystals. For all these systems exhibit 'phase transitions', as a function of temperature, that change the emergent symmetry at the bulk material level. As we saw, liquid crystals possess molecules with a high degree of anisotropy – one can think of them as molecular 'rods' or 'needles'. At high temperatures, the thermal fluctuations at the molecular level are large enough to maintain complete randomness in the distribution of orientations of the molecular rods. At the macroscopic level the material is a fluid – with no translational structure or preferred orientation, even though there is very local orientation ('short-range order') in the immediate vicinity of a single molecule, since near neighbours tend to be correlated in their orientation. However, as the temperature is lowered, the thermal fluctuations become weaker, and the local interactions between rods that tend to align them are able to drive a transition of the entire system. In the resultant 'nematic' phase of the liquid crystal, a preferred orientation appears at the bulk level. Although there is no spatial order so that the system is still a fluid in its propensity to flow, there is orientational order (experimentally apparent, for example, in the optically polarising properties of the fluid).

When the liquid crystal system is described in terms of the emergent field of orientational order, a mathematical similarity to phase transitions in superconducting materials becomes apparent. The general structure of the formalism invokes a field of coarse-grained local 'order parameter' $S(\mathbf{r})$. In the case of a liquid crystal, S describes the degree of local orientational molecular order; in the case of a superconductor, it measures the strength of the superconducting wave function. The free energy becomes a functional of the field $S(\mathbf{r})$:

$$F\big(S(\mathbf{r}),T\big) = \int \Big\{ a_0(T) + a_2(T)\big[S(\mathbf{r})\big]^2 + a_3(T)\big[S(\mathbf{r})\big]^3 + a_4(T)\big[S(\mathbf{r})\big]^4 + K(\nabla S)^2 \Big\}dV \quad (5)$$

Symmetry considerations can eliminate some of the orders of expansion of the field in this expression – for example, in two dimensions the nematic liquid crystal order parameter must satisfy the symmetry $S \to -S$, so a_3 can be set to zero. The 'Landau-Ginzburg' form of (5) was first written down as an ansatz for a coarse-grained free energy in terms of averaged quantities. Later a formal procedure enabled the coarse-graining to be computed from a fine-grained Hamiltonian.

The phenomenology of (5) is very rich, identifying similar phenomena in magnetism, liquid crystals and particle theory. To take one example, when the symmetry $S \to -S$ applies, a 'second-order phase transition' appears at the temperature for which $a_2 = 0$. When $a_2 > 0$ the equilibrium value for the field S is zero everywhere, but when $a_2 < 0$, the free energy is minimised at $S = \sqrt{(-a_2/2a_4)}$. Below the transition point (which might be the onset of a finite magnetisation or a finite orientation of a liquid crystal field), the field grows continuously in strength. The form of F at temperatures above and below the transition is shown in Figure 20.2.

When spatial structure is added to the theory (for example, the last term in (5) penalises spatial gradients in the order parameter), the appearance of 'textures' and 'defects' can be understood. The underlying symmetry of the system, expressed by zero field at high temperature, is at low temperature recovered by an unbiased choice of direction in which the symmetry is broken locally. This naturally gives rise to spatial structures in which the direction of the liquid crystalline order is different at different spatial locations. To accommodate such spatial variation of local orientations, very localised 'topological defects' arise – points at which the orientation field is mathematically singular and cannot be defined. Figure 20.3 illustrates four possible such topological defects in terms of the local behaviour of the orientation field around them and an experimental micrograph of a nematic liquid crystal that contains instantiations of these objects.

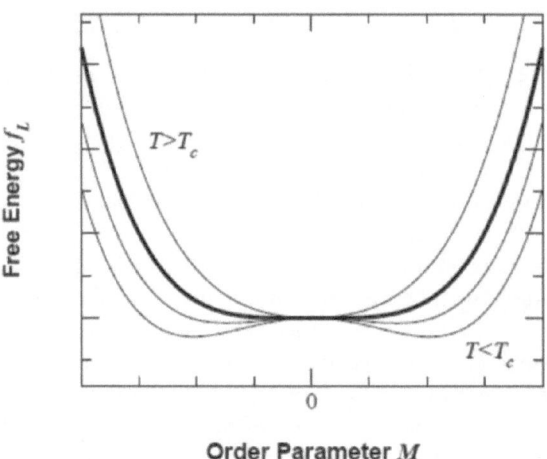

Figure 20.2 The Landau free energy (5) for a field undergoing a continuous phase transition

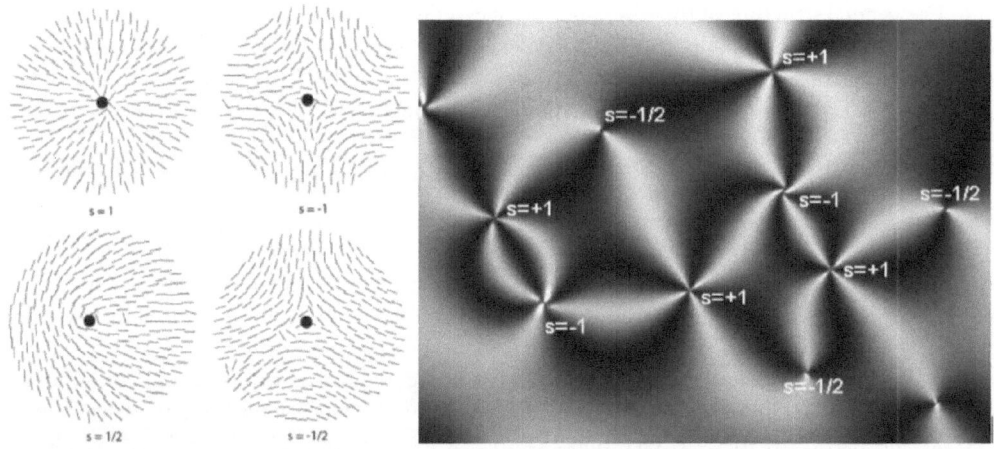

Figure 20.3 (a) Nematic liquid crystal defects of four different orders; (b) a spatial defect structure containing the examples in (a) under polarised light microscopy

[Courtesy of B. Senyuk, University of Colorado]

The nematic liquid crystal, in which the only order parameter is the mean molecular orientation, is the simplest of a very complex family of materials. 'Smectic' liquid crystals introduce one-dimensional (layered) order in addition to the orientational order (so breaking spatial symmetry), and 'cholesteric' liquid crystals introduce the characteristic of twist and a breaking of mirror symmetry. The defect structures of these higher-order liquid crystals are correspondingly richer. They also possess similarity mappings with completely different physical systems. So, for example, an analogue of a vortex phase in a (Type II) superconductor that supports a lattice of magnetic field lines was predicted in smectic cholesteric liquid crystals and later found experimentally as their 'twisted grain boundary' phase.

Experimental commonalities

It would be wrong to suggest that the coherence of soft matter as an emergent discipline is generated entirely by its underpinning theoretical ideas, although such conceptual cross-ties are essential. There is also a strong family of experimental tools that reflect the universality within the materials themselves, and especially their meso-structural properties. Characterising the nano-scale structures within polymeric, colloidal or membrane-dominated materials relies on both electron and optical microscopies (see Figure 20.3). However, material opacity and complexity mean that spatial information is often acquired more effectively by scattering techniques. Rather than probe spatial structures directly, recording the variation of density at different *positions*, scattering of light or, more often, x-rays, responds to the averages of density waves of different *wavelengths* within the materials (one might speak loosely and say that scattering is the Fourier transform of microscopy).

Neutron scattering is especially associated with soft matter for a number of reasons. Neutrons possess a wavelike nature by virtue of quantum mechanics, and as neutral particles, are not affected by the charged electrons of atomic and molecular orbitals. They can therefore penetrate into the bulk of any matter whose nuclei scatter them weakly enough. The existence of different isotopes of elements gifts another precious tool to the neutron scatterer from soft matter: careful synthetic chemistry may incorporate different isotopes into selected parts of the molecular components of a material. Although such 'isotopic labelling' does not change the chemical behaviour of the molecules, thus leaving the structure unchanged, the strength with which different isotopes scatter neutrons can differ markedly. The most common secondary isotope of hydrogen – deuterium – scatters neutrons, but with the opposite phase. So with careful quantitative matching, hydrogen-containing molecules can be made to 'disappear' from a neutron-scattering experiment. In this way, the spatial structuring of components of a soft matter system may be visualised. For example, polymers composed of blocks, themselves comprising segments of different chemistry (we recalled a commercial example in the two-phase material SBS earlier) may self-assemble into periodic structures so that sub-chains of the same chemistry are in mutual contact. Scattering from labelled polymers can pick out the spatial patterns of just one component and by repeated application can provide very rich and selective spatial information on self-organised structures in soft matter.

There is an appealing and direct link between scattering experiments and common theoretical framings of soft matter systems. In the absence of crystalline order, the expressions for the density of free energy (in terms of density, or concentration of one or more components) of polymers, liquid crystals or colloids at the coarse-grained level must obey translational symmetry. They are also typically short range and so contain powers of the density and its spatial gradients and higher derivatives; equation (5) is an example, with the field $S(r)$ now interpreted as a density. If this local variable is also the density of relevant *scatterers* within the experimental framework, then the scattering function (the Fourier transform of the scattering density) can be calculated as a thermal average directly, where the spatial gradient terms become simply algebraic in the scattering function. The simplest example is the very common Ornstein-Zernicke function for a field whose free-energy density is quadratic in density and its gradient:

$$F\big(\rho(r),T\big) = \frac{1}{2}\int\left\{a_2(T)\big[\rho(r)\big]^2 + K(\nabla\rho)^2\right\}dV \qquad (6)$$

In terms of the scattering intensity $I(k) = <\overline{\rho(k)\rho(-k)}>$ predicted experimentally (k is both the experimental 'scattering vector' and the Fourier transform conjugate variable for r) from this density field, the form is simply 'read off' from (6) once the 'recipe' Gaussian integrals are applied:

$$I(k) = \frac{1}{2\pi} \frac{V\rho^2}{\left(a_2(T) + Kk^2\right)} \tag{7}$$

Scattering experiments possess a pleasingly direct access onto the coarse-grained, mesoscopic Hamiltonians of soft matter. The presence of thermal disorder, spatial symmetry and local structure of interactions means that the natural language for theory is in the scattering function of the density variables, not in their spatial correlation functions. Spatial structures are better thought of as consequences of, rather than fundamental to, the physics. So to complete our example, the density correlation function corresponding to the general quadratic and local expression for the free energy of a scalar field in (6) that has the scattering function of (7) takes the 'Yukawa' form, of a screened Coulomb field:

$$I(r) \equiv \rho(0)\rho(r) = \rho^2 \frac{e^{-r/\zeta}}{(r/\zeta)} \tag{8}$$

where the screening length $\zeta = \sqrt{(K/a_2)}$ reflects the interplay of energies from density and its gradient, manifest in the Hamiltonian (6) in terms of the Fourier variables. Near a critical point where two mixed polymers of different chemical character begin to demix, large and long-range fluctuations in composition are modelled by this form of correlation.

The feature of slow dynamical structures within soft matter motivates a further set of techniques to identify and characterise, common to soft matter, and which focus on the mechanical properties of the material. Furthermore, there are complex connections between the structural changes at the nanoscale and the resultant dynamical properties of the bulk material. When these appear in macroscopic flow or deformation, the corresponding techniques fall under the banner of 'rheology'. Careful application of different geometries of material deformation – shear (equivalent to 'lubrication flow') and uniaxial extension (equivalent to 'fibre spinning flow') and at different rates – can identify processes of structural reorganisation at a range of timescales. The mechanical stress on the material surface is typically measured at the same time through sensitive transducers. Since stress has many components, corresponding to forces on the material surfaces in both parallel and normal directions, the rheology of soft matter is often especially rich. There is a historical and conceptual resonance with scattering experiments here: Weissenberg, who invented one of the first rheometers for soft matter, called it a 'rheogoniometer', deliberately alluding to the detector device used in scattering experiments. He conceived of rheology as a sort of scattering experiment – samples 'received' a range of geometries of strain and 'scattered' a range of geometries of stress (McLeish 2008).

Examples will serve to illustrate the point. The uncross-linked melt of polymers, which flows as a liquid at long timescales, mimics a cross-linked gel when mechanical oscillation is performed on it at high enough frequencies. This is because the polymer chains are subject to *topological* constraints – they are unable to pass through each other, even if they are not chemically linked. So in the melt form, the dynamical reconfigurations that a single chain would rapidly adopt through random thermal motion are strongly suppressed. Since these are precisely the reconfigurations that would be responsible for the relaxation of mechanical tension created by extending or deforming the chains through flow, they will retain these strained configurations for extended periods until much slower, topologically allowed modes of motion are able to re-equilibrate the ensemble. In this case, de Gennes and Edwards together contributed insights that identified the predominant slow mode of structural relaxation in dense, entangled polymer systems. Each individual polymer chain is constrained by topological constraints from its neighbours to a tube-like path that follows its own contour. It renews its (random-walk) configuration via the only non-constrained mode of Brownian motion – the one that carries its centre of mass in diffusive dynamics along this path. This is the motion de Gennes termed 'reptation' (Doi and Edwards 1086) and is described in more detail later. When the flow rate is increased, the topological constraints between chains are removed by convection as well as by the diffusional reputation, thus endowing the material with a 'shear-thinning' property (the apparent viscosity reduces with flow rate).

A range of colloidal fluids with very dense packing exhibit an opposite effect: while fluids at low flow rates, they may suffer a strong increase in the effective viscosity as the imposed flow rate increases, leading in some cases to a liquid-to-solid like transition. As in the dense polymer case, this solidification is a macroscopic consequence of microstructural processes. But here, rather than a moderation due to flow, as in the (topological) interactions of polymer chains, attractive colloidal particles suffer an increased rate of collision in a shear flow and can build up temporary chains of particles under compressive load that can no longer disassemble through diffusion. Such 'jamming transitions' constitute a highly cooperative non-equilibrium change of state and give rise to remarkable bulk phenomena. An example is the extreme, but temporary, thickening of cornflour paste made with very little water or milk.

The discovery that polymer and colloid chemists, theoreticians thinking about the same systems, and experimentalists exploring the simultaneous measurement of macroscopic and microscopic structure in complex fluids all faced common challenges and had developed similar tools proved a significant cohesive force in the constitution of the new field of soft matter. Ideas from one material-dependent subfield nourished others, and continue to do so.

Multiple realisation and top-down causation – examples from polymer physics

From what has been said already, there is clearly a rich field of exemplars in soft matter physics that constitute relevant source material for current discussions within the philosophy of science. In particular, soft matter physics indicates why the level of physical structure, energy or coarse-graining to which is attributed the label 'fundamental', or assigned causal powers, is up for debate (Anderson 1972). In this section we re-examine two examples from the case of polymer physics that explore two important ideas for the philosophy of emergence: multiple realisation and top-down causation.

Multiple realisation

A central theme in the discussion of the tension between emergence and reductionism has been the notion of multiple realisability (Bickle 2016). Originally adopted by Putnam and Fodor in the context of multiple realisability of mental states by physical ones, it has since been recognised as a pervasive yet contentious challenge to reductionism (Aizawa and Gillett 2009). The claim against reductionism is that the observation that all mental (or otherwise defined high-level or 'special science' states) are multiply realised renders implausible the reductionist requirement that all high-level states, including their causal relations, map onto physical (low level) states via bridge laws that preserve causal relations for every realised kind.

The idea is normally discussed in the context of the contested reduction of the mental to the physical, but examples from soft matter, underpinned as they are by statistical mechanics, provide rather cleaner examples of multiple realisation that are more straightforward to discuss (in that all the science is understood at both levels) than the mental. We saw earlier (the path integral expression of equation 4) that the statistical physics of a polymer chain is expressed mathematically and explicitly as a sum over the multiple realisations of its end-to-end vector. One emergent quantity that can be derived from the integral is the effective force-displacement relationship $f(R)$ of this vector between the chain endpoints. From standard statistical mechanics:

$$f(R) = -kT\nabla_R \ln Z(R,T) = -\frac{3kT\,R}{2b^2} \tag{9}$$

The Hookean linear spring behaviour of the polymer is an emergent property of the entire chain, predicated *on the multiple realisability of the microstates alone.* Here the emergent physics at the level of the polymer chain (its 'rubber elasticity') depends on the many realisations of the chain configuration and nothing else. Furthermore, causal powers of the chain itself (its role in sustaining elastic waves in a cross-linked polymer rubber, for example) depend upon this chain-level property, not the time-dependent microphysics of the chain segments. This is because, through the thermal coupling to the environment accounted for in the canonical ensemble within which the path integral is performed, the dynamics of the microstates are (explicitly) non-causal, while the dynamics at the level of the Hookean chains are.

Top-down causation – long-range topological order in polymer rings

The topological physics that constrains the snake-like 'reptation' diffusion of polymers in the melt phase, albeit for kinetic, not thermodynamic, behaviour, becomes much stronger in the case of polymer rings. This fascinating soft matter system has emerged both experimentally and theoretically to generate a rich and puzzling ensemble of materials from the liquid to the solid. In these cases the polymer chains have no ends at all, but are synthesised as connected rings. The richness in the multiple possibilities for the macroscopic behaviour of a system of many such rings arises from the multiplicity of topologies that the rings may enjoy with respect to each other.

Figure 20.4 demonstrates the variety of topologies of a single ring (4b) and multiple rings (4c and d) in contrast to the temporary interactions of non-cyclised chains (4a). The unlinked system of 4b has remarkable properties (Everaers 2004) – the constraints on the chains generated by the trivial topology (all linking numbers are zero) becomes, via its large restriction on entropy in the partition function (multiple path integral) of the system, a huge 'topological pressure' on the chains. They collapse, as a consequence, into much smaller dimensions that would be the case for free rings. Furthermore, the emergent material is a fluid, with no memory of its initial configuration and with each polymer able to diffuse arbitrarily long distances. The linked melt (4c), by contrast, generates a solid at the macroscopic level, even though the polymer chains possess as

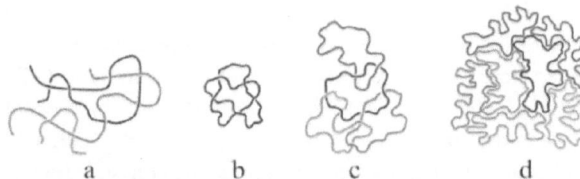

a   b   c   d

Figure 20.4 Possible configurations of polymer chains in different topological states in a melt of other chains; (a) linear chains in Gaussian interpenetration; (b) a single ring-chain self-knotted (the trefoil knot); (c) a melt of Gaussian ring-chains with typically complex topological invariants leading to an emergent network; (d) melt of rings in trivial topology in crumpled globule states.

much local mobility as they do when unlinked. The long-range topological order between the linked rings both localises their mobility and creates an emergent elastic solid from the ensemble.

From the point of view of emergence and causation, this system bears strong structural similarities to the fractional quantum Hall effect (FQHE), discussed recently by Lancaster and Pexton (2016). High-level and causative degrees of freedom (those of the elastic solid) emerge because of the long-range topological constraints. Without knowing the linking numbers of the chains, constituting a set of variables to be specified in addition to the molecular-scale configurations, the trajectory of the system at the coarse-grained scale is not possible to compute. Furthermore, the causative macroscopic trajectory is explored through multiple realisations of the low-level molecular (chain-linked) variables, which are themselves not causal (they are coupled to an external heat bath which exchanges heat with them in the canonical way). The topological variables are not an epistemological convenience in this case – a charge that can be made of the coarse-grained Fourier variables of theories such as the Landau-Ginzburg Hamiltonian of equations (5) and (6). They appear as ontologically constitutive of the emergent material, be it the gel-like solid of the linked system, the viscoelastic fluid of the unlinked melt, or any one of the infinite family of anomalously viscoelastic fluids that lie between these two limits.

Soft matter and biology – emergence from looped causation

The classical structures of soft matter – macromolecules, membranes, self-assembled nanoscopic systems – are all recruited by biology to serve the structural and functional substrate of life. In a surprising recent development in the history of science, these features, together with the experimental and theoretical developments discussed in this chapter, have furnished the scientific community with the latest example of a series of translations from the physical to the biological sciences of both people and ideas.

Inasmuch as the scientific structures apply, but with careful and specific modifications, then so do the philosophical applications. In particular, the self-replicating, evolving and functional aspects of biological organisms recruit, but also extend, soft matter materials. As a consequence, the illustrations of emergent variables, top-down causation and non-locality become even stronger in the biological than in the non-living context. There are several reasons for this, whose provenance is the special additional features that apply to biological soft matter. Principally these are:

1   The ubiquitous context of evolution, a context of both temporal and spatial non-locality which also provides a pressure towards optimisation of replication within niches at every level of an organism.
2   The 'active' properties of biological matter, in contrast to the 'passive' equilibrium or out-of-equilibrium behaviour of non-living soft matter, endow the colloidal, polymeric or self-assembled structures with the capability to metabolise ('burn fuel') and via entropy production globally to sustain highly non-equilibrium occupation of phase-space.
3   The duality of genotype and phenotype: so every biological structure is implicit in another biological structure (the DNA of the relevant sections of the genome) in an information-processing sense.

These three additional dimensions to the science of soft matter have generated in turn their own programmes of research, including non-living models of active matter (Ramaswamy 2010), computational coarse-graining of genotype–phenotype maps and both experimental and computational evaluations of evolution (Khatri *et al.* 2009). These have now entered the graduate physics curriculum in many leading research institutions (Nelson 2004).

To exemplify the way that biological physics is linked to and emerges from soft matter physics, we briefly look at one example which draws heavily on the macromolecular topology of rings.

Entangled DNA and topoisomerases

A remarkable application of polymer physics came to light in the theoretical examination of bacterial cell division. Since the early work of de Gennes, Doi and Edwards (Doi and Edwards 1986) motivated by the phenomenon of viscoelasticity in concentrated polymer solutions and melts, we have understood that the principle underlying physics is that of the *topology* of strings in random, fluid configurations. Rather than attractive or repulsive interactions between molecules, as we saw earlier, it is the uncrossability of two 1-dimensional objects embedded in a 3-dimensional space that endows the system with very slow dynamics. For in order that the coarse-grained fluid composed of many overlapping polymer molecules may flow, the molecular chains must themselves repeatedly reconfigure themselves, adopting new neighbours and leaving old ones. This process cannot be achieved by simple convection with the flow, as the molecular chains cannot pass through each other.

In synthetic polymeric fluids, the dominant diffusive process is the only one not inhibited by these topological constraints: the diffusion and convection of chains along their own contours. It is the chain ends, not subject to the same topological constraints as the inner chain segments, which allow new configurations to be adopted. As described earlier, this one-dimensional contour diffusion was named *reptation* by De Gennes, who was reminded of a snake-like crawling (Figure 20.5). He showed, since confirmed by many experiments, that the timescale for reconfiguring a single polymer molecule by the reptation process scales as the third power of its molecular weight. Since macromolecular chains can reach very high molecular weights, these times can become (in molecular terms) extremely long, even seconds or minutes.

This topological slowing down of the dynamics of diffusion and flow appears in biological contexts as well. The most extreme example is the requirement of the separation of daughter strands of bacterial DNA into the two new daughter cells. In bacteria, DNA is not confined to the ordered structures of chromosomes, but is much more randomly distributed through the organism. So when it divides into two strands prior to cell division, the two macromolecules thus created are in a highly entangled state. If they were to disentangle by reptation, or even by forced

Figure 20.5 Schematic of the reptation of a polymer chain (solid curve) entangled with neighbours (dots intersecting the plane) and moving in an effective tube (dashed lines)

diffusion along the contour length determined by the topological uncrossability constraints, the timescale for cell division would be astronomically long. Instead, one of a family of enzymes known as topoisomerases (Roca 1995) performs local breaking and recombination of DNA strands at points at which two strands meet together with the enzyme. During the process, the unbroken strand is passed though the nick in the other strand before it is healed and the enzyme releases from the two strands. The action of the topoisomerase is to change the topological state of the two strands. Essentially, this is an *active* process – the information of the earlier state is erased, and as shown by Landauer (1961), this must be accompanied by the dissipation of heat from a fuel source or another source of free energy. The topoisomerase oxidises the common biochemical fuel of ATP into ADP for each information-processing event.

Among many remarkable aspects of this near-miraculous example of evolved molecular engineering is that a small fleet of topoisomerase II molecules are able to resolve the topological constraints that would otherwise inhibit DNA segregation on cell division. This is all the more surprising since the crossing-over events are local but need to respond to a direction of topological complexity (the strands need to move from higher degrees of entanglement to lower in order to separate) which is defined only globally. It is not, for example, possible to decide whether two loops of string are knotted by examining them locally, but only from their global configurations. The current hypothesis for the mechanism of communication from the global topology to the local activity of the enzyme draws on statistical mechanics. There small biases in thermal fluctuations will tend to explore less constrained states slightly more frequently than more. So if the complex of topoisomerase and the two strands of DNA is sensitive to the bias in these fluctuations, represented as an attempt frequency to cross or to escape, after many such encounters, there will be a drift in the very high dimensional space of DNA topologies toward simpler entangled states.

The 'top-down' causative role of long-range topology in the physics of string-like structures is directly connected to and analogous with the synthetic case of ring polymers. Here, as we saw, the two extreme states in which no ring molecule is linked with any other and in which they all are constitute an emergent liquid and solid, respectively. This is true in spite of the fact that all local physics is identical in the two cases. Intermediate topological states tune continuously between the liquid and solid states via an unusual type of percolation transition. Usually for any liquid-to-solid transition, this is second order, not first order, in the control parameter (the mean linking number of the ring molecules).

Topology is defined in all these systems only globally and in terms of the coarse-grained variables of the complete molecular paths. Furthermore, it furnishes an additional set of state variables themselves undefined at the atomistic level, yet which are highly determinative of the future evolution and macroscopic properties of the systems. In the biological case, the long-range topology is not only causatively operative on the macromolecular chains themselves and the microstates to which they are restricted, but onto the function of the smaller, topoisomerase molecules as well. Their enzymatic function is controlled, through biases in dynamic fluctuations locally, from the long-range topological relationships between crossing DNA strands, and is therefore another example of a physical instantiation of multiple realisability.

The role of evolution is evident from the highly specific function of the topoisomerase, which depends delicately on its internal fluctuations, hydrophobic and charged interactions, and a delicately tuned affinity for DNA that is enhanced when strands cross. The selection of protein structures over very long periods of evolutionary time has found solutions to the entanglement problem of extraordinary sophistication. We might note here that a different strategy seems to have been adopted by eukaryotes, in which DNA packing in the chromosome is affected in such a way that entanglements are highly suppressed (Mirny 2011).

Biological lessons

The potential for strongly emergent physics within soft matter seems to be recruited by biological systems ubiquitously. Perhaps this is not surprising, given our prior experience of candidates for top-down causation within living organisms. What may be more surprising is the relatively low level at which examples are already multiply far removed from considerations of mind and cognition.

The example earlier has allowed us to follow in detail the way that long-range (e.g. topological) physics is differentiated from the merely coarse-grained and leads to a strong, rather than weak, notion of emergence. Similarly the phenomenon of multiple realisability serves to ground a strongly emergent ontology for realised material form and function in addition to the weak coarse-graining of low-level descriptions.

The approach of taking a physical perspective onto biological matter additionally illustrates the unboundedness of physics from any special scale of length or energy. Rather, it locates the 'physical' at the set of fundamentally causal variables, which themselves may simultaneously occupy multiple length scales (and in biology unvaryingly do).

Finally, the biological context seems to provide a unique theatre for the emergence of loops of causality between different levels of coarse-graining. To follow our example, the top-down causation generated by the long-range entanglement topologies are in this case implemented through protein structures themselves coded for by the DNA polymers themselves. The direct physical, contextual emergence of biological mesostructures, such as the entanglements of prokaryotic DNA, is a consequence of local molecular structures in the gene. These in turn are consequences of the large-scale assembled structures that they code for. Another example is furnished by the structure of cell membranes, themselves active in bringing membrane-resident proteins together for enzymatic or signalling activity, while the membranes themselves are coded for at a lower level.

The development of 'biological physics' in the sense of a truly new sub-discipline at the same level as 'chemical physics' shows every sign of being at a very early stage (McLeish 2011 and other articles in the same issue). Beyond the mere application of theoretical and experimental ideas from physics *to* biological systems, the maturation of the field would fill out with more examples of new physics motivated by the interdisciplinary encounter. The original work on active matter and a possible field of the physics of evolution seem promising in this regard.

Conclusions

Soft matter science constitutes a rich exemplar for both historians and philosophers of science. Although this chapter has focussed on the philosophical material relevant for the debate around emergence, the historical context is relevant as well, as it traces an inherently interdisciplinary encounter of physics, chemistry, materials science and, latterly, biology. To take one example, there are deep mathematical relationships, through the structure of field theory, between the statistical mechanics of soft matter and quantum field theory which have been historically operational in the development of the field.

Soft matter physics has clarified the salient difference between coarse-graining and non-locality and has identified examples of causal variables which, while 'fundamental' in at least the same sense by which atoms are 'fundamental', are non-reducible, have mathematical description and have causal power. The field also illustrates how the philosophical notion of 'multiple realisation' carries essential physical concomitant structures, strongly exemplified by the statistical mechanics of polymers. Material properties such as polymer elasticity and membrane curvature emerge from (thermally generated) multiple realisability.

Biological matter is amenable to a perspective from soft matter science and suggests additional components (evolutionary context, active matter, self-coding) that create new possibilities for causal connection between variables at different length scales that form self-sustaining loops. Not present in non-living soft matter, these long-range/short-range connected causal structures may provide another route to define what we mean by 'living'.

Acknowledgements

I would like to thank Robin Findlay Hendry, Alex Carruth, Mark Pexton, Tom Lancaster and others in the Durham Emergence project and the John Templeton Foundation for support. Helpful conversations are also acknowledged with George Ellis, Jennifer Wilson, Carl Gillett and Robert Bishop.

Note

1 The long-running Gordon Conferences on 'Polymer Physics' and 'Condensed Matter Physics', for example, still continue, but spawned a new biennial series on 'Soft Condensed Matter Physics' from 2009, see www.grc.org/conferences.aspx?id=0000050

References

Aizawa, K. and C. Gillett (2009), "The (Multiple) Realization of Psychological and Other Properties in the Sciences", *Mind and Language*, 24, 181–208.

Anderson, P.W. (1972), "More is different", *Science*, 177, 393–396.

Batterman, R. (2002), *The Devil in the Details: Asymptotic Reasoning in Explanation, Reduction, and Emergence*, Oxford University Press, Oxford.

Bickle, J., "Multiple realizability", *The Stanford Encyclopedia of Philosophy* (Spring 2016 Edition), Edward N. Zalta (ed.), URL = <http://plato.stanford.edu/archives/spr2016/entries/multiple-realizability/>.

de Gennes, P.-G. and J. Prost (1995), *The Physics of Liquid Crystals*, Clarendon Press, Oxford.

Doi, M. and S.F. Edwards (1986), *The Theory of Polymer Dynamics*, Clarendon Press, Oxford.

Duran, J. (2000), *Sands, Powders and Grains*, Springer-Verlag, New York.

Ellis, G.F.R. (2012), "Top-down causation and emergence: some comments on mechanisms", *Interface Focus*, 2, 126–140.

Everaers, R., S.K. Sukumaran, G.S. Grest, C. Svaneborg, A. Sivasubramanian and K. Kremer. (2004), "Rheology and microscopic topology of entangled polymeric liquids", *Science*, 303, 823.

Feynman, R.P. and A.R. Hibbs (1965), *Path Integrals and Quantum Mechanics*, McGraw Hill, New York.

Jones, R.A.L. (2002), *Soft Condensed Matter*, Oxford University Press, Oxford.

Khatri, B.S., T.C.B. McLeish and R.P. Sear (2009), "Statistical mechanics of convergent evolution in spatial patterning", *PNAS*, 106, 9564–9569.

Landauer, T. (1961), "Irreversibility and heat generation in the computing process", *IBM Journal of Research and Development*, 5(3), 261–269.

Lancaster, T. and M. Pexton (2016), "Reduction and emergence in the fractional quantum Hall state", *Studies in History and Philosophy of Modern Physics*, 52, 343–357.

Larson, R.G. (1999), *The Structure and Rheology of Complex Fluids*, Oxford University Press, Oxford.

Lekkerkerker, H.N.W. and R. Tuinier (2011), *Colloids and the Depletion Interaction*, Springer Lecture Notes in Physics 833, Springer, Dordrecht.

McLeish, T.C.B. (2008), "Molecular polymeric matter, Weissenberg, Astbury and the pleasure of being wrong", *Rheologica Acta*, 47, 478–489.

McLeish, T.C.B. (2011), "'Physics met biology, and the consequence was . . .', in Studies in history and philosophy of biological and biomedical sciences, Ed. D. P. Rowbottom", *Studies in History and Philosophy of Science C*, 42, 190–192.

McLeish, T.C.B., M.E. Cates, J.S. Higgins and P.D. Olmsted (2003), "Slow dynamics in soft matter: Introduction", *Philosophical Transactions of the Royal Society A*, 361 (1805), 637–639.

Mirny, L.A. (2011), "The fractal globule as a model of chromatin architecture in the cell", *Chromosome Research*, 19, 37–51.

Nelson, P. (2004), *Biological Physics: Energy, Information, Life*, W.H. Freeman, New York.

Ramaswamy, S. (2010), "The mechanics and statistics of active matter", *Annual Review of Condensed Matter Physics*, 1, 323–345.

Roca, J. (1995), "The mechanisms of DNA topoisomerases", *Trends in Biochemical Sciences*, 20(4), 156–160.

Rubinstein, M. and R.C. Colby (2003), *Polymer Physics*, Oxford University Press, Oxford.

Safran, S.A. (2003), *Statistical thermodynamics of surfaces, interfaces and membranes*, Westview Press, Boulder, CO.

Weare, D. and S. Hutzler (1999), *The Physics of Foams*, Oxford University Press, Oxford.

Witten, T. A. and P. A. Pincus (2004), *Structured Fluids: Polymers, Colloids, Surfactants*, Oxford University Press, Oxford.

21

EMERGENCE IN NON-RELATIVISTIC QUANTUM MECHANICS

Stewart Clark and Iorwerth Thomas

Introduction

Materials are a collection of atoms, which consist of atomic nuclei surrounded by electrons, and these interact with each other electrostatically by means of quantum mechanics via Coulomb's law. These are the "parts" that make up materials. One conception of emergence is that new properties arise out of combinations of these parts, so that a material (solid, liquid, molecule, etc.) has properties that the parts do not. In this chapter we will discuss examples of how properties emerge from the parts when described using only non-relativistic quantum mechanics. This is often called a *first-principles* approach. One of the most challenging problems in condensed matter physics today is predicting the properties of materials from first principles, that is, from knowledge of the nuclei, electrons and their interactions, but without the requirement of additional experimental information. Can we take a take a simple description of a material and give insight into its emergent properties? Can we predict the properties of everyday materials and objects around us, both simple and complex, ranging from metals to glasses, to complex technological materials and biological systems?

Materials can be described, in principle, by the Schrödinger equation (Schrödinger 1926). Quantum mechanics provides a reliable way to calculate what electrons and atomic nuclei do in any situation. The behaviour of electrons, in particular, governs most of the properties of materials. This is true for a single atom, or for assemblies of atoms in condensed matter, because quantum mechanics describes, explains and allows us to define a rich variety of properties, for example, chemical bonds. Therefore, being able to solve the Schrödinger equation would give us a great deal of information about material systems. However, there are two issues here: (i) it is a tremendously difficult task mathematically and computationally to solve the Schrödinger equation for anything but the simplest cases; and (ii) if we could, the amount of data and information that is gained is huge, and so it would be a technological challenge to find emergent properties of materials from these data. Exact, closed-form (pencil-and-paper, mathematical) solutions of the Schrödinger equation exist only for a single electron in exceptionally simple situations. The problem of interacting electrons in condensed matter physics, a manifestation of the *many-body problem*, is the defining challenge of the subject.

This chapter is concerned with how we can deal with these issues. The discussion will allow us to comment on emergent properties that arise from consideration of non-relativistic quantum mechanics. Specifically, we will describe some examples of emergent properties that can be found by investigating solutions of the Schrödinger equation for realistic, complex systems.

The difficulty in solving the Schrödinger equation for materials, and analysing the data to identify emergent properties, could be called the *fundamental problem of condensed matter physics* – as regardless of whether you are a friend of emergence or the most stringent reductionist, it is the reality described by the Schrödinger equation that forms the backdrop of the debate. Discussions of both reduction and emergence (weak or strong, ontological or epistemological) often confuse not only the differences between quantum and classical physics but also the real difficulties that arise in attempting to solve the Schrödinger equation for systems containing many particles. Clarifying how quantum mechanics can give directly accurate (exact) properties of materials is surely important if the scientifically informed debate is to progress, and we attempt to do so here by means of examples while making use of the bare minimum of mathematical formalism.[1]

Much of, in fact almost all of, chemistry could be considered an emergent property of the Schrödinger equation if the *chemical bond* is emergent, as this is one way to describe the properties of a material. In a basic course in chemistry one finds terms such as *covalent bond, ionic bond* and *metallic bonding*. A little further into the subject, and we come across terms such as *hydrogen bond, molecular bonding* and *van der Waals bond*. These alone give us insight into some of the properties of materials. Covalently bonded materials have a high melting point, are good thermal insulators, are bad electrical conductors and are brittle. In materials with metallic bonding, we observe a high melting temperature, good electrical and thermal conductivity and find the materials are ductile. Ionic bonds yield materials with a high melting point, good electrical insulation when solid but conducting well in the melt and so on. So given a description of bonding types, we immediately infer some bulk properties of materials.

There are, however, at least two problems with this approach. First, these concepts are purely qualitative and second, the definitions of such bonds are actually very badly defined: there are many (in fact, most) materials' bonding that is either intermediate between various pairs of these loosely defined categories or that have a mixture of these bonds. Additionally, one usually infers the type of bonding in a material from its properties, rather than the other way round. So the question arises: Given only fundamental principles of physics, can we do better? Can we find a model in which the chemical bond is a well-defined property rather than a conceptual one imposed after properties are known? Will other properties emerge from such formalism?

Yes: using quantum mechanics. However, it is only in recent years, with the onset of fast computers, advanced theory and algorithmic methods, that it has become possible to use quantum mechanics in such a quantitative manner to describe real materials.

Non-relativistic quantum mechanics and Schrödinger's equation

First we will introduce enough basic quantum mechanics that we can demonstrate the practical problems in applying the formalism and how these can be obviated. We will then step through some examples to demonstrate emergent properties in real materials using only quantum mechanics; from simplicity, through complexity and back to emergent simplicity again.

The *wavefunction* of a material system, such as a solid, liquid or molecule, is a mathematical representation describing the potential results of a measurement on a system. Its square gives us a measure of the probability of finding particles in certain locations. If we have N particles (usually the electrons and atomic nuclei in a material) which are labelled with positions. $r_1, r_2, \ldots, r_N$ then the wavefunction is written in full as

$$|\Psi\rangle = |\Psi(r_1, r_2, \ldots, r_N)\rangle,$$

that is, it is a function of all of the positions. Given that a macroscopic piece of material contains on the order of 10^{26} atoms, it is immediately apparent that the wavefunction is a function in a huge number of dimensions. This high dimensionality is fundamentally the reason why solving its governing equation, the Schrödinger equation, computationally is far from trivial.

The Schrödinger equation describes the evolution of the wavefunction through time and is typically written

$$-i\frac{\partial}{\partial t}|\Psi\rangle = \hat{H}|\Psi\rangle,$$

or in a time-independent form

$$E|\Psi\rangle = \hat{H}|\Psi\rangle.$$

The left-hand side of this equation tells us how the wavefunction changes with time (in the time-dependent form) or the possible values of energy (in the time-independent form). The right-hand side describes the interactions of the particles and their surroundings. The symbol, $\hat{H}$ known as the Hamiltonian of the system, consists of the sum of the kinetic energy (due to movement) and potential energy (due to interaction with other particles or with external fields). The operation of $\hat{H}$ on $|\Psi\rangle$ gives us the possible values of the total energy, E, of the system. The term $-i\partial / \partial t$ is the time evolution operator. On the left side of the equation, the operation of this on the wavefunction gives us the rate of change of the wavefunction with time. Taken together, the Schrödinger equation tells us that the rate of change with time of the wavefunction at a given instant is related to the structure of the energy levels of the system.

More generally, a *measurement* on a system is mathematically described by an operator $\hat{O}$ that acts on the wavefunction and gives us the result of the measurement and the likelihood of observing that result. That is, a given outcome of a measurement is indeterministic, with the probability of a possible outcome given by a weighting derived from the wavefunction.

The apparent simplicity of the Schrödinger equation is deceptive, concealing a number of snares. First, consider what is physically measurable in the Schrödinger equation. The two main "objects" in this equation are the energy and wavefunction, but neither of these are experimentally measurable quantities. Conversely, although time is experimentally measurable, there is no operator that acts on a wavefunction to give the time. There is therefore some disconnect between experimental measurement and the mathematical statement of the Schrödinger equation. Second, the number of dimensions in the wavefunction equals the number of degrees of freedom of the electrons. As noted earlier, macroscopic pieces of material contain on the order of 10^{26} electrons, and it is simply not computationally feasible to handle such objects. That aside, we have to do some work on these objects to obtain the results of the experiment. In summary, the Schrödinger equation contains all the information needed to describe non-relativistic physical systems, but for realistic systems a full description in terms of equation is often analytically intractable as there is *too much* information. We will return to this point later.

As is common in problems involving many particles (including classical many-body problems), finding emergent properties often gives us insight into why things behave as they do that are not apparent from the constituent particles and their interactions. A simple analogy of emergent properties is seen by considering a flock of birds: if we take that a bird flies and introduce some simple rules (the behaviour of the flock causing individual birds to turn in a given direction, or climb, or dive, etc.), then the properties of the flock follows. We have something similar in the quantum system: the moving particles interact via a simple rule (Coulomb's law). The Schrödinger equation is a shorthand mathematical formulation of these rules from which all material properties emerge.

Solving the Schrödinger equation

In general, it is not yet possible to solve the Schrödinger equation in its entirety either analytically or computationally. However, a combination of recent theoretical and computational advances allows us to make considerable headway towards this and hence to determine emergent properties of materials. Here we will briefly mention the two methods required for solution: computer simulation (Clark et al. 2005) and density functional theory (Kohn and Sham 1965).

In simulation, one builds a model of a real system and uses it to explore the system. The model is mathematical, and the exploration is done using a computer, but in many ways simulations share the same goals as the experiment. However, in a simulation there is absolute control and access to detail, the ability to compute almost any observable quantity, and, given enough computer power, accurate answers for the model. Our model in the current case is the Schrödinger equation. This is vastly ambitious, because our goal is to model real systems using no approximations whatsoever.

Given unlimited computational power (both in memory and speed), the accurate, numerical solution to the full Schrödinger equation can be computed. However, we have relatively limited computational hardware, even with today's fastest supercomputers, and the solution of the Schrödinger equation is still many, many orders of magnitude away from being possible. The theoretical physics tool that we also need is *density functional theory* (DFT) which takes a remarkable approach to solving this complex mathematical and computational problem. DFT is both a profound, exact theory of interacting electrons and a practical prescription for calculating a solution. The idea is that it can be shown that the *wavefunction is not necessary* and we do not need to handle this huge-dimensional object. DFT says that this fundamental part of the Schrödinger equation can be bypassed and, instead, DFT aims to calculate the *electron density*, which is the probability of finding an electron at a particular point in space. This is a three-dimensional object, not a 10^{26} dimensional object. DFT puts this electron density on a firm physical footing from a quantum mechanical point of view, and, from the electron density, we are able to define and calculate many emergent properties of materials using only fundamental ideas in quantum mechanics.

DFT tell us that when a system of interacting particles is in its lowest energy (ground) state, the density distribution of those particles in space uniquely determines the properties of the system. A functional (a function of a function) for the energy of the system can be written in terms of the density distribution that will determine the exact ground state density and energy. Therefore, a particular ground state energy and density distribution will be associated with a particular arrangement of atoms in the functional. The ground state density distribution will be that which globally minimises the energy functional, and the ground state energy will be that global minimum. Having bypassed the wavefunction and targeting the electron density directly means that we have reduced the dimensionality of our problem from $\sim 10^{26}$ to 3. The problem is then computationally tractable. DFT therefore allows one to calculate the exact ground state properties of any interacting many-body object by transforming it into this more tractable system.

An important approximation that is made in most density functional calculations is known as the Born-Oppenheimer Approximation. In the limit where the masses of nuclei are much larger than the electron mass, it is possible to separate out the portion of the system that corresponds to the fast-moving electrons from that of the slow-moving atomic nuclei. This means that the behaviour of the electrons is only dependent on the positions of the ions, not their dynamics. We can then "clamp" the atoms in place while solving the Schrödinger equation for the electrons, and this will give us the ground state energy for that ionic configuration. Born-Oppenheimer is an excellent approximation for cases where nuclear motions are much slower than the rate at which electrons move. This is due to the large differences in electronic and nuclear masses, the nuclear mass being thousands of times larger than the electronic mass. Hence, for a given instantaneous set of nuclear positions, the electrons will essentially always be in their lowest energy state. Therefore, in many cases, separating the degrees of freedom between electrons and nuclei is an incredibly good approximation.

Born-Oppenheimer allows us, for example, to compare the ground state energies of different static structural configurations with the same chemical formula and determine which is the most stable. Suppose that there are several possible stable molecular structures for a given set of ions. If we were to solve the Schrödinger equation for this collection, given arbitrary starting positions and away from the Born-Oppenheimer limit, we would find that we obtain a superposition of these stable molecular structures. By working in the Born-Oppenheimer limit, we have *selected* one of these structures. At first sight, this might seem to be an artificial, ad hoc intervention, but note that we have done nothing that nature does not itself do. In reality, the constituent ions would be interacting with the surrounding environment, and it is this interaction, which we have not so far considered, that would draw the nuclear wavefunction towards a form approximating that of the Born-Oppenheimer Approximation while realising one of the possible molecular structures. A close examination of the nature of this interaction between the molecular system and its environment is thus needed if emergentist accounts of molecular structure are to be sustained.

Emergence of physical properties

Loosely speaking, many properties can "emerge" from the Schrödinger equation, some more easily than others. (One could ask if *all* properties of materials can be derived[2] from it.) In this section we detail a few interesting examples of properties seen to emerge from the Schrödinger equation using methods such as density functional theory.

Pauli exclusion: Before attempting to derive emergent properties from the Schrödinger equation, let's examine an interesting property that we can obtain simply from the wavefunction. A fairly naïve, semi-classical picture of an atom is that of electrons in orbit around the nucleus – essentially a mini solar system. Unlike a solar system, the orbits are only allowed at particular distances (actually energies), and only one electron is allowed in the orbit (we're ignoring a property called *spin* here, which complicates this argument without adding to the result). This property: *only one particle is allowed in each orbit* is called the *Pauli exclusion principle*. We will show that the less-than-obvious quantum result of the Pauli exclusion principle drops out of the wavefunction.

Electrons are indistinguishable, that is, you can't tell one from another. If we have only two electrons in a system, its wavefunction can be written $\left| \Psi\left(r_1, r_2\right) \right\rangle$. If we swap the two electrons, then the wavefunction is $\left| \Psi\left(r_2, r_1\right) \right\rangle$. Now since electrons are indistinguishable, are these two wavefunctions equal? As noted earlier, we use operators on wavefunctions to make changes, so let's call this swap $\hat{S}$. Mathematically we can write the description of swapping electrons as

$$\hat{S}\left| \Psi\left(r_1, r_2\right) \right\rangle = x\left| \Psi\left(r_2, r_1\right) \right\rangle.$$

We have inserted the unknown x here, as we don't know if the swap changes the wavefunction. However, if we swap them back, then we return to where we started; two swaps restores the system to its original state, so we do know that

$$\hat{S}^2\left| \Psi\left(r_1, r_2\right) \right\rangle = 1\left| \Psi\left(r_1, r_2\right) \right\rangle$$

so we can immediately read off that $\hat{S}^2 = 1$, hence $\hat{S} = \pm 1$. Therefore, on swapping particles the wavefunction can do one of two things – remain unchanged ($\hat{S} = 1$) or change sign ($\hat{S} = -1$).

This gives us two classes of particles; the particles with $\hat{S} = 1$ we call bosons and the particles with $\hat{S} = -1$ we call fermions. Electrons are in the second class of particles, where the wavefunction changes sign on swapping. Let's now attempt to put both electrons at the same position (call that position r) so we know that on swapping them

$$\left|\Psi(r,r)\right\rangle = -\left|\Psi(r,r)\right\rangle$$

that is, when the particles are at the same place, then the wavefunction must be equal to negative itself. The only function that this holds for is zero, hence the probability of finding two fermions (for example, electrons) at the same place is zero. This is the Pauli exclusion principle, an emergent property, arising out of the wavefunction of a system that has no classical analogue. Returning to atoms, it is the exclusion property that determines the emergent orbital-like picture (essentially only one electron can exist in each "orbit").

Colour: Why does a material have a particular colour? This is a straightforward question but with an answer deeply rooted in quantum mechanics. In isolated atoms, electrons occupy different energy levels. Due to the Pauli exclusion principle, it is only possible for electrons with different (up and down) *spins* to occupy the same energy level. In periodic, crystalline solids, these energy levels are replaced with bands of closely packed energy levels whose relative location and form are determined by the solution of the Schrödinger equation for that material. If an electron is excited from a filled level to an empty one by absorbing energy (usually a packet of light called a photon), then we have an electron in an *excited state*. However, not all photons are absorbed by the material, because electrons cannot absorb energies of arbitrary energies – electrons can only be in specific energy states. So to move an electron from one allowed state to another allowed state requires a photon whose energy is the difference in energy of those two states.

White light contains photons of all energies[3] in the visible spectrum. When this light shines on an object, it interacts with the electrons in the material. Photons with energies that correspond to the difference in electron energy levels get absorbed, and electrons are promoted from the lowest energy states to higher energy states. The remaining light is reflected from the object, and that is what reaches our eyes. The colour we see corresponds to the light that is not absorbed. For example, if light towards the red end of the spectrum (red, orange, yellow) is absorbed, then light towards the blue end is reflected, and that is the colour we see the object to be.

Hence the explanation of colour is quantum mechanical in nature; there is no classical description that explains colour. We can, in principle, solve the Schrödinger equation for an object and determine the energies of the electrons. The differences in these energies correspond to photon absorption, leaving non-absorbed photons to be reflected off of the object and some reach our eyes. Colour is an emergent property of the Schrödinger equation.

Of course one might ask, what happens to the energy of the absorbed photon? The electron which absorbs it can release the energy in a number of ways, the most common being transferring that energy to the kinetic energy of the atom – it makes the atom move or vibrate. The technical name for the quantum of motion of atoms in a material is a *phonon*. The kinetic energy of atoms is directly related to temperature. If the average kinetic energy of the atoms is given by E_{KE} then the temperature of the system is $T \sim E_{KE}/k_B$ where k_B is known as Boltzmann's constant. The energy of absorbed photons gets translated into atomic motion and the material heats up. This is why darker materials (more light absorbed, less light reflected) get warmer than lighter-coloured materials.

The reverse of this process is also interesting. If we take a material and heat it up, then the kinetic energy of the atoms increases, and this energy can be transferred to the electrons and excite them into higher energy states. The electrons then fall back to their ground state and a photon is emitted. This is the process that causes a material when heated to glow red, then as it gets hotter orange, and then white (all colours are emitted). Even hotter, such as in some stars, the peak of energy moves off to the high-energy end of the spectrum and the star is blue. This can all be determined from the Schrödinger equation, and so, closely related to colour is the emergent property of *temperature*.

Chemical bonds: It is normal for chemists to discuss the properties of materials derived from knowledge of their chemical bonding (Winter 1994). As described earlier, various bonding types exist: (i) *covalent* bonding where electrons are shared between pairs of neighbouring atoms which creates the intuitive "ball and stick" picture of atoms bonded into molecules and solids; (ii) *ionic* bonding where one atom gives up one or more (negative) electrons to another atom, leaving the donor atom positive and the receiving atom negatively charged. This allows the application of Coulomb's law to explain attraction between the oppositely charged species; and (iii) *metallic* bonding, where some loosely bound electrons move in a negatively charged electron sea, held together by the positive ions. This electron sea is responsible for electrical conductivity.

Are these ideas of bonding emergent from the Schrödinger equation? Yes, but unfortunately in not such an elegant manner as the properties described earlier. The ideas of particular bonding types are somewhat idealised models and, in general, the electronic structure of materials doesn't often lie fully within one bonding type, but is an incomplete mixture of bonding types. Let's take the isolated atomic constituents of a material and solve the Schrödinger equation for each one of the isolated atoms. We then know what the wavefunction associated with each atom is and from this can construct a wavefunction associated with each electron in each one of the atoms. Let these wavefunctions be denoted by $|\varnothing_n\rangle$ where n labels each of the electronic states in each of the atoms. We can then express the solution to the Schrödinger equation of the solid or molecule in terms of these

$$|\Psi\rangle = \sum_n c_n |\varnothing_n\rangle$$

where c_n is a number.[4] Loosely speaking, the square of the number c_n is the proportion of the electron density of the whole system that is found in atomic state $|\varnothing_n\rangle$. We therefore know the equivalent atomic occupations that most closely resemble reconstructing the entire system from atomic orbitals. Using this, we can construct the number of electrons associated with each atom, and if we know the number of electrons on each atom, we know its charge and hence how *ionised* it is. This represents a direct measure of the strength of ionic bonds in the material.

For covalent bonds, instead of expressing the complete wavefunction as an expansion of atomic orbitals, we express it as an expansion of orbitals that represent sharing between pairs of atoms. We then know the amount, directionality and occupation of covalent bonding in the material. Metallic bonding also proceeds similarly, but with the complication that states localised around atoms are found to be partially occupied, containing, on average, fractions of electrons. This is because some of the electronic charge is smeared out over the entire material representing the metallic sea of electrons that essentially form the metallic behaviour.

In realistic systems, when one evaluates the wavefunction and expresses it in the earlier form, it is usual to find that such idealised pictures of covalent, ionic or metallic bonding individually don't represent the electron density. This is because in real systems bonding tends to be a mixture of these different types, and one has to say, for example, a bond is partially ionic and partially covalent. The emergent picture of bonds becomes less clear when we start to classify how much of different types of bonds form the bonds in a material. We end with classifications such as strongly covalent, weakly ionic, semi-metallic and semiconducting, to name a few. These can all be classified by the proportions of types of bond in the material (and also using the energy difference between the ground state of the system and the energy of the first excited state).

Bonding might appear a somewhat artificial emergent property derived from the Schrödinger equation. However, there are some fundamental reasons why such bonding forms a good viewpoint of the electronic structure of materials and leads to the deeper and qualitative understanding of their properties. Quantum mechanics tells us that electrons being shared between atoms, or transferred from one atom to another, can reduce the energy. This is because the lowest energy state of atoms is one in which electrons pack in a certain order. By forming materials by grouping large numbers of atoms, the electrons still try to form the lowest energy states by sharing themselves or moving between atoms, and the manner in which they do this (transferring/sharing) is exactly what we call a chemical bond. So in essence, a chemical bonding is a consequence of the Schrödinger equation and hence an emergent property of it, and these bonds allows us a simple but powerful picture of materials at the atomic level.

Phonons: In a material the constituent atoms have kinetic energy, which are directly related to the temperature of the material. Once a regular crystal structure has been made, quantised lattice vibrations called phonons (Born and Huang 1954) may be excited in it. The mechanism is analogous to a vibrating string; specific vibrational frequencies are allowed. Phonons, which are a collective excitation of atomic positions, play an important role in the thermal properties of material properties, since they conduct heat alongside electrons, and also play an important role in the formation of Cooper pairs in the BCS theory of superconductivity (Bardeen et al. 1957).

We can obtain phonons from the Schrödinger equation. Starting from a very simple case we can build up the picture. A mass hanging on a spring will vibrate at a given frequency, and the stiffer the spring, the higher the vibrational frequency. Similarly, a taut string will vibrate at a given frequency, and the more the tension on the string, the higher the frequency. Let the mass or string extend from its equilibrium length by an amount, x, and let the tension in the string or strength of the spring be k. Then the energy and displacement are related by Hooke's law $E = \frac{1}{2}kx^2$. The second derivative of this equation with respect to x gives us k, thus $k = d^2E / dx^2$.

Returning to the atomic systems of materials, the Schrödinger equation gives us the energy, E and x represents the displacements of atoms from their equilibrium sites under excitation of atomic vibrations. It is a technicality to take the second derivative of the Schrödinger equation and a complex computational task to solve such an equation, but within density functional theory this has been done (Refson et al. 2006), and essentially the spring constants k can be evaluated. The vibrational frequencies are given by $\omega = \sqrt{k / m}$ where m is atomic mass. So atomic vibrations can be evaluated from the Schrödinger equation, leading to thermodynamic quantities (Kittel and Kroemer 1980) such as vibrational entropy or Gibbs free energy.

Higher-order derivatives can also be calculated which go beyond what is known as the harmonic approximation and leads to temperature-dependent quantities such as thermal expansion, thermal conductivity and spectroscopic features such as those of Raman spectroscopy. Therefore, temperature, heat and thermodynamic properties of materials are emergent from the Schrödinger equation.

Magnetism: In the previous discussion we mentioned one of the properties of fundamental particles such as electrons. It is known as the spin and has units of angular momentum. To get some insight as to what this is, let's return to the solar system cartoon of the atom. In this picture, an electron's orbit of the atomic nucleus would represent the "year" (a planet orbiting a star, one planet-year is the time taken for one orbit). The spin can be thought of as the "day", the time for the object to rotate on its axis once. As all electrons are indistinguishable, the "day" is a particular value; essentially the electron carries a fixed, intrinsic value of angular momentum that we call its spin (for an electron this value of angular momentum is $\hbar/2$ where $\hbar$ is Planck's constant as found in the Schrödinger equation; it is the quantised, fundamental unit of angular momentum).

Angular momentum is a vector quantity, which means it points in a given direction. The electron spin pointing in a certain direction is the basis for magnetism; the electron carries a magnetic field, with a north pole and a south pole. In many materials electrons tend to pair up with one electron having its spin pointing in one direction (often called spin-up) and the other pointing in the opposite direction (spin-down). Such materials are generally called non-magnetic, as the magnetic fields from the electron spins exactly cancel. However, in some materials, the lowest energy configuration occurs when the spins on the electrons are not all equally paired up, leaving a net local electron spin. These are commonly described as *magnetic materials.* Can we get this from the Schrödinger equation? Yes, but we have to be more precise with our language and concept of a wavefunction: the wavefunction is not a function of just the positions of the electrons but also their spin:

$$|\Psi\rangle = \left|\Psi\left(r_1, s_1, r_2, s_2, \ldots, r_N, s_N\right)\right\rangle,$$

where s labels the spin on the electrons. The Schrödinger equation becomes more complicated to solve, but the principle of emergent quantities that can be obtained from it still holds.

The magnetic properties of a material arise from the interaction of many spins located on ions composing the crystal lattice of the material. We can define an exchange constant J along pathways between pairs of neighbouring spins, which is the energy needed to flip one spin into the opposite direction. We use the solution of the Schrödinger equation (often in the form of density functional theory) to calculate the relative energies of spin configurations, which allow us to directly calculate the spin-flipping J energies. These allow us to build a model of a magnetic material with spins being placed at regular lattices. This is known as an *Ising model* for the case where the spins are only allowed to take two directions, up and down. For so-called Heisenberg systems, the spins are allowed to rotate in any direction in space, but the principle is the same. This allows us to state a very simple, but very powerful, model of magnetism:

$$E = J \sum_{\{i,j\}} s_i \cdot s_j$$

where the sum is over pairs of neighbouring spins. We can then introduce temperature and obtain the temperatures at which various spin configurations occur. Although we have only introduced a simple, idealised model, it is used only by way of example to illustrate how magnetic properties emerge from non-relativistic quantum mechanics. Such models can be made more detailed and used to obtain significant insight into the other emergent properties of magnetic materials (Landau 1969).

Conclusions

We have outlined a minimum number of concepts in quantum mechanics, allowing a description of sample emergent properties. Although some concepts in quantum mechanics arise in a probabilistic manner, it does not follow that all properties are probabilistic. A large number of material properties exist that follow directly from the Schrödinger equation that are definite in value. Although the locations of electrons in a material can only be determined with a probability, the properties of materials are often determined not by the location of electrons but by their energy. The Schrödinger equation gives us energies which give rise directly to properties such as colour, temperature and magnetism. Even given the probabilistic nature of the position of electrons, we can evaluate types of bonding, which gives us great insight into properties of materials.

Finally, we conclude with some speculation. We have demonstrated that starting from nuclei, electrons and their interactions, we are able to obtain properties of materials that are not properties of the "parts". These are emergent properties, and we would likely call this "weak" emergence. However, we are not aware of any properties that have been *demonstrated* to be emergent in any other sense, which we could call "strongly" emergent. Are there any (non-relativistic) properties of materials that cannot be obtained from solving the Schrödinger equation? Essentially, how likely is the reductionist point of view of the world? Here we have described a small number of properties but could have included electrical conductivity, melting points, mechanical strength, viscosity, speed of sound in a material, specific heat and so on. So what can we not get, and are there any "strongly" emergent properties? There is much debate on this, but we note that, historically, as soon as we discover a mechanism of how a particular property of a material works in the quantum mechanical realm, we quickly find out how to extract it from the Schrödinger equation. An interesting example is superconductivity. A mechanism for superconductivity was discovered in the 1950s (the BCS mechanism), which involved electrons and phonons coupling together. That allowed the calculation of superconducting properties such as transition temperatures. In the 1980s a new and different class of superconductivity was discovered (called high-T_c superconductivity). The mechanism for this remains unknown, but is it hidden in the Schrödinger equation? Probably.

Notes

1 It is useful to highlight the acute disjunction between the classical and quantum depictions of the world around us. Conceptually, the quantum realm is quite alien compared to the domain of classical physics: notions such as entanglement and non-locality have no clear equivalents in classical mechanics, and discussions of reduction and emergence must, if they are to take science seriously, avoid simply assuming that there is no real difference between the two. There is no interpretation of quantum mechanics in which all the weirdness "goes away" – at best, one only replaces one kind with another.
2 There is some debate (Clark 2017) as to whether this is the case and whether some properties of materials are not derivable properties of a quantum treatment of systems, causing debates on top-down versus bottom-up causation and divisions into definitions of strong and weak emergence.
3 The energy of a photon determines many of its properties. For example, if a photon has an energy, E, it has a frequency, f, related by $E = hf$, where h is known as Planck's constant. If c is the speed of light, then the wavelength, λ, of the photon is given by $\lambda = c/f$.
4 Note that there are different ways to combine these single electron states that add unnecessary detail to this picture, so we will, without loss of generality, continue with this straightforward combination.

References

Bardeen, J., L. N. Cooper and J. R. Schrieffer (1957) "Theory of superconductivity", *Physical Review* 108: 1175.
Born, M. and K. Huang (1954) *Dynamical theory of crystal lattices*, Oxford: Oxford University Press.
Clark, S. J. and T. Lancaster (2017) "The use of downward causation in condensed matter physics" in *Philosophical and scientific perspectives on downward causation*, Routledge studies in contemporary philosophy, New York and Oxford: Tailor & Francis.
Clark, S. J., M. D. Segall, C. J. Pickard, P. J. Hasnip, M. J. Probert, K. Refson and M. C. Payne (2005) "First principles methods using Castep", *Zeitschrift für Kristallographie* 220: 567.
Kittel, C. and H. Kroemer (1980) *Thermal physics*, New York: W. H. Freeman Press.
Kohn, W. and L. J. Sham (1965) "Self-consistent equations including exchange and correlation effects", *Physical Review* 140: A1133.
Landau, L. D. (1969) *Collected papers*, Moscow: Nauka.
Refson, K., P. R. Tulip and S. J. Clark (2006) "Variational density functional perturbation theory for dielectrics and lattice dynamics", *Physical Review* B 73: 155114.
Schrödinger, E. (1926) "An undulatory theory of the mechanics of atoms and molecules", *Physical Review* 28: 1049.
Winter, M. J. (1994) *Chemical bonding*, Oxford: Oxford University Press.

22

THE EMERGENCE OF EXCITATIONS IN QUANTUM FIELDS

Quasiparticles and topological objects

Tom Lancaster

Introduction

Quantum field theory (QFT) is one of the most successful and rigorously tested areas in all of physics. Central to this discipline is the notion of a *particle* as an excitation in a field. In fact, the central idea of QFT may be stated as: *all particles are excitations in a quantum field* (Lancaster and Blundell 2014). QFT is the only known method of combining quantum mechanics and special relativity. Moreover, as soon as quantum mechanics and relativity are combined, it becomes impossible to deal with single particles, and we are forced to consider the physics of collections of many particles. As a result, QFTs describe not only the relativistic realm of the standard model of particle physics but also the multiparticle realm of condensed matter physics, including the sub-disciplines of (quantum mechanical) hard matter and (the classical statistics of) soft matter. This variety allows a number of insights into cases where candidate emergent phenomena might be found. In this chapter we discuss the emergence of the properties of interacting particles, followed by the emergence of qualitatively new forms of particles. Our aim is to track how *interactions* lead directly to the physical masses and charges of particles in QFT and how they bring about states of matter that support the existence of novel forms of particles.

A particle in classical mechanics is known in Russian, very appropriately, as a *material point* (Landau and Lifshitz 1976). That is, a massive object localized at a point in real space with no extension in space. In conventional quantum mechanics (QM), where position-momentum uncertainty limits our simultaneous knowledge of a particle's position and momentum, it is often more convenient to describe a particle using a basis of states where particles are eigenstates of the momentum operator, rather than the position operator. The quantum particle is then an object defined by its well-defined momentum, with its position in real space being subject to uncertainty. In QFT, in addition to this choice of how to describe particles, their properties are encoded in a small number of parameters known as *coupling constants*, which could include the particle's mass m and charge q. These feature in the *Lagrangian*, which is a mathematical expression giving the difference between kinetic and potential energy densities, and which is built from fields $\phi(x)$. The Lagrangian defines a quantum field theory such that we often speak of "a theory", meaning the state of affairs described by a particular Lagrangian. An example Lagrangian density describing scalar fields is $\mathcal{L} = \frac{1}{2}\partial_\mu\phi^2 - \frac{m^2}{2}\phi^2 - \frac{g}{4!}\phi^4$, that is, a fairly simple polynomial expression in the field and its derivatives. Here, the parameter m will turn out to give the mass of the particle excitations, and the parameter g (which is rather like a charge) will tell us how strongly the particles interact.

In practice, QFT involves the process of writing down a Lagrangian, built from classical fields, and then *quantizing* it. Technically, quantization involves setting up commutation relations between the fields in a theory and then expanding the fields in terms of operators that create and destroy particles. The aim of quantization is to express the properties of a system (such as its total energy, or total momentum) in terms of the number of particle excitations present. This procedure is most easily carried out for so-called non-interacting theories. Such theories are frequently mathematically solvable and, as the name suggests, have properties expressible in terms of particles that don't interact with each other. One might think of the particles in a non-interacting theory as passing, ghost-like, through one another, rather than undergoing scattering collisions with one another (as interacting particles do). Of course, real systems involve interactions (we would not be able to measure or experience them in any way if they didn't), and when interactions are included in the description there are frequently two dramatic consequences to our analysis: (i) the theory is often no longer exactly solvable and must be approximated somehow; and (ii) the interactions have a profound effect on the properties of the particles themselves. It is this latter point, and the possibility of emergent properties that it presents, with which this chapter is concerned.

In order to understand the properties of particles in interacting theories, we commonly employ a thought experiment. This involves focusing now on the *Hamiltonian*[1] (rather than the Lagrangian, to which it bears a straightforward mathematical relationship) and splitting up this Hamiltonian into a sum of non-interacting (H_0) and interacting (H') parts: $H = H_0 + \lambda H'$. We then imagine what happens to a non-interacting theory (with the Hamiltonian H_0) when we slowly turn on the interaction Hamiltonian (H'). To do this we simply multiply it by the "turning-on" function λ that starts vanishingly small and slowly grows to unity, at which point the previously non-interacting theory becomes an interacting one that describes a bubbling cauldron of interactions.

When λ is zero the non-interacting system described by the theory is simple. The ground state of the theory is known as a vacuum (mathematically given the symbol $|0\rangle$ to represent a state in which there is nothing at all). The excitations from this ground state are particles $|p\rangle$, labelled by their momentum. These are known as non-interacting, or *bare*, particles and have the masses and charges that may be read off directly from the Lagrangian. In order to create an excitation mathematically, we use the particle creation operator $\hat{a}_p^\dagger$ which acts on the vacuum to create a particle with momentum p thus: $\hat{a}_p^\dagger|0\rangle = |p\rangle$.

As we turn on the interactions the ground state vacuum changes from $|0\rangle$ into another state we will call $|\Omega\rangle$. If we attempt to excite the interacting system using our operator $\hat{a}_p^\dagger$ we now excite particles that may interact with each other, causing more particles to be created or annihilated. The original particle may become lost in the havoc, and it is legitimate to ask whether it is even *possible* to have particles at all in an interacting system.

It will turn out that particles of a kind are still possible in the presence of interactions: they are called *quasiparticles*. These quasiparticles have different properties to their non-interacting relatives. The process of changing the properties of a particle (such as its mass or charge) as we turn on interactions is known as *renormalization*.[2]

Renormalization in a metal

An influential picture of the example of renormalization of electrons in a metal was formulated by Lev Landau and is known as Landau Fermi Liquid theory (Lifshitz and Pitaevskii 1980; Anderson 1984). It has been called the standard model of condensed matter physics. The quasiparticles in Landau Fermi Liquid theory are slightly different from the field theory quasiparticles described thus far (we will call them Landau quasiparticles to avoid confusion), but the lack of mathematical machinery needed to understand Landau's picture makes it useful for gaining an insight into the subject. In Landau's picture we pay particular attention to the turning on of the interactions.

The non-interacting model of a metal is the *Fermi gas*. This is a state of matter formed by trapping non-interacting electrons in a rigid box. Although the electrons in this model do not interact with each other via the Coulomb force, they must obey Fermi statistics and the Pauli exclusion principle, which ensures that electrons (which are identical Fermi particles) may not doubly occupy momentum states. Electrons therefore stack up in energy, up to the so-called Fermi energy. As we imagine turning on Coulomb interactions, the particles of the Fermi gas change their energy, with the gas evolving into the interacting Landau Fermi liquid. This is known as the *adiabatic turning on* process (Anderson 1984) and provides us with a picture of how particle properties emerge as interactions are turned on.[3]

The vital thing here is that all of the singly occupied states in the gas become singly occupied states in the liquid. We call the object occupying the states in the gas a bare particle and in the liquid that emerges from the gas, a Landau quasiparticle. Despite the fact that the energies change as interactions are turned on, there is no ambiguity about which bare particle has evolved into which Landau quasiparticle, which is to say that the energy levels should not cross during the turning-on process, as shown in Figure 22.1(a). This will be the case since

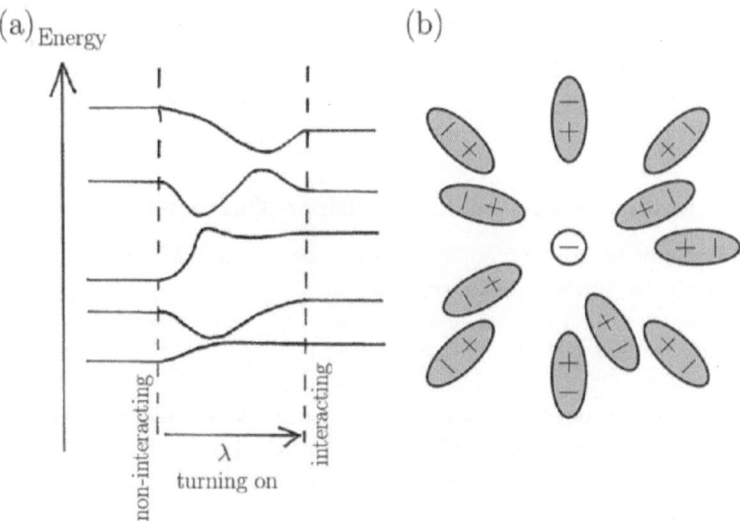

Figure 22.1 (a) The turning-on process in the Landau Fermi liquid evolves non-interacting electrons into Landau quasiparticles. The key feature is that the energy levels change, but never cross, as the interactions are turned on. (b) The change in particle charge on renormalization, viewed as a classical picture of a bare charge being screened by a vacuum made of dipoles.

[Adapted from Lancaster and Blundell 2014]

the Fermi gas and the Landau Fermi liquid share the same symmetry (i.e. the interactions do not cause a symmetry-breaking phase transition, a topic to which we return later). In the absence of a phase transition the slow turning-on process provides a one-to-one correspondence between Landau quasiparticles and free particles. This state of affairs is known as *adiabatic continuity*.

Landau's quasiparticles are different in some ways from QFT quasiparticles. In QFT, the ground state $|\Omega\rangle$ is the interacting vacuum which contains no quasiparticles. In Landau's theory, the liquid state contains the same number of particles as did the Fermi gas from which it evolves. This guarantees charge conservation, as long as Landau quasiparticles carry the same charge as conventional electrons. However, although Landau quasiparticles have the same charge as bare electrons, the change in energy levels that occurs leads to the particles taking on a new "effective" mass m^*. (That is, we account for the change in energy of the particle that results from the interaction by saying that its mass has changed.)

A valuable insight into the properties of the Landau Fermi liquid is found by looking at the details of the interaction of one electron with all of the others as the interactions are turned on. We therefore imagine introducing an extra electron in an unoccupied state. Interactions have the effect of causing some other electron to be scattered from its momentum state. This latter electron will therefore have a finite *lifetime*. However, the scattering process is strongly constrained by energy-momentum conservation and the Pauli exclusion principle. It may be shown that the lifetime is long (compared, say, to the timescale for turning on the interactions) only for those electrons with energies very near the Fermi energy, since in that case, there are so few available vacant states to accommodate the scattered electrons, they tend not to scatter at all. Thus, it is meaningful to speak about Landau quasiparticles *only* if they have energies near the Fermi energy, as these have a long lifetime against decay. At lower energies, well below the Fermi energy, the status of these excitations is less clear: although the energy levels are filled, the particles are scattering between states very rapidly compared to the slow timescale of the turning-on process and so enjoy only a fleeting existence. Put less technically, only the most energetic electrons qualify as quasiparticles, since the less energetic ones fail to survive for long enough to qualify as particles (where "long enough" is judged against the imagined time taken to turn on the interactions).

The formalism of making quasiparticles in field theory

We give here a brief flavour of how quasiparticles are dealt with mathematically. The process of renormalization may be summed up conceptually as

(Quasiparticle) = (bare particle) + (interactions)

As Philip Anderson points out (Anderson 1963; Anderson 1984), there is nothing guaranteeing the existence of quasiparticles in an interacting system. Returning now to QFT, in order to see where the quasiparticles arise in field theory, we may try to create a particle operationally, by using the creation operator $\hat{a}_p^\dagger$ on the interacting vacuum $|\Omega\rangle$. The result is not only the expected particle with momentum p but also lots of other particles with momenta that add up to p. That is

$\hat{a}_p^\dagger |\Omega\rangle$ = (single particle part) + (multiparticle parts)

It is also possible that we only approximately create the expected single particle and actually make an unstable narrow wavepacket known as a resonance. The consequence is that the particle-like resonance takes on a finite lifetime τ. With all of these particles flying around, we measure the likelihood of the occurrence of quasiparticles by defining the *quasiparticle weight* Z_p Formally speaking, Z_p gives us the overlap of the states of matter we have created and the particle we intended to create. We may say that if $Z_p \neq 0$, then we have quasiparticles in the system. This allows us, mathematically, to write the dressing-up process as the passage from a non-interacting particle with wavefunction $e^{-i\,p.x}$ to a quasiparticle with wavefunction $Z_p e^{-i\,p.x-t/\tau}$, that is, in becoming a quasiparticle, the bare particle's amplitude is reduced by a factor Z_p and takes on a finite lifetime τ.

When interactions between particles are turned on, the most obvious consequence is that particles scatter from each other. This scattering is encoded in QFT using *perturbation theory* and *Feynman diagrams*, which provide a shorthand for calculations (Lancaster and Blundell 2014). As stressed earlier, the interactions also cause renormalization of the particle properties, which can also be encoded mathematically in the following manner. For a particle to propagate in a system, we define the *propagator G* as the quantum amplitude for a particle to be created at point y and be detected at x. One possibility is that a particle might propagate from y to x without being impeded in any way. The amplitude for this is given by the non-interacting propagator G_0. If the particle suffers an interaction at some intermediate position with an amplitude V then the amplitude becomes $G = G_0 + G_0 V G_0$. If there is a possibility of a second interaction then the sum becomes $G = G_0 + G_0 V G_0 + G_0 V G_0 V G_0$, and so on. Feynman diagrams allow us to turn this infinite series of interactions into a process of cartooning, allowing the depiction of the nature of each interaction V. Since both the cartoons and the algebra represent an infinite, geometric series, these may be summed to give

$$G = G_0 + G_0 V G_0 + G_0 V G_0 V G_0 + \ldots = \frac{G_0}{1 - V G_0} = \frac{1}{G_0^{-1} - V}$$

For the simplest quantum field theory (the scalar field theory), the non-interacting propagator G_0 may be shown to be given by $G_0 = i / \left(p^2 - m^2 \right)$. This implies that the interacting propagator is given by

$$G = i / \left(p^2 - m^2 - V \right).$$

Comparing G and G_0, we see that the scattering process that the particle undergoes has had the effect of shifting its mass from m to $\left(m^2 + V \right)^{1/2}$. It is in this way in QFT that the interactions cause the mass of a particle to be altered.

The means by which interactions cause a change in the charge of a particle may be understood in a similar way to the argument presented earlier. In this case the particle of interest is a photon, which transmits the electrostatic force between charges. Quantum electrodynamics (QED) is the QFT that describes the interaction between electrons and photons. QED allows a photon to instantaneously turn into an electron–positron pair and then return to being a photon. This transformation forms the content of the interaction V discussed in the previous example, with interactions from several instances of V corresponding to the photon repeatedly turning into electron–positron pairs. In this case, however, a more intuitive picture of the charge is possible in terms of classical screening [Figure 22.1(b)], where the electron–positron pair may be thought of as a polarized dipole. The bare electron is then viewed as being surrounded by these dipoles. An attempt to measure the electronic charge from a large distance measures not only the negative charge of the electron but also the combined effect of this and the screening supplied by all of the surrounding dipoles. Viewed macroscopically, it is as if the vacuum itself has become a polarized dielectric, which attenuates the charges of electrons.

Lessons from our case study discussion of metals can be carried over to the consideration of QED and quasielectrons more generally. In the metal, we are able to distinguish between the behaviour of the electron outside of the metal and the electron quasiparticle that exists within the metal. This allows a story to be told about the dressing up of a bare electron into a quasielectron via the electron–electron interactions. The vacuum for the metal in field theory is the Fermi gas of electrons. Quasiparticle excitations then exist as excitations out of that gas. QED is different, in that it is impossible to take an electron outside of the QED vacuum, since this would imply removing electrons from our universe. So while it is difficult to tell the same story about dressing a QED electron in QED interactions, we are left with an idea that the QED electron is itself a quasiparticle, whose properties emerge from interactions with the interacting vacuum of the universe.

Infinities

Renormalization provides an account of masses and charges being altered by interactions with the vacuum. It is especially important in the case of many field theories (including QED) where a troubling aspect of the theory arises: the presence of infinities. In fact, renormalization is often presented as a method of removing infinities from a theory. However, it should be borne in mind that renormalization is a necessary step in formulating a field theory. It is required to take interactions into account and is independent of the mathematical presence or absence of any infinities (Weinberg 1995; Lancaster and Blundell 2014).

Mathematically, infinites are found in many theories and result from attempts to sum behaviour over all length scales (or equivalently, over all momentum scales). They can be avoided if we admit that there is a minimum length scale in any practical case below which we don't have access. (See later for a discussion of this.) Technically one can then identify where this minimum length scale, known as a cut-off, occurs in the theory and then add extra terms to the Lagrangian (known as counter-terms) in order to remove them. The consequence of this strategy is to shift the values of the mass and the charge of the theory. The theory is therefore saved mathematically via the removal of infinities, with the shifts (i.e. renormalization) in particle properties a by-product.

We can trace the origin of the shifts in particle properties as follows. We start doing QFT by writing down a Lagrangian that (we assume) accounts for the properties of some area of physics. This includes parameters such as the masses and charges of particles. If the theory has interactions, we usually can't solve it, and so we rely on perturbation theory to account for the properties of a system. This involves expanding a series about the values of the parameters m and g that we believe happen to encode the particle properties. However, it turns out we were wrong! We were doing the wrong perturbation theory as we were expanding about the wrong masses and wrong couplings. We had asked a nonsensical question and ended up with an infinite (and therefore meaningless) answer. The couplings we should have been using are those found experimentally in nature. Renormalization is not, therefore, a formal, mathematical exercise in hiding infinities; it is an exercise in shifting the parameters that describe particle properties so that they describe nature as we experience it. For many theories, the magnitudes of the shifts in particle properties are infinite. This implies that the bare values of mass m and coupling g are infinite, and we shift from them by an infinite amount to obtain a sensible (finite) value. This is certainly a troubling consequence of renormalization that has caused practitioners of QFT much pain and uncertainty. However, the instructive use of a set of techniques known as renormalization group analysis has provided a great deal of insight into the origin and explanation of these infinities.

Renormalization group

A valuable insight into the meaning of renormalization in QFT is found in renormalization group (RG) analysis, discussed elsewhere in this volume (Blundell 2019). RG has been the subject of much discussion in the philosophical emergence literature in the context of its usefulness in describing phase transitions (see e.g. Batterman 2001; Morrison 2012), where its use in identifying critical points and analysing universality classes of models allows an insight into multiple realizability in condensed matter physics. For our purposes, RG allows us an insight into how the dressing-up process of renormalization affects particle properties through the notion of *scale*.

In examining quasiparticles, rather than correct for infinities as is done in renormalization, in RG the cut-off is used as a handle that allows us to vary the length scale of interest in a problem. Specifically, the physical content of the cut-off is that it provides a length below which all details of the physics are irrelevant to the problem at hand. Consider, for example, some atoms in a box. If we are interested in sound waves, then the length scale of interest is the wavelength of sound (centimetres), and the cut-off might be a few microns. If we're interested in the physics of the electrons in the atoms of the gas, then the length scale of interest will be the size of an atom, and the cut-off should be set to the size of a nucleus. The renormalization group analysis consists of asking how the coupling constants (masses and charges again) vary as we change the value of this cut-off.

For theories of condensed matter physics, the motivation for varying the cut-off is to examine the limit of behaviour measured by experiment. Solid materials are generally macroscopic, and so the length scale of interest in condensed matter physics is that in which, on the scale of the underlying atoms, length tends to infinity. Examining the variation of the coupling constants in this limit then allows insights into the qualitative and quantitative details of behaviour.[4] The size of a coupling constant will change as we interrogate the system on different length scales. The result is that certain interactions will be important in a particular limit, while others will die away.

Although this all sounds rather artificial, it is a real possibility in experiments involving particle collisions, where the momentum involved in the collision, and hence the length scale probed, can be varied. This, in turn, provides an insight into the particle "dressing-up" scenario that we have described earlier. High energy and momentum corresponds to small length scales, so high-momentum collisions allow us to probe electronic charges at very low length scales. It is found that the charge on an electron appears to increase at these small probing distances. This corresponds to our measuring the electronic charge inside the screening cloud of polarized electron–positron dipoles [described earlier and shown in Figure 22.1(b)]. As the probe momentum decreases (i.e. length scale increases), the probing distance is greater, there is more screening between the bare particle and the probe and we correspondingly measure a smaller charge. One consequence of this is that many of the "fundamental" physical constants of nature, such as electronic charge, should not be regarded as constants. They depend on the length scale at which we measure them, and this is a consequence of the quasiparticle view of particles.

Finally, it is worth noting that a further consequence of the RG view of nature is that all theories (described by a Lagrangian in the sense noted earlier) are *effective theories*. They may contain terms that operate at momentum scales so large that we never detect them. As a result, the physics we have is a low-energy approximation to a more complete field theory of nature. This feature has led to philosophers such as Jonathan Bain to suggest that effective theories provide an explanation of emergent phenomena in systems described by field theories (Bain 2013).

Summary: emergence of particle properties in QFT

The quasiparticles concept implies that the properties of particles (their masses and other couplings) take on values by virtue of the interaction of the fields in a theory. These interactions are themselves mediated via particles, so we might say that the many-particle nature of QFT is what gives rise to the couplings we observe. It is perhaps in this sense that particles' properties should be thought of as emergent. Moreover the ability to describe and predict the details of the dressing-up process makes this a rather weak form of emergence when compared to some of the more dramatic phenomena outlined later, which resist such a description. Whether we regard the bare particles of a theory realistically depends, according to Sidney Coleman (Coleman 2018), on "how weird you're willing to believe the world is". That is, one can imagine an extreme realist position where the electronic charge is infinite, but we never observe this infinity due to the spontaneous existence of particles that screen the infinite charge.

From the point of view of renormalization group analysis, the emergence of particle properties is bound up in notions of scale. Specifically, different properties emerge at different scales. We implicitly accept that a theory will have a realm of applicable scales, and insight is provided through considering how the theory must change to remain consistent as the scale of interest is varied. The RG picture gives us a clue as to the reality of the dressing-up process that turns particles into quasiparticles. It tells us that the electronic charge depends on the momentum used to investigate it. If one attempts to investigate the electronic charge at sufficiently small distances (using very high momentum probes), then the measured charge of the electron is actually larger. This makes sense in the context of the quasiparticle view of the world as a particle being dressed in interactions.

Broken symmetry and topological objects

A very clear link between particle properties and emergence is found in the much-discussed topic of broken symmetry (Blundell 2019). In fact, Philip Anderson discusses this topic in terms of a breakdown of adiabatic continuity (Anderson 1984). In the case of the metal, discussed earlier, turning on interactions causes particle properties to vary smoothly. This might be characterized in terms of a weak form of emergence, where we follow the gradual change in particle properties as interactions are dialled up. However, in the case where turning on interactions causes the particle energy levels to cross, there is a breaking of symmetry. The result is that the particles effectively lose their identity in the resulting reconstruction of a system. The properties of the new excitations realized in the broken symmetry state might then have a claim to be realized by a stronger form of emergence.

In QFT we think of the production of a particle as involving the excitation of the ground state. Breaking a symmetry in QFT is characterized in terms of a dramatic change in the potential energy density (which is part of the Lagrangian describing the system) as shown in Figure 22.2. When a symmetry is broken, the nature of the vacuum state itself is reconstructed, with the twofold consequence that (i) the properties of the particles change and (ii) qualitatively new varieties of particles are possible.

The first of these features is due to the details of the change in the shape around the minima of the potential being reflected in a change in the Lagrangian parameters, from which the particle properties follow. The second of these is more dramatic and reflects the fact that the new potential landscape allows new forms of excitation within it.

Qualitatively new particles arise, for example, when a *continuous symmetry* is broken. In this case, new forms of particle excitation are possible, whose very existence could not be supported

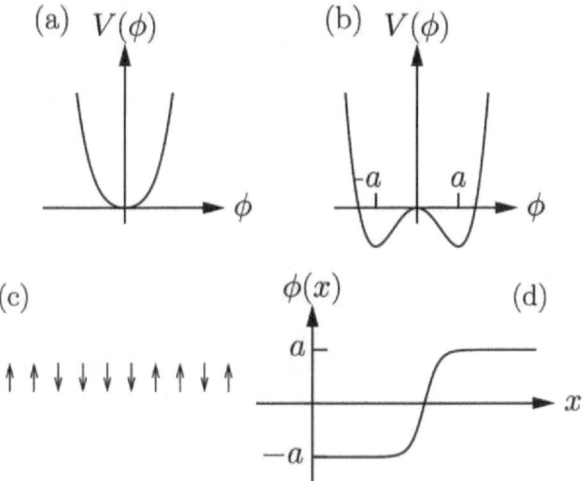

Figure 22.2 The dramatic change in potential energy density caused by the breaking of a symmetry. (a) Before symmetry breaking; (b) after symmetry breaking. (c) A model of a one-dimensional magnet with spins (arrows) arranged randomly. (d) The kink in a field theory can be thought of as an ordered state where arrows [represented here by a field $\phi(x)$] go through a transition as a function of distance x, from being aligned downwards (on the left) to upwards (on the right).

by the non-broken symmetry state. An example is found by examining the freezing of a liquid into a crystalline solid. The solid is a state of broken *translational* symmetry. On average, the density of atoms is the same at every point in a liquid. In a solid, atoms are arranged into a periodic array and are only found at regular distances from other atoms. The solid supports the existence of a new particle excitation: the phonon. Phonons are massless excitations, which means they can be created with arbitrarily low energy. In addition, the number of phonons is not a conserved quantity. In general, the emergence of massless particles on symmetry breaking is not too mysterious – it simply reflects the fact that the broken symmetry potential can support a different manner of excitation from the non-broken symmetry potential. At an atomic level, the phonon quasiparticle owes its existence to all of the coordinates of the underlying crystal lattice changing. Phonons are therefore sometimes called quanta of lattice vibrations. As a result of this rather extreme reliance on the existence of the ordered crystal, the phonon can't exist outside of the solid.

There is still another class of particle-like objects that exist in theories with broken symmetry: these are *topological objects*. The key point to these objects is that they owe their existence to a system-breaking symmetry in different ways in different spatial locations. The simplest example is a magnet with spins (i.e. magnetic moments, or arrows) arranged along a line [Figure 22.2(c)]. The non-broken symmetry system is disordered, with the spins arranged randomly pointing up or down. (In the language of potential energy [Figure 22.2(a)], the system sits in the minimum of the potential well.) In breaking symmetry in this system, the spins must collectively align up or down. [This corresponds to the system choosing to sit in one of the two minima of the double well potential in Figure 22.2(b).] However, a more interesting configuration is possible if half of the spins on the left of the chain order down, while the other half align up. This leads to a transition region in the middle of the chain known as a *kink* [Figure 22.2(d)]. The kink is yet another form of particle-like excitation. This object may be shown to exist with a finite energy above the ground state. It can be moved around, rotated and translated just like any other particle. The main difference is that it is a spatially extended object, in a sense quite unlike the spatial extension of the quantum particle localized in a momentum state.

Kinks certainly exist in nature, most obviously in magnets where they are known as domain walls and are used extensively in magnetic memory applications. The kink is a good example of the general concept of a *topological object*. Topological objects are (roughly speaking) a sort of untieable knot, in that it is impossible to remove them from a system without incurring an enormous energetic cost. These objects are therefore stable. It is impossible to deform the kink shape without costing the system a (semi-) infinite amount of energy, since this would involve lifting half of the kink shown in Figure 22.2(d) over the potential barrier shown in Figure 22.2(b). Like many particles, kinks carry a conserved quantity. However, this is not the usual conserved charge (guaranteed by symmetry due to Noether's theorem). Rather, it is a topological charge, independent of the geometry of space-time. [Technically, the quantity does not involve the metric $g_{\mu\nu}$ (Lancaster and Blundell 2014).]

Kinks exist as a result of the one-dimensional physics of a chain of spins [Figure 22.2(c)]. In two dimensions we have a new topological object: the vortex, which looks rather like a whirlpool in a liquid. Again, this is formed by considering configurations that exist in different broken symmetry ground states in different regions of space. In three dimensions the analogous particle-like object is the monopole (also known as a hedgehog, due to its resemblance to that mammal, when the field is drawn as arrows). In addition, there exist many more examples of objects whose stability is due to topological considerations, which emerge on symmetry breaking and which deserve serious consideration as particles.[5] In summary, the topological objects we have described are a species of excitation since they have a finite energy above the system's ground state. They can exist only by virtue of the system breaking a symmetry. They are necessarily extended over space, so are rather different from classical point-like particles, and so one might be tempted to classify them as non-fundamental. They do, however, have many of the fundamental symmetry properties of particles (they can be translated and transformed in the same way that particles can), so bear a very strong resemblance to simple particles. Finally, just like the new particles that exist only by virtue of symmetry breaking, the topological objects lose their meaning in a system without broken symmetry and therefore have the same emergent character.

Conclusions

We conclude with some observations about the emergent nature of quasiparticles. There are many ways to characterize weak and strong forms of emergence (Lancaster and Pexton 2015), invoking, for example, (i) a failure of explanatory reduction; (ii) the appearance of new entities; (iii) novel systematic properties; and (iv) a failure of mereological supervenience, leading to the necessity of considering whole-system properties.

Perhaps the key factor in all of the examples presented has been interactions. A system without interactions might support particle excitations, but these will not be detectable as they pass through each other (and everything else). As soon as interactions are added to a system, they change the properties of the particles. In fact, the emergence, existence and demergence of a particle are quite dramatic in Landau's quasiparticle picture. Assuming that no symmetry is broken, the properties of the particles can be shown to evolve continuously from the non-interacting theory upon turning on the interactions. We might characterize this as the emergence of the particle properties. If we introduce a new quasiparticle into the metal, then the length of time it will retain its identity as a particle is a continuous function of its energy. Too far from the Fermi energy, and it ceases to exist as a particle over the lifetime of an observation, instead being consumed by the liquid-like system. Its existence becomes longer the closer it gets in energy to the Fermi energy. The short lifetimes of those excitations with energies well below the Fermi energy cause them to lose their meaning as particles and instead exist as slightly ill-defined parts of a whole metallic system. It is worth noting in this context that there are also more exotic cases of quasiparticles, such as a phenomenon known as the fractional quantum Hall effect (Lancaster and Pexton 2015), where if we introduce an extra electron to the system, it will fall apart into the allowed quasiparticle excitations of the quantum Hall fluid. This is particularly dramatic in this case since the fractional quantum Hall quasiparticles carry fractions of electronic charge. There is a sense in which the electron gives up its properties to the fractional quantum Hall fluid, where they reappear as fractionally charged quasiparticles.

Interactions can also lead to the spontaneous breaking of symmetry. Symmetry breaking leads to the possibility of new forms of particles, such as the massless particles like the photon, and the massive W and Z particles in the standard model of particle physics. In the broken symmetry case we also have a new class of excitations that have a topological character such as kinks or vortices. Both of these types of broken symmetry particles might be considered under each of the characterizations of emergence given earlier: (i) A failure of explanatory reduction occurs due to a singularity in the free energy on symmetry breaking. (More specifically, if you use perturbation theory to perturb around the symmetric ground state, you won't derive the existence of the broken symmetry particles, nor the topological particles, which are fundamentally non-perturbative. If we were to heat the system up and regain the symmetry, then these particles would cease to exist.) (ii) These quasiparticles are qualitatively new excitations that could not exist before symmetry breaking. (iii) They exist by virtue of a systematic restructuring of a system on symmetry breaking.

However, what all of the quasiparticles we have considered have in common (i.e. Landau/field theory quasiparticles and broken symmetry quasiparticles) is that their emergence follows from (iv): the necessity of considering whole-system interactions. This is most obvious for the collective excitations, such as the phonon, for example, which represent a change in coordinates of all of the atoms making up a crystal. However, there is a sense in which it is true for all quasiparticles since, in all cases, the renormalization of particles involves them dressing themselves in interactions, which are collective expressions of the excitations of the vacuum of the whole system. Finally, the theme of internal interactions leading to constraints on the whole of a system seems to be common in describing many of the candidates for emergent phenomena examined within the physical sciences, both in this volume and elsewhere (e.g. Clark and Lancaster 2017) and has become an increasingly pervasive theme in modern condensed matter physics research, where notions of fields, topology and emergence are now integral concepts.

Notes

1 The Hamiltonian and related concepts are discussed in the chapter by Clark and Thomas in this volume (Clark and Thomas 2019).
2 A distinction may be made between two sorts of elementary excitation of a system. The first are called *collective excitations* and correspond to a change in coordinates of all particles in a system (and we discuss these later in this chapter). The second are the quasiparticles described here. Examples of the latter include the quasielectrons in a metal.
3 This is not as artificial as it might sound. In some materials, interactions can be tuned by applying hydrostatic pressure to the solid system, for example.
4 It is also possible to ask about the very low length scale regions of the theory, which corresponds to high-energy measurements carried out by particle physicists, for example.
5 Interestingly, we can also go further and investigate these structures in time rather than space. The kink in time is known as an *instanton* and has found uses mathematically in solving tunnelling problems.

References

Anderson, P.W. (1963) *Concepts in Solids: Lectures on the Theory of Solids.* Singapore River Edge, NJ: World Scientific.

Anderson, P.W. (1984) *Basic Notions of Condensed Matter Physics.* Boulder: Westview.

Bain, J. (2013) "Emergence in Effective Field Theories" *European Journal for Philosophy of Science* 3: 257–273.

Batterman, R.W. (2001) *The Devil in the Details: Asymptotic Reasoning in Explanation, Reduction and Emergence.* Oxford: Oxford University Press.

Blundell, S.J. (2019) "Phase Transitions, Broken Symmetry and the Renormalization Group" in *The Routledge Handbook of Emergence* (Eds. Sophie Gibb, Robin Findlay Hendry, and Tom Lancaster) London: Routledge.

Clark, S.J. and Lancaster, T. (2017) "The Use of Downward Causation in Condensed Matter Physics" in *Philosophical and Scientific Perspectives on Downward Causation* (Eds. Michele Paolini Paoletti and Francesco Orilia) New York: Routledge.

Clark, S.J. and Thomas, I.O. (2019) "Emergence in Non-Relativistic Quantum Mechanics" in *The Routledge Handbook of Emergence* (Eds. Sophie Gibb, Robin Findlay Hendry, and Tom Lancaster) London: Routledge.

Coleman, S. (2018) *Physics 253: Quantum Field Theory: Lectures by Sidney Coleman* available from: www.physics.harvard.edu/events/videos/Phys253 (accessed 05/06/18).

Lancaster, T. and Blundell, S.J. (2014) *Quantum Field Theory for the Gifted Amateur.* Oxford: Oxford University Press.

Lancaster, T. and Pexton, M. (2015) "Reduction and Emergence in the Fractional Quantum Hall State" *Studies in History and Philosophy of Science Part B: Studies in History and Philosophy of Modern Physics* 52: 343–357.

Landau, L.D. and Lifshitz, E.M. (1976) *Mechanics.* Oxford: Pergamon.

Lifshitz, E.M. and Pitaevskii, L.P. (1980) *Statistical Physics, Part 2.* Oxford: Pergamon.

Morrison, M. (2012) "Emergent Physics and Micro-Ontology" *Philosophy of Science* 79: 141–166.

Weinberg, S. (1995) *The Quantum Theory of Fields*, volume 1. Cambridge: Cambridge University Press.

23

EMERGENCE

A personal perspective on a new paradigm for scientific research

David Pines

Introduction

Twenty years ago, in the physical sciences, the particle physics community was focused on a reductionist "top-down" paradigm whose success was embodied in the standard model of particle physics, in which one wrote down a model Hamiltonian for the so-called elementary particles – quarks and gluons – and pursued its consequences. There was heady talk of arriving at a "Theory of Everything" based on what was regarded as the fundamental approach to understanding the universe in which we live.

A contrary view had been expressed by P.W. Anderson in his seminal 1972 *Science* article on quantum matter, "More Is Different" (Anderson, 1972), in which he focused on broken symmetry and hierarchical organization and argued that these significant phenomena would never appear in a reductionist approach. Anderson wrote:

> The ability to reduce everything to simple fundamental laws does not imply the ability to start from those laws and reconstruct the universe. In fact, the more the elementary particle physicists tell us about the nature of the fundamental laws the less relevance they seem to have to the very real problems of the rest of science, much less to those of society.
>
> *(393)*

> The behavior of large and complex aggregates of elementary particles, it turns out, is not to be understood in terms of a simple extrapolation of the properties of a few particles. Instead, at each level of complexity entirely new properties appear, and the understanding of the new behaviors requires research which I think is as fundamental in its nature as any other.
>
> *(393)*

The past twenty years have seen a major shift away from the reductionist paradigm for carrying out scientific research to one that focuses on emergence and emergent behavior, terms that interestingly are not to be found in the Anderson article. When electrons or atoms or individuals in societies interact with one another or their environments, the collective behavior of the whole is strikingly different from that of its parts. We now call this resulting behavior emergent. *Emergence* thus refers to collective phenomena or behaviors in complex adaptive systems that are not present in their individual parts. The study of emergent behavior at every level is as fundamental as work on the laws that govern the behavior of quarks and gluons.

In the study of inanimate and living matter, the Institute of Complex Adaptive Matter (ICAM), a distributed institute with its home on the Web (http:icam-i2cam.org), has played a significant role in bringing about this change. ICAM researchers focus on identifying the organizing principles responsible for the emergent behavior observed in the laboratory and, importantly, bring this research to the attention of the broader scientific and educational community.

In this chapter, I give a personal perspective on the role played by ICAM, which began in 1998 with six branches and now has twenty-six branches in North America and twenty-three branches in Asia, Europe, South America, and Israel. ICAM's message about emergence is communicated through a program of workshops, fellowships and exchange programs, and a significant outreach program of science, education, and engagement activities (Web 1). I describe the impact of a paper that served as ICAM's initial emergence manifesto, "The Theory of Everything", by me and R.B. Laughlin (Laughlin and Pines, 2000a) and the ICAM outreach efforts that led to the development of an online science museum devoted to emergent behavior (http://emergentuniverse.org). I then consider the role played by emergence in an online course, "Physics for the Twenty-First Century", and an emergence-based university course, "Gateways to Emergent Behavior in Science and Society". I conclude by describing a recent effort to teach eighth graders to "Think Like a Scientist", which uses a new approach, emergent scientific thinking, to introduce them to a broad spectrum of emergent behaviors.

ICAM begins

ICAM's origins may be traced to a 1996 conversation over lunch in which Zachary Fisk and I were discussing ways in which the many disparate research efforts on materials at Los Alamos might be more successful. We realized that finding a common theme would be a step forward, and I suggested that "complex adaptive matter" might offer that theme, since complex adaptive systems had proved to be a unifying theme for the Santa Fe Institute. We defined complex adaptive matter as the study of the generally unpredictable emergent behavior brought about by the interactions of the component parts found in living and inanimate matter. A succinct description of the results of those interactions had been provided earlier in Anderson's "More Is Different".

Zachary and I then proposed the idea to a number of senior Los Alamos National Laboratory officials, whose response was on the whole positive, sufficiently so that some two years later, in December 1998, the Center for Materials Research at LANL organized what proved to be the founding workshop for ICAM. Would-be participants in the workshop were asked to submit brief statements on the potential of such an institute, such that even before the workshop convened, it was already clear that a number of leaders in the condensed matter and biological physics communities were supportive of our idea of forming a distributed institute to study emergent behavior in matter.

ICAM began operating as an independent unit of the University of California, located at Los Alamos, reporting to the UC Office of the president, on March 4, 1999, with founding meetings in Oakland, California, of its board of trustees, chaired by Zachary Fisk (Florida State) and Alexandra Navrotsky (UC Davis), and its Science Steering Committee, chaired by Robert Laughlin (Stanford) and Peter Wolynes (UIUC).

It began as a joint project of the University of California's Office of the President (UCOP) and Los Alamos National Laboratory with support and oversight being provided by Robert Shelton, the UCOP Vice-President for Research, and Bill Press, the LANL Deputy Laboratory Director. Don Parkin, leader of LANL's Center for Materials Science, and I were asked to become

ICAM's founding co-directors. I accepted a five-year appointment as a LANL staff member to help ICAM emerge.

The first ICAM workshop was on "Adaptive Atoms in Physics, Chemistry, Biology, and the Environment". Co-chaired by Daniel Cox (UC Davis), Zachary Fisk, Andrew Shreve, and John Kazuba (Los Alamos), the workshop examined a broad range of systems in which adaptive atoms (atoms whose valency is determined through their interaction with their environment) play a key role in bringing about their observed emergent behavior. The participants concluded that there was so much to be gained by bringing physical scientists, biologists, and environmental scientists together to tackle problems at the frontier in this field that the study of adaptive atoms and their role in electron transfer reactions, broadly defined, could be regarded as a grand scientific challenge for the ICAM scientific community. The workshop soon gave rise to ICAM's first scientific paper (Cox, 1999) and to some seven nascent research groups – scientists from different fields and institutions who sought bridging support to initiate collaborations on problems identified at the workshop. The workshop also served to stimulate the formation of ICAM nodes in this and related areas at UC Davis, Rice University, Oxford University, and Los Alamos. This first success was soon followed by the workshop on "Mesoscopic Organization in Matter" described later.

The Theory of Everything

On January 1, 2000, *Proceedings of the National Academy of Sciences* (PNAS) published an article written by Bob Laughlin and David Pines, "The Theory of Everything" (Laughlin and Pines, 2000a), that, *inter alia*, served as an *emergent manifesto* for ICAM. Although Bob and I had been collaborating on bringing ICAM into existence during the previous two-year period, it had not occurred to us to write a paper about emergent behavior. Matters changed in March 1999, when my newly diagnosed cancer was being treated at UCSF, and I realized that by scheduling an early morning dosage of radiation, I could be free to do science the rest of the day. I wrote to Bob asking whether I might spend a few days each week during the seven-week course of treatment visiting him at Stanford and received a positive response.

As the visits took place, we realized we had a mutual antipathy to the then-popular idea that out of the reductionist approach of particle physics there might emerge a "Theory of Everything". We shared as well the belief that it was timely to call the attention of the scientific community to the importance of starting with an emergent paradigm in which the key ingredient was the realization that we live in an emergent universe in which the interactions between the fundamental components lead to unpredicted emergent behavior at every scale.

In "The Theory of Everything", we demolished the idea that there could ever be a theory of everything in which the behavior of complex adaptive matter could be derived from a set of "fundamental laws" in the traditional reductionist sense. We introduced the concept of *protectorates* to characterize emergent physical phenomena governed by emergent rules that are insensitive to microscopics.

There are many of these protectorates, with the first to be recognized being the Landau Fermi liquid, the state of matter represented by conventional metals and normal ^{3}He, the lighter-weight isotope of helium. Landau realized that the existence of well-defined fermionic quasiparticles at a Fermi surface was a universal property of such systems that is independent of microscopic details. Landau eventually abstracted this to the more general idea that low-energy elementary excitation spectra were generic and characteristic of distinct stable states of matter. Other important quantum protectorates include conventional superconductors; superfluid Bose liquids such as ^{4}He; and the newly discovered atomic condensates, band insulators, ferromagnets, antiferromagnets, and the quantum Hall states. The low-energy excited quantum states of these systems are particles in exactly the same sense that the electron in the vacuum of quantum electrodynamics is a particle. These quantum protectorates, with their associated emergent behavior, provide us with explicit demonstrations that the underlying microscopic theory can easily have no measurable consequences whatsoever at low energies. The nature of the underlying theory is unknowable until one raises the energy scale sufficiently to escape protection.

As Laughlin and Pines tell us,

> [L]iving with emergence means, among other things, focusing on what experiment tells us about candidate scenarios for the way a given system might behave before attempting to explore the consequences of any specific model. This contrasts sharply with the imperative of reductionism, which requires us never to use experiment, as its objective is to construct a deductive path from the ultimate equations to the experiment without cheating. But this is unreasonable when the behavior in question is emergent, for the higher organizing principles – the core physical ideas on which the model is based – would have to be deduced from the underlying equations, and this is, in general, impossible. Repudiation of this physically unreasonable constraint is the first step down the road to fundamental discovery.
>
> *(2000: 30)*

Perhaps the best way to get a sense of the continuing impact of the article is to go to Google Scholar. Here one finds that each year during the past decade some forty authors have cited "The Theory of Everything". If 2015 is a reasonable sampling, some 25% of their papers deal with physics and about the same number deal with the philosophical issues relating to emergence; the remainder cover a remarkably broad area of science that includes as a few unusual examples biological autonomy, business ecosystem dynamics, mental states as emergent properties, interferometric probes of Planckian quantum geometry (Kwon, 2015), crime as a complex system, "cupping" therapy, proteins as nanomachines, materials aging at the mesoscale, quantized orbits in weakly coupled Belousov-Zhabotinsky reactors (Weiss and Deegan, 2015), agent-based modeling, language as a values-realizing activity, and biological emergence and inter-level causation.

The range of topics suggests that our message about emergence and protected behavior is spreading throughout the world of science and is turning up in some quite unexpected places.

Workshop II: "Mesoscopic Organization in Matter"

Chaired by Robert Laughlin, Peter Wolynes, David Pines, and Alexander Balatsky (Los Alamos), a second ICAM workshop brought together thirty-three senior scientists, three postdocs, and three graduate students for an in-depth discussion of "Mesoscopic Organization in Matter". "Mesoscopic" describes a length scale intermediate between microscopic and the human scale, and organization of matter on the mesoscopic scale is a fascinating example of the emergent behavior found in both living and inanimate matter. There was unanimous agreement among the participants that the mesoscopic world is a key frontier in science, and that one of the grand scientific challenges in the study of matter is establishing the existence or non-existence of mesoscopic protectorates, rules of organization of matter on this scale that transcend details. A second grand scientific challenge identified at the workshop was establishing the connection in biology between structure, energy landscapes, dynamics, and function that make possible the predictive design and synthesis of biomolecules. There was agreement that sharp disciplinary boundaries are counterproductive in pursuing these challenges and that great and largely untapped research opportunities lie at the interface between the cultures of the physical and biological sciences.

Immediately after the workshop a group of us spent four days at my home working on the draft of a paper that would tell a PNAS audience about what we had learned at the workshop. "The Middle Way", by R.B. Laughlin, D. Pines. J. Schmalian, B. Stojkovic, and P.G. Wolynes (Laughlin et al., 2000b), was intended to provide an in-depth overview of mesoscopic organization in matter. (The "Buddhist" title was proposed by Wolynes.) We suggested to PNAS that it appear as a companion article to "The Theory of Everything", and the two papers appeared together in PNAS in its inaugural issue for this century.

The workshop also led to nine requests for bridging support to assist in the formation of multidisciplinary, multi-institutional research groups devoted to examining specific aspects of emerging macroscopic order in both living and quantum matter and to the establishment of ICAM research nodes at UCSB, UCLA, UCSD, UCB, UIUC, Stanford, Rutgers, Iowa State, and the University of Chicago.

Workshop III: "Designing Emergent Matter"

"Designing Emergent Matter", the title of the third ICAM workshop, focused on the extent to which one could use well-established organizing principles for emergent behavior in matter to design new materials. It brought together chemists, biological physicists, and quantum physicists who discussed state-of the-art approaches toward achieving this goal. What made it different from earlier ICAM workshops was the inclusion of two major science writers among the invitees, Sandy Blakeslee and George Johnson of the *New York Times*. Sandy published a contemporaneous article about the workshop that conveyed to the general public the challenges of designing emergent matter. George waited some months before producing, in December 2002 for the *New York Times*, an account of the tensions in the scientific community associated with transition underway from reductionism to emergence.

The *New York Times* on emergence vs. reductionism

What George Johnson succeeded in doing in his lengthy and lucid article was to set the stage for a global debate on whether the reductionist paradigm would be overtaken by an emergence-based paradigm. In the physical sciences the hard-core reductionists were to be found in the theoretical particle physics community, who firmly believed in a top-down approach to understanding their field and, by inference, all of physics. Looking back, it appears likely that at this time, in 2002, their view was accepted by a significant portion of the physics community.

The minority view in the physics community – that we live in an emergent universe – was largely espoused by theorists who studied quantum matter, a field in which experiment seemed to provide an unending set of unexpected and unpredicted behaviors. (After all, as early as 1972 the battle was joined by Phil Anderson in "More Is Different.", in which he argued that the study of quantum matter [and matter more generally] was every bit as fundamental as the study of quarks and gluons.)

Those interested in science communication quickly picked up on the debate in progress. The day after George's essay was published, I was invited by Ira Flatow, host of the NPR show *Science Friday*, to debate Leon Lederman on the topic. As it turned out, in the *Science Friday* debate, Lederman was joined by a leading particle theorist, Chris Quigg. I like to think that I held my own against the two of them.

Bob Laughlin and I received an invitation from one of the leading scientific literary agents to write a book expanding the views we had expressed in "The Theory of Everything". Laughlin and I did start collaborating on a book, but it soon became clear that Bob's literary voice was sufficiently different from mine that it was better for him to write it alone. His book, *A Different Universe*, was published in 2004 (Laughlin, 2005).

The 115 materials and the two-fluid model

As part of its continuing focus on emergent behavior in quantum matter, ICAM held a workshop in December 2002 on some remarkable materials, the heavy electron family $CeMIn5$, where M could be Co, Rh, or Ir. At low temperatures, as the external pressure was varied, these "115" materials could be antiferromagnetic, superconducting, or both. At high temperatures they behaved like a collection of interacting local moments, but as the temperature was lowered the hybridization of the local moments against a background of conduction electrons led to the emergence of an itinerant heavy component with masses that could become as large as 200 free electron masses. It was this mixture of local moments and heavy electrons that displayed the remarkable low temperature behavior noted earlier.

But how to sort this out? The answer came shortly after the end of the workshop, as Zachary Fisk; his postdoc, Satoru Nakatsuji; and I were discussing experiments they had carried out by doping the cerium sites with lanthanum. We realized that a consistent phenomenological account of their results could be obtained with a two-fluid model in which an order parameter described the emergence of heavy electrons. What we were doing was carrying out the first step toward a physical understanding of the emergence of heavy electrons in these and other heavy electron materials.

What makes this interesting to those who are not quantum theorists is that this initial phenomenology was followed by another fifteen years of work developing the phenomenological two-fluid model in much more detail and using it to achieve a better understanding of the emergent behaviors turning up experimentally, while still not being able to solve the problem microscopically. This research can serve as a proof of concept for what can be achieved at a phenomenological level with emergent behavior.

From quantum criticality to "Music of the Quantum"

Among the fascinating emergent behaviors found in correlated electron materials is the quantum-critical behavior found when there exist two competing states of matter near absolute zero. This leads to a quantum-critical point that marks a transition from one state to the other and produces fluctuations that are called quantum critical. ICAM held a workshop on this topic at Columbia University in March 2003.

One of the workshop organizers, Piers Coleman, saw this as providing a unique opportunity to inform the greater New York City community about ICAM and emergence. His idea was to hold a combined musical and scientific event that would be of interest to both the scientific and musical communities in New York. He persuaded his brother, Jaz, a talented musician and composer, to write a piece, "Music of the Quantum", for the occasion, and I was able to obtain funding for the event.

Expectations for the concert and popular lectures on March 22 were high, but the timing was terrible, since that was the day the Gulf War began. As such, there was no possibility of coverage in the major newspapers, many potential listeners stayed home to follow the latest news of the war, and concert impact on the city was minimal. Those who were able to attend were rewarded by the world premiere of a science-inspired composition and two excellent popular lectures by Bob Laughlin and Piers Coleman.

Fortunately, the event was recorded and posted online at http://musicofthequantum.rutgers. edu, to include not only a recording of the concert but also a number of interviews on emergent behavior with scientists who came to the workshop. I strongly encourage readers of this chapter to relive the event online.

ICAM and emergence go global

The NSF held a competition in 2004 for awards to three institutions to establish international materials institutes. ICAM was encouraged to enter the competition and emerged as one of those selected. It used the award to establish I2CAM, the International Institute for Complex Adaptive Matter, which soon had as many branches abroad as it did on home soil. During the next decade (the period over which such NSF support was available) I2CAM was able to expand its emergence-based research, educational, and outreach program to some eighteen countries in Europe and Asia. The global expansion was so successful that, three years after the end of NSF support., ICAM has twenty-six North American branches, eleven Asian branches, eight European branches, two South American branches, and one branch each in the Middle East and Australia. Many of these latter branches are consortia of several institutions, so that, in fact, some forty-nine scientific institutions outside the United States are members of ICAM. This significant global presence translates into a major expansion in the number of scientists who have come to appreciate and embrace the emergent paradigm.

Estimating the impact of ICAM workshops on emergent thinking

State-of-the art workshops are a remarkably effective way of communicating new science and new ways of doing science. Not only are the participants informed, but as they return home, and in the months and years that follow, they spread the word through organizing new research initiatives and giving talks at meetings. The founding 1998 ICAM workshop and the three that followed involved some 200 participants. A conservative estimate of their impact would be to assume that each participant in turn introduced five students or colleagues to emergent thinking during the first three years of ICAM's existence. If we assume that during this period, "The Theory of Everything" had some 500 attentive readers, that the George Johnson piece in the *New York Times* added another 500 new readers, and that the *Science Friday* piece on the topic added another 1000 scientific listeners, we arrive at an estimate of some 3000 scientists whose thinking about emergent behavior may have begun to change. If then each of these scientists during the following decade influenced the thinking of say, 5 other scientists, we arrive at an impact factor of 75, from the 200 founding participants to some 15,000 emergence-oriented scientists today. This is likely a significant underestimate, as some fifty additional ICAM workshops were held and a number of other efforts, described later, also began to have an impact.

Emergentuniverse.org

The most ambitious emergence-based outreach program undertaken by ICAM was to inspire and find support for the establishment of a virtual science museum, emergentuniverse.org, that was aimed at an audience of millennials who are not scientists. ICAM was able to find the perfect person to carry out this ambitious project, Suzi Tucker, who had changed careers from being a successful statistical theorist with a tenured professorship to become a scientific exhibit and web designer.

Her museum is divided into three main parts: (1) an introduction to emergence, with examples ranging from fish schooling to cars in traffic to the "Game of Life"; (2) "Resistance is futile" – an introduction to superconductivity, the poster child for emergent behavior in quantum matter; and

(3) an examination of Alzheimer's disease, from its origin in the formation of amyloid plaques in the brain to its onset and consequences. Among the many segments on the website are a flash dance group illustrating a key aspect of superconductivity, the highly correlated motion of electron pairs, and a moving illustration of the changes in personality that accompany Alzheimer's disease.

One can keep track of the number of visitors to emergentuniverse.org. As of January 2018, in its ninth year, it has had some 182,000 visitors. It lifetime may be limited, as it is based on flash technology, which is being abandoned by the industry. So ICAM and its outreach-based spin-off, described next, are exploring what might be done to preserve this online museum and develop a version addressed to eighth graders.

"Physics for the 21st Century"

I turn now to the role played by emergence in three educational experiments. "Physics for the 21st Century" (Web2) is an online course that explores the frontiers of physics. The eleven units, accompanied by videos, interactive simulations, and a comprehensive Facilitator's Guide, work together to present an overview of key areas of rapidly advancing knowledge in the field, arranged from the subatomic scale to the cosmological. The goal is to make the frontiers of physics accessible to anyone with an inquisitive mind who wants to experience the excitement, probe the mystery, and understand the human aspects of modern physics.

The course has been distributed free of charge on the Web since 2010. It was designed by Harvard Professor of Astronomy and Physics Christopher Stubbs, with units developed by a distinguished group of physicists from Harvard and other top universities and research centers. Produced by the Science Media Group at the Harvard-Smithsonian Center for Astrophysics and funded by Annenberg Media, it is intended to open the doors to an exciting world of ideas, to help bridge the gap between what is being taught in high school and college and what is exciting physics researchers.

Two of its eleven units in the course are devoted to emergent behavior: "Emergent Behavior in Quantum Matter", which I wrote and is based on a course with that title that I taught for a number of years at UC Davis, and "Biophysics", written by Princeton University Professor of Physics Robert W. Austin. Taken together, these provide a unique introduction for high school students and their teachers to emergent behavior in quantum and living matter.

"Gateways to Emergent Behavior in Science and Society"

A course with the title "Gateways to Emergent Behavior in Science and Society" was taught at UC Davis in 2012. The aim of the course was to provide students with the tools they need to develop an emergent perspective on problems in science and society by focusing on gateways to emergent behavior we have identified in the physical and biological sciences and on gateways that have been proposed for solving some of our major societal problems.

This new course was an experiment, stressing interdisciplinary and integrative learning at the upper undergraduate and graduate level. There are several aspects to the experiment: the topic of emergence as a unifying principle bringing together students in different sciences; the emphasis on emergent global problems and the science needed to assure clean, secure, and sustainable energy and food supplies to power and stabilize our world; the integration of high-profile guest lecturers who will also participate in other campus activities; and grading based on a website created by each student. This experiment will set the stage for other innovative, rigorous, and adventurous intellectual experiences for UC Davis students.

The course was organized and co-taught by David Pines and Alexandra Navrotsky (Distinguished Interdisciplinary Professor of Ceramic, Earth, and Environmental Chemistry at UC Davis).

They enlisted the help of a number of internal and external distinguished colleagues who are each world leaders in their respective fields and invited them to spend two days on campus, during which they gave a major public lecture that formed part of the course and met with our students.

The first guest lecturer was Ralph Cicerone, the then-president of the National Academy of Sciences. He was followed by Peter Littlewood, a past director of Argonne National Laboratory; University of Colorado physicist Ivan Smalyukh; MacArthur Fellow Shawn Carlson; the ecologist Simon Levin of Princeton University; Peter Smerud, the director of the Wolf Ridge Environmental Learning Center; the earth scientist Sue Kieffer of Princeton University; and the Carnegie Institution geophysicist, Robert Hazen. Their lectures were videotaped and may be viewed online (Web3).

Think Like a Scientist

Emergence is reaching a much younger audience through Think Like a Scientist (TLS), a global movement that recognizes, connects, and expands the efforts of individuals and organizations seeking to stimulate *all* students to think like scientists in the emergent universe in which they live. It is convening teams of engaged scientists, innovative educators, citizen science and science museum leaders, and game designers who use emerging technologies to create resources and activities that are designed to inspire and teach middle school students to think like scientists, provide teachers with new ways of thinking about the science they teach, and connect engaged teachers and students to mentors from the working scientist community.

To bring emergent behavior to an eighth-grade audience, TLS is focusing on *emergent scientific learning*, defined as a multifaceted approach using emerging technologies to learn about science. In emergent scientific learning, because of the many possible interactions and synergies between these different modes, the whole becomes much more than, and different from, its parts.

To help identify the skills of Thinking Like a Scientist, TLS has introduced *TLS/Concepts*, a summary of the skills that go into TLS. As may be seen next, emergence and emergent scientific thinking play a key role among these thirteen steps that enable an eighth grader to think like a scientist.

1 **Be curious – inquire!** Never accept that something is unexplainable. We just don't have the answer yet. Do not think there are black boxes; these are just boxes that we don't yet understand.
2 **Learn to look for patterns** in your data, whether they are obtained on the Web or by recording the results of your own hands-on experiments.
3 **Carry out experiments** or use the Web to study the result of using different probes to make measurements and ask how the results might be connected.
4 **Collaborate and communicate** through participation in group work on projects.
5 **In the emergent universe** in which we live, "More is different" because the interactions between people and matter and the environment in which they exist lead to unexpected consequences and unpredictable behavior at every scale. The whole is different from its parts, as everyone who has ever baked a cake knows full well.
6 **Know how to get science-based facts, experiments, observations, and data from the Web.**
7 **Know how to analyze and interpret data.**
8 **Begin to understand systematic and statistical errors in your data.**
9 **Begin to devise scenarios and logical alternatives to explain your observations.** For example, what is connected to what and how your observations change with the environment in which experiments are carried out.

10 **Understand that making mistakes is part of doing experiments** or carrying out observations, and mistakes can also advance science.

11 **Be skeptical** – Ask if the new fact you were told is science-based. Consider how the parts of the system interact – can you do an experiment, apply a mathematical principle, engineer a solution, or integrate technology to show that it is true?

12 **Be a complex problem solver,** recognizing that almost all problems are emergent in that there is no single cause and therefore no single solution. Therefore, to make progress, you need to experiment – try several solutions at once and then determine from observations which combination works best.

13 **Connect** – ask whether there are connections between different kinds of emergent collective behavior (e.g., birds flocking, fish schooling) or between the results of an experiment/ observation and one of the big ideas in science.

In collaboration with the Smithsonian Institution's Science Education Center, TLS established ATLAS (Always Thinking Like a Scientist) as "A Journey into Your Emergent Universe". The first semester of ATLAS/Aspen, an experimental out-of-school program in which high school students act as lead explorers who expose a group of middle school students to ideas, concepts, and methods that help them learn to think like scientists, has just been completed. As measured by the enthusiasm displayed by the students in their eagerness to continue being part of ATLAS/ Aspen, the experiment was an outstanding success.

Conclusion: emergent strategies

For the layman, perhaps the key takeaway that comes from recognizing that you live in an emergent universe is the realization that any problem you face, whether at home or in society, is likely to be emergent. This implies that it has no single cause and no single solution. So your emergent strategy is to experiment with many solutions at once, look for emergent synergies among these, and take advantage of these as they emerge.

For the scientist, it means building a physical understanding of your field based on experiment and using a phenomenological approach to explaining the emergent behaviors you study. Indeed, in many, if not most, cases, a detailed phenomenological description may be the best one can do. Recognizing this is the case requires abandoning the reductionist dream of being able to develop a microscopic model that captures all aspects of the emergent behavior one is studying and represents a key step toward making progress in understanding it.

Acknowledgments

I thank Zachary Fisk, Robert Laughlin, Alexandra Navrotsky, Don M. Parkin, and Peter Wolynes for the central role they played in starting ICAM, and Gregory Boebinger, Paul Chaikin, Daniel Cox, Piers Coleman, Laura Greene, Peter Littlewood, Jose Onuchic, Arthur Ramirez, and Rajiv Singh for their subsequent ICAM leadership roles.

References

Anderson, P. W. (1972) "More is different", *Science* 177: 393–396.

Cox, D. L., Singh, R. R. P. and Pati, S. K. (1999) "A new mechanism for long-range electron transfer in biological systems".

Kwon, O. (2015) "Interferometric probes of Planckian quantum geometry", PhD Thesis. Available from https://search.proquest.com/docview/1722273157?pq-origsite=gscholar (accessed 04/06/18).

Laughlin, R. B. (2005) *A Different Universe*, New York: Basic Books.

Laughlin, R. B. and Pines, D. (2000) "The theory of everything", *PNAS* 97: 28–31.

Laughlin, R. B., Pines, D., Schmalian, J., Stojković, B. P. and Wolynes, P. (2000a and 2000b) "The middle way", *PNAS* 97: 32–37.

Web1: http://icam-i2cam.org/index.php/outreach (accessed 04/06/18).

Web2: http://seminars.uctv.tv/Host.aspx?hostID=41 (accessed 04/06/18).

Web3: www.learner.org/courses/physics/ (accessed 04/06/18).

Weiss, S. and Deegan, R. D. (2015) "Quantized orbits in weakly coupled Belousov-Zhabotinsky reactors", *EPL (Europhysics Letters)* 110: 60004.

24

EMERGENCE AND REDUCTIONISM

An awkward Baconian alliance

Piers Coleman

This chapter discusses the relationship between emergence and reductionism from the perspective of a condensed matter physicist. Reductionism and emergence play an intertwined role in the everyday life of the physicist, yet we rarely stop to contemplate their relationship: indeed, the two are often regarded as conflicting worldviews of science. I argue that in practice, they complement one another, forming an awkward alliance in a fashion envisioned by the philosopher-scientist, Francis Bacon. Looking at the historical record in classical and quantum physics, I discuss how emergence fits into a reductionist view of nature. Often, a deep understanding of reductionist physics depends on the understanding of its emergent consequences. Thus, the concept of energy was unknown to Newton, Leibniz, Lagrange or Hamilton, because they did not understand heat. Similarly, the understanding of the weak force awaited an understanding of the Meissner effect in superconductivity. Emergence can thus be likened to an encrypted consequence of reductionism. Taking examples from current research, including topological insulators and strange metals, I show that the connection between emergence and reductionism continues to provide a powerful driver for frontier scientific research, linking the lab with the cosmos.

Introduction: reductionism and emergence

Reductionism is the marvelous idea that as we take matter apart to its smallest constituents and understand the laws and forces that govern them, we can understand everything. This bold idea traces back to Greek antiquity and has served as a key inspiration in the natural sciences, particularly physics, up to the present day. Emergence, by contrast, is the intriguing idea that as matter comes together, it develops novel properties and unexpected patterns of collective behavior.[1] This is something that scientists have long understood intuitively – we observe emergence all around us – from snowflakes floating on a cold day, the pull of a mundane refrigerator magnet, a flock of geese flying overhead and, for those of us who have seen it, the magic of a levitating superconductor and life in all its myriad forms. These are all examples of natural science that that are not self-evident, linear extrapolations of the microscopic laws and that often require new concepts to be understood.

Emergence and reduction are sometimes regarded as opposites. The reductionist believes that all of nature can be reduced to a "final theory", a viewpoint expressed beautifully in Steven Weinberg's *Dreams of a Final Theory* (Weinberg: 1992). Whereas reductionism is an ancient concept, the use of the word *emergence* in the physical sciences is a comparatively recent phenomenon, dating back to the highly influential article by Philip W. Anderson, entitled "More Is Different" (Anderson: 1972). In this highly influential work, Anderson put forward the idea that each level in our hierarchy of understanding of science involves emergent processes and that moreover, the notion of "fundamental" physics is not tied to the level in the hierarchy. Yet despite the contrast between these two viewpoints, neither repudiates the other. Even the existence of a final theory does not mean that one can go ahead and calculate its consequences "ab initio". Moreover, the existence of emergence is not a rejection of reductionism and in no way implies a belief in forms of emergence which can never be simulated or traced back to their microscopic origins.

In this chapter I present a pragmatic middle ground: arguing that reductionism and emergence are mutually complementary and, quite possibly, inseparable. Sometimes methods and insights gained from a reductionist view, including computational simulation, do indeed enable us to understand collective emergent behavior in higher-level systems. However, quite frequently an understanding of emergent behavior is needed to gain deeper insights into reductionism. The ultimate way to gain this deeper insight is through experiment, which reveals the unexpected consequences of collective behavior among the microscopic degrees of freedom. In this way, the empirical approach to science plays a central role in the intertwined relationship between emergence and reductionism. This connection between reductionism, emergence and empiricism lies at the heart of modern physics.

Here, I will illustrate this viewpoint with examples from the historical record and also some current open challenges in condensed matter physics.

The Baconian view

Four hundred years ago, the Renaissance philosopher-scientist, Francis Bacon, championed a shift in science from the top-down approach favored in classical times to the empirically driven model that has been so successful up to the current day. In 1620, Bacon wrote (Bacon: 1620)

> There are and can exist but two ways of investigating and discovering truth. The one hurries on rapidly from the senses and particulars to the most general axioms, and from them, as principles and their supposed indisputable truth, derives and discovers the intermediate axioms. And this way is now in fashion. The other derives axioms from the senses and particulars, rising by a gradual and unbroken ascent, so that it arrives at the most general axioms last of all. This is the true way, but as yet untried.
>
> *Francis Bacon,* Novum Organum, Book 1, Aphorism **XIX**, (1620)

Bacon argues for an integrated experimental-theoretical approach to science, in which progress stems not from imposing the most general axioms, but by using experiment and observation without preconception, as guidance to arrive at the "most general axioms". Bacon's approach is not an abandonment of reductionism, but a statement about how one should use experiment and observation to arrive there. The Baconian approach however leaves room for surprises – for discoveries which are unexpected "collective" consequences of the microscopic world, consequences which often shed new light on our understanding of the microscopic laws of physics. Bacon's empirically driven approach plays a central role in the connection between emergence and reductionism.

A.   *The incompleteness of classical mechanics*

Modern education teaches classical mechanics as a purely reductionist view of nature, yet historically it remained conceptually incomplete until the nineteenth century, 200 years after Newton and Leibniz. Why? Because the concept of energy, a reductionist consequence of classical mechanics, could not be developed until heat was identified as an emergent consequence of random thermal motion. This example helps us understand how emergence and reductionism are linked via experiment.

Energy is most certainly a reductionist consequence of classical mechanics: if the force on a particle is given by the gradient of a potential, as it is for gravity, $\vec{F} = -\vec{\nabla} V$, then from Newton's second law of motion, $\vec{F} = m\dfrac{d\vec{v}}{dt}$, one can deduce the energy $E = \frac{1}{2}mv^2 + V$ is a constant of motion. Moreover, this reductive reasoning can be extended to an arbitrary number of interacting particles. Yet although Newton and Leibniz understood the motion of the planets, understanding that was considerably sharpened by Lagrange and Hamilton, the concept of energy was unknown to them.

Gottfried Leibniz had intuitively identified the quantity mv^2 (without the half) as the *life force* ("*vis viva*") of a moving object, but he did not know that it was the conserved counterpart of momentum ("*quantitas motus*"). Lagrange (Lagrange: 1788) introduced the modern notation $T = \sum_j \frac{1}{2} m_j v_j^2$, with the factor of 1/2, and he certainly knew that $T + V$ was conserved, provided both are time-independent, a point later formulated as a consequence of time-translation symmetry by Emmy Noether in the twentieth century.[2] Yet still, the concept of energy had to wait a full two centuries after Newton. From a practical point of view, momentum is a vector quantity which is manifestly conserved in collisions, so that a macroscopic momentum can never dissipate into random microscopic motion. By contrast, energy as a scalar quantity inevitably transforms from manifest bulk kinetic energy into microscopic motion. Without an understanding of heat, kinetic energy appears to vanish under the influence of friction.

In Munich in 1798, the Colonial American–born royalist, inventor and physicist, Benjamin Thompson (Count Rumford), carried out his famous experiment demonstrating that as a cannon is bored, heat is produced. He wrote afterwards (Rumford: 1798) that

> It appears to me to be extremely difficult, if not quite impossible, to form any distinct idea of any thing, capable of being excited and communicated in the manner the Heat was excited and communicated in these experiments, except be it MOTION.
> *Benjamin Count of Rumford,* Phil. Trans. Roy. Soc. London **88**, p99 (1798)

Thompson's identification of heat as a form of motion eventually put an end to the "caloric" theory of heat as a fluid, clearing a conceptual log jam that had prevented progress for two centuries.

From our twenty-first-century hindsight, it seems almost inconceivable that several generations of physicists would miss energy conservation. Yet nothing in science is ever simple. It was certainly known to Louis Lagrange, and to William Rowan Hamilton after him, that the "Hamiltonian" $H = T + V$ is constant, provided that T and V have no explicit time dependence, but the notion of the universally conserved quantity we now call energy is completely absent from their theoretical treatises. In his treatise of 1835 Hamilton (1835) in which William Hamilton introduces the concept of phase space and modern Hamiltonian dynamics, he explicitly comments that H is constant because $dH/dt = 0$ (Equation 31 in Hamilton (1835)), but the significance of this constancy is not discussed, and Hamilton simply refers to it by its symbol, "H". In fact, though the word *energy* was most probably first introduced by Thomas Young in 1802 (Young: 1807), the common usage of this concept had to wait until the middle of the nineteenth century.

The modern reductionist might argue that the early Newtonian physicists were just not reductionist *enough*! Perhaps, had they been so, they would have realized that the conservation law known for simple systems would apply microscopically throughout macroscopic objects.

Yet, historically, until it was clear that heat was a form of random motion, this connection was not made.

Newton, Leibniz, Lagrange and Hamilton were the greatest minds of their generation – they believed fundamentally in the power of reductionism, yet they failed to make the link. Would a modern reductionist, without modern hindsight, have fared any better? The fact is that the concept of energy was hidden from the most brilliant minds of the era and was not unlocked from its reductionist origins until physicists had understood one of the most remarkable emergent consequences of classical mechanics: heat. Classical mechanics thus provides a beautiful illustration of the intertwined relationship of reductionism and emergence.

B. *Darwin-Maxwell-Boltzmann*

Biologists trace the idea of *emergence* back to Charles Darwin's *On the Origin of Species*, and the use of the term in science began in biology. Already, in the nineteenth century, scientists struggled with the relationship between emergence and reductionism. In *On the Origin of Species*, Darwin writes (Darwin: 1859)

> whilst this planet has gone cycling on according to the fixed law of gravity, from so simple a beginning endless forms most beautiful and most wonderful have been, and are being evolved.
>
> *Charles Darwin,* On the Origin of Species, *p. 490 (1859)*

Here one glimpses in Darwin's writings the idea that emergence and reductionism are connected. Around the same time that Charles Darwin was writing his opus, a young James Clerk Maxwell was trying to work out how Newton's laws could give rise to Saturn's rings. To describe the rings, Maxwell constructed what was, in essence, an early model for his theory of atomic motion. In his prize essay on the theory of Saturn's rings, Maxwell (Maxwell: 1859) wrote

> We conclude, therefore, that the rings must consist of disconnected particles; these may be either solid or liquid, but they must be independent. The entire system of rings must therefore consist either of a series of many concentric rings, each moving with its own velocity, and having its own systems of waves, or else of a confused multitude of revolving particles, not arranged in rings, and continually coming into collision with each other.
>
> *James Clerk Maxwell,* On the Stability of Saturn's Rings. *p. 67 (1859)*

Maxwell understood that the properties of Saturn's rings were a collective consequence of collisions between its constituent particles. Later, when he moved from Aberdeen to London, he used the astronomic inspiration from Saturn's rings as a model to develop his molecular theory of gases. At a time where the concept of an atom was as controversial as modern string theory, his particulate model for Saturn's rings provided a valuable launching pad for his derivation of the kinetic theory of molecular motion.

Maxwell, and Boltzmann after him, realized the importance of the Baconian approach to science – and in particular, that the collective motion of particles required new statistical approaches, inspired by observation and experiment. Here's a quote from Boltzmann in the early twentieth century (Boltzmann: 1905):

> We must not aspire to derive nature from our concepts, but must adapt the latter to the former. . . . Even the splitting of physics into theoretical and experimental

is only a consequence of methods currently being used, and it will not remain so forever.

Ludwig Boltzmann, Populäre Schriften, *p. 77 (1905)*

Boltzmann pioneered a reductionist explanation of thermodynamics and the field of statistical mechanics, yet it is clear he was a strong believer in the importance of an empirically based approach.

From the Angstrom to the micron

The vast discoveries in physics during the twentieth century, from the discovery of the structure of the atom, to relativity and quantum mechanics, to the successful prediction of antimatter from relativistic quantum mechanics and the discovery of gauge symmetries that lie behind the standard model of particle physics, are a monumental tribute to the power of reductionism (Pais: 1986). Today, the well-known extensions of this frontier lie in the puzzles of dark matter and dark energy, the observation of gravity waves, the confirmation of the Higg's particle in the standard theory and string theory with its prediction of 10^{500} alternate multiverses (Weinberg: 2005). The excitement of this frontier is widely shared with society, for instance, in Stephen Weinberg's *Dreams of a Final Theory*, Brian Greene's *Elegant Universe* (Greene: 1999) and Hawking's *A Brief History of Time* (Hawking: 1988). These expositions capture the beauty and romance of discovery while giving rise to a popular, yet false, impression that the frontier of science is purely reductionist and that the frontiers lie exclusively at the extreme sub-quark scale, the Planck mass and the first moments of the Big Bang.

Yet this is only one element of today's physics frontier: we need only to look just below the limits of the optical microscope and classical engineering, at scales of an order of a micron, to find remarkable emergent physics that we barely begin to understand. This is the view expounded by Philip W. Anderson (Figure 24.1) in his highly influential article "More Is different" (Anderson: 1972).

Anderson's article, and his subsequent writings, helped to crystallize the idea of emergence in the physical sciences. The concept of emergence re-invigorated the field of solid-state

``The behavior of large and complex aggregations of elementary particles, it turns out, is not to be understood in terms of a simple extrapolation of the properties of a few particles. Instead, at each level of complexity entirely new properties appear, and the understanding of the new behaviors requires research which I think is as fundamental in its nature as any other.''

Philip. W. Anderson
from ``More Is Different'', p 393, (1972).

Figure 24.1 Philip W. Anderson. Anderson introduced the concept of emergence into condensed matter physics in his influential "More Is Different" article.

[Source: musicofthequantum.rutgers.edu]

physics, prompting the field to redefine itself under the broader title "condensed matter physics".

The terrestrial counterpart to the multiverse of string theory is the periodic table. While there are only ninety-two stable elements, quantum mechanics and chemistry mean that each new compound provides a new universe of collective behavior. As we go from elements to binary, tertiary and quaternary compounds, out towards the organic molecules of life, the number of unique combinations exponentiates rapidly. It is this emergent multiverse that provides the back-drop for quantum materials, biology, life and all its consequences.

On the length-scale of atoms, an Angstrom (10^{-10}m), we understand pretty much everything about the motion of electrons and nuclei. This motion is described in terms of the many-body Schrödinger equation, which describes the system in terms of a wave, described by the many-body wavefunction $\Psi(1, 2, 3 \ldots N)$, where $1, 2, \ldots$ denote the coordinates of the particles. The squared magnitude of this wave provides the probability of finding the particles at their respective coordinates,

$$p(1, 2 \ldots N) = |\Psi(1, 2 \ldots N)|^2, \tag{1}$$

and in principle, with a few caveats, all the statistics of the particle motion, momentum, energy, fluctuations, correlations and response can be determined from Ψ. One important aspect of this description is its wave character, reflected by the phase of the wavefunction. When we add particles together, their waveforms overlap and interfere with each other, so that unlike classical systems, the probability distribution of the sum is not the sum of its parts:

$$|\Psi_A + \Psi_B|^2 \neq |\Psi_A|^2 + |\Psi_B|^2. \tag{2}$$

This is part of the answer to something chemists know intuitively: that when one combines elements together, the compound that forms is utterly different from a simple mixture of its components. The other important aspect is that the wavefunction depends on a macroscopically huge number of variables – classically, a system of N particles requires $3N$ position and momen-tum variables – a quantity that is itself huge; yet quantum mechanically, the number of variables required to describe a wavefunction is an exponential of this huge number.

As we scale up from the Angstrom to the micrometer ($1\mathring{A} = 10^{-10}$m, 1μm $= 10^{-6}$m), a mere four orders of magnitude, matter acquires qualitatively new properties. The particles come together to form crystals: this we can understand classically. However, the electron waves that move throughout these immense periodic structures interfere with each other, and this interfer-ence endows matter with remarkable new properties, hardness, rigidity, magnetism, metallicity, semiconductivity and superfluidity, phase transitions, topology and much, much more. To take an example proposed by Anderson (Anderson: 1972), on the scale of the nanometer, the motion of electrons in metallic gold is identical to that in niobium or tin. Yet on scales of a micron, electrons in niobium and tin correlate together into Cooper pairs to form superconductors that expel magnetic fields and levitate magnets. Niobium and tin are examples of low-temperature superconductors, requiring the extreme low temperatures of liquid helium to cool them to the temperatures where they conduct without resistance, but today physicists have discovered new families of "high-temperature superconductors" that only require liquid nitrogen, and there is a dream that room-temperature superconductivity might occur in hitherto undiscovered com-pounds. Yet superconductivity is just a beginning, for already by the micron, life develops. The organism *Mycoplasma mycoides*, found in the human gut, forms self-reproducing cells of 250 nm in diameter (Kuriyan *et al.*: 2013). While we more or less understand the physics of Cooper pairs in periodic, equilibrium superconductors, we are far from understanding the emergent physics of life that develops on the same scale in aperiodic, non-equilibrium structures. This lack of understanding occurs despite our knowledge of the microscopic, many-body Schrödinger equation, and it is this realization that prompts us to appreciate emergence as a complementary frontier (Laughlin *et al.*: 2000). It prompts us to pose the question:

What are the principles that govern the emergence of collective behavior in matter?

IV. A selective history of emergence and reductionism in condensed matter physics

Condensed matter physics is rife with historical examples of intertwined reductionism and emergence, with the one providing insights into the other (Figure 24.2). One of the things we learn from these examples is that fundamental physics principles are not tied to scale: that while insights from the cosmos influence our understanding in the lab, equally, understanding of emergent principles gleaned from small-scale physics in the lab has given us extraordinary new insights into the early universe and the sub-nuclear world.

To illustrate this interplay between emergence and reductionism, let us look at some examples. Quantum condensed matter physics arguably began with Albert Einstein's 1906 proposal (Einstein: 1907) that the concept of quanta could be extended from light to sound. In the previous year he had proposed the idea of quanta, or *photons*, to interpret Planck's theory of blackbody radiation (Einstein: 1905). By proposing that light is composed of streams of indivisible quanta of energy $E = hf$, where $h = 6.626 \times 10^{-34}$ Js is Planck's constant and f is the frequency, Einstein was able to inject new physical insight into Planck's earlier work, and using it, he could make the link between blackbody radiation and the photoelectric effect. By 1906 he saw that he could take the idea one step further, proposing that analogous sound quanta occur in crystals. By treating a crystal as an "acoustic blackbody", Einstein was able to develop the theory of the low-temperature specific heat capacity of diamond as it drops below the constant value ("Dulong and Petit's law")

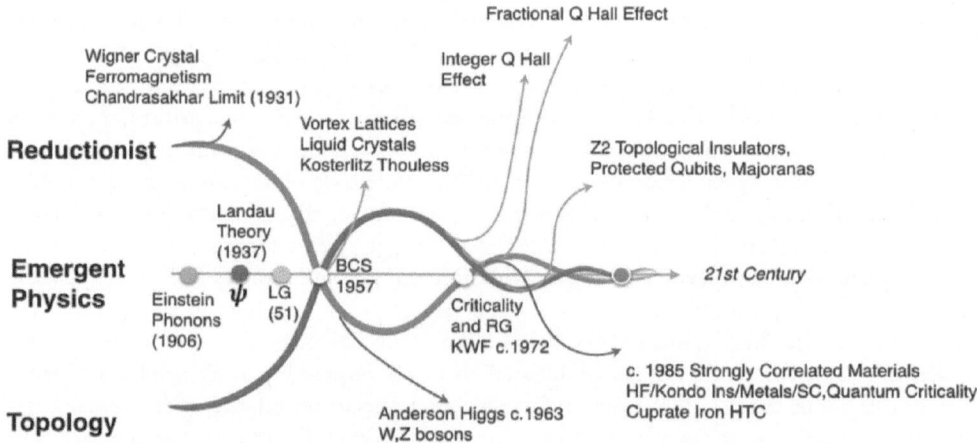

Figure 24.2 Schematic timeline illustrating developments in condensed matter physics over the past century. The three arrows show developments following a reductionist, emergent and topological track.

predicted by classical equipartition. Einstein's work in 1905 and 1907 are remarkable examples of high-grade phenomenology – driven by experiment and careful physical reasoning. Moreover, Einstein's "phonons" as we now call them, are emergent quanta of the solid state: the result of the quantization of the collective motion of a macroscopic crystal.

Another early idea of emergence in physics is Landau's order parameter theory of phase transitions, developed in 1937 (Landau: 1937). Here he introduced the key concepts of an order parameter and spontaneously broken symmetry: the main idea is that the development of order at a phase transition can be quantified in terms of *order parameter ψ*, which describes the development of a macroscopic property, such as a magnetization ($\psi = M$) or an electric polarization ($\psi = P$). With a very simple phenomenological theory, Landau showed how to use this concept to describe phase transitions, without reference to the microscopic origin of the order parameter or the mechanism by which it developed. In Landau's theory, close to a second-order phase transition, the dependence of the bulk free energy $F[\psi]$ on the order parameter is given by

$$F[\psi] = a\left(T - T_c\right)\psi^2 + b\psi^4 + O\left(\psi^6\right) \tag{3}$$

where a and b are positive constants, T is the temperature and T_c is the critical temperature. For $T > T_c$, the free energy is a minimum at $\psi = 0$, but for $T < T_c$, it develops two "broken symmetry" minima at $\psi = \pm\left[(a/2b)\left(T_c - T\right)\right]^{1/2}$ (Chaikin and Lubensky: 1995; Coleman: 2016). The important point about Landau theory is that it describes a universal property of matter near a phase transition, independently of the microscopic details of the material. Thirteen years later in 1950, Ginzburg and Landau (Ginzburg and Landau: 1950) showed how a more detailed version of Landau's theory, or "Ginzburg Landau theory", in which $\psi(x)$ is a complex order parameter with spatial dependence, could provide a rather complete macroscopic description of superconductors accounting for the expulsion of magnetic flux and the levitation of magnets a half-decade before the Bardeen-Cooper-Schrieffer (BCS) microscopic theory of the same phenomenon.

Yet condensed matter physics could not have developed without reductionism (Pais: 1986). With the arrival of Heisenberg's matrix mechanics in the 1920s, it became possible to attempt a first-principles description of quantum matter. Suddenly, phenomena such as ferromagnetism that were literally impossible from a classical perspective could be given a precise microscopic description, and these phenomena could be linked in a reductionist fashion to the equations of quantum mechanics. The idea that electrons are probability waves, described by Schrödinger's equation, led to the notion of Bloch waves: electron waves inside crystals. The idea of antimatter, predicted by Paul Dirac using his relativistic theory of electrons (Dirac: 1931), had its direct parallel in condensed matter physics in Peierls' and Heisenberg's concept of "hole" excitations in semiconductors (Heisenberg: 1931; Hoddeson *et al.*: 1987). Landau and Néel extended Heisenberg's ideas of magnetism to predict antiferromagnetism, first observed in the 1950s, while Wigner used reductionist principles to predict that electrons would form "Wigner crystals" at low densities, a remarkable result not confirmed until the 1980s. Quantum mechanics also enjoyed application in the new realm of astrophysics, most dramatically in Subrahmanyan Chandrasakhar's theory of stellar collapse (Chandrasekhar: 1984). By combining classical gravity with the statistical (quantum) mechanics of a degenerate fluid of protons and neutrons, Chandrasakhar was able to predict that beyond a critical mass, stars would become unstable and collapse. The critical Chandrasakhar mass M of a star,

$$M \approx M_P\left(\frac{M_P}{m_p}\right)^2 \tag{4}$$

is given in terms of the proton and the Planck mass, m_p and $M_P = \left(\dfrac{hc}{G}\right)^{\frac{1}{2}}$, respectively. Chandrasakhar's formula, built on principles designed to understand the terrestrial statistical mechanics of electrons, is the first time that gravity and quantum mechanics come together in a single expression.

Yet the fully reductionist revolution of quantum mechanics ran out of steam when it came to understanding superconductivity: the phenomenon whereby metals conduct electricity without resistance at low temperatures. Some of the greatest minds of the first half of the twentieth century, Bohr, Einstein (Sauer: 2008), Bloch, Heisenberg and Feynman (Schmalian: 2010), attempted microscopic theories of superconductivity, but without success. In 1957, the reductionist and emergent strands of condensed matter physics came together in a perfect storm of discovery, with the development of the Bardeen-Cooper-Schrieffer (BCS) theory of superconductivity (Bardeen *et al.*: 1957). On the one hand, it required a reductionist knowledge of band theory and the interaction of electrons and phonons; it also took advantage of the new methods of quantum field theory, adapted from the theory of quantum electrodynamics by early pioneers such as Fröhlich, Gell-Mann and Hubbard. On the experimental front, it required the discovery of the Meissner effect: the expulsion of magnetic fields that occurs when a metal becomes superconducting; it also built strongly on the phenomenological ideas of London, Landau and Ginzburg, Pippard and Bardeen; finally, it required stripping the physics down to its bare minimum, in the form of a minimalist model now known as the "BCS model". The important point is that rather than attempting a fully reductionist description of the combined electron–lattice and electron–electron interactions, which led to something far too complicated to be solved in one go, Bardeen, Cooper and Schrieffer captured the combined effects of these phenomena in terms of a simple low-energy attractive interaction between pairs.

BCS theory had many further ramifications: pairing was generalized to the nucleus, where it led to an understanding of the stability of even-numbered nuclei; it led to the prediction of superfluidity in neutron stars and He-3. Most unexpectedly, it opened up new perspectives on broken symmetry that inspired Anderson and then Higgs and others to identify a mechanism for how gauge particles acquire mass that we now call the "Anderson-Higgs mechanism" (Anderson: 1963; Higgs: 1964). At a time when particle physicists had almost abandoned field theory, the new success in superconductivity provided a case study of field theory in action that stimulated a resurgence of interest in field theory in particle physics, leading to electro-weak theory (Witten: 2016). Indeed, key elements of electro-weak theory can be understood as a simple two-component spinorial extension of Landau Ginzburg theory, and from this perspective, the weak force in nuclear particle physics can be understood as a kind of cosmic Meissner effect that expels the W and Z fields from our universe.

A second example of the intertwined nature of reductionism and emergence is provided by the theory of critical phenomenon, a revolution in understanding of phase transitions that occurred a decade after BCS theory, between 1965 and 1975 (Domb: 1996). From the 1960s, physicists were increasingly aware of a failure in the classical theory of phase transitions, based on the work of Van der Waals, Landau and others, which was unable to describe the observed properties of second-order phase transitions. Experiments and Onsager's solution to the two-dimensional Ising model showed that phase transitions were characterized by unusual, indeed, universal, power-law behavior. For example, the magnetization of a ferromagnet below its critical temperature develops with a power-law $M \propto (T_c - T)^\beta$. Landau's theory predicts $\beta = 1/2$, yet in three-dimensional Ising ferromagnets, $\beta = 0.326$. Moreover, the unusual critical exponents were found to occur in a wide variety of different phase transitions, exhibiting the phenomenon of "universality".

To understand this discrepancy required a revolution in statistical mechanics, involving new, high-precision measurements of phase transition; it meant borrowing methods that had been developed to control or "renormalize" divergences in particle physics, but it also involved developing new ideas about how physics changes and scales with size. Today these ideas are captured by a "scaling equation" that describes the evolution of a Hamiltonian H with length scale L. Schematically, such scaling equations are written as

$$\frac{\partial H}{\partial Log[L]} = \beta[H], \tag{5}$$

where H is the Hamiltonian; L represents some kind of minimum cut-off length scale to which the Hamiltonian applies; and $\beta[H]$, the function that describes how $H[L]$ depends on length scale, is called the "beta function". The culmination of this work in Fisher and Wilson's "epsilon expansion" showed how to calculate scaling behavior using a beautiful innovation of following physics as a function of dimension d (Wilson and Fisher: 1972). Remarkably, for the simplest models, the classical theories of phase transitions worked in dimensions above $d = 4$. Wilson and Fisher showed that a controlled expansion of the critical properties could be developed in terms of the deviation from four dimensions $\varepsilon = 4 - d$. The Fisher-Wilson theory is a theory of an emergent phenomenon, yet it draws on methodologies from reductionist quantum field theory.

V.   Two examples from current physics

The convective exchange of reductionist and emergent perspectives continues to drive current developments in condensed matter physics. I'd like to touch on two active examples: research into *topological* properties of quantum matter and the mystery posed by the discovery of classes of phase transitions at absolute zero, which radically transform the electrical properties of conductors into *strange metals*.

A.   *A topological connection*

One of the most remarkable developments has been the discovery of a topological connection to emergence (Hasan and Kane: 2010; Moore: 2010). Topology describes global properties of geometric manifolds that are unchanged by continuous deformations. For instance, a donut can be continuously deformed into a one-handled mug: the presence of the hole, or the handle, is topologically protected, and we say they have the same topology. Mathematics links the differential geometry of two-dimensional manifolds to the topology:

differential geometry $\leftrightarrow$ topology

via the Gauss-Bonet theorem,

$$\frac{1}{4\pi} \int kdA = (1 - g)$$

which relates the area integral of the curvature to the number of handles or the *genus g* of the surface. Topology is a kind of mathematical emergence: a robust property that depends on the global properties of a manifold.

The rise of topology in condensed matter physics involved a marvelously tortuous path of discovery. While the microscopic physics is a reductionist consequence of the band theory of insulators developed in the 1930s, the discovery of a topological connection had to await another half-century, culminated in the discovery of a new class of band insulator, the "topological insulator". One of the remarkable properties of topological matter is that the surface remains metallic. The 2016 Nobel Prize in Physics was awarded to Haldane, Kosterlitz and Thouless for their early contributions to topology in condensed matter physics.

Topological structures in physics can develop in both real space and momentum space. An example of the first kind of topology is vortices in a superfluid. In a superfluid, the phase $\phi(x)$ of the complex order parameter $\psi(\vec{x}) \propto e^{i\phi(x)}$ is a smooth function of position, and in passing around a closed path the order parameter must change smoothly and come back to itself so that the change in the phase must be an integer multiple n of 2π, $n \times 2\pi$. The integer n describes the quantization of circulation in a superfluid, first predicted by Onsager and Feynman.

A second kind of topology involves the wavefunction of electrons, in which a non-trivial topological configuration constitutes a new kind of "topological order":

differential geometry of the wavefunction $\leftrightarrow$ topological order

Topological order is distinct from broken symmetry, and it manifests itself through the formation of gapless surface or edge (2D) excitations around the exterior of an otherwise insulating state. The first example of such topological order is the quantization of the Hall constant in two-dimensional electron gases, according to the relationship

$$\rho_{xy} = \frac{1}{\nu} \frac{h}{e^2}$$

where the Hall resistivity, $\rho_{xy} = V_H/I$, is the ratio of the transverse Hall voltage V_H to the current I and ν, an integer associated with the topology of the filled electron bands (Thouless et al.: 1982); one of the manifestations of this effect is the formation of ν "edge states" which propagate ballistically around the quantum Hall insulator.

Microscopically, this topology is determined by way the phase of the electron wavefunction twists through momentum space, which is given by a quantity called the "Berry connection" associated with the filled electron bands, given by

$$\vec{A}(k) = -i \sum_{m=1,N} \langle u_{m,k} | \nabla_k | u_{m,k} \rangle$$

where $u_{m,k}$ is the Bloch wavefunction of the mth filled electron band at momentum $\mathbf{k}$. The Berry connection $\vec{A}(k)$ plays the role of an emergent vector potential: a momentum space analog of the electromagnetic field. The corresponding magnetic flux, or "Berry curvature", $\kappa_k = \vec{\nabla} \times \vec{A}(k)$ plays the same role as the curvature in the Gauss-Bonnet theorem, and the integral of this curvature over momentum space gives the integer "Chern number",

$$\nu = \frac{1}{2\pi} \int \kappa_k d^2 k. \tag{6}$$

Later in the 1980s, Duncan Haldane showed that such topological order could occur without a net magnetic field (Haldane: 1988). Haldane's 1987 theoretical model had a honeycomb structure (Figure 24.3a). Fifteen years later, the discovery of a 2D carbon structure, "graphene", with an uncanny resemblance to Haldane's model, inspired Charlie Kane and Eugene Mele (Kane and Mele: 2005) to propose that topological order would develop in graphene without any magnetic field (Figure 24.3b). The key to their idea was "spin–orbit" coupling – an internal magnetic coupling between the spin and orbital motion of electrons. Kane and Mele recognized that spin–orbit coupling allows spin-up electrons to create a magnetic field for spin-down electrons, and

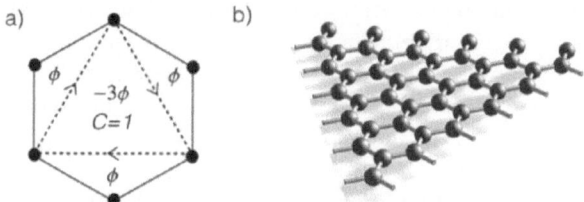

Figure 24.3 (a) Haldane's tight-binding model after (Haldane: 1988) on a honeycomb lattice used to show that topological Chern insulators can form without a net magnetic field. (b) Graphene, which together with Haldane's model, provided stimulus for the discovery of topological insulators.

vice versa, creating two separate versions of the quantum Hall effect, one for spin-up and one for spin-down electrons. The resulting edge states carry spin, forming an early version of the modern topological insulator, the "spin-Hall insulator".

Although the spin–orbit coupling in real graphene turned out to be too weak to give rise to a topological insulator, the idea held and was confirmed by experiment (Bernevig *et al.*: 2006; König *et al.*: 2007) and later generalized to the three-dimensional topological insulators (Fu and Kane: 2007; Fu *et al.*: 2007; Roy: 2009). The current view is that spin–orbit coupling changes the topology of an insulator by inducing a crossing between the unoccupied conduction and occupied valence bands. Such crossings can only take place at certain allowed high symmetry points in momentum space defined by the crystal symmetry, and when they do, they change the topology. Like the braiding of a ribbon, where an odd number of twists produces a non-trivial configuration or Möbius strip, in insulators, an odd number of band crossings leads to a "strong topological insulator" (STI) with conducting surface states.

From a fully reductionist viewpoint, one might wonder why the topological revolution did not occur along with the development of electron band theory, from which it can be deduced. Indeed, one of the early pioneers of band theory, the co-inventor of the transistor, William Shockley (Shockley: 1939), came remarkably close. Yet new emergent principles, while traceable back to their microscopic origins, required the experimentally inspired development of new concepts. We see here a close analogy with the 200-year delay in the discovery of energy as a consequence of Newtonian mechanics.

B. Strange metals

As a counterpoint to the discovery of topological insulators, I'd like to say a little about how our understanding of metals appears to be on the verge of radical change. The foundations of the modern theory of metals were established not long after the discovery of the electron, at the turn of the twentieth century by Paul Drude. One of the main ideas of Drude's theory of metals is that electrons defuse through a metal due to their scattering from imperfections and vibrations. The resulting "transport relaxation time", τ_{tr}, governs most aspects of the electron transport. The arrival of quantum mechanics in the 1920s led to a major upheaval in the understanding of the electron fluid. In particular, electrons, as identical quantum particles, were found to obey the Pauli exclusion principle, which prevents more than one of them occupying the same eigenstate. This individualism causes electrons to fill up momentum space to higher and higher momentum states up to some maximum momentum, the Fermi momentum. The occupied states at this maximum momentum define a Fermi surface in momentum space, and almost all the action in a metal involves electrons at the Fermi surface.

Yet when the dust of quantum mechanics settled, Drude's picture had survived almost intact: in particular, the concept of a transport relaxation time could be extended to describe the scattering of electrons at the Fermi surface by disorder and mutual interactions, leading back to Drude's diffusive electron transport picture. One of the consequences of this robustness is that one can measure the resistivity, Hall constant and dependence of its resistivity on a magnetic field, the so-called "magneto-resistance", to check if these quantities scale with the scattering time, τ_{tr}, in the way predicted by Drude's theory. Although the rate at which electrons scatter is temperature dependent, various ratios appearing in the transport theory are independent of the scattering rate and become temperature independent. One well-known consequence of Drude's theory is that the cancelation between the scattering rate associated with the Lorentz force cancels with the scattering rate due to the electric force so that the ratio of the two, determined by the Hall constant $R_H = V_H/I$ is temperature independent. Another consequence is a scaling law known as Kohler's law. In Drude's theory the resistivity R is proportional to the scattering rate $R \propto \tau_{tr}^{-1}$, whereas the magneto-resistivity grows with the square of the angle of deflection (Hall angle) of the electrical current in a field, $\Delta R / R \propto \Theta_H^2$. Now the Hall angle depends on the product of the cyclotron precession frequency and the scattering rate, $\theta_H = \omega_c \tau_{tr}$ so that $\Delta R / R \propto \theta_H^2 \sim \tau_{tr}^2$, which when combined with the resistivity, leads to Kohler's rule $\Delta R/R \propto R^{-2}$. This scaling relation works remarkably well for a wide range of simple metals, vindicating Drude's theory.

Of course, quantum mechanics does have radical consequences for metals. For example, disorder can cause electron waves to *Anderson localize* (Abrahams *et al.*: 1979; Anderson: 1958), completely stopping electron diffusion to produce an insulator. The many-body version of this phenomenon, *many-body localization* (Nandkishore and Huse: 2015), is of great current interest. Another radical consequence that I want to discuss now is the formation of *strange metals*. Over the past three decades, experiments have revealed a new class of "strange metal" which deviates from Drude's theory in a qualitative way. This unusual metallic behavior tends to develop in metals that are close to instability. When the interactions are increased inside a metal, through the effect of pressure, chemistry or external fields, the metal can become unstable, giving rise to *quantum phase transition* into an ordered state, such as magnetism. Such instabilities occur at an absolute zero, where there are no thermal fluctuations to drive a phase transition. Instead, the phase transition is driven by quantum zero-point fluctuations, and it is thought that these fluctuations play a role in transforming the electron fluid, causing the resulting conductor to deviate qualitatively from Drude's behavior.

The most famous strange metals are the high-temperature cuprate superconductors (Chien *et al.*: 1991; Takagi *et al.*: 1992), but similar behavior is also seen in their low-temperature cousins, the family of heavy electron superconductors known as "115" superconductors (Nakajima *et al.*: 2004), and most recently, in artificially constructed two-dimensional electron gases (Mikheev *et al.*: 2015), which are not superconducting. High-temperature cuprate superconductors lose their resistance at temperatures as high as 90 K, high enough to be able to use liquid nitrogen to cool them into the superconducting state. But above these temperatures, they are equally remarkable, for they exhibit a resistivity that is linear up to very high temperatures $R(T) \propto T$ (Figure 24.4a). In fact, this linearity can be traced back to a Drude scattering rate $\Gamma_{tr} = \tau_{tr}^{-1}$ that is proportional to the temperature, given approximately by $\Gamma_{tr} \sim \dfrac{k_B T}{\hbar}$. The timescale $\tau_{tr} \sim \dfrac{\hbar}{k_B T}$ is sometimes called the "Planck time", because it is the timescale derived from combining the energy–time uncertainty relationship $\Delta E \Delta \tau \sim \hbar$ with the Boltzmann energy $\Delta E \sim k_B T$. This simple scaling of the scattering rate with the temperature is very unusual, and in a typical metal the scattering rate has a much more complicated dependence on temperature, on disorder and on the coupling to vibrations of the crystal. Perhaps the strangest aspect of these metals is their departure from Drude's behavior in a magnetic field because the scattering response to the

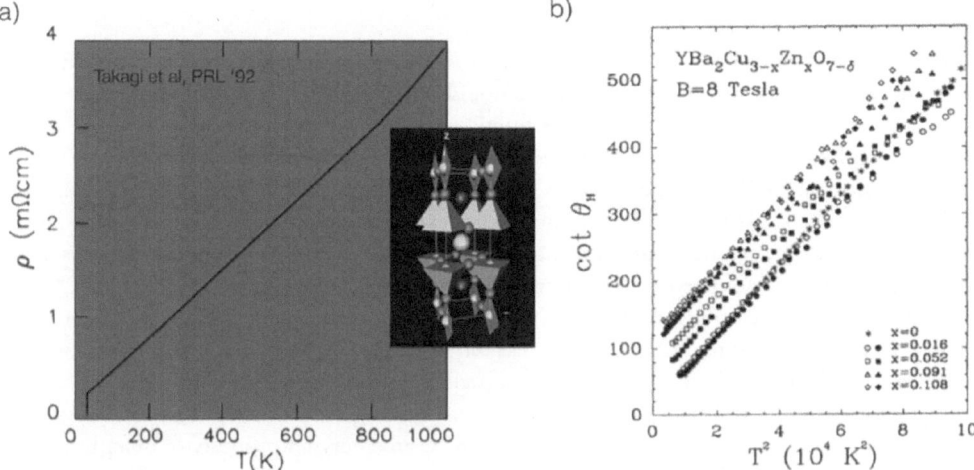

Figure 24.4 Strange metals. (a) Linear resistivity of the high-temperature superconductor $La_{2-x}Sr_xCuO_4$ (x = 0.15) (adapted with permission from H. Takagi *et al*, Phys. Rev. Lett. 69, 2975 (1992) (Takagi *et al.*: 1992)), showing the remarkable linear resistivity up to 1000K, indicating that the electrical current relaxation rate $\Gamma_{tr} \propto T$ is proportional to the temperature. (b) Quadratic temperature dependence of the Hall angle in a cuprate superconductor (reprinted with permission from T. R. Chien, *et al.*, Phys. Rev. Lett. 67, 2088 (1991)(Chien *et al.*: 1991)), indicating that Hall currents in these strange metals exhibit a decay rate $\Gamma_H \propto T^2$. The appearance of two relaxation timescales in a simple conductor poses a challenge to our current understanding of metals.

Lorentz force, measured in a magnetic field, is qualitatively different from the response to a pure electric field. Whereas the linear resistivity gives a scattering rate Γ_{tr} proportional to temperature, the magneto-resistivity and Hall resistivity give a scattering rate that is quadratic in the field $\Gamma_H \propto \theta_H^{-1} \propto T^2$ (Figure 24.4b). Summarizing

$$\Gamma_{tr} \propto T, \qquad \Gamma_H \propto \frac{T^2}{W},$$

where W is a scale that governs the decay of Hall currents. The presence of these qualitatively different scattering rates leads to a strongly temperature-dependent Hall constant and a "modified Kohler's rule", whereby the magneto-transport scales with the square of the Hall angle, rather than the square of the conductivity:

$$R(T) \propto \tau_{tr}^{-1}, \qquad \frac{\Delta R}{R} \propto \theta_H(T)^2$$

This behavior is not unique to cuprate superconductors, and it has also been observed in certain heavy fermion superconductors (Nakajima *et al.*: 2004), which are low-temperature cousins of the cuprate superconductors, and in low-dimensional oxide interfaces (Mikheev *et al.*: 2015), but which have quite different microscopic chemistry and structure. These results taken together suggest that a fundamentally new kind of metal has been discovered, one that may require a new conceptual framework for interacting electrons. Unlike the new developments in our understanding of insulators, the emergent framework for understanding strange metals is still very much in the early days of discovery.

Conclusion

This chapter has illustrated examples of emergence in condensed matter physics, seeking to highlight the close interdependence of a reductionist and emergent approach. Perhaps the most exciting aspect of this linkage is that it may provide a way to accelerate the way we solve major problems in physics and the natural sciences. While reductionism provides the mathematics and the computational tools to tackle complex problems and gain new insight into emergence, at the same time, it is likely that the importance of understanding of physics in the lab, particularly emergent physics between the Angstrom and the micron, will, as it has in previous centuries, yield important insights into our reductionist understanding.

As in previous generations, condensed matter physicists are looking to the tools of particle physics, such as the holographic principle (Zaanen *et al.*: 2015), to make new progress on the many-body problem, while in a similar vein, particle physicists and cosmologists are looking to emergence and condensed matter for inspiration. One of the prevalent ideas for unifying gravity and quantum mechanics is that space-time itself may be an emergent property of quantum gravity on scales beyond the Planck length (Seiberg: 2006). Another area of activity is the problem of dark matter. For example, recently Verlinde (Verlinde: 2016) has suggested that the dark matter problem may be a consequence of an emergent aspect of gravity in which the unseen gravitating force inside galaxies is not interpreted as a cloud of particles, but as a kind of gravitating condensate.

These developments tempt us to speculate whether our current understanding of quantum mechanics might parallel that of classical mechanics, which remained incomplete 200 years after *Principia*, because one of its key emergent consequences, heat, prevented an understanding of energy. Perhaps, in a similar fashion, ninety years after Heisenberg, Schrödinger and Dirac, a more complete understanding of quantum mechanics might await new perspectives from emergence.

I should like to thank Natan Andrei, Stephen Blundell, Tom Lancaster and Jan Zaanen for stimulating discussions related to this work. This work was supported by NSF grant DMR-1309929.

Notes

1 For a careful discussion of the definition of emergence in physics, see for example (Kivelson and Kivelson: 2016).
2 Lagrange writes in his treatise *Méchanique Analitique* "In effect the integral $T + V$ = constant follows when T and V have no t dependence" ("*En effect, l'intégrale $T + V$ = const, ayant nécessairement lieu, puisque T & V sont fonctions sans t*") (Lagrange: 1788).

References

Abrahams, E., P. W. Anderson, D. C. Licciardello, and T. V. Ramakrishnan (1979), "Scaling Theory of Localization: Absence of Quantum Diffusion in Two Dimensions," *Physical Review Letters* 42, 673–676.

Anderson, P. W. (1958), "Absence of Diffusion in Certain Random Lattices," *Physical Review Letters* 109, 1492–1505.

Anderson, P. W. (1963), "Plasmons, Gauge Invariance, and Mass," *Physical Review Letters* 130 (1), 439–442.

Anderson, P. W. (1972), "More Is Different," *Science* 177, 393.

Bacon, Francis (1620), "Novum Organum," Book I, Aphorism XIX.

Bardeen, J., L. N. Cooper, and J. R. Schrieffer (1957), "Theory of Superconductivity," *Physical Review* 108 (5), 1175–1204.

Bernevig, B. Andrei, Taylor L. Hughes, and Shou-Cheng Zhang (2006), "Quantum Spin Hall Effect and Topological Phase Transition in HgTe Quantum Wells," *Science* 314 (5806), 1757–1761.

Boltzmann, Ludwig (1905), *Populäre Schriften* (Leipzig: J. A. Barth).

Chaikin, P. M., and T. C. Lubensky (1995), *Principles of Condensed Matter Physics* (Cambridge: Cambridge University Press), See for example: 151–154.

Chandrasekhar, S. (1984), "On Stars, Their Evolution and Their Stability," *Reviews of Modern Physics* 56, 137–147.

Chien, T. R., Z. Z. Wang, and N. P. Ong (1991), "Effect of Zn Impurities on the Normal-State Hall Angle in Single-Crystal YbBa2Cu3−xZnxO7−Δ," *Physical Review Letters* 67, 2088.

Coleman, Piers (2016), *Introduction to Many Body Physics* (Cambridge: Cambridge University Press), See for example: 357–415.

Darwin, Charles (1859), *On the Origin of Species by Means of Natural Selection*, Chap. 14 (London: John Murray).

Dirac, P. A. M. (1931), "Quantised Singularities in the Quantum Field," *Proceedings of the Royal Society A* 133, 821.

Domb, Cyril (1996), *The Critical Point: A Historical Introduction to the Modern Theory of Critical Phenomena* (Boca Raton, FL: CRC Press).

Einstein, A. (1905), "Concerning an Heuristic Point of View towards the Emission and Transformation of Light," *Annalen der Physik* (Leipzig) 17, 132.

Einstein, A. (1907), "Planck's Theory of Radiation and the Theory of the Specific Heat," *Annalen der Physik* 22, 180.

Fu, L., and C. L. Kane (2007), "Topological Insulators with Inversion Symmetry," *Physical Review* B 76 (4), 45302.

Fu, Liang, C. L. Kane, and E. J. Mele (2007), "Topological Insulators in Three Dimensions," *Physical Review Letters* 98, 106803.

Ginzburg, V. L., and L. D. Landau (1950), "On the Theory of Superconductivity," *Zhurnal Éksperimental'noĭ i Teoretícheskoĭ Fiziki* 20, 1064.

Greene, Brian (1999), *The Elegant Universe: Superstrings, Hidden Dimensions and the Quest for Ultimate Reality* (New York: W. W. Norton).

Haldane, F. D. M. (1988), "Model for a Quantum Hall-Effect without Landau-Levels–Condensed-Matter Realization of the Parity Anomaly," *Physical Review Letters* 61 (18), 2015–2018.

Hamilton, W. R. (1835), "Second Essay on a General Method in Dynamics," *Philosophical Transactions of the Royal Society* 125, 95–144.

Hasan, M. Z., and C. L. Kane (2010), "Colloquium: Topological Insulators," *Reviews of Modern Physics* 82, 3045.

Hawking, Steven (1988), *A Brief History of Time: From the Big Bang to Black Holes* (New York: Bantam Dell Publishing).

Heisenberg, W. (1931), "The Pauli Exclusion Principle," *Annalen der Physik* (Leipzig) 10, 888.

Higgs, Peter W. (1964), "Broken Symmetries and the Masses of Gauge Bosons," *Physical Review Letters* 13 (16), 508–509.

Hoddeson, Lillian, Gordon Baym, and Michael Eckert (1987), "The Development of the Quantum-Mechanical Electron Theory of Metals: 1928–1933," *Reviews of Modern Physics* 59, 287–327.

Kane, C. L., and E. J. Mele (2005), "Z2 Topological Order and the Quantum Spin Hall Effect," *Physical Review Letters* 95, 146802.

Kivelson, Sophia, and Steven A. Kivelson (2016), "Defining Emergence in Physics," *Quantum Materials*), 16024.

König, Markus, Steffen Wiedmann, Christoph Brüne, Andreas Roth, Hartmut Buhmann, Laurens W. Molenkamp, Xiao-Liang Qi, and Shou-Cheng Zhang (2007), "Quantum Spin Hall Insulator State in HgTe Quantum Wells." *Science* 318, 766–770.

Kuriyan, John, Boyana Konforti, and David Wemmer (2013), *The Molecules of Life: Physical and Chemical Principles* (Garland Science).

Lagrange, Joseph-Louis (1788), *Mécanique Analytique* (Paris: Hachette livre, BNF), 258 pp.

Landau, L. D. (1937), "Theory of Phase Transformations," *Physikalische Zeitschrift der Sowjetunion* 11 (26), 545.

Laughlin, R. B., D. Pines, J. Schmalian, B. P. Stojkovic, and P. Wolynes (2000), "The Middle Way," *Proceedings of the National Academy of Sciences* (USA) 97.

Maxwell, James Clerk (1859), *On the stability of Saturn's Rings* (Cambridge, London: Macmillan and Co).

Mikheev, Evgeny, Christopher R. Freeze, Brandon J. Isaac, Tyler A. Cain, and Susanne Stemmer (2015), "Separation of Transport Lifetimes in SrTiO3-Based Two-Dimensional Electron Liquids," *Physical Review* B B 91 (16), 165125.

Moore, Joel E. (2010), "The Birth of Topological Insulators," *Nature* 464 (7286), 194–198.

Nakajima, Y., K. Izawa, Y. Matsuda, S. Uji, T. Terashima, H. Shishido, R. Settai, Y. Onuki, and H. Kontani (2004), "Normal-State Hall Angle and Magnetoresistance in Quasi-2D Heavy Fermion CeCoIn5 Near a Quantum Critical Point," *Journal of the Physical Society of Japan* 73, 5.

Nandkishore, Rahul, and David Huse (2015), "Many-Body Localization and Thermalization in Quantum Statistical Mechanics," *Annual Review of Condensed Matter Physics* 6, 15–38.

Pais, Abraham (1986), *Inward Bound: Of Matter and Forces in the Physical World* (Oxford: Oxford University Press).

Roy, Rahul (2009), "Z2 Classification of Quantum Spin Hall Systems: An Approach Using Time-Reversal Invariance," *Physical Review B* 79, 195321.

Rumford, Benjamin Count of (1798), "An Inquiry Concerning the Source of the Heat Which Is Excited by Friction," *Philosophical Transactions of the Royal Society* 88, 80–102.

Sauer, Tilman (2008), "Einstein and the Early Theory of Superconductivity, 1919–1922," *Archive for the History of Exact Sciences* 61, 159–211.

Schmalian, Joerg (2010), "Failed Theories of Superconductivity," arXiv:1008.0447 in *Bardeen, Cooper and Schrieffer: 50 Years*, eds L. Cooper and D. Feldman (Singapore: World Scientific), 41–55.

Seiberg, Nathan (2006), "Emergent Spacetime," arXiv:hep-th/0601234.

Shockley, William (1939), "On the Surface States Associated with a Periodic Potential," *Physical Review* 56, 317–323.

Takagi, H., B. Batlogg, H. L. Kao, J. Kwo, R. J. Cava, J. J. Krajewski, and W. F. Peck (1992), "Systematic Evolution of Temperature-Dependent Resistivity in La2−xSrxCuO4," *Physical Review Letters* 69, 2975–2978.

Thouless, D. J., M. Kohmoto, M. P. Nightingale, and M. Den Nijs (1982), "Quantized Hall Conductance in a Two-Dimensional Periodic Potential," *Physical Review Letters* 49 (6), 405–408.

Verlinde, Erik P. (2016), "Emergent Gravity and the Dark Universe," arXiv.org 1611.02269.

Weinberg, Steven (1992), *Dreams of a Final Theory: The Scientist's Search for the Fundamental Laws of Nature* (New York: Vintage Books).

Weinberg, Steven (2005), "Living in the Multiverse," arXiv hep-th, 0511037.

Wilson, Kenneth G., and Michael E. Fisher (1972), "Critical Exponents in 3.99 Dimensions," *Physical Review Letters* 28, 240–243.

Witten, Edward (2016), "Phil Anderson and Gauge Symmetry Breaking," in *PWA90: A Lifetime of Emergence*, eds. P. Chandra, P. Coleman, G. Kotliar, Ph. Ong, D. L. Stein and C. Yu (World Scientific), 73–90.

Young, Thomas (1807), "A Course of Lectures on Natural Philosophy and the Mechanical Arts," *Lecture* 6, 41.

Zaanen, Jan, Yan Liu, Ya-Wen Sun, and Koenraad Schalm (2015), *Holographic Duality in Condensed Matter Physics* (Cambridge: Cambridge University Press).

25

THE EMERGENCE OF SPACE AND TIME

Christian Wüthrich*

Research in quantum gravity strongly suggests that our world is not fundamentally spatiotemporal, but that spacetime may only emerge in some sense from a non-spatiotemporal structure, as this chapter illustrates in the case of causal set theory and loop quantum gravity. This would raise philosophical concerns regarding the empirical coherence and general adequacy of theories in quantum gravity. If it can be established, however, that spacetime emerges in the appropriate circumstances and how all its relevant aspects are explained in fundamental non-spatiotemporal terms, then the challenge is fully met. It is argued that a form of spacetime functionalism offers the most promising template for this project.

Space and time, it seems, must be part and parcel of the ontology of any physical theory, of any theory with a credible claim to being a *physical* theory, that is. After all, physics is the science of the fundamental constitution of the material bodies, their motion in space and time, and indeed of space and time themselves. Usually implicit, Larry Sklar (1983) has given expression to this common intuition:

> What could possibly constitute a more essential, a more ineliminable, component of our conceptual framework than that ordering of phenomena which places them in space and time? The spatiality and temporality of things is, we feel, the very condition of their existing at all and having other, less primordial, features. . . . We could imagine a world without electric charge, without the atomic constitution of matter, perhaps without matter at all. But a world not in time? A world not spatial? Except to some Platonists, I suppose, such a world seems devoid of real being altogether.
>
> *(45)*

The worry here, I take it, goes beyond a merely epistemic concern regarding the inconceivability of a non-spatiotemporal world; rather, it is that such a world would violate some basic necessary condition of physical existence. It is contended that space and time partially ground a material world. The alternative to a spatiotemporal world, it is suggested, is a realm of merely abstract entities.[1] Part of what it means to be 'physically salient' (Huggett and Wüthrich 2013) is to be in space and time. In other words, what it is to give a physical explanation of aspects of our manifest world is, among other things, to offer a theory of how objects are and move in space and time.

However, it turns out that physics itself may lead us to the conclusion that space and time are not part of the fundamental ontology of its best theories. Just as the familiar material objects of our manifest world arise from qualitatively rather different basic constituents, spacetime may emerge only from the collective action of fundamental non-spatiotemporal degrees of freedom. Spacetime, in other words, may exist merely 'effectively', just as many salient aspects of our physical world, such as temperature, pressure, or liquidity.

The present chapter investigates this possibility. To this end, I precisify the situation as I see it (§1) and illustrate it with concrete cases (§2). In §3, I articulate and defend the position of 'spacetime functionalism' as the appropriate position to take vis-à-vis the suggestion that spacetime is emergent. Conclusions are in §4.

1. The emergence of space and time in fundamental physics

There are two areas in physics in which space or spacetime may emerge from something qualitatively rather different: wave function monism in quantum mechanics and quantum theories of gravity. The former case concerns a particular interpretative issue in non-relativistic quantum mechanics. Although this in itself does not solve the measurement problem, part of such a solution will be to get clear on the ontological status of the wave function. A central part of the formalism of quantum mechanics, the wave function mathematically expresses the state of a quantum system as the system's configuration. Governed by the Schrödinger equation, it is a function whose domain is a mathematical space whose dimensions are the chosen degrees of freedom (and time), normally the system's position in space. If the system at stake consists in N particles moving in three-dimensional space, then the wave function's domain is the $3N$-dimensional configuration space.

All interpretations of quantum mechanics are faced with the question of the ontological status of the wave function, and I will not be concerned with this general problem here. Suffice it to mention that it has been proposed (Albert 1996) that the wave function is the *only* thing in the ontology of quantum mechanics. This position – 'wave function monism' – faces the challenge of explaining the features of the manifest world, which seem very distant from the object described by the wave function. The issue, then, is to understand how three-dimensional physical space and what goes on in it emerges from the wave function in its high-dimensional configuration space. Thus, wave function monism must recover the spatiality of our physical world and its three-dimensionality. David Albert (2015) defends a functionalist response on behalf of the wave function monist: ordinary objects in three-dimensional space are first 'functionalized' in terms of their causal roles, that is, reduced to a node in the causal network of the world; then it is argued that the wave function dynamically enacts these causal relations and, more generally, these functional roles precisely in a way which gives rise to the empirical evidence we have and to the manifest world more generally.[2]

The second area concerns quantum theories of gravity.[3] Although general relativity stands unrefuted as our best theory of gravity and hence of spacetime, it assumes that matter exhibits no quantum effects. Since this assumption is false, it must ultimately be replaced by a more fundamental, that is, more accurate and more encompassing, theory of gravity which takes the quantum nature of matter into full account. There are many proposals for how to articulate such a quantum theory of gravity, and unfortunately no empirical constraints to guide the search other than those confirming our currently best theories, such as the standard model of particle physics and general relativity. Saving concrete examples for §2, an analysis of different research programmes into quantum gravity reveals that in almost all of them, spacetime is absent from the fundamental ontology. More precisely, the ontology of these theories seems to consist of physical systems with less-than-fully spatiotemporal degrees of freedom. In other words, the structures postulated by these theories lack several, or most, of those features we would normally attribute to space and time, such as distance, duration, dimensionality, or relative location of objects in space and time. In short, spacetime, either as it figures in general relativity or phenomenologically in the experience of the world, is *emergent* (Huggett and Wüthrich forthcoming).

In both cases, we witness the emergence of physical space or spacetime from something nonspatial or non-spatiotemporal.[4] However, there are also significant disanalogies. For starters, space and time play very different roles in quantum mechanics: while position is an observable with a corresponding operator, time is the dynamical parameter entering the wave function and hence the Schrödinger equation. Departing from this difference between space and time, Alyssa Ney (2015, §7) notes that the wave function monist has the dynamical evolution of the wave function available for their functionalist reconstruction project, which is generally not the case in quantum gravity, where time is part of the spacetime structure that evaporates at the fundamental level. Ney argues that this difference is responsible for the differential success of the two cases in dealing with the threat of empirical incoherence. A theory is *empirically incoherent* just in case its truth undermines conditions necessary for its empirical confirmation. Ney maintains that since empirical confirmation is ineliminably diachronic but not spatial, fundamental time – and change – is essential for saving a theory from being empirically incoherent in a way that space is not. She concludes that, unlike wave function monists, quantum theories of gravity without time face imminent empirical incoherence. However, it is not clear on what basis such a distinction can be made; as long as space and time, which both seem necessary for empirical confirmation, can be shown to emerge at the appropriate scales of human science, the threat is averted in either case (Huggett and Wüthrich 2013). Be this as it may, it is clear that in quantum gravity it is *spacetime*, and not mere *space*, which is emergent rather than fundamental, and this constitutes the first important disanalogy.

There is a second crucial disanalogy. While it is mandatory for any interpretation of quantum mechanics to pronounce itself on the ontological status of the wave function, the content of that pronouncement need not be that the wave function gets reified and admitted to the ontology. Instead, one can opt for a 'primitive ontology' of local 'beables', that is, basic entities in space and time (Allori et al. 2008). This primitive ontology comprises the basis which makes up the entire world, including empirical evidence, and is postulated prior to an interpretation of the mathematical formalism. Only once the primitive ontology is set up, we check whether the formalism of the theory commits us to further entities. Among Bohmians, for instance, it is then popular to interpret the wave function as part of the *nomological*, rather than the *ontological*, structure of the world (Miller 2014; Callender 2015). While wave function monism is thus clearly optional, it seems as if the fundamental absence of space and time from the ontologies of quantum gravity is not a matter of interpretation, and thus unavoidable. Furthermore, it is generic, that is, present in most or all approaches to quantum gravity, as can be seen, among others, from the examples to be discussed in §2.

Turning to quantum gravity for the remainder of the chapter then, what exactly do we mean when we say spacetime 'emerges' from some non-spatiotemporal reality? Before anything else, it should be emphasized that quantum gravity is very much a work in progress, and hence we cannot hope to command a conclusive understanding of the situation just yet. In that sense, the term 'emergence' serves as placeholder for a relation, the investigation of which is part of the very project of quantum gravity: it is, as it were, a working title for that relation.

Nevertheless, there are a few things we can presently say, however provisional they may be. First, what are the relata of the 'emergence' relation? What we are ultimately interested in are two aspects of the emergence of spacetime. First, the relationship between general relativity and the quantum theory of gravity must be understood. So it designates a relation between *theories*. Second, it also refers to a relation that is supposed to generically hold between the physical structures of quantum gravity and relativistic spacetimes. Let us discuss these two aspects in turn.

Qua relation between theories, emergence designates an asymmetric relation obtained between general relativity and a quantum theory of gravity capturing several important aspects. First, it is a relation of *relative fundamentality*, which partially orders scientific theories (Wüthrich 2012b, §2). A putative theory of quantum gravity is more fundamental than general relativity, although it may not be fundamental *tout court* (Crowther and Linnemann forthcoming). Second, it should be expected to be a *broadly reductive* relation, both in the ontological and the explanatory sense. The ontology of general relativity should be expected to depend on, and hence be reducible to, that of a quantum theory of gravity. The scientific explanations at the level of general relativity should ultimately also be understood in terms native to the more fundamental theory, although this may be less obvious than for the ontology. This reducibility currently remains a promissory note. Although not strictly necessary, it would constitute a central part of an explanation in quantum gravity for the predictive successes (and failures, should there be) of general relativity. In this sense, reduction can be seen as a methodological directive in explicating the relation of emergence in the present case. This means that the form of 'emergence' at stake cannot be the strong form, which excludes reducibility (Chalmers 2006). Finally, although broadly reductive, the relation should also turn out to be somewhat *corrective* in the sense that the more fundamental theory should deliver predictions which are measurably improved over those of general relativity.

Qua relation between the entities or structures postulated by the theories at the two levels, emergence is an equally asymmetric relation. Unlike the case of relating theories, here it will generally be *many-to-one*, relating a potentially very large number of fundamental structures to one and the same relativistic spacetime. This is to be expected, as some minor differences at the fundamental level should not lead to any discernible difference at the emergent level: recombine the quantum-gravitational structure slightly, and the same relativistic spacetime will be its best approximation at the effective level. In other words, fundamental structures multiply realize the less fundamental structure, though the fundamental structures are not different in kind. A second marked – and important – difference to relating theories is the fact that in general, there will be some model of the quantum theory of gravity, that is, some physically possible fundamental structures, which will not give rise to anything like spacetime. Just as elementary particles may fail to combine to give rise to a carbon molecule, or a liquid, or a life, fundamental structures in quantum gravity may inauspiciously assemble such that *no effective spacetime* results. Thus, the fundamental theory should be expected to contain some models which do not resemble our actual world at all.

Philosophers may complain that this use of 'emergence' is unjustified in the context of quantum gravity, particularly in light of the expected reducibility. I will not quarrel about words here, but hasten to add that such protest misses the fact that 'emergence' here is supposed to express precisely the dual aspects of *dependence* and *independence* traditionally associated with emergence. On the one hand, it captures the anticipated complete ontological and, to some degree explanatory, dependence of the level of general relativity on the more fundamental one of quantum gravity and its structures. On the other hand, it enounces the fact that the spacetime structures are qualitatively distinct from those found at the more fundamental level. This qualitative independence can be cashed out in different currency, with varying strength. At its strongest end, it asserts full autonomy of the levels and assumes, for example, the form of non-supervenience, or at least of irreducibility. Even in its weakest form, however, it gives voice to the novelty involved in the contention that the fundamental structures of quantum gravity are relevantly non-spatiotemporal. That such novelty in itself need not entail some irreducibility but is commensurate with an underlying reductive relation is a widespread view in philosophy of physic, and has been given an explicit statement and defence by Jeremy Butterfield (2011). Motivated by effective field theory, Karen Crowther (2016, particularly Ch. 2) similarly argues that of the many concepts of emergence on the market, those operative in quantum gravity should be considered compatible with reduction. In recent work (Crowther 2018), she convincingly contends that it is the *reductive aspect* which should be considered necessary for the relation between a quantum theory of gravity and general relativity, and emergence – if it obtains at all – does so merely non-essentially and in ways that fully depend on the details of the approach to quantum gravity taken.

If it does obtain, the emergence of spacetime (particularly in so-called 'canonical' quantum gravity) can come in grades or 'levels' of severity, or so Daniele Oriti (forthcominga) suggests. At the mild end of the spectrum, we find the apparent disappearance of (space and) time exemplified by moving to a quantum description of the gravitational field and thus of the chronogeometric aspects of spacetime. At the next step removed, the fundamental structure may consist in 'pre-geometric' degrees of freedom, which are not merely in a quantum superposition, but whose eigenstates do not afford a chronogeometric interpretation at all. At the final level, these pre-geometric degrees of freedom may dynamically combine into different continuum phases, only some of which are chronogeometric or spatiotemporal. In this case, there are phase transitions from one phase to another, including, for example, a transition from a non-spatiotemporal or pre-geometric one to a spatiotemporal or geometric phase – a 'geometrogenesis'.

2. Concrete examples of spacetime emergence

The severity and the characteristics of the disappearance of spacetime differ from one approach to quantum gravity to another. This section illustrates both of these dimensions of spacetime emergence. To this end, I will discuss causal set theory as well as loop quantum gravity and some of its extensions. It should be noted that the emergence, and hence non-fundamentality, of spacetime is a rather generic feature of quantum gravity.[5] In this sense, causal set theory and loop quantum gravity are merely representative examples of this.

Causal set theory starts out from a central result in general relativity according to which the causal structure determines, up to a so-called 'conformal factor', the geometry of a relativistic spacetime. This result motivates causal set theory to postulate a discrete causal ordering as capturing the basic structure – a 'causal set'. The causal ordering is a partial order formed by a fundamental relation of 'causal precedence' on a set of otherwise unspecified elementary events, which is designed to be the discrete analogue of the causal (lightcone) structure of relativistic spacetimes. See Figure 25.1 for an illustration of the resulting structure. As indicated in this figure, a relativistic spacetime may emerge from the fundamental causal set structure. A necessary condition for such emergence is that there exists a map ϕ from the elements of the causal set to spacetime events such that the causal structure is preserved: if an elementary event p causally precedes another, q, then its image under ϕ must be in the causal past of $\phi(q)$, as in Figure 25.1. This condition, however, is far from sufficient, as generically causal sets give rise only to highly deformed spacetimes. The standard remedy is to impose additional conditions whose task it is to rule out most of these unwieldy structures. The additional conditions are dynamical rules, which a causal set must follow as it grows from a finite past. Furthermore, since the causal set is a discrete structure and relativistic spacetimes are continua, emerging spacetimes cannot exhibit interesting features below the characteristic scale of the discrete structure of the causal set at stake. If it did, then the underlying causal set would not have the resources to ground them. Finally, a given causal set should, at least to a good approximation, give rise to no more than one emergent spacetime. This is imposed to ensure that while a single emergent entity can arise from distinct

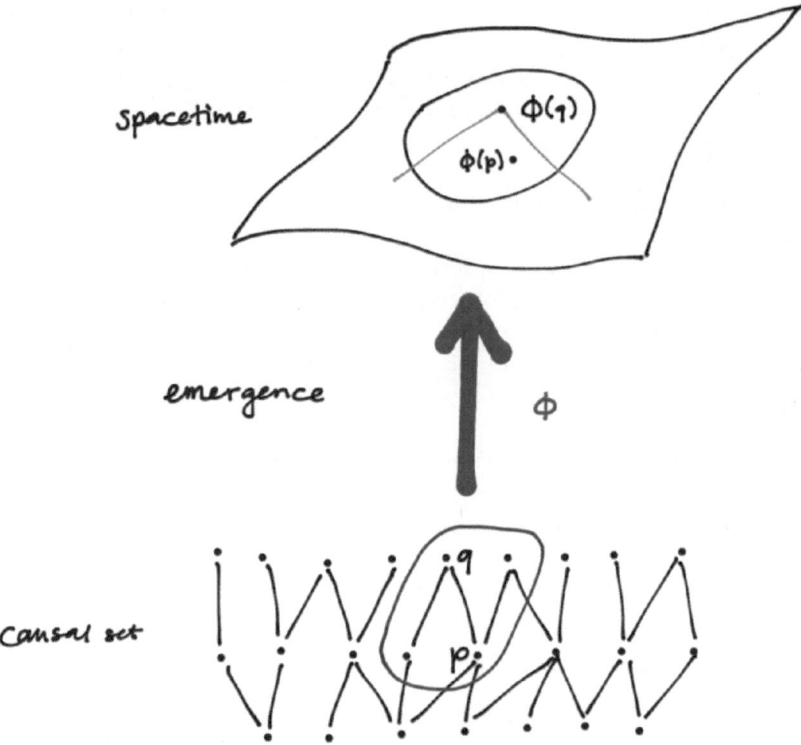

Figure 25.1   A causal set and an emergent spacetime related by a map ϕ.

fundamental structures, one and the same fundamental structure cannot help but give rise to one specific emergent entity, if it does at all. The relation has to be many-to-*one*, not many-to-many. Call this demand the 'unique realization requirement'.

The research programme of causal set theory remains incomplete to date in a number of respects. First, the unique realization requirement has not been shown to be satisfied in causal set theory. That it is satisfied is called the 'Hauptvermutung', or main conjecture, by advocates of the research programme. Its unproven status marks one of the programme's main gaps. Second, the theory is so far entirely classical and as such cannot claim to meet the basic demand for a quantum theory of gravity. At the very least, it would have to offer a way of accommodating the quantum nature of matter. So far, it does not.

Let us compare causal set theory and general relativity in terms of emergence as discussed in the previous section. It is clear that causal set theory is a candidate for a more fundamental theory than general relativity. Second, the ambition of the research programme is to provide a reduction of general relativity, at least in the functional sense to be explicated in Section 3. Should the reduction succeed, it will clearly be ontological and, again at least in the functional sense, explanatory. Third, although this can only be judged with the complete theory in place, the functional reduction of the relevant quantities is likely to result in small corrections for measurable spatiotemporal quantities, such as spatial distances and temporal durations. Furthermore, advocates of causal set theory claim that it predicts quantities, such as the cosmological constant, which remains a free parameter that has to be set by hand in general relativity.[6]

Turning to the relation between the structures rather than the theories, we do find that many distinct causal sets should give rise to one relativistic spacetime, and so the latter are multiply realizable by causal sets. Second, there are (too) many causal sets from which nothing spatiotemporal emerges: unless substantive dynamical constraints are added to the theory, most causal sets do not look like our world at all.

Up to a few provisos, then, let us accept that many of the conditions for there to be the right kind of dependence of the general relativity on causal set theory are satisfied. But what about the *independence*, or novelty, of the higher level from the lower level? Relativistic spacetimes (or general relativity) are independent from causal set (theory) insofar as the fundamental structures are not spatiotemporal, or at least not fundamentally so. As only the relation of causal connectibility is fundamental, a 'spacelike' cross-section of a causal set encompassed no relations among the basal elements whatsoever (cf. Figure 25.1). Hence, what would be the best candidate to correspond to space is completely unstructured, and so *a fortiori* does not have topological or metrical structure as we would expect a space to have.

Although time is really only emergent qua aspect of the emerging spacetime structure, the causal relation ordering the fundamental structure bears clear similarities to time: it looks like a B-theoretic, discrete, relativistic version of an asymmetric temporal precedence relation. Once the full quantum character of the fundamental structure is implemented, however, one should expect to see these similarities erode and the independence become even more salient. Before I will comment on how the non-trivial task of recovering relativistic spacetimes can be accomplished, let me turn to loop quantum gravity, a case where we have at least a sketch of a quantum theory and where consequently the gap between the fundamental structures and spacetimes is wider.

Loop quantum gravity also starts out from what it takes to be the central insights of general relativity and applies a standard quantization procedure in order to arrive at a quantum theory of gravity. Whatever the details of the quantization, the research programme aims at the most conservative way to transform the dynamical geometry of general-relativistic spacetimes into quantum properties of a quantum system, relying on techniques that have proved successful on other occasions. The result, so far, is a quantum theory describing the fundamental structures that give rise to spacetime, yet are so different from spacetime. These structures are spin networks, that is, discrete structures consisting of nodes and connecting edges (Figure 25.2). Both the nodes and the edges have additional properties expressed by half-integer spin representations besides their connection in the network – hence 'spin network'.

Before I address the question of time and dynamics, how should we interpret these spin networks? First, there is a sense in which they are 'spatial': they result from a quantization of geometric structures that are standardly assumed to capture 'space'. Second, the spin representations sitting on the nodes and the edges are eigenvalues of operators, which admit a natural geometric interpretation themselves: they represent three-dimensional volumes and two-dimensional surfaces, respectively. According to this interpretation, we thus obtain a compelling granular picture of spin networks: the spin representations on the nodes depict the size of the volume associated with that 'atom' of space, and the spin representations on the edges give the size of the area of the mutual surface of the two atoms connected by the edge. Thus, edges naturally express 'adjacency' or contiguity between the 'grains' of the discrete structure.

However, the physical interpretation of spin networks as straightforwardly characterizing the elementary building blocks from which space is formed by joining the blocks together by the thread of adjacency is limited in two ways. First, one finds a peculiar form of 'non-locality' (Huggett and Wüthrich 2013): parts of the spin network that may be connected by an edge and thus are fundamentally adjacent may end up giving rise to parts of emergent spacetime which are spatially very distant from each other as judged by the distances operative at the level of emergent spacetime (the blue regions in Figure 25.2 try to illustrate this non-locality). 'Non-local' behaviour like this

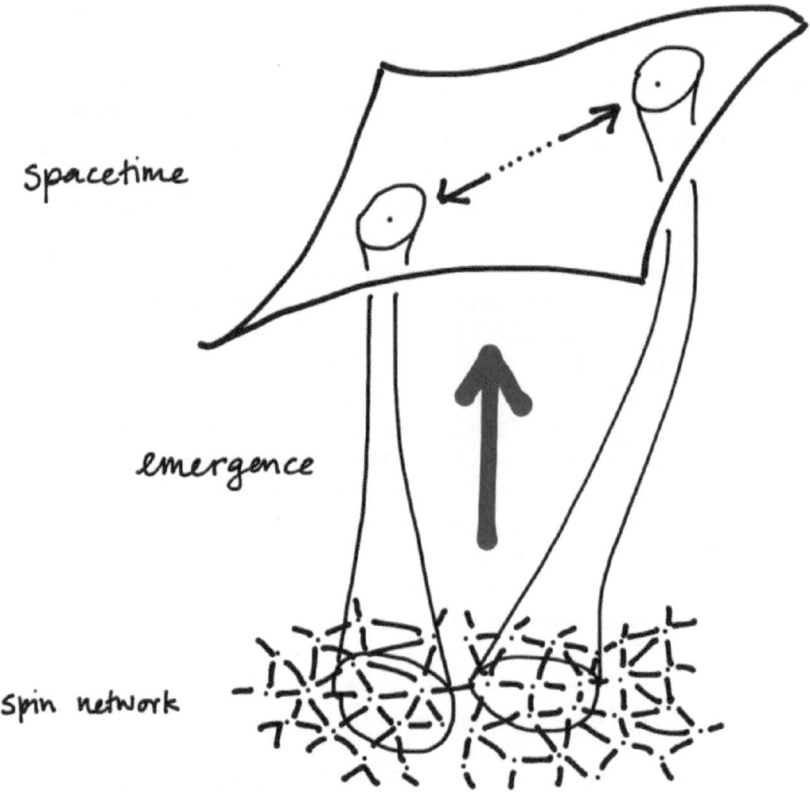

Figure 25.2 A spacetime emerging from a spin network.

cannot be too rampant, for if it were, then the spin network in question would have given rise to a different spacetime, one which better reflects the fundamental locality structure.

Second, it should be noted that the spin networks form a basis in the Hilbert space of the fundamental quantum systems at stake. If the state of such a system is in an exact spin network state, then the earlier geometric interpretation works well (modulo the 'non-local' behaviour). However, in general the state of the system will be in a superposition of states in the spin network basis and thus not have any determinate geometric properties. Perhaps the state snaps into an eigenstate of the geometric operators upon measurement, but then the measurement problem of quantum theory rears its complicating head. Presumably, the measurement problem will be even less tractable in the present context, where we are dealing with a quantum system which is supposed to underwrite space(time) – all of it.

What about the temporal aspect of spacetime? First, loop quantum gravity is a canonical approach to quantum gravity and as such faces the ominous 'problem of time' (Earman 2002; Maudlin 2002; Huggett et al. 2013), better called 'the problem of change'. This problem arises in canonical approaches: if we follow a standard way of interpreting physical theories, no genuinely physical magnitude can ever change – they are all 'constants of the motion'. Following this line of thinking, then, the structures loop quantum gravity aims to quantize appear 'frozen' in time. Furthermore, if one analyses what ought to be the central dynamical equation (the so-called 'Wheeler-DeWitt equation'), it appears as if no reference to time is made – rather odd for a dynamical equation expressing how a physical system is supposed to evolve over time! Various strategies exist to avoid or dissolve the problem, but the general conclusion seems to be that there is nothing naturally temporal in the fundamental structures described by loop quantum gravity.

In sum, then, the structures postulated by loop quantum gravity are non-spatiotemporal, arguably more so than causal sets. Relative to loop quantum gravity, general relativity enjoys a higher degree of novelty and thus independence than it did vis-à-vis causal set theory. Having said that, the fundamental structures described by some approaches to quantum gravity live on yet higher rungs of the ladder of increasing non-spatiotemporality as proposed by Oriti (forthcominga). For instance, group field theory (Oriti 2014; Huggett forthcoming), a 'second quantization' of loop quantum gravity, proposes fundamentally algebraic, rather than primarily geometric, relations among the basic constituents. The structures can be in different phases, just like gases and condensates, and only when they 'transition' into a spatiotemporal phase do they form structures such as spin networks that can give rise to the geometry of spacetime. Such a phase transition can obviously not be a dynamic process in time – it is a process 'prior' to there being time as it were. Instead, it is captured by a renormalization group flow (Oriti forthcomingb).

Turning to the dependence aspects, loop quantum gravity is more fundamental and more encompassing than general relativity, with a relation between the theories that is broadly reductive. It potentially also offers deeper explanations and corrections of general relativity. For instance, although the expectation values of sufficiently large regions of space tend very strongly to the classical values, for sufficiently small regions, there ought to be noticeable differences between the values as predicted by loop quantum gravity and by general relativity. Furthermore, the measurement outcomes can only assume discrete values in loop quantum gravity. Perhaps closer to practise, loop quantum gravity promises to deliver explanations of phenomena connected with black holes or the very early universe, which remained unexplained in general relativity. As for the relation between the structures postulated by the theories at the two levels, we find again an instance of multiple realizability, as well as fundamental structures which fail to give rise to spacetimes.

3. Spacetime functionalism

To date, no research programme in quantum gravity has managed to fully recover general relativity and its spacetimes. Not only does this remain a central explanatory task of a new scientific theory unfulfilled, but the threat of empirical incoherence as articulated earlier also looms over quantum gravity. The details are too rich and too technical to review here,[7] but let me finish by sketching in broad philosophical outline what the task involves.

First and foremost, this will involve the scientific work establishing a rather general mutual dependency under variations of the physics at the fundamental level and the spacetime physics at the phenomenal level. Will this suffice to discharge the task and to dispel the threat of empirical incoherence? According to the *spacetime functionalist*, it does. Recently risen to popularity,[8] functionalism asserts that in order to secure the emergence of spacetime from non-spatiotemporal structures in quantum gravity, it suffices to recover only those features of relativistic (or indeed phenomenal) spacetimes which are functionally relevant in producing empirical evidence. Specifically, it is not necessary to recover the full manifold structure and the full metric of general relativity. If this is right, the threat of empirical incoherence is *ipso facto* averted.

In order to execute the functionalist programme in quantum gravity, two steps would thus be necessary and sufficient: first, the spatiotemporal properties or entities are 'functionalized' or 'functionally reduced', that is, they are given a definition in terms of their functional roles; second, an explanation is provided as to how the fundamental entities or structures found in quantum gravity successfully instantiate these functional roles. The functional roles that need to be filled include, among others, spacetime (relative) localization, spatial distances, and temporal durations;

the fundamental structures which enact these roles are parts of causal sets, or parts of spin network states, as the case may be. Lam and Wüthrich (2018, §§4–5) sketch in more detail how this functionalist endeavour can be implemented in causal set theory and in loop quantum gravity.

It should be clear that multiple fundamental structures can then, in principle, realize the relevant spacetime roles. This very liberal functionalist attitude raises concerns about the metaphysical robustness of the proposal. For instance, in analogy to the philosophy of mind, one may worry that the spacetime functionalist cannot muster the resources to account for the 'spacetime qualia', that is, the qualitative aspects of spacetime – the 'spatiality' and the 'temporality' of our experienced world – which seem phenomenologically vital and hence ontologically ineliminable. To make the objection salient, imagine the case of 'spacetime zombies'. Suppose there are two worlds exactly alike in all non-spatiotemporal respects, with one of the worlds also containing exemplifications of spatiotemporal properties, which the other one lacks. Although their fundamental physics and all the functionally realizable spatiotemporal appearances are precisely the same, this latter world lacks genuinely spatiotemporal properties and instead only contains 'spacetime zombies'. If this world is indeed numerically distinct from the first and metaphysically possible, it seems as if a merely functionalist treatment of spacetime does not command the resources to distinguish between the two. Hence, spacetime functionalism is false.

A more persuasive form of the worry has been raised by Ney (2015). In the context of quantum gravity, even though spacetime and spatiotemporal aspects of its occupants may be functionally reduced to fundamental, non-spatiotemporal physics, the objection goes, such functional reduction is insufficient in demonstrating how the fundamental structure does, in fact, *constitute* spacetime and its spatiotemporal occupants. Similarly, David Yates (forthcoming) argues that functionalism is insufficient to solve the challenge of empirical incoherence, as our concepts for spatiotemporal properties are non-transparent in a fundamentally non-spatiotemporal world and functionalism cannot overcome this non-transparency.

Against these detractors, Lam and Wüthrich (2018) argue that there remains no task left undone once a research programme can show 'how the fundamental degrees of freedom can collectively behave such that they appear spatiotemporal at macroscopic scales *in all relevant and empirically testable* ways' (10, emphasis in original). If this is right, then however counterintuitive a fundamentally non-spatiotemporal world may appear, it arises as a live possibility – one perhaps borne out in contemporary fundamental physics. This possibility also suggests that the widespread and intuitive 'constitutionalism', that is, the view according to which our experienced world is ultimately constituted by, or built up from, elementary building blocks or atoms of spatiotemporal 'beables', may not offer an adequate understanding of the truly puzzling situations we presently face in fundamental physics. Instead, a functionalism devoid of constitutionalism and hence lighter on metaphysical prejudice is recommended.

4. Conclusion and outlook

Despite robust appearances to the contrary, research in quantum gravity suggests that the world we inhabit may not be fundamentally spatiotemporal. Philosophers should pay attention to these developments, as they have potentially profound implications for the philosophy of space and time, and beyond. If borne out, the absence of spacetime at the fundamental level forces at least a reconsideration of the traditional debate between substantivalism, relationalism, and structuralism. If substantivalism is reduced to the assertion that spacetime is an independent substance, and as such in no way derivative, then a substantivalist position could hardly be maintained. But relationalism does not obviously fare any better: although spacetime indeed seems to be derivative, it is not clear that it is derivative on material entities, let alone on the spatiotemporal relations in which these entities stand. The fundamental structures of quantum gravity are not obviously material, and they fail to exemplify spatiotemporal relations as directly as standard relationalism presupposes. The traditional position which is most naturally adapted to the present context is structuralism (Wüthrich 2012a), although, of course, the relevant structure in our ontology will not be straightforwardly spatiotemporal as is again assumed in the traditional formulation of spacetime structuralism.

Although the non-spatiotemporality varies in degree and character among the different research programmes, as was illustrated in §2, the absence of spacetime from the ontology of quantum theories of gravity raises worries about the empirical coherence of these theories and, more generally, about their adequacy as replacements of general relativity. In order to resolve these concerns, however, it suffices to establish how spacetime emerges in the appropriate way from the fundamental non-spatiotemporal structures. This task consists in both technical and philosophical challenges that must be met in order to arrive at a satisfactory understanding. In §3, I have suggested that a form of spacetime functionalism may furnish a key element to this project. But much work – including philosophical – remains to be done!

Notes

* I thank Robin Findlay Hendry and Tom Lancaster for their insightful and challenging comments on an earlier draft of this chapter. This work was partly performed under a collaborative agreement between the University of Illinois at Chicago and the University of Geneva and made possible by grant number 56314 from the John Templeton Foundation, and its content is solely the responsibility of the author and does not represent the official views of the John Templeton Foundation.
1 The defenders of the claim that the world is purely abstract, formal, or mathematical – as opposed to *partly* abstract, formal, or mathematical – are usually referred to as 'Pythagoreans', rather than as 'Platonists'.
2 Ney (2015) offers the most up-to-date evaluation of functionalism in the context of wave function monism.
3 See the collection by Oriti (2009) for a useful overview of approaches to quantum gravity.
4 Although, of course, the fundamental structures or their states may well be described by representations which inhabit a *mathematical* space, such as a configuration space or a Hilbert space.
5 This is the case even in string theory (Witten 1996; Huggett 2017) and condensed matter approaches to quantum gravity (Bain 2013).
6 Sorkin (1991) takes the cosmological constant to result from statistical effects resulting from the large number of discrete bits. The 'prediction' of the cosmological constant is severely limited in several ways: first, it is off by at least an order of magnitude; second, it relies only on the scale of the fundamental discreteness, and not on any specifics of causal set theory; third, this scale is assumed to be the Planck scale, and thus also just put in 'by hand'.
7 For those details, see Wüthrich (2017), Huggett and Wüthrich (forthcoming).
8 Knox (2013) offers a statement and defence of spacetime functionalism in the context of classical gravity in terms of inertial frames, although it only receives this label in Knox (2014). Lam and Wüthrich (2018) apply the idea to the context of non-spatiotemporality in quantum gravity. Functionalism also offers an important perspective for wave function monism; cf. e.g. Albert (2015, Ch. 6), Ney (2015), Lam and Wüthrich (2018, §2).

References

Albert, David Z. Elementary quantum metaphysics. In James T. Cushing, Arthur Fine, and Sheldon Goldstein, editors, *Bohmian Mechanics and Quantum Theory: An Appraisal*, pages 277–284. Kluwer, Dordrecht, 1996.

Albert, David Z. *After Physics*. Harvard University Press, Cambridge, MA, 2015.

Allori, Valia, Sheldon Goldstein, Roderich Tumulka, and Nino Zanghì. On the common structure of Bohmian mechanics and the Ghirardi-Rimini-Weber theory. *British Journal for the Philosophy of Science*, 59:353–389, 2008.

Bain, Jonathan. The emergence of spacetime in condensed matter approaches to quantum gravity. *Studies in History and Philosophy of Modern Physics*, 44:338–345, 2013.

Butterfield, Jeremy. Emergence, reduction and supervenience: A varied landscape. *Foundations of Physics*, 41:920–959, 2011.

Callender, Craig. One world, one beable. *Synthese*, 192:3153–3177, 2015.

Chalmers, David J. Strong and weak emergence. In Philip Clayton and Paul Davies, editors, *The Re-Emergence of Emergence*, pages 244–254. Oxford University Press, New York, 2006.

Crowther, Karen. *Effective Spacetime: Understanding Emergence in Effective Field Theory and Quantum Gravity*. Springer, Cham, 2016.

Crowther, Karen. Inter-theory relations in quantum gravity: Correspondence, reduction, and emergence. *Studies in History and Philosophy of Modern Physics*, 63:74–85, 2018.

Crowther, Karen and Niels Linnemann. Renormalizability, fundamentality and a final theory: The role of UV-completion in the search for quantum gravity. *British Journal for the Philosophy of Science*, forthcoming.

Earman, John. Thoroughly modern McTaggart: Or what McTaggart would have said if he had read the general theory of relativity. *Philosophers' Imprint*, 2(3), 2002.

Huggett, Nick. Target space ≠ space. *Studies in History and Philosophy of Modern Physics*, 59:81–88, 2017.

Huggett, Nick. Spacetime 'emergence'. In Eleanor Knox and Alastair Wilson, editors, *Companion to the Philosophy of Physics*. Routledge, London, forthcoming.

Huggett, Nick, Tiziana Vistarini, and Christian Wüthrich. Time in quantum gravity. In Adrian Bardon and Heather Dyke, editors, *A Companion to the Philosophy of Time*, pages 242–261. Wiley-Blackwell, Chichester, 2013.

Huggett, Nick and Christian Wüthrich. Emergent spacetime and empirical (in)coherence. *Studies in History and Philosophy of Modern Physics*, 44:276–285, 2013.

Huggett, Nick and Christian Wüthrich. *Out of Nowhere: The Emergence of Spacetime in Quantum Theories of Gravity*. Oxford University Press, Oxford, forthcoming.

Knox, Eleanor. Effective spacetime geometry. *Studies in History and Philosophy of Modern Physics*, 44:346–356, 2013.

Knox, Eleanor. Spacetime structuralism or spacetime functionalism? *Manuscript*, 2014.

Lam, Vincent and Christian Wüthrich. Spacetime is as spacetime does. *Studies in History and Philosophy of Modern Physics*, 64:39–51, 2018.

Maudlin, Tim. Thoroughly muddled McTaggart: Or how to abuse gauge freedom to generate metaphysical monstrosities. *Philosophers' Imprint*, 2(4), 2002.

Miller, Elizabeth. Quantum entanglement, Bohmian mechanics, and Humean supervenience. *Australasian Journal of Philosophy*, 92:567–583, 2014.

Ney, Alyssa. Fundamental physical ontologies and the constraint of empirical coherence: A defense of wave function realism. *Synthese*, 192:3105–3124, 2015.

Oriti, Daniele, editor. *Approaches to Quantum Gravity: Toward a New Understanding of Space, Time and Matter*. Cambridge University Press, Cambridge, 2009.

Oriti, Daniele. Disappearance and emergence of space and time in quantum gravity. *Studies in History and Philosophy of Modern Physics*, 46:186–199, 2014.

Oriti, Daniele. Levels of spacetime emergence in quantum gravity. In Christian Wüthrich, Baptiste Le Bihan, and Nick Huggett, editors, *Philosophy beyond Spacetime*. Oxford University Press, Oxford, forthcominga.

Oriti, Daniele. The Bronstein hypercube of quantum gravity. In Nick Huggett, Keizo Matsubara, and Christian Wüthrich, editors, *Beyond Spacetime: The Foundations of Quantum Gravity*. Cambridge University Press, Cambridge, forthcomingb.

Sklar, Lawrence. Prospects for a causal theory of space-time. In Richard Swinburne, editor, *Space, Time, and Causality*, volume 157 of *Synthese Library*, pages 45–62. Berlin: Springer, 1983.

Sorkin, Rafael D. Spacetime and causal sets. In J. C. d'Olivo, E. Nahmad-Achar, M. Rosenbaum, M. P. Ryan, L. F. Urrutia, and F. Zertuche, editors, *Relativity and Gravitation: Classical and Quantum*, pages 150–173. World Scientific, Singapore, 1991.

Witten, Edward. Reflections on the fate of spacetime. *Physics Today*, pages 24–30, April 1996.

Wüthrich, Christian. The structure of causal sets. *Journal for General Philosophy of Science*, 43:223–241, 2012a.

Wüthrich, Christian. When the actual world is not even possible. *Manuscript*, 2012b.

Wüthrich, Christian. Raiders of the lost spacetime. In Dennis Lehmkuhl, Gregor Schiemann, and Erhard Scholz, editors, *Towards a Theory of Spacetime Theories*, pages 297–335. Birkhäuser, Basel, 2017.

Yates, David. Thinking about spacetime. In Christian Wüthrich, Baptiste Le Bihan, and Nick Huggett, editors, *Philosophy beyond Spacetime*. Oxford University Press, Oxford, forthcoming.

PART 4

Emergence and the special sciences

26

DIGITAL EMERGENCE

Susan Stepney

Introduction

Aristotle (1924) has one of the earliest descriptions of what we now call emergent systems:

> things which have several parts and in which the totality is not, as it were, a mere heap, but the whole is something beside the parts.

This is now commonly phrased as *the whole is more than the sum of its parts*, yet this formulation misses the essence of the original (or, at least, its translation): "*something beside* the parts". That is, the whole is *qualitatively* different from its parts, rather than having a mere quantitative difference. The whole is *other than* the sum of its parts. Anderson (1972) pithily captures this essence in the title of his solid-state physics paper "More Is Different". He goes on to say:

> the behavior of large and complex aggregates of elementary particles, it turns out, is not to be understood in terms of a simple extrapolation of the properties of a few particles. Instead, at each new level of complexity entirely new properties appear, and the understanding of the new behaviors requires research which I think is as fundamental in its nature as any other.
>
> *(p. 393)*

The whole is not a "mere heap", because its parts have organisation, relationships, form patterns and exist in a context. For a more detailed discussion, see Stepney et al. (2006).

Here we explore this view that emergence is the result of a quantitative change becoming a qualitative one in the digital, computational domain. In this domain, what is it that is "more", and what constitutes a "difference"?

In classical computing, that of abstract Turing machines and von Neumann hardware architectures (such as are mostly realised by everyday PCs, tablets and smartphones), there are essentially only two things we can have more of: space (memory) and time (number of operations). Computational power, complexity and capability come from these two parameters.

First we investigate how space can contribute to emergence, through a range of concepts: the *amount* of memory and how that memory can be *chunked* into higher-level concepts, from simple data structures to entire virtual machine layers. Some properties and functionality of the resulting computation can be readily reduced to the underlying structures; others are more truly emergent, in that they are global properties of the entire system.

Next, we investigate the effect of time on emergence, through the effect of iteration: repeating the same process over and over from a different starting point at each step. This can give rise to a whole zoo of emergence, linked to communication between components of the overall system.

Space

More memory

One of the simplest computational structures is the finite-state machine (FSM). The system has a finite number of states and a state transition function defining which state to move to next, given a current state and an input. An FSM starts in its initial state, and at each step the function determines which state to move to next, based on the next input item. If the system ends in a final state when it processes its final input, it "accepts" the complete input string, or else it "rejects" it (see Figure 26.1). The amount of memory (holding the state transition table and remembering the current state) is finite, bounded and fixed. A computational step cannot introduce a new state or new transition. Because of this restriction, only certain sets of input strings can be accepted (the "regular languages"). For example, FSMs cannot recognise arbitrarily deeply nested matched parentheses, such as $(((. . .)(()))())$ where the ". . ." can have further parentheses. This means they cannot be used to recognise input strings that comprise valid computer programmes written with arbitrarily deeply nested parenthesised expressions.

A pushdown automaton has more memory; it includes an unbounded stack, where symbols can be pushed onto the top of the stack and popped off the stack, and where the top symbol can be used as input to the decision of which transition to take. (Think of a stack of dinner plates of different colours, say. *Pushing* is adding a new plate on to the top of the stack. *Popping* is removing the plate currently on the top of the stack. The colour of the plate currently on the *top* of the stack can be observed in order to make a decision. The stack is *unbounded* in that, no matter how big the stack is, more plates can always be pushed onto it.) So the system can have memory of how it got to its current state. Pushdown automata can accept more sets of input strings (the "context-free languages"), including arbitrarily deeply nested matched parentheses (see Figure 26.2).

Turing machines (TMs) add further memory: instead of a stack, which can be accessed only from the top, a TM has an unbounded *tape*, which can be accessed anywhere on its length. This makes them computationally more powerful still. Turing machines are believed to be the computationally most powerful devices possible: no further adding of memory (under certain physically plausible restrictions) can result in a more computationally powerful device.

So we see how the addition of more memory, from finite and bounded (FSMs), through finite, unbounded stacks (pushdown automata), to finite, unbounded and accessible anywhere

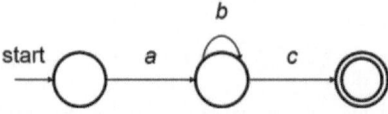

Figure 26.1 A finite state machine to recognise strings of the form ab^nc, that is, *ac, abc, abbc, abbbc,* . . . The start state is shown. Input of an *a* moves the system to the next state. There, input of any number of *b*s leaves the system in that state. Input of a *c* moves the system to the final accepting state, indicated by the double circle. Input of any other character at each stage leads to the string being rejected.

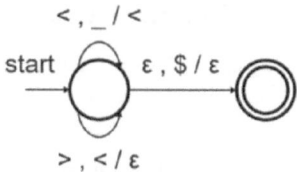

Figure 26.2 A pushdown automaton to recognise nested matched '<' and '>' brackets. The labels on the arcs represent: (input symbol, top of stack before/top of stack after). The start state is shown; the stack is initially empty (represented as $). In that state, input of a < is allowed, no matter what is on the stack (either the empty stack, $, or a <, jointly represented as an underscore); the system stays in that state and the < is pushed onto the stack. Input of a > requires there to be a corresponding < on the top of the stack; the system stays in that state and the < is popped off the stack (represented by being replaced with the "empty" symbol ε). If the end of the string has been reached (represented by the "input" of the empty symbol) and the stack is empty (is $), then the system moves to the final accepting state. Any other input attempts result in the string being rejected. The stack is unbounded in depth: any number of <s can be pushed before the first > is seen.

on the tape (TMs), leads to the emergence of qualitatively different computational power of the machines. More memory means different computational power.

For an excellent and readable introduction to computability and complexity, see Harel & Feldman (2004). For more technical details of computational automata, see any standard textbook on the theory of computation, for example, Sipser (1997).

Chunking memory

Although formal computational theory allows the basic memory to store arbitrary finite symbol sets, in practice, the symbol set has two elements: the binary digits (bits) 0 and 1. These are chunked in various ways to form higher-level components.

At the hardware level, bits are chunked into longer lists of bits: 8 bits chunked to form a byte, 32 or 64 bits to form a word. The hardware is designed to act on these chunks as a single unit.

At the software level, bits are chunked into data structures, components that make sense at the programming level. The hardware-based chunks are used to represent primitive data values: bits to represent boolean values, bytes to represent characters, words to represent integers and floating-point values. Operations are made available to manipulate these values: logic operations on booleans, comparison operations on characters, arithmetic operations on integers and floats.

These primitive values are then chunked into higher-level data structures: stacks, lists, trees, graphs, records, dictionaries, databases – again, with accompanying operations that act on the values. These values with their operations provide the semantics of what the bits represent.

In these higher-level structures, not all underlying bit patterns are "well formed" higher-level values, and the operations provided cannot create ill-formed bit patterns. This can be thought of as a simple form of downward causation (where higher levels of a system have a causal influence on lower levels): here some collection of bits is being used in a particular high-level context, and as a result can have only the particular "well-formed" subset of all the possible values they could take outside that context.

These higher-level structures can be generic and independent of their constituent types. For example, a list may be a list of characters, of integers, of lists and so on. The list operations work only at the list level, and so the semantics of being a list can be thought of as an emergent property, independent of the underlying component properties.

Software levels

This section provides several disparate examples of where emergence can be considered to occur in software systems at a variety of levels of abstraction.

Virtual machines

The virtual machine is an important architectural concept in computer design. It is a software implementation of a computer that runs on another (virtual or real) machine. For example, a Microsoft Windows virtual machine can be implemented to run on top of a Linux machine, looking like a Windows computer, "hiding" the underlying Linux layer from the user. There may be several layers of virtual machines, each hiding the underlying layers. It is not possible when using such a machine to tell if it is virtual or real (except maybe from its speed of execution) or to tell the nature of any underlying layers. As Laughlin (2005) points out, the properties of an emergent layer are insensitive to certain details of the underlying substrate, so you cannot draw conclusions about such details from them. The virtual machine is an engineered emergent layer.

Non-functional properties

In software development, there are *functional* properties (what the system does, for example, add numbers together, or search a database, or animate a graphic on the screen) and *non-functional* properties (how well the system does it, for example, how fast, or how accurately, or how smoothly). The non-functional properties generally cover such things as security, usability, performance, accessibility, maintainability and reliability (often referred to *en masse* as "the ilities").

There is a more technical definition: functional properties are those preserved by the technique of *refinement* (de Roever & Engelhardt 1998), an essentially reductionist approach to software development that takes high-level data structures and operations (for example, a data structure representing a mathematical set and operations representing set union and intersection) and refines them to lower-level counterparts with the same properties but executable on the platform of choice (for example, representing a set of items as a list of items and implementing the union and intersection operations using list manipulation operations). Refinements can be performed on individual parts and aggregated to form a refinement of the composite system.

Refinement also provides the link between the detailed data structures and the abstract values they represent through a relation called *finalisation*. A further representational step provides the link between the abstract values and their physical instantiation (Horsman et al. 2014). This representational step can be pointed to as the place where the meaning of a physical instantiation emerges: the pattern of pixels that is "10" may mean the decimal value "ten", the binary value "two", the bitstring "one-zero" or something else entirely, depending on the context in which it is viewed.

Non-functional properties are not preserved by refinement because they are properties of the entire system and cannot be decomposed into properties of its components in this way. Non-functional properties are emergent properties of the entire system, not of any of its parts.

Bugs

Certain classes of bugs in a system may also be considered (undesired) emergent properties of the way the entire system works, rather than due to a localisable error.

Some of these are of the "more is different" kind of emergent bug: a system may work perfectly until a certain input, or resource demand, tips over a particular threshold. Examples include resource starvation (where there are in principle sufficient resources in total, but they are being used by other parts of the system and are never allocated to one particular part that needs them to make progress); race conditions (where a part of the system attempts to access a high-level construct that is actually only partway through being manipulated at the lower level); and feature interaction (where functions that independently work clash when all are brought together).

In some cases the "downward causation" – requiring the underlying bits to act as if they were parts of higher-level constructs – fails to ensure the required constraints. Examples include such bugs as buffer overflows (where a high-level construct is too large to fit in the space provided and overflows into surrounding low-level memory, where it gets interpreted as a different higher-level construct) and memory leaks (where low-level memory resources used to support a high-level construct are not released back for use when the higher-level construct is deleted, leading to a gradual exhaustion of memory).

In other cases the device might work as designed, but alternative undesired meanings emerge if it is viewed in an unconventional manner (Clark et al. 2005). Examples include side channel attacks (Kocher 1996) of cryptographic systems, where measuring the physical time that different operations take to execute can be used to discover encryption keys.

Iteration

Iteration, when coupled with sufficient memory, is a major source of digital emergence. With iteration, the system is in a particular state and an operation is applied repeatedly, each time to the modified or updated state. This can be thought of as a simple feedback process: the "output" of one iteration is fed back as the "input" of the next. The simplest example of iteration is a counter: the current state is the count n, and the operation is to increment the count, $n := n+1$. Applied repeatedly, the operation results in the counter continually increasing in value.

Logistic map example

The logistic map (May 1976) is defined by the iteration $x := \lambda*x*(1-x)$, where x is a real number between 0 and 1, and λ is a real-valued parameter between 1 and 4. For some values of λ, the value of x settles down (after an initial transient behaviour) to a regular cycle; for other values of lambda, x exhibits chaotic behaviour (see Figure 26.3).

The change between periodic behaviour and chaotic behaviour as λ changes can be thought of as an emergent behaviour: there appears to be nothing in the individual values to signal such a drastic change in global behaviour. Crutchfield (1994) analyses this behaviour by building statistical models of x's behaviour for different values of λ. In the periodic regions, and also in the chaotic regions, these models can be captured by (stochastic) finite-state machines. But in the transitional regions, the "edge of chaos" (Langton 1990) regions, no FSM can capture the behaviour, as it has structure on all temporal scales, and a pushdown automaton is needed.

Mandelbrot set example

Consider the iteration $z := z*z + c$, where z and c are complex numbers and the initial value of z is 0. For some values of c, the magnitude of z diverges to infinity as the iteration progresses; for others, it stays bounded. When the values of c for which it stays bounded are plotted in the

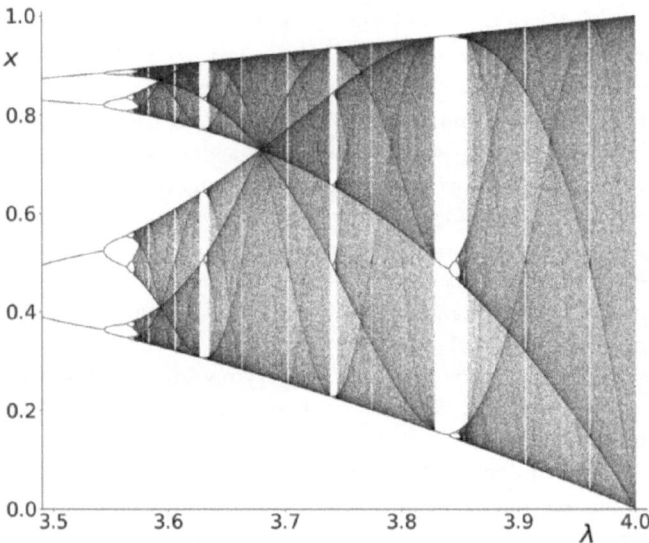

Figure 26.3 The logistic map shown for values of λ between 3.5 and 4. When λ = 3.5, x settles down to a cycle of period 4. (This plot does not show any initial transient behaviour of x.) At λ = 3.54409 . . ., there is a "period doubling" transition, to period 8. As λ increases, there are further period doublings. When λ = 3.56994 . . . the behaviour of x becomes chaotic. As λ increases further, "islands" of odd periodicity occur: for example, at λ = 3.828427 . . . an island of period 3 appears and then period–doubles to further chaos.

complex plane, the famous image of the Mandelbrot set emerges. Colours are often used to represent how quickly divergent values of z are discovered by iteration; the black centre is the set itself.

Here each pixel (value of c) is iterated independently. The image emerges from the underlying structure of complex numbers and the fact that the neighbourhood of the iterated values matches the neighbourhood of the pixels. One pattern (the black centre) emerges from the binary mathematical distinction between divergent and non–divergent initial conditions; another pattern (the swirling colours surrounding the centre) emerges from the iterative process itself used to compute which values are divergent.

This iterative algorithm produces an ever-better approximation to the Mandelbrot set: at each iteration more points are discovered to have diverged, and hence can be coloured as not members of the set. The set emerges as the infinite limit of the computation (see Figure 26.4).

Cellular automaton example

A cellular automaton (CA) comprises a grid or lattice of cells, each of which contains a copy of an FSM and has a current state. At each iteration, the state of each cell is updated based on the FSM rule, its current state and the current state of its neighbouring cells. Depending on the specific FSM transition rule and the initial state, CA can exhibit complex behaviour. Here the emergence is not the "final" state of the system, but rather the dynamic pattern of states exhibited as the rule iterates.

Conway's "Game of Life" CA (Gardner 1970) is the best known of these. This uses a 2D square grid of cells, each of which can be "alive" (black) or "dead" (white). The neighbourhood of a cell comprises its eight immediately surrounding cells. If a cell is currently dead, then it

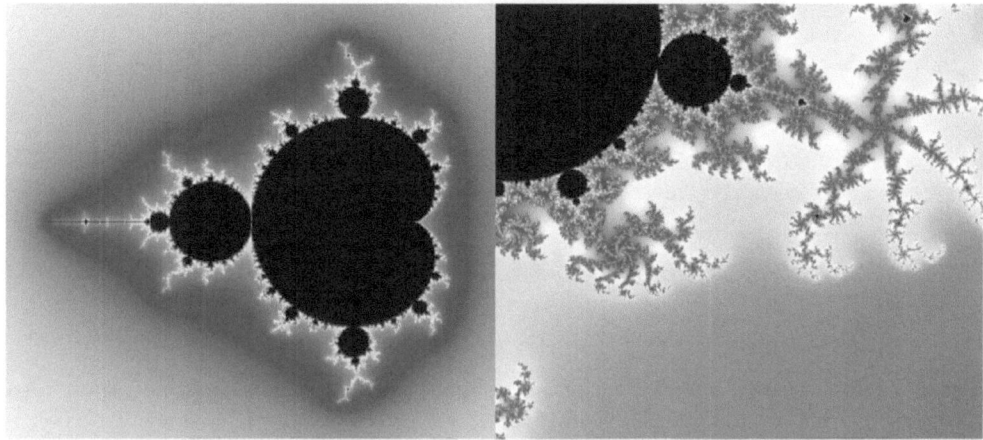

Figure 26.4 The Mandelbrot set (left), and a zoomed-in view (right). The set itself is the black area. The greyscale bands around the outside are a function of the algorithm used to detect divergence: different greys represent different numbers of iterations needed before divergence is detected. Other algorithms for calculating the set can result in different visual patterns.

[Figures drawn using www.jakebakermaths.org.uk/maths/mandelbrot/canvasmandelbrotv12.html]

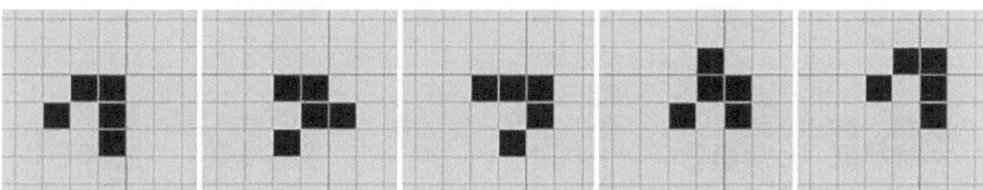

Figure 26.5 Game of Life. A glider, a pattern that moves one square up and to the right over four timesteps.

[Figures drawn using Golly from golly.sourceforge.net]

becomes alive in the next step if it has exactly three live neighbours; otherwise, it stays dead. If a cell is alive, it stays alive if it has two or three live neighbours; otherwise, it dies.

With these very simple rules, Life exhibits a whole zoo of emergent behaviours, from small oscillating patterns, to gliders and spaceships (patterns that move across the grid, see Figure 26.5), glider guns (patterns that emit gliders, see Figure 26.6) and even an emulation of a Turing machine (Rendell 2002), itself built from gliders and other lower-level emergent processes. It is even possible to emulate the Game of Life in the Game of Life, where each high-level cell, or "meta-pixel", is implemented by a large square of low-level cells. The emulation is larger and runs slower than the underlying substrate.

Life exhibits *sensitive dependence on initial conditions*: a single cell change can change the entire behaviour of a pattern. This is demonstrated clearly by the behaviour of the 12 different "pentominoes", initial patterns of five connected live cells (see Figure 26.7). Eleven of these become still lifes or small oscillators within 10 timesteps. The "r-pentomino", however, does not stabilise until 1103 timesteps, during which time it emits six gliders.

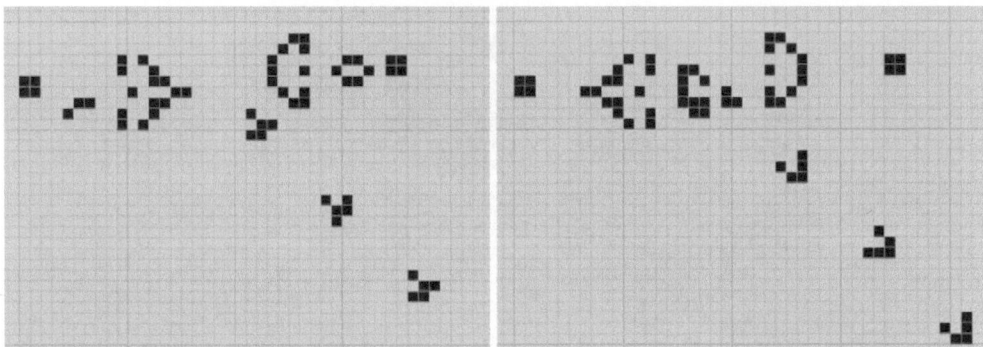

Figure 26.6 Game of Life. Gosper's glider gun, an oscillating pattern that emits a stream of gliders; it has period 30; here two states are shown 15 timesteps apart.

[Figures drawn using Golly from golly.sourceforge.net]

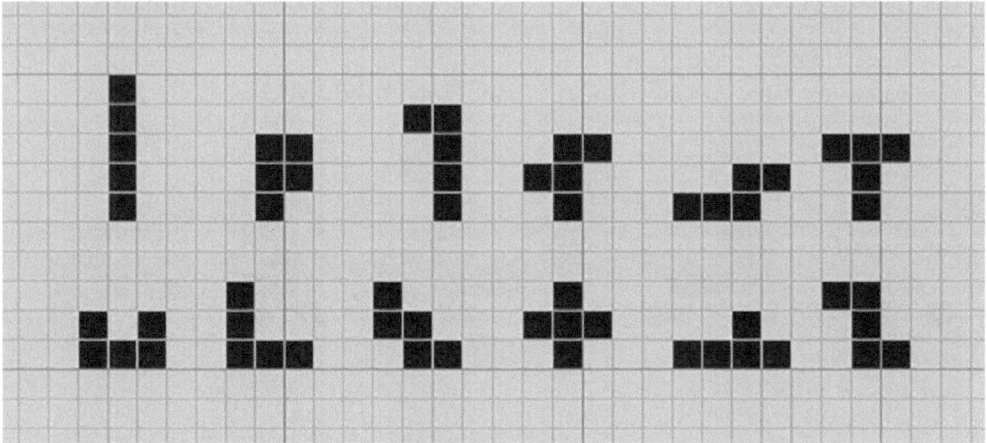

Figure 26.7 Game of Life. The 12 essentially different pentominoes (ignoring reflections and rotations). The "r-pentomino" is shown on the top row, fourth from the left.

[Figure drawn using Golly from golly.sourceforge.net]

In CAs the emergent forms are supported by the underlying communication between neighbouring cells: information, instantiated in emergent patterns, can propagate across the grid.

Flocking

Another form of iteration of multiple agents with communication between them is Reynolds' (1987) "boids" flocking simulation. Here each agent obeys three simple rules (don't collide, match velocities, stay close to others), the detailed instantiation of which depends on the positions and velocities of all the other agents. As these rules are iterated through time, the individual boids exhibit natural-looking flocking behaviours.

As with the CAs, the emergence arises from iterating the same rule over multiple communicating agents: the iteration allows small changes, and the communication allows those small changes to be coordinated.

Evolutionary algorithms

The next level of iteration with communication includes feedback. Agents may still behave and interact, as with the boids, but now the evolutionary part of the algorithm adds generational selection of the fittest agents (relative to some extrinsically defined fitness measure that evaluates their performance) and then reproduces these selected agents, with some variation, to produce the next generation of potentially fitter agents. So the computational agents can adapt and evolve to perform their imposed task better. Again there is communication between agents in the same generations of their fitness (via the selection process) and between generations from parent to child of their genome.

The emergence here is of a more open-ended form (Banzhaf et al. 2016), because the underlying system of agents and rules can change, leading potentially to new forms of emergence. Such open-endedness can be selected for explicitly by using "novelty search" (Lehman & Stanley 2011), where fitness is determined by being different from what has gone before. In this way, novel and more complex forms emerge.

Developmental algorithms

Another iterative form of algorithm is based on biological processes of growth, development or morphogenesis. Rather than a population of agents competing, as in evolutionary algorithms, here we have a population of component agents cooperating and communicating by sensing their local environment to build an emergent adult or mature form.

One simple form of such developmental systems are Lindenmayer systems (Prusinkiewicz & Lindenmayer 1990), a form of generative grammar or rewrite system. Symbols in strings are rewritten in parallel, based on simple grammar rules. At each rewrite iteration the string may be interpreted as a graphical program, and the sequences of images so defined form a view of the developing system.

More sophisticated iterative forms, where the agents can be more complex than stateless symbols and can divide and differentiate, akin to cells in a growing organism, can also be designed to support emergent adult states. Alternatively, the agents may be more akin to termites, building a system by laying down (simulated) material in a manner dependent on the local context, but not themselves growing or changing.

These systems rely on the same concept: with each iteration the components modify themselves or the system based on their local context; that modification changes the context so that on the next iteration the next context-sensitive modifications can be different. After multiple iterations, a complex form can emerge.

Evo-devo algorithms

A further form of iteration happens when such evolutionary algorithms and developmental algorithms are combined. This combination yields "evo-devo" algorithms, so called by reference to the analogous "evo-devo" branch of biology (Hall 2003).

With evo-devo algorithms, we have two nested iterative loops: a developmental loop (from embryo to adult) inside a generational evolutionary loop. Small evolutionary changes in a genome can potentially lead to large developmental changes in the associated phenotype, leading to the possibility of more open-ended emergent novelty.

Iteration is the continued application of a rule to a state. But in biology, the rules themselves may be encoded in the state, in the form of protein machines encoded in the DNA. We can develop computational analogues and start iterating the rules, either implicitly, by having them encoded in the state (Clark et al. 2017), or explicitly, by iterating rules that modify the rules. This leads to the potential for true open-ended emergence *in silico*.

Summary

Digital computational systems exhibit many forms and levels of emergence. In some cases, the emergence itself is directly engineered: sophisticated data types, virtual machines, non-functional properties. In some cases, that engineering is defective and the emergence manifests as an undesired bug, often resulting from the desired downwards causality being ineffective. In some cases the system is engineered so that open-ended novelty can emerge; in such systems some emergent bugs may be accepted as felicitous, if unexpected, manifestations of the desired open-endedness.

References

Anderson, P.W., 1972. More Is Different. *Science*, 177(4047), pp. 393–396.

Aristotle, 1924. *Metaphysics, Book VIII (translation by W. D. Ross)*, Oxford: Oxford University Press.

Banzhaf, W., Baumgaertner, B., Beslon, G., Doursat, R., Foster, J.A., McMullin, B., de Melo, V.V., Miconi, T., Spector, L., Stepney, S. & White, R. 2016. Defining and Simulating Open-Ended Novelty: Requirements, Guidelines, and Challenges. *Theory in Biosciences*, 135(3), pp. 131–161.

Clark, E.B., Hickinbotham, S.J. & Stepney, S., 2017. Semantic Closure Demonstrated by the Evolution of a Universal Constructor Architecture in an Artificial Chemistry. *Journal of the Royal Society, Interface*, 14:20161033.

Clark, J.A., Stepney, S. & Chivers, H., 2005. Breaking the Model: Finalisation and a Taxonomy of Security Attacks. *Electronic Notes in Theoretical Computer Science*, 137(2), pp. 225–242.

Crutchfield, J.P., 1994. The Calculi of Emergence: Computation, Dynamics, and Induction. *Physica D*, 75, pp. 11–54.

de Roever, W.-P. & Engelhardt, K., 1998. *Data Refinement: Model-Oriented Proof Methods and Their Comparison*, Cambridge: Cambridge University Press.

Gardner, M., 1970. The Fantastic Combinations of John Conway's New Solitaire Game "Life". *Scientific American*, 223(4), pp. 120–123.

Hall, B.K., 2003. Evo-Devo: Evolutionary Developmental Mechanisms. *The International Journal of Developmental Biology*, 47(7–8), pp. 491–495.

Harel, D. & Feldman, Y., 2004. *Algorithmics: The Spirit of Computing*, 3rd ed., Boston, MA: Addison-Wesley.

Horsman, C., Stepney, S., Wagner, R. C. and Kendon, V., 2014. When Does a Physical System Compute? *Proceedings of the Royal Society A*, 470(2169), p. 20140182.

Kocher, P., 1996. Timing Attacks on Implementations of Diffie-Hellman, RSA, DSS, and Other Systems. In *CRYPTO'96*. LNCS. Berlin: Springer, pp. 104–113.

Langton, C.G., 1990. Computation at the Edge of Chaos: Phase Transitions and Emergent Computation. *Physica D. Nonlinear Phenomena*, 42(1–3), pp. 12–37.

Laughlin, R.B., 2005. *A Different Universe: Reinventing Physics from the Bottom Down*, New York: Basic Books.

Lehman, J. & Stanley, K.O., 2011. Abandoning Objectives: Evolution through the Search for Novelty Alone. *Evolutionary Computation*, 19(2), pp. 189–223.

May, R.M., 1976. Simple Mathematical Models with Very Complicated Dynamics. *Nature*, 261(5560), pp. 459–467.

Prusinkiewicz, P. & Lindenmayer, A., 1990. *The Algorithmic Beauty of Plants*, Berlin: Springer.

Rendell, P., 2002. Turing Universality of the Game of Life. In A. Adamatzky, ed. *Collision-Based Computing*. Berlin: Springer, pp. 513–539.

Reynolds, C.W., 1987. Flocks, Herds and Schools: A Distributed Behavioral Model. In *Proceedings of the 14th Annual Conference on Computer Graphics and Interactive Techniques*. SIGGRAPH '87. ACM, pp. 25–34.

Sipser, M., 1997. *Introduction to the Theory of Computation*, Worcester: PWS Publishing.

Stepney, S., Polack, F.A.C. & Turner, H.R., 2006. Engineering emergence. In *11th IEEE International Conference on Engineering of Complex Computer Systems (ICECCS'06)*. IEEE, pp. 89–97.

27

EMERGENCE IN CHEMISTRY

Substance and structure

Robin Findlay Hendry

Introduction

Chemistry has a history in the emergence debate. Even before the term 'emergence' acquired its modern philosophical meaning, John Stuart Mill cited chemical compounds as the bearers of emergent properties that could not be predicted from those of their constituent elements. Mill's successors in the philosophical tradition that Brian McLaughlin has called 'British Emergentism', including C.D. Broad, followed him (McLaughlin 1992). Chemistry is not now widely cited as a rich source of candidate examples of emergence. The debate has moved on in two ways. First, the emergence debate has moved beyond the epistemic criteria for emergence they applied, such as predictability. Second, many philosophers and scientists would no doubt agree with McLaughlin's judgement that the advent of quantum mechanics in the 1920s, and the explanatory advances that came in its wake, rendered the British Emergentists' central claims about chemistry 'enormously implausible' (McLaughlin 1992, 89).

There should be no doubt about how important the scientific advances were that unified chemistry and physics in the twentieth century. During the nineteenth century, chemists had established that chemical substances are composed of a finite number of more basic substances: the chemical elements. To account for the existence of isomers – distinct chemical substances which are composed of the same elements in the same proportions – chemists developed theories of structure. The structures were the different *ways* in which the same elements are combined within them. In the early twentieth century, investigations of atomic structure offered a physical account of the nature of the elements: they are essentially classes of atoms which are alike with respect to their nuclear charge, but which may differ with respect to their mass. G.N. Lewis then offered an account of how bonds are realised by shared electrons. Although this identification produced many novel insights into the mechanisms of chemical reactions, especially in organic chemistry, Lewis' work was (so the story goes) in its turn superseded by the advent of quantum mechanics.

In what follows I will argue that chemistry is a far more fertile ground for emergence than many scientists and philosophers allow. Chemistry studies substances, accounting for their chemical and spectroscopic behaviour in terms of their structure. There is scope for seeing substances themselves as emergent, and bearing emergent causal powers, with respect to the populations of molecular species of which they are composed. Structures themselves are emergent from dynamical processes within molecular populations: different structures emerge at different scales of energy, time and length, and there is no reason to privilege processes at higher energies or shorter lengths – or timescales. Finally, there are no purely scientific reasons to believe that molecular structure is determined to exist by the quantum mechanics of systems of electrons and nuclei. The emergence of quantum chemistry – the application of quantum-mechanical principles to systems of electrons and nuclei – was not a deductive affair and offers no grounds for ruling out emergence in chemistry. Structures arise and persist within special dynamical conditions. Quantum mechanics provides not a derivation of structure, but a precise identification of the conditions under which it can exist.

Emergence and reducibility

In the most abstract terms, emergence is dependent novelty: A emerges from B when it is dependent and yet also novel or autonomous with respect to B. More specific varieties of emergence result when particular kinds of dependence and novelty are filled in. Thus, for instance, A might depend on B in the sense that it could not exist without B. This kind of weak ontological dependence is compatible with A being novel or autonomous with respect to B in a number of ways. Perhaps A's existence cannot in principle be predicted on the basis of B; or perhaps B can give rise to A only in the presence of certain substantive conditions, so that the mere existence of B does not entail A's existence.

There is also a long tradition according to which emergence is contrasted with reducibility; hence, reduction is also an important notion to consider. Twentieth-century discussions of emergence concentrated on intertheoretic reducibility, conceived of as the derivability of chemical theories from more fundamental physical theories (Nagel 1979). Intertheoretic reduction clearly excludes emergence: A cannot be novel or autonomous with respect to B if it is entailed by it. Yet intertheoretic reduction is a tall order. Applying physical theories such as quantum mechanics to chemical systems such as atoms and molecules yields complicated and computationally intractable equations, so chemists and physicists introduce models and approximations that simplify the exact equations or replace them entirely. Hence, the failure of strict intertheoretic reduction might be explained in two different ways. Reductionists tend to see the failure of derivability as expressing our limited mathematical abilities; models and approximations are proxies for the exact laws, with no independent explanatory power (Hendry 1998).[1] In contrast, emergentists tend to see the failure of derivability as showing the need to identify substantive further conditions for the existence of the phenomena they are attempting to explain. When a special science such as chemistry applies general physical theories to the study of emergent phenomena, the construction of physical models explaining these phenomena will take the form of a synthesis of physical and special-science principles, rather than a derivation of one from the other.[2]

Since reductionists and emergentists may agree that explanatorily relevant facts about B cannot be derived from theories concerning A, yet disagree on whether this is a sign of A's ontological autonomy, how might we express their disagreement? Ideally it should involve contrasting views of reality – different metaphysical accounts of how the world is, expressed in terms that transcend any claims about logical relationships between theories. In the metaphysics of mind, strong emergence has come to be associated with the existence of novel causal powers. Our understanding of the ontological relationship between chemistry and physics can learn from this and then feed back into metaphysics and the philosophy of mind, showing how the existence of novel causal powers is much less exotic or mysterious than many philosophers take it to be (for argument see Hendry 2006a, 2010a).

Chemical substances are emergent

Many philosophers might reject the idea that there can be strong emergence in chemistry because they think that the reducibility of chemical entities and properties to physical entities and properties, or their identity with physical entities and properties, has been established and expressed in such theoretical identities as 'water is H_2O'. The argument is supposed to be that 'water is H_2O' should be read as 'water = H_2O'. To be H_2O *just is* to be composed of H_2O molecules. This settles the emergence debate, because if the contents of a particular jug have any causal powers in virtue of being water (e.g. the power to quench thirst or to dissolve salt), then those contents have those powers in virtue of their being composed of H_2O molecules. From a range of different perspectives, the claim that 'water is H_2O' has been challenged by Barbara Malt (1994), Paul Needham (2000, 2002, 2011), Jaap van Brakel (2000), Michael Weisberg (2006) and Hasok Chang (2012). I would not challenge the identity, so long as it is properly construed. I have argued elsewhere that chemical substances are individuated by their microstructural properties and relations (see Hendry 2006b), that the identity of water with H_2O can be said to have been discovered[3] and that a substance's structure at the molecular scale *is what makes it the substance that it is* from a chemical point of view (see Hendry 2016a). However, I will argue that it is a straightforward misreading of 'water is H_2O' to think that it establishes the reducibility of water, or some chemical analogue of the mind–brain identity theory.

Historically, the microstructural identity of water was established via a number of distinct steps, the first being the compositional claim that water is a compound of hydrogen and oxygen. Eighteenth-century chemists did not content themselves with giving hypothetical explanations of water's behaviour in terms of its elemental composition. Rather, they took care to establish it experimentally, taking known weights of water, decomposing them into hydrogen and oxygen, weighing the separate elements to establish that their combined weights were (roughly) the same as those of the decomposed water, then recombining them, recovering close to the original weights of water. Later, in the nineteenth century, they introduced quantitative compositional formulae which represented the proportions between the constituent elements: H_2O in the case of water.[4] Finally, although such atomist interpretations were controversial for much of the nineteenth century, the compositional formulae came to be interpreted as embodying molecular facts, at least in some cases: for water, this meant that its characteristic molecule contains two atoms of hydrogen and one of oxygen. All this should be salutary for materialist philosophers of mind who would wish to use parallels between 'water is H_2O' and 'pain is c-fibres firing' as a guide in developing a materialist theory of the mind. Establishing that water is H_2O was a detailed process, whose first step involved the analysis of water into its proposed constituents and then a re-synthesis from them. It does not seem unreasonable to withhold one's assent to 'pain is c-fibres firing' until something analogous has been achieved.

That point made, how should 'water is H_2O' be construed? To survey the alternatives, we need to identify the relata (water and H_2O) and the relation between them.[5] First consider the relata: following Paul Needham, we can treat 'water' and 'H_2O' as predicates, or more likely a range of predicates. We can discuss whether these predicates correspond to properties later. What does it mean to say that something is water? First note that some substance names – 'ice' or 'diamond', for instance – refer only to specific states of aggregation: the name determines whether the relevant stuff is solid, liquid or gas. Other substance names are used independently of the state of aggregation, which must be added explicitly if it is to be specified, as in 'liquid nitrogen' or 'solid carbon dioxide'. 'Water' has a phase-neutral use, in which we may ask (for instance) how much of it there is in the solar system.[6] A comprehensive answer will include the solid water in the polar icecaps of various planets,[7] liquid water in their seas (perhaps only in the case of the Earth), water vapour in their atmospheres and isolated water molecules strung out in interplanetary space. In the case of hydrogen, one would have to include the large quantities present as plasma in the interior of the sun. Clearly, nothing of interest depends on whether one allows the phase-neutral scientific usage or insists on the supposed 'ordinary language' usage (I use scare quotes because I am highly sceptical that there *is* an ordinary language usage that is consistent enough to be said to have an extension). From the chemists' point of view, since there is something important that all water's different states of aggregation share, it makes sense to have one name for all these forms.

Now consider 'being H_2O'. People who know little of chemistry may take this simply to be a molecular condition (something like 'being composed of H_2O molecules'), but in general a chemical formula need not convey much information at the molecular level: it may, for instance, specify just the elemental composition of a substance, which may be shared by more than one substance. The formula 'C_2H_6O', for instance, applies both to ethanol (often written CH_3CH_2OH) and dimethyl ether (sometimes written CH_3OCH_3), which are distinct compounds with very different physical and chemical properties. So, we must ask, is 'H_2O' intended to specify the molecular make-up of water, or merely its elemental composition?

Finally, we come to the relation itself. It is well known that 'A is B' bears interpretation in terms of either identity or predication. In the present case two such interpretations suggest themselves: clearly 'water' and 'H_2O' are not the same predicate, though they may correspond to the same property. A weaker interpretation involves a relation of coextension or containment between the two predicates or properties: all A is B. If a necessity operator is envisaged (and on my view, one is required), then the source of the necessity is important. On the strongest microstructural essentialist view, which I would endorse, the relationship could be put in one of two ways: (identity) to be water is to be composed of H_2O molecules; (coextension) necessarily, all samples of water are samples of stuff composed of H_2O molecules, with the necessity in question being full metaphysical necessity. Putting this all together, 'water is H_2O' could mean either (i) 'to be water is to be made up of two parts of hydrogen to one part (by equivalents) of oxygen'; (ii) 'to be water is to be composed of H_2O molecules'; (iii) 'every sample of water is made up of two parts of hydrogen to one part (by equivalents) of oxygen'; and (iv) 'every sample of water is composed of H_2O molecules'. Different versions of (iii) and (iv) also result if modal operators are appended, and if one attends to the source of such modality (see van Brakel 2000).

Even if one takes the *strongest* essentialist reading, according which to be water *is to be* H_2O, then on the only scientifically plausible reading of what it is to be H_2O, reductionism does not follow. Why? Hilary Putnam once said that the extension of 'water' is 'the set of all wholes consisting of H_2O molecules' (Putnam 1975, 224). If a 'whole' is taken to be a mereological sum, or the result of any other composition operation in which the components are assumed to survive, this is straightforwardly false according to chemistry. Being a whole that consists of H_2O molecules may well be sufficient to be a quantity of water, but it is not necessary. Pure liquid water contains other things apart from H_2O molecules: a small but significant proportion of H_2O molecules (at room temperature, about 1 in 10^7) dissociate (or rather self-ionise) forming H_3O^+ and OH^- ions:

$$2H_2O \rightleftharpoons H_3O^+ + OH^-$$

Furthermore, H_2O molecules are polar and form hydrogen-bonded chains which are similar in structure to ice. One might regard the ionic dissociation products and chains as impurities, but the presence of these charged species is central to understanding water's electrical conductivity. Since chemists regard the electrical conductivity they measure as a property *of pure water*, it seems gratuitous for philosophers to interpret it instead as a property of an aqueous solution of water's ionic dissociation products. Looked at this way, liquid water can at best be considered to be composed of some diverse and constantly changing population of species at the molecular scale, including H_2O molecules, H_3O^+ and OH^- ions and various oligomolecular species. Can we defend the claim that water is H_2O? Yes, by considering water in all its forms to be the substance brought into being by interactions among H_2O molecules (see Hendry 2006b).

One way to synthesise all this is to regard being water as a distinct property that both molecular species and macroscopic bodies of stuff can have. H_2O molecules have it merely by virtue of being H_2O molecules. Larger bodies of stuff get it by being composed of (possibly diverse) populations of molecular species of kinds which are produced when H_2O molecules interact. Given the assumption that every part of water is water, this means that molecular species (such as H_3O^+ and OH^- ions) can be water by virtue of being part of a diverse population of molecular species which is produced when H_2O molecules interact. Hence, they acquire the property of being water by association. There is nothing strange in this. If we consider the protons in water to be part of the water, they acquire their wateriness by association too.

In a less exciting sense, wateriness is therefore an emergent property because nothing below a particular size (that of an H_2O molecule) can be water on its own account, and some smaller fragments acquire the property by association.[8] But that doesn't tell us whether being water is a *strongly* emergent property, that is, whether or not being water confers additional causal powers. This is where the standard argument I mentioned earlier comes in, except we can now see that it runs into difficulty. Consider all the different kinds of thing that, we have agreed, count as quantities of water, from mereological sums of water molecules to steam, liquid water and (the various forms of) ice. Trivially, a mereological sum of water molecules is no more than the sum of its parts. Any powers it has are acquired from its constituent H_2O molecules. But it has no bulk properties, so there is no distinction to be made between its molecular and its bulk properties. Steam, liquid water and (the various forms of) ice do have bulk properties, each bearing distinct sets of properties produced by the distinct kinds of interactions between their parts. Wherever there is significant interaction between the H_2O molecules, there is scope for that interaction to bring new powers into being. This is particularly obvious if that interaction includes self-ionisation and the formation of oligomers: the excess charge of solvated protons can be transported across a body of liquid water without the transport of any matter to carry it, via what is called the Grotthuss mechanism. This, in fact, is why water conducts electricity so well, unlike other, similar hydrides. The power to conduct electricity is not possessed by any sum of (neutral) H_2O molecules. The mechanism by which that power is exercised requires some part of the molecular population to be charged. It therefore depends on a feature of a diverse *population* of molecular species.

The reductionist will say at this point that the water can only acquire its causal powers from its parts and interactions between them. So, no novel causal powers have been introduced. The strong emergentist will ask why, when it is being decided whether they are novel, the powers acquired only when the molecules interact are already accounted for by the powers of H_2O molecules. If the rule is that any power possessed by any molecular population produced by any interaction between H_2O molecules is included, and we know this rule to apply independently of any empirical information we might ever acquire about what water can do and how it does it, then it seems that we know *a priori* that there will be no novel causal powers. The strong emergentist is within her rights to insist on a good scientific argument for this claim before accepting it into the debate. This does not, of course, mean that the strong emergentist wins the argument by default: only that in the absence of a specific *scientific* argument, the reductionist and the strong emergentist conclude this discussion honours even. Anti-reductionists need not fear theoretical identities, and should even learn to love them.

Structure and scale

Structure is central to chemistry: substances are named and classified by chemists in ways that depend entirely on their structure at the molecular scale, and structure grounds our understanding of their chemical reactivity and spectroscopic behaviour. In an important sense, the identity of a substance depends on its structure at the molecular scale (Hendry 2016a).

To say that a chemical substance has a 'structure' at the molecular scale might suggest to some philosophers that physical forces hold atoms and ions in static relative spatial positions. This may be why the term 'arrangement' has such currency in debates between philosophers concerning the metaphysics of composition (see for instance van Inwagen 1990; Sider 2013). However, objects at the molecular scale *cannot* be static because of zero-point motion: a minimal random motion associated with the lowest possible energy state of a quantum-mechanical system. A proper understanding of structure at the molecular scale must be consistent with the fact that atoms and ions are always in motion, even at absolute zero. Above absolute zero one must add thermal fluctuations: if the parts of a structure exhibit relative positions, such as the sodium and chloride ions in the lattice of common salt, these must be equilibrium positions, around which the various parts move, like a pendulum bob. Motion in a physical system depends on the energy it contains, so the ionic motions in a salt lattice will increase as its temperature increases (until, of course, the lattice breaks down). This dependence means that the structure of an ionic lattice is relative to temperature, and therefore energy (for a more extended discussion of all these points see Hendry forthcoming).

A well-known discussion of ice provides another example of the scale-relativity of lattice structure. Eisenberg and Kauzmann (1969, 150–152) point out that H_2O molecules in ice undergo vibrational, rotational and translational motions, with the molecules vibrating much faster than they rotate or move through the lattice. At very short timescales (shorter than the period of vibration), the structure of ice is a snapshot of molecules caught in mid-vibration. It will be disordered because different molecules will be caught in slightly different stages of the vibration. As timescales get longer, the structure averages over the vibrational motions and then (at yet longer scales) the rotational and translational motions. This yields successively more regular but diffuse structures. Therefore

> the term 'structure' can have three different meanings when applied to a crystal such as ice. The meaning depends on whether one considers a time interval short compared to the period for an oscillation (τ_v), or an interval longer than the period of an oscillation but less than the time for a displacement (τ_D), or an interval considerably longer than the displacement time.
>
> *(1969, 151)*

Chemists describe organic substances in terms of what I have called bond structure, which is distinct from the kind of structure exhibited by such ionic lattices as solid sodium chloride (see Hendry 2016b). This slightly complicates matters. Consider ethane (C_2H_6), in which a single bond links two carbon atoms whose remaining valences are used up by six hydrogen atoms. The geometry of the carbon atoms is roughly tetrahedral, the H-C-H angle being around 109 degrees. Further geometrical details of the ethane molecule are complicated by the fact that the two methyl groups rotate quite freely around the single bond between the two carbon atoms. Given the bond angles between the hydrogens attached to each carbon atom, one can distinguish two different geometrical configurations (or conformations) that ethane can take: staggered ethane, in which the hydrogen atoms are offset, and eclipsed ethane, in which they are aligned. These two different conformations can be represented by Newman and sawhorse projections, as in Figure 27.1.

The staggered conformation is of slightly lower energy than the eclipsed conformation, so rotation around the C-C bond is not entirely free. Nevertheless ethane, as Bassindale colourfully puts it, 'can be thought of in terms of two linked CH_3 propellers, with each CH_3 rotating rapidly' (1984, 25), and the eclipsed conformation now appearing as three small regularly occurring

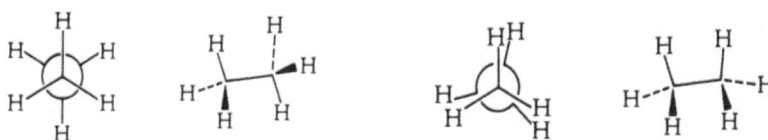

Figure 27.1 Newman and sawhorse projections of staggered ethane and eclipsed ethane, from Bassindale 1984, 50. Staggered ethane is represented by the two projections on the left, eclipsed ethane by the two on the right.

hindrances, passed over by the propellers like bumps in the road. Because ethane is always in motion, it presents distinct geometrical structures at different timescales: according to Bassindale 'measurement techniques with relatively long timescales would show an averaged geometry for ethane, but very short-timescale techniques could observe ethane in a staggered conformation' (1984, 51). Ethane's bond structure is an invariant feature throughout these motions. It is a complex structural property in common to all the different conformations.[9]

A substance may also present distinct structures at different length scales. Seen from afar, a steel bar looks homogeneous, with homogeneity being a null kind of structure. Closer in, there are imperfections: discontinuities giving rise to a grain structure at the millimetre scale. Within the grains are atomic lattices. These differences of structure across scales are reflected in physical explanations of steel's behaviour at the various scales (see Batterman 2013).

This suggests that chemical substances and other materials cannot be said to have a single structure, instead exhibiting a variety of them, each emerging at different scales of length, time and energy. It may be tempting to privilege the structures presented at close range, and at shorter timescales, over those which are observed at long range and longer timescales. One might even doubt that the longer-range and timescale structures are really distinct at all if they are just dim impressions of, or averages over, the close-range or short-timescale structures. If one takes a series of snapshots at a higher frequency, one could reconstruct the average one would see at a lower frequency.[10] Yet there is a counterfactual difference: many different series of snapshots could have given rise to the same average, and which particular series was seen is irrelevant. Put another way, the averaged picture is what one sees at the particular frequency that is appropriate to it. In a slightly different context (a discussion of boundaries), Achille Varzi goes a step further than reduction, arguing that a boundary seen from afar must be illusory, because it disappears on closer approach:

> It is true that I had the impression of seeing the shoreline of Long Island from my plane; but it is also true that when you actually go there, ground-level, things look very different. What looked from the air like a sharp line turns out to be an intricate array of stones, sand, algae, piers, boardwalks, concrete blocks, musk sediments, marshy spots, putrid waters, decayed fish.
>
> *(2011, 139)*

Moreover if a coastline is identified with the 'water/sand interface', '[t]hat boundary is constantly in flux, and it is only by filtering it through our cognitive apparatus – it is only by interpolating objects and concepts – that a clear-cut line will emerge' (2011, 139). I don't think one should dismiss the lower-resolution views. Even if the straightness disappears when one gets closer, a beach that looks long and straight from the air *really does* present a long straight boundary to incoming water waves, producing plane waves by reflection. Concentrating only on the view from close up, one may literally lose this bigger picture. The shape of the beach, like the structure of a molecule, depends on the scale at which one is considering it. The correct scale may not be a choice, but is fixed for us by some salient process, such as the formation of water waves (for the beach) or interaction with radiation (for molecules). These are matters of physical interaction, not of perception or conception.

A second aspect of the scale relativity of structure is that relationships of structural sameness and difference vary across different scales. This variation can occur at two different levels: in the way that molecules interact to form macroscopic substances and in the structural distinctness of the molecules themselves. At the level of the substances, consider Louis Pasteur's achievement in separating by hand crystals of the L- and D-forms of sodium ammonium tartrate, obtained from a racemic solution (an equal mixture of the two). This is a famous exemplar of structural explanation in science, and its experimental demonstration for the L- and D-forms are enantiomers: structures which are mirror images but which cannot be superimposed on each other. It is less well known that had Pasteur attempted the separation at a higher temperature than he did, he would likely have failed, because above 26°C the L- and D-salts form a single racemic crystal (Kauffman and Myers 1975).[11] How the two different species aggregate to form macroscopic crystals depends on temperature: it is a scale-dependent process.

For an example of the timescale dependence of structural sameness and difference in a single molecule, consider substituted biphenyls, which contain pairs of benzene rings connected by a single bond (see Figure 27.2).

When the hydrogen atoms in the benzene rings are substituted by functional groups X and Y in the four positions shown, the possibility of a new form of stereoisomerism arises – the molecule can in principle exist in two enantiomeric forms – but the isomerism is interestingly temperature dependent. In general, single carbon–carbon bonds, like the one connecting the two benzene rings, allow free rotation of the groups they connect. If the groups X and Y are relatively small (e.g. single atoms such as hydrogen or fluorine), then that rotation will be relatively unrestricted, though there will be some stearic hindrance: groups X and Y will bump past each other, like the hydrogens in ethane. In such cases the two enantiomers will not be separable at room temperature because they will interconvert. There are not two enantiomers, just two enantiomeric conformations of a single structure. For larger X and Y, perhaps the enantiomers can be separated, but they racemize rapidly. For really bulky groups such as $-NO_2$, or $-COOH$, the interaction between them will constitute a barrier to the rotation – the two enantiomers will be separable and will racemize only slowly. So for any given substituent groups X and Y, the physical distinctness of the enantiomers disappears above a characteristic temperature. Hence, the structural sameness and difference of molecular species is also a temperature-dependent (and therefore energy- dependent) phenomenon. The structures of molecules are dynamic configurations which are 'frozen in' at the particular energy scales at which they can exist. Each biphenyl enantiomer breaks the underlying symmetry of the situation, a symmetry which, as we have seen, reasserts itself sooner or later.

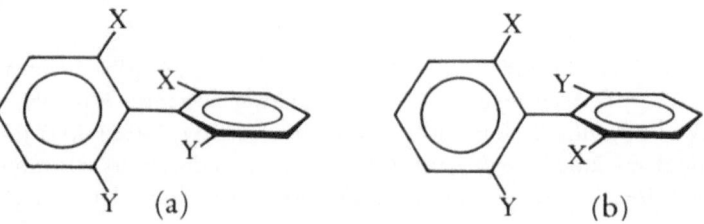

Figure 27.2   Biphenyl atropisomers, from Bassindale 1984, 58.

The idea that structure is frozen in when a system lacks energy could be applied to the whole of chemistry, which studies states of matter whose behaviour is dominated by the particular kinds of interactions that occur between atoms and ions: nuclei which are at low enough temperatures to have captured electrons and exhibit the stable electronic structures which give rise to the formation of chemical bonds. Once we go beyond the energy scales at which stable atoms and ions can exist, chemistry as we know it ceases to happen.

Emergence of molecular structure

A detailed examination of how molecular structures are explained within quantum mechanics puts pressure on the idea that they can be said to be derived from the basic laws of that theory. Quantum mechanics describes many-body systems of electrons and nuclei in terms of the Schrödinger equation, whose solutions correspond to the possible stationary states of the system. Electronic and nuclear motions are first separated, on account of the very different rates at which they move, and respond to external interactions. This is the adiabatic approximation, which yields wavefunctions for two coupled systems (of electrons and of nuclei), dynamically evolving in lockstep. When modelling a molecular structure, the nuclei are then typically localised, corresponding to the equilibrium positions of the known structure. Density-functional theory (DFT) replaces the 3N-dimensional electronic wavefunction with a three-dimensional electron density function (it can be shown that this can be done without approximation). According to the Hellmann-Feynman theorem, the overall force on a nucleus in the system is determined by the electron density, so effectively the nuclei are being pushed around by their interactions with the electrons.[12] The structure is then explained by showing that it corresponds to a configuration in which the energy is a minimum (i.e. the forces on the nuclei are effectively zero).

Two things are worth noting. Firstly isomers, which are molecules in which the same types of atoms are bonded together in different ways, share their molecular Schrödinger equations, the starting point of the earlier explanation. Thus, ethanol CH_3CH_2OH and dimethyl ether (CH_3OCH_3) share the same Schrödinger equations, as do enantiomers such as L- and D-tartaric acid (see Sutcliffe and Woolley 2012). The starting point of the explanation, the molecular Schrödinger equation, does not respect the differences between isomers. What results from localising the nuclei *does* respect these differences. Hence, in localising the nuclei in positions which correspond to the different isomers, these differences are being put in by hand. Thus, the different structures, their different symmetry properties and the different causal powers they ground are effectively introduced as unexplained explainers. As noted, the reductionist will see this as a pragmatic move: we cannot directly solve the molecular Schrödinger equations and so must introduce approximations whose results should not be seen as having any independent explanatory power. But it is hard to see these moves as mere approximations because they so radically transform the physics of the problem. This transformation is not a mere approximation since it transforms the scope of the equation, and it seems hard to argue that the approximations have no independent explanatory power. The molecular structures could not, *in principle*, appear without them.

A second point concerns how physical approximations work. The reductionist assumes that they have no independent explanatory power, because they are proxies for the exact equations: anything that could be explained using an approximate model could be explained using the exact equations. This would be a reasonable thing to conclude if the conditions which define the model could be derived from the exact equations or shown to be metaphysically unimportant in some way. One might, for instance, regard the necessary conditions merely as initial or boundary conditions and no more interesting than the 'auxiliary assumptions' that generate the Duhem-Quine problem. The story goes like this: quantum mechanics (QM) implies the existence of molecular structure (MS) only in conjunction with statements describing the necessary boundary and initial conditions (BIC). Thus, QM and BIC imply MS. This is all correct: adiabaticity and nuclear localisation might plausibly be thought of as boundary conditions, while the choice of nuclear positions looks like an initial condition. But this is no help to the reductionist who wants to see this explanation as a derivation from quantum mechanics, because as we have noted, exactly meeting the adiabaticity and nuclear localisation conditions is impossible for any quantum system. Thus, the conjunction of QM and BIC is, in some important sense, incoherent.

It is less misleading to put the point as follows: the mathematics provides only what one might call a *dynamical consistency proof*: the conditions which define the model *could not* hold exactly in any situation in which the exact equations hold,[13] but the two kinds of systems will evolve dynamically in approximately similar ways for some given level of accuracy and over the timescales relevant to the calculation. As we have seen, the molecular structure calculations described earlier assume dynamical conditions – the adiabatic separability of electronic and nuclear motions and the localisation of the nuclei – which *could not* hold exactly in any quantum system. All that can be assumed therefore is that a quantum-mechanical system of electrons and nuclei will display approximately similar dynamics to the model. No derivation of the model dynamics from the exact equations has been provided, nor even a demonstration of their logical consistency. All that the mathematics provides is that approximations introduced in the model can be neglected for a given level of accuracy over relevant timescales.

In joint work with Robert Schoonmaker (Hendry and Schoonmaker forthcoming), rather than treating adiabatic separability and nuclear localisation as mere approximations, we interpret them as substantive special assumptions about dynamical interactions within a many-body system of electrons and nuclei. If it can be shown that these conditions are necessary for molecular structure, the emergentist will regard them as necessary conditions for molecular structure itself to emerge (for the moment, I leave aside the question of whether these conditions are sufficient for the emergence of molecular structure).

As already noted, these conditions radically transform the dynamical behaviour of quantum systems and the scope of the equations that describe them. Adiabatic separability makes the overall energy of the electrons and nuclei a function of the nuclear configuration so that the dependence of energy on nuclear positions can be mapped by a potential energy (PE) surface (or rather as a hypersurface). This is not a global assumption (it depends on adiabatic separability), and a system of electrons and nuclei will not have a global PE surface: PE surfaces are not foliated, and near where they cross, the adiabatic separability of nuclear and electronic motions breaks down (see Lewars 2011, chapter 2). The effect of nuclear localisation is just as radical and interesting, for it suppresses quantum statistics. In general, any quantum system of electrons and nuclei must obey nuclear permutation symmetries: the overall wavefunction must be symmetric (for bosons) or antisymmetric (for fermions). In a quantum system with a molecular structure, the nuclei can be localised by their interaction with the electrons (whose joint quantum-mechanical state also reflects their interactions with the other nuclei) so that it makes a negligible difference to the dynamical evolution of the system whose statistics is assumed to apply: whether the overall wavefunction is symmetric (Bose-Einstein statistics), antisymmetric (Fermi-Dirac statistics) or indeed *asymmetric* (classical statistics) with respect to nuclear permutation. Interaction with the rest of the system effectively transforms the nuclei from quantum entities into classical objects. It should be emphasised that neither of these conditions is necessary for bonding as such: condensed matter physicists study systems such as metals and superconductors in which there is bonding (since they form cohesive materials), but in which nuclear and electronic motions are not adiabatically separable, and in which the nuclei are not localised, in the sense that quantum statistics must be taken into account in describing their structure. These conditions are necessary only for the kind of structure that is describable in terms of the classical chemical structures developed in organic chemistry during the nineteenth century. Interestingly there are molecules, such as protonated methane, in which some of the nuclei are localised but others are not. The permutation symmetries of two non-localised protons in protonated methane are therefore expressed in the dynamical behaviour of the molecule (see Hendry and Schoonmaker forthcoming).

These two conditions – adiabatic separability of nuclear and electronic motions and nuclear localisation – are not, however, sufficient for the emergence of classical molecular structure. Some molecules, such as cyclobutadiene, tunnel between two different structures, each of which is expressed in the molecule's interaction with radiation: in infrared (IR) spectra the molecule exhibits square symmetry, while higher-frequency x-ray diffraction catches it in the rectangular states between which it tunnels (see Schoonmaker, Lancaster and Clark 2018). This is another expression of the scale relativity of structure. It should be emphasised that tunnelling between different structures is the normal quantum-mechanical behaviour, but the dynamical behaviour of many molecules can be understood in terms of a single classical structure. Hence, dynamical restriction to a single structure is a third necessary condition for the classical kind of structure that is exhibited by many organic molecules.

Conclusion

In this chapter I have explored the positive reasons that chemistry provides to see substances and their structures as emergent. There were three levels of emergence: how substance properties, and substances themselves, emerge from the diverse populations of molecular species which compose them; how the structures of molecular species emerges in a scale-relative manner from their characteristic motions; and how the emergence of molecular structure requires three necessary conditions governing dynamical interactions between electrons and nuclei. Each case exemplifies a characteristic feature of emergence: that the existence of the emergent requires distinct kinds of cooperation and organisation between its parts.

Too often, scientists and philosophers approach relationships between physics and the special sciences via 'one-size-fits-all' interpretations, including various kinds of ontological reductionism combined with instrumentalist interpretations of the special sciences, according to which they provide no more than useful fictions that stand in proxy for the computationally intractable fundamental equations of physics. These approaches are inadequate in so many ways. They are insensitive to the important differences between the various special sciences, one especially salient distinction being that between sciences which are commensurable with physics (such as chemistry and condensed matter physics) and those which are not, because they involve function or teleology (such as biology and psychology). They can also be epistemically closed to evidence against them, like conspiracy theories. To everything I have said, the ontological reductionist can respond, quite correctly, that it is all perfectly consistent with the ontological reducibility of chemistry. The solipsist can make the same response to the evidence of their senses. A better understanding of the situation requires us to weigh the evidence *for* ontological reducibility, as against weaker kinds of dependence.

Acknowledgements

I am grateful to Stephen Blundell, Alisa Bokulich, Alex Carruth, Nancy Cartwright, Stewart Clark, Tom Lancaster, Tom McLeish and Robert Schoonmaker for many helpful conversations on the topic of this chapter and for critical comments on earlier drafts. Probably none of them would agree with all of it, and some of them would agree with very little.

Notes

1 Reductionists should also accept the possibility that non-derivability reflects a need to revise currently accepted physical principles.
2 Linus Pauling took just this view of quantum chemistry (see Hendry 2012).
3 An issue also discussed by Joseph LaPorte (2004) and Michela Massimi (2012).
4 Note that the proportions were not between the weights of the elements, but between equivalents; thus, 'water is H_2O' represents the fact that it contains twice as many equivalents of hydrogen as of oxygen, although the oxygen–hydrogen weight ratio in water is more like 8:1.
5 In the following discussion I am indebted in many ways to Paul Needham (2000, and many conversations), although we are in clear disagreement on some of the issues.
6 Note that various chemical processes produce and consume water. Hence, the total amount of water in the solar system will vary, so the question must be asked with reference to some period of time.
7 In fact John Finney (2004) identifies 16 distinct structures for ice, which form under different thermodynamic conditions.
8 In case the claim that 'every part of water is water' sounds odd, consider that there is nothing in pure water that isn't water. It may become more acceptable if one considers chemical substances to be modes in the medieval sense: ways that matter can be. All the parts of a pure sample of water are acting together to be water. Wateriness is a property that matter may acquire and lose. I am grateful to Alisa Bokulich for pressing me on this point.
9 For the distinction between bond structure (which the different conformations share) and geometrical structure (with respect to which they differ), see Hendry (2016b), where I also argue that neither kind of structure is more fundamental than the other kind.
10 I am grateful to Robert Schoonmaker for helpful discussion on these points.
11 I am grateful to John Hudson for this reference.
12 Although note that, since DFT folds interactions with other nuclei into the electron density, each of the nuclei is really being pushed around by its interactions with the rest of the system.
13 In a similar fashion, under the assumption that the sun has a finite inertial mass, no Newtonian solar system could display Keplerian orbits. Nevertheless, astronomers down the ages have chosen planetary orbits from among the conic sections, which assume that the sun is at rest.

References

Bassindale, Alan 1984 *The Third Dimension in Organic Chemistry* (Chichester: John Wiley and Sons).
Batterman, Robert W. 2013 'The tyranny of scales' in R.W. Batterman (ed.) *The Oxford Handbook of Philosophy of Physics* (Oxford: Oxford University Press), 255–286.
Chang, Hasok 2012 *Is Water H_2O? Evidence, Realism and Pluralism* (Berlin: Springer).
Eisenberg, D. and W. Kauzmann 1969 *The Structure and Properties of Water* (Oxford: Oxford University Press).
Finney, John L. 2004 'Water? What's so special about it?' *Philosophical Transactions of the Royal Society B* 359, 1145–1165.
Hendry, Robin Findlay 1998 'Models and approximations in quantum chemistry' in Niall Shanks (ed.) *Idealization in Contemporary Physics: Poznan Studies in the Philosophy of the Sciences and the Humanities* 63 (Amsterdam/Atlanta, GA: Rodopi, 1998), 123–142.
Hendry, Robin Findlay 2006a 'Is there downward causation in chemistry?' in Davis Baird, Lee McIntyre and Eric Scerri (eds.) *Philosophy of Chemistry: Synthesis of a New Discipline*, Boston Studies in the Philosophy of Science Volume 242 (Dordrecht: Springer), 173–189.
Hendry, Robin Findlay 2006b 'Elements, compounds and other chemical kinds' *Philosophy of Science* 73, 864–875.
Hendry, Robin Findlay 2010 'Ontological reduction and molecular structure' *Studies in History and Philosophy of Modern Physics* 41, 183–191.
Hendry, Robin Findlay 2012 'Linus Pauling (1901–1994)' in Robin Findlay Hendry, Paul Needham and Andrea I. Woody (eds.) *Handbook of the Philosophy of Science 6: Philosophy of Chemistry* (Amsterdam: Elsevier), 151–157.
Hendry, Robin Findlay 2016a 'Natural kinds in chemistry' in Grant Fisher and Eric Scerri (eds.) *Chapters in the Philosophy of Chemistry* (Oxford: Oxford University Press), 253–275.
Hendry, Robin Findlay 2016b 'Structure as abstraction' *Philosophy of Science* 83 (2016), 1070–1081.

Hendry, Robin Findlay forthcoming 'Structure, scale and emergence' in *Studies in History and Philosophy of Science* (in press).

Hendry, Robin Findlay and Robert Schoonmaker forthcoming 'The emergence of the chemical bond' (in preparation).

Kauffman, George B. and Robin D. Myers 1975 'The resolution of racemic acid: A classic stereochemical experiment for an undergraduate laboratory' *Journal of Chemical Education* 52, 777–781.

LaPorte, Joseph 2004 *Natural Kinds and Conceptual Change* (Cambridge: Cambridge University Press).

Lewars, Errol G. 2011 *Computational Chemistry: Introduction to the Theory and Applications of Molecular and Quantum Mechanics*, Second Edition (Dordrecht: Springer).

Malt, Barbara 1994 'Water is not H_2O' *Cognitive Psychology* 27, 41–70.

Massimi, Michela 2012 'Dwatery ocean' *Philosophy* 87, 531–555.

McLaughlin, Brian P. 1992 'The rise and fall of British Emergentism' in A. Beckermann, H. Flohr and J. Kim (eds.) *Emergence or Reduction? Essays on the Prospects for Non-Reductive Physicalism* (Berlin: Walter de Gruyter), 49–93.

Nagel, Ernest 1979 *The Structure of Science* (London: Routledge and Kegan Paul).

Needham, Paul 2000 'What is water?' *Analysis* 60, 13–21.

Needham, Paul 2002 'The discovery that water is H_2O' *International Studies in the Philosophy of Science* 16, 205–226.

Needham, Paul 2011 'Microessentialism: What is the argument?' *Noûs* 45, 1–21.

Putnam, Hilary 1975 'The meaning of "meaning"' in his *Mind Language and Reality* (Cambridge: Cambridge University Press), 215–271.

Schoonmaker, Robert, Tom Lancaster and Stewart Clark 2018 'Quantum-mechanical tunneling in the automerization of cyclobutadiene' *The Journal of Chemical Physics* 148, 104109.

Sider, Theodore 2013 'Against Parthood' in Karen Bennett and Dean Zimmerman (eds.) *Oxford Studies in Metaphysics* (Oxford: Oxford University Press), Volume 8, 237–293.

Sutcliffe, B. and R.G. Woolley 2012 'Atoms and molecules in classical chemistry and quantum mechanics' in Robin Findlay Hendry, Paul Needham and Andrea I. Woody (eds.) *Philosophy of Chemistry* (Amsterdam: Elsevier), 387–426.

van Brakel, Jaap 2000 *Philosophy of Chemistry* (Leuven: Leuven University Press).

van Inwagen, Peter 1990 *Material Beings* (Ithaca, NY: Cornell University Press).

Varzi, A.C. 2011 'Boundaries, conventions and realism' in J.K. Campbell, M. O'Rourke and M.H. Slater (eds.) *Carving Nature at Its Joints: Natural Kinds in Metaphysics and Science* (Cambridge, MA: MIT Press), 129–153.

Weisberg, Michael 2006 'Water is not H_2O' in Davis Baird, Lee McIntyre and Eric Scerri (eds.) *Philosophy of Chemistry: Synthesis of a New Discipline*, Boston Studies in the Philosophy of Science Volume 242 (Dordrecht: Springer), 337–345.

28

EMERGENCE IN BIOLOGY

From organicism to systems biology

Emily Herring and Gregory Radick

Introduction

The idea that a causal system as a whole can have features that no one studying the component parts in isolation would have guessed predates the twentieth century. But the widespread use of the term "emergence" as a name for this idea does not. It derives from a 1923 book about evolution by the British scientist and philosopher Conwy Lloyd Morgan, *Emergent Evolution* (Morgan, 1923). Nowadays, Morgan's term leads two lives in biology. One is as the familiar label for an important but, from the perspective of conventional science, unchallenging class of phenomena. Richard Dawkins, the great publicist in our day for a disenchanted, matter-in-motion understanding of Darwinian nature, describes the human capacity to act against the interests of our selfish genes as an emergent property of brains that evolved in the service of those genes (Dawkins, 2017, pp. 3, 39–40). The term's other life – our main concern here – is more philosophically colourful and, for the likes of Dawkins, disreputable. Following Morgan's lead, emergence has been a perennially attractive option for biologists and others seeking a middle path between the extremes of what Morgan's generation called "mechanism" – the view that life is nothing but (and so fully reducible to) complex machinery – and "vitalism" – the view that life is something more than (and so not fully reducible to) complex machinery. Emergence in this anti-reductionist sense has overlapped untidily with a range of anti-reductionist stances and schools, from organicism in the first third of the twentieth century to systems biology in the first third of the twenty-first century. Even a brief history of emergence in biology will, of necessity, be a history of anti-reductionist tendencies in post-1900 biology more broadly.

Before and around organicism

Debating the autonomy of biology

The nature of biology's connection with the physical sciences and, relatedly, the status of biology as a science in its own right were common themes in the reflective writings of early twentieth-century biologists. One stimulus was the apparent success of mechanical explanation in the rapidly professionalizing biology of the later nineteenth century, notably in embryology, physiology, biochemistry, animal behaviour and – most famously – evolution. (It was in this period that Lamarckism, with its emphasis on animal activity as an evolutionary agent, acquired an anti-reductionist character that it retains to this day.) Another was the growing popularity, from 1900, of the mathematical Mendelian theory of heredity, widely touted as biology's atomic theory. In a 1911 address, the German-born American physiologist Jacques Loeb declared that "ultimately life, i.e., the sum of all life phenomena, can be unequivocally explained in physico-chemical terms" (Loeb, 1912, p. 3). A choice seemed to be forced. Either life was nothing more than physics and chemistry, in which case the science of biology was effectively an immature branch of the physical sciences, or life was not just physics and chemistry, in which case biology was a truly autonomous science, standing on its own feet. But what, metaphysically considered, did those feet stand upon? Did autonomy require the positing of irreducibly biological and even teleological forces, along the lines of the *entelechies* invoked by the embryologist Hans Driesch to explain development (Driesch, 1900)? If so, then, for Loeb and like-minded thinkers, the price of autonomy seemed to be vitalism, and so a turning away from anything recognisable as modern science since the reforms of Bacon and Descartes.

In Britain, several biologists sought to articulate a third way. They defended the idea that biology was undergoing – or needed to undergo – some sort of theoretical and methodological redefinition which would establish its autonomy by enabling biologists to explain the phenomena of life and mind on distinctively biological terms, the better to meet the overriding scientific goal of giving a correct account of nature and its workings (Thomson, 1911; Johnstone, 1914; Darbishire, 1917). These efforts often involved severe criticisms of Mendelian genetics and the neo-Darwinian theory of natural selection, viewed as the most prominent biological manifestations of reductive mechanism. After World War I, resistance to a fully physicalized biology grew stronger, thanks to developments outside as well as inside biology. Outside biology, mechanism and reductionism in some quarters became associated with the evils of war (Harrington, 1996), while the rise of the quantum theory within the physical sciences was seen as discrediting the old clockwork image of deterministic nature. Within biology, meanwhile, the search for alternatives drew inspiration from several different sources. We shall touch on three that took shape around the same time as organicism. One was Morgan's evolutionary biology of emergence. The other two arose in association with resurgent romantic holism and a new metaphysics of life-as-process. (The term "holism," it should be noted, is a product of the same period and conversation, coined by the South African statesman and philosopher Jan Smuts in a book on evolution; see Smuts, 1926.)

Resurgent romantic holism

A great deal of early twentieth-century organism-centred biology in Germany and Austria (Harrington, 1996), as well as in Germanophile British and American biology (Esposito, 2013), continued a tradition of romantic holism tracing back to Johann Wolfgang von Goethe's appropriation of the philosophy of Immanuel Kant. Goethe, inventor of morphology, was particularly impressed with Kant's definition of organisms as self-generating, self-organising wholes. For Kant, human understanding of organisms required them to be considered teleological entities, with the activities of the parts subservient to the needs of the whole. ("There will never be a Newton of a blade of grass" became Kant's great one-liner against mechanistic biology.) The enduring legacy from this post-1900 boom in Kantian-Goethean biology is the Scottish morphologist D'Arcy W. Thompson's *On Growth and Form*, first published in 1917. Although the book's emphasis on geometry and physics made Thompson a grassblade-Newton (or perhaps grassblade-Maxwell) hero to later generations of reductionists, Thompson saw himself as striking a blow for the view that, as he wrote, when it comes to organisms,

> the whole is not merely the sum of its parts. It is this, and much more than this. For it is . . . an organization of parts, of parts in their mutual arrangements, fitting one with another in what Aristotle calls "a single and indivisible principle of unity."
>
> *(Thompson, 1942, p. 1019, quoted in Esposito, 2013, p. 78)*

A new metaphysics of life-as-process

The French philosopher Henri Bergson's metaphysically rich interpretation of biological evolution, developed in *L'Evolution Créatrice* (Bergson, 1907), had enormous appeal across professional biology, from the neo-Darwinian end (Gayon, 2008) through to the neo-Lamarckian end (Herring, 2016). Bergson introduced what he called an "image" (Bergson, 1907, p. 258), the *élan vital*, to characterize the history of life as a dynamic, unitary, self-creating whole, consisting of successive divisions stemming from the same original impulse and pursuing a common progressive tendency. Bergson's philosophy was profoundly anti-mechanistic and anti-deterministic, centring on the idea that the evolution of life is, in its details, an inherently unpredictable (and therefore, for Bergson, non-teleological), internally driven movement of complexification, in the direction of the development of more and more sophisticated minds. The British philosopher Alfred North Whitehead acknowledged Bergson's influence on his own process philosophy (Whitehead, 1929, p. xii). Criticising mechanistic trends in science, Whitehead proposed that scientists should concentrate on "multi-perspectival networks of relationships," rather than on "the behaviour of aggregated atomic units" (Peterson, 2014, p. 286). His philosophy of the organism (Whitehead, 1920, 1925) went on, as we shall see, to inspire the British organicists of the 1930s.

Emergent evolution

Favourably disposed towards Bergson's and Whitehead's anti-mechanistic metaphysics (Blitz, 1992, p. 91), Conwy Lloyd Morgan came to similar views independently through reflections that began for him as a student of Thomas Huxley's. Contra Darwin, Huxley thought that nature occasionally makes leaps. Morgan turned that teaching into a rule of method for comparative psychology, "Morgan's canon," first promulgated in the 1890s, and commanding investigators not to attribute cognitively rich powers to the animal mind (i.e. powers of reason) when more humble powers (i.e. trial-and-error learning) will serve, on the view that reason-and-language-enabled human minds are different in kind from the minds of our evolutionary kin. The doctrine of emergence was Morgan's canon generalized to the whole of evolution, indeed the whole universe, understood as a system in which, level by level, spirit manifests in new and ever more complex forms, first of matter, then of life, then finally of mind (Radick, 2004; Morgan, 1923).[1] "The whole doctrine of emergence," wrote Morgan, "is a continued protest against mechanical interpretation, and the very antithesis to one that is mechanistic" (Morgan, 1923, pp. 6–7). Behind that protest lay not just Huxleyan biology but, among other sources, Herbert Spencer's evolutionary philosophy, which likewise emphasized progression towards complexity and distinguished three levels successively attained by universal evolution: the "inorganic," the "organic" and the "super-organic" (Blitz, 1992). For Morgan – and also for another of the era's philosophers of life and mind, the less well-remembered Leonard Hobhouse – appreciation of the qualitative novelty of the emerging levels went along with a need to recognize that methods of investigation appropriate for one level could not be transferred to a higher level (Radick, 2017).

Organicism

Early organicism

Energized by this ambient, multi-source biological anti-mechanism, the Scottish physiologist John Scott Haldane, in lectures published the same year as of *On Growth and Form*, proposed his own solution to the shortcomings of both mechanistic and vitalistic views. Pointing out that biological phenomena such as self-regulation arose from interactions between the different parts and levels of the organism, Haldane argued that organisms themselves need to be studied as complex wholes, irreducible to the physico-chemical elements composing them. He insisted that this view did not amount to a disguised form of vitalism and explicitly excluded explanations relying on immaterial internal forces. In a footnote, he suggested a name for this organism-focused doctrine: "organicism" (Haldane, 1917, p. 3).[2] Soon others would publish along similarly organicist lines, including the American biologist William Emerson Ritter, who laid out the principles for his own "organismal conception of life" in a two-volume work (Ritter, 1919); the Scottish biologist E.S. Russell, who defended "the validity and independence of biological laws over against the laws of physics and chemistry" (Russell, 1924, p. 41); and the English philosopher Joseph Henry Woodger, author of *Biological Principles* (Woodger, 1929), which went on to be a "cornerstone theoretical text of the 'third way'" (Peterson, 2016, p. 62). These organismic thinkers all shared a commitment to viewing the organism as a hierarchical complex whose properties at any given level of organization emerged from interactions among the parts at a lower level. By the 1930s, a younger generation of self-identified organicists was building upon the foundations put in place by the previous generation.

The Theoretical Biology Club

A prominent member of this younger generation was the English biochemist Joseph Needham, who had converted from reductionism to third-way thinking in the late 1920s through correspondence with Woodger. Along with the Edinburgh-based evolutionary embryologist Conrad Hal Waddington, Needham and Woodger attempted to launch an organicist research programme – an ambition recently given a boost by the experiments of the Berlin embryologist Hans Spemann, who had shown, with German embryologist Hilde Mangold, that certain areas of the embryo (what he called the "organizer centres") induced the organization of the adjacent areas (Spemann and Mangold, 1924). The challenge of understanding these organizer centres attracted the new British organicists, who saw in the problem a potentially tractable version of the larger problem of understanding the hierarchical self-organizing properties of life. In 1931, Needham and his wife, the biochemist Dorothy Needham, deciding to join the organizer "scientific gold rush," moved to Berlin, where they became Waddington's neighbours (Peterson, 2016, pp. 93, 103). By the end of the year, the Needhams and Waddington were back in Britain and eager to pursue their organismic research. The following spring, the first meeting of the Theoretical Biology Club took place in Woodger's summer cabin. In addition to the Needhams, Woodger and Waddington, the crystallographer J.D. Bernal and the mathematician Dorothy Wrinch attended.

The mix of disciplines reflected and expressed a shared commitment to trying to understand organisms holistically. Other sources of unity included a common left-wing politics and, notwithstanding the professional success of many, a sense of outsider, even outlaw, status. They were unashamedly intellectual, drawing not just on Whitehead's process philosophy but also the logical positivism of the Vienna Circle, along with the latest work in symbolic logic and mathematics. In their meetings, held once or twice a year over a period of six years, they discussed problems such as Spemann's organizer, geometrical patterning in biology, philosophy of science and language, as well as socialism (for a more detailed account see Abir-Am, 1987). Over the course of the club's existence, Needham and Woodger published their most important works in organismic biology (Needham, 1932, 1936; Woodger, 1930a, 1930b, 1931), while the Needhams and Waddington were pursuing their organizer research, making Cambridge a centre of research in experimental embryology until the late 1930s (Peterson, 2014, p. 291).

From 1934, Joseph Needham began negotiations with the New York–based Rockefeller Foundation to obtain funding for a laboratory in Cambridge, in the hope of institutionalizing the club's research programme. A first proposal, seeking funds for seven interlinked research facilities devoted to embryology in all its aspects from the experimental to the psychological, was deemed too ambitious, with the multidisciplinary aspect of the project viewed as a handicap (Peterson, 2016, p. 118). Needham subsequently scaled down his requests, but these less ambitious proposals too were rejected. Although some historians suggest that this failure can be put down to the increasing success of mechanistic approaches in biology (Bowler, 2001, pp. 174–175) – and indeed, it was at just this time that the Rockefeller Foundation began promoting the new science of molecular biology – the foundation's growing awareness of how little support the club had in Cambridge itself was at least as important. The club's socialist-communist ties did nothing to help its cause (Peterson, 2016, p. 119).

Organisms as systems: Ludwig von Bertalanffy

Britain was not the only country in which third-way solutions were being proposed in this period. In Austria, the embryologist Ludwig von Bertalanffy developed his own "organismic" theory of life. Often cited as a founding figure for systems theory and systems biology, Bertalanffy always acknowledged his debt to another Austrian biologist, Paul A. Weiss, who used the notion of organisms as systems to criticize Loeb's reductionist approach to animal behaviour (Weiss, 1925). Learning about British organicism, Bertalanffy in the late 1920s started corresponding and collaborating with Woodger. Their discussions led to a translated and revised edition of Bertalanffy's 1928 manifesto for a "critical theory of morphogenesis" (Bertalanffy, 1928). Published in 1933 under the title *Modern Theories of Development, An Introduction to Theoretical Biology*, it argued that biology was in need of a "theoretical clarification" (Bertalanffy, 1933, p. 4) that would distinguish it from a simple collection of empirical facts and give it the status of a full-grown theoretical science with its own laws. To this end, Bertalanffy propounded what he called "an *organismic* or *system theory* of the organism," which subscribed neither to "machine theory" nor to a vitalistic "mystical entelechy" (Bertalanffy, 1933, pp. 177–178). Eschewing an analytico-summative manner, by which the organism was analysed into its separate parts (cells or characters, for instance) and studied as nothing more than the sum of these parts, Bertalanffy proposed considering the organism as a system, that is, as a constellation of elements mutually interacting in order to maintain a state of dynamic equilibrium. His "system theory" of the organism aimed to provide the framework to study "the forces immanent in the living system itself" (Bertalanffy, 1933, pp. 177–178) – that is, the interactions between the parts of the system which brought about the hierarchical order characteristic of the organism.

The politics of organicism

In the German-speaking world, the political appropriations of organismic and holistic theories served a radically different ideological agenda than that of the socialist organicism of the Theoretical Biology Club. The British organicists of the left used analogies between organism and society in order to argue for a socialist ideal of cooperation in the interest of the parts and the whole, and often viewed the third-way synthesis between mechanism and vitalism as a form of Marxist dialectic. In Germany and Austria during the same period, by contrast, holistic ideals were instead being used to serve the fascist and racist agenda of National Socialism, with mechanistic science increasingly disparaged as the work of Jews, and comparisons between state and organism served the purpose of asserting the superiority of the whole by negating the power of its individual parts (Harrington, 1996, p. 175). Whether it was out of conviction, opportunism or (most likely) both, Bertalanffy joined those thinkers who explicitly linked their organismic biology to totalitarianism. While he apparently voiced disapproval of Nazi policies at first, Bertalanffy ended up joining the Nazi Party in 1938, after the annexation of Austria by Hitler. Academic promotion followed in short order (Drack et al., 2007, pp. 361–362; cf. Burkhardt, 2005, ch. 5, for the comparable case of Konrad Lorenz, whose science of ethology was likewise pitched between mechanism and vitalism).

After the Second World War, the term "organicism" became increasingly associated with discredited ideology rather than creditable science. One of the unfortunate effects was to obscure the diversity and even real achievements of the emergentist, holistic, organicist biology of the interwar years. As we have seen, although it did not last long, the left-leaning Theoretical Biology Club went some way toward showing what such a biology would look like in practice. But if the term "organicism" did not survive, the programme did, as well as key proponents such as Bertalanffy. Before he could be prosecuted during the denazification process after World War II he emigrated to Britain and then Canada, where he began promoting what he now labelled "general systems theory." This time his ideas were met with greater institutional enthusiasm, as systems thinking and mathematized models more generally came to be seen increasingly as promising ripostes to the challenges posed by new forms of post-war reductionism.

Developments after the Second World War

The triumph of the new reductionism

In the 1940s, discoveries in biochemistry later regarded as foundational for molecular biology, such as Beadle and Tatum's one-gene-one-enzyme hypothesis (Beadle and Tatum, 1941), strengthened confidence in reductionist quarters that biology could be unified from the bottom up, that is, from the simple to the complex, from molecule to organism. The post-war period generally saw reductionist approaches in biology go from strength to strength. One source of this recovered momentum was the intellectual migration of accomplished physicists into biology, among them George Gamow and Erwin Schrödinger (Schrödinger, 1944), thanks in part to the attractiveness of the exciting new problems and in part to the unattractive weaponization of physics during the war (Morange, 1994, pp. 92–93). Major achievements flowed swiftly, most famously the discovery of the double-helical structure of DNA, the chemical basis of the gene, by James Watson and Francis Crick (another ex-physicist) in 1953; Crick's formulation of the "central dogma of molecular biology" in 1957;[3] and the relating of DNA's structure to its function via the working out, by the mid-1960s, of the genetic code, linking nucleotide sequences in genes to amino acid sequences in proteins.

Watson and Crick famously claimed to have discovered not just the structure of a molecule but "the secret of life." That bold statement has since been understood as a calculated coup against their opponents in the context of a cultural war, with the materialist and reductionist Watson and Crick hoping this would be the final blow against their holist and spiritualist adversaries (Bud, 2013). And indeed, the publicizing in the 1960s of DNA-centric views by science popularizers such as Isaac Asimov (Asimov, 1960) helped to spread the idea that the organism is nothing more than the sum of its molecular components or the sum of the effects of its discrete genes. Even so, aspirations for a different model for biological thinking, one that would endeavour to come up with new ways to apprehend the complexity of life, not only survived but thrived across the same period and beyond.

Taking complexity seriously

For some, a science that could crack the problem of organized complexity in biology – a problem that molecular biology seemed to them utterly incapable of addressing, let alone solving – was the key not just to understanding life but to understanding more or less everything. They found an influential spokesman in Bertalanffy, whose project for a general systems theory started reaching wider audiences in the 1950s and 1960s. In calling it a "general" theory, he meant it. In Bertalanffy's words, his general system theory was "a general science of 'wholeness' which up till now was considered a vague, hazy and semi-metaphysical concept. In elaborate form it would be a logico-mathematical discipline, in itself formal but applicable to the various empirical sciences" (von Bertalanffy, 1969, p. 37). Those sciences included particle physics, psychology, the social sciences and potentially even disciplines such as history. Such radical inclusiveness did not at all suggest that different levels of the hierarchic structure of the universe could all be collapsed into one another. On the contrary, the study of organization would likely bring the recognition of new types of laws.

> To arrive at such organizational laws . . . we need on the one hand, empirical investigation and definition in each case and at each particular level. On the other hand we need a general conceptual framework which transcends that of traditional science.
>
> *(von Bertalanffy, 1969, p. 59)*

Bertalanffy was far from alone. In the 1940s and 1950s, other influential attempts were made to construct mathematical models and laws which would apply to different types of systems and were thought to have the potential to unify science. These included the Belgian chemist (and future Nobelist) Ilya Prigogine's study of non-equilibrium thermodynamics (Prigogine, 1947); the American mathematician Norbert Wiener's cybernetics (Wiener, 1948), which studied the transfer of information in self-regulating systems; and the beginnings of what would develop, in later decades, into catastrophe theory, chaos theory and complexity theory. A reviewer for the first volume of the *Yearbook of the Society for the Advancement of General Systems Theory* characterized ambitions along these lines as the search for "potentially universal methods" (Whyte, 1956, p. 171). As the title of that volume suggests, the Bertalanffyan end of the search was especially successful institutionally. In 1954, he founded the Society for the Advancement of General Systems Theory[4] alongside the neurophysiologist and behavioural scientist Ralph Gerard and the economist Kenneth Boulding. Among the aims of the society in their initial programme were to "investigate the isomorphy of concepts, laws and models in various fields, and to help in useful transfers from one field to another," as well as to "promote the unity of science through improving communication among specialists."

Anti-reductionist evolution: a return for emergence

Another flank in the post-war rise of reductionist biology was the Neo-Darwinian theoretical enterprise that Julian Huxley named "The Modern Synthesis" (Huxley, 1942): the union of Mendelian genetics and Darwinian natural selection. Nevertheless, emergentism soon found its way back into evolutionary thinking, as a way of blunting the edge of the idea that evolution was nothing more than the outcome of changes in the frequencies of randomly mutating genes. In France, Pierre Teilhard de Chardin, a controversial but very influential French Jesuit palaeontologist, gave emergent evolution a theological spin. In his posthumously published *Phenomenon of Man* (Teilhard de Chardin, 1955), he depicted evolution as a universal movement of complexification and the universe as a hierarchical system composed of emergent levels: "Pre-Life," "Life" (the biosphere) and "Human Consciousness" (the "noosphere" – a term borrowed from the Ukrainian chemist Vladimir Verdnasky). The ultimate development of evolution would be "the Omega Point," a point of divine convergence of all spheres of consciousness. Despite being accused of producing nothing more than pseudo-scientific mystical nonsense by two Nobel Prize laureates, Peter Medawar and Jacques Monod (Medawar, 1961; Monod, 1970), Teilhard de Chardin had a strong following not only among French Neo-Lamarckians such as Albert Vandel and Pierre-Paul Grassé but also among the architects of the Modern Synthesis, notably Julian Huxley and Theodosius Dobzhansky, who, in the later stages of their careers developed emergentist visions of evolution, inspired by his ideas.

The specifically Lloyd Morgan version of "emergence" talk and thinking re-entered anti-reductionist discussions in the 1950s and 1960s via another one of Teilhard de Chardin's readers, the British-Hungarian chemist and polymath Michael Polanyi, author of the philosophy-of-science classic *Personal Knowledge* (1958). According to Polanyi, the Neo-Darwinian theory of evolution carried a "fundamental vagueness" (Polanyi, 1962, p. 383), arising from its reductionist definition of life in terms of physics and chemistry. Polanyi posited a teleological and irreducible "ordering principle" or an "orderly innovating principle" to explain evolution as the emergence of radical and irreducible novelties. Thanks to Polanyi's advocacy, in that book and elsewhere (see, e.g., Polanyi, 1968), the doctrine of emergence once again became a lively resource for anti-reductionist biologist thinkers, as one can see in the writings in this period of the Cambridge-based Quaker ethologist William Thorpe (Radick, 2017).

Systems biology

In 1966, the Serbian scientist Mihajlo D. Mesarovic, at the time head of the Systems Engineering Group at Case Western Reserve University in Cleveland, organized a symposium themed around systems thinking. This symposium was the third on the subject of systems, but the first focussing specifically on systems theory in relation to biology. In the introduction to the proceedings of the symposium (Mesarovic, 1968), Mesarovic put systems at the centre of biological inquiry. "The fundamental question for the community of biologists is whether an explanation on the systems theoretic basis is acceptable as a true scientific explanation in the biological inquiry" (p. 76), he wrote, adding: "If the answer to the question of the acceptance of systems-theoretic explanations in biology is in the affirmative (as I contend) then . . . [we will have] a field of systems biology with its own identity and in its own right" (p. 77). In 1968, a reviewer of the proceedings represented systems biology as a new form of holistic reaction against the reductionist excesses of molecular biology, predicting that system-theoretic concepts were destined to play an important role in the future of biology and recommending his readers familiarize themselves with these concepts (Rosen, 1968, with Mesarovic quoted on p. 34).

That same year, in the Austrian village of Alpbach, the British-Hungarian man of letters and Cold Warrior Arthur Koestler gathered scientists from various disciplines for a symposium to reflect on ways of going "Beyond Reductionism," aiming to re-examine what he called the "totalitarian claims of the Neo-Darwinian orthodoxy" and its molecular-biological outrider (Koestler and Smythies, 1969, p. 1). Among the invitees were several figures already mentioned, including Waddington, Thorpe and the systems theorists Bertalanffy and Weiss. Both of the latter represented themselves as opposed to the reductionism of molecular approaches, with Weiss speaking in favour of a more interdisciplinary biological science (Koestler and Smythies, 1969,

p. 3), seeking a more synthetic (i.e. less atomistic and analytic) understanding of organisms. Notwithstanding the like-mindedness of the symposiasts in so many ways, however, no consensus was reached of what it meant to be an anti-reductionist biologist (Stark, 2016).

Over the course of the next decades, new ways of modelling biological systems emerged, from the biochemical systems theory and metabolic control theory that developed in the 1960s and 1970s through to the first *in silico* models of cells and viruses. At the same time, new prospects for advancing systems-oriented biology arose from the increasing powers of biologists, both to gather vast amounts of data – thanks especially to the automated genome-sequencing techniques developed in the 1990s and by the early 2000s – and to process that data using increasingly powerful computers. Key aspects of systems thinking have since been popularized in science books aimed at the general public (see, e.g., Noble, 2006). But there have also been major disciplinary developments within professional biology. "Systems biology" is increasingly a fully institutionalized, specialized branch of science, with its own journals, societies, textbooks and university departments with teaching programmes. On the website of the Department of Systems Biology at Harvard Medical School, we meet the following message of welcome:

> Systems biology is the study of systems of biological components, which may be molecules, cells, organisms or entire species. Living systems are dynamic and complex, and their behavior may be hard to predict from the properties of individual parts. To study them, we use quantitative measurements of the behavior of groups of interacting components, systematic measurement technologies such as genomics, bioinformatics and proteomics, and mathematical and computational models to describe and predict dynamical behavior. Systems problems are emerging as central to all areas of biology and medicine.

A lot of the past century's anti-reductionist buzzwords are present, and indeed systems biologists mostly, if not exclusively, identify as anti-reductionists (Calvert and Fujimura, 2009). Is anti-reductionist biology so configured headed for an institutionally secure future? The history we have reviewed here suggests caution. As we have seen, "organicism" was a name devised and adopted by biologists who viewed their discipline as in crisis because torn between two impossible-to-reconcile extremes, vitalism and mechanism. Likewise, for some at least, "systems biology" has come to designate less a new solution to the old problems than a new label for them. It is "the name of the crisis," as one systems biologist remarked; "it's the name of the fright that every one's gone into about having all the pieces and still not knowing how biology works" (Calvert and Fujimura, 2009, p. 48).

Notes

1 In his 1922 Gifford lectures "Emergent Evolution," Morgan explicitly framed his conception of emergence in terms of higher and lower levels (Morgan, 1923).
2 Haldane added that this term was not original to him but had previously been used to describe the theories of Xavier Bichat, Karl Ernst von Baer and Claude Bernard.
3 The central dogma held that the transfer of information was strictly unidirectional from DNA to RNA to protein.
4 The society is now called the International Society for the Systems Sciences.

References

Abir-Am, P. G. 1987. The Biotheoretical Gathering, Trans-Disciplinary Authority and the Incipient Legitimation of Molecular Biology in the 1930S: New Perspective on the Historical Sociology of Science. *History of Science*, 25(1): 1–70.

Asimov, I. 1960. *The Intelligent Main Guide to Science, Vol. II: The Biological Sciences.* New York: Basic Books, Inc.

Beadle, G. W. and Tatum, E. L. 1941. Genetic Control of Biochemical Reactions in Neurospora. *Proceedings of the National Academy of Sciences of the United States of America*, 27(11): 499–506.

Bergson, H. 1907. *L'Evolution Créatrice.* Paris: Félix Alcan.

Bertalanffy, L. v. 1928. *Kritische Theorie der Formbildung.* Berlin: Gerbrüder Borntraeger.

Bertalanffy, L. v. 1933. *Modern Theories of Development: An Introduction to Theoretical Biology.* London: Oxford University Press.

Bertalanffy, L. v. 1969. *General System Theory: Foundations, Development, Applications.* New York: G. Braziller.

Blitz, D. 1992. *Emergent Evolution: Qualitative Novelty and the Levels of Reality.* Dordrecht: Springer Science.

Bowler, P. J. 2001. *Reconciling Science and Religion: The Debate in Early Twentieth-Century Britain.* Chicago: The University of Chicago Press.

Bud, R. 2013. Life, DNA and the Model. *British Journal for the History of Science*, 46(2): 311–334.

Burkhardt, Richard W., Jr. 2005. *Patterns of Behavior: Konrad Lorenz, Niko Tinbergen, and the Founding of Ethology.* Chicago: University of Chicago Press.

Calvert, J. and Fujimura, J. H. 2009. Calculating Life? A Sociological Perspective on Systems Biology. *EMBO Reports*, 10(1S): S46–S49.

Darbishire, A. D. 1917. *An Introduction to a Biology: And Other Papers.* New York: Funk and Wagnalls.

Dawkins, R. 2017. *Science in the Soul: Selected Writings of a Passionate Rationalist.* London: Bantam.

de Chardin, Teilhard. 1955. *Le Phénomène humain.* Paris: Seuil.

Drack, M., Apfalter, W. and Pouvreau, D. 2007. On the Making of a System Theory of Life: Paul A Weiss and Ludwig von Bertalanffy's Conceptual Connection. *The Quarterly Review of Biology*, 82(4): 349–373.

Driesch, H. 1900. Die isolirten Blastomeren des Echinidenkeimes. *Archiv für Entwicklungsmechanik der Organismen*, 10(2–3): 361–410.

Esposito, M. 2013. *Romantic Biology, 1890–1945.* New York: Routledge.

Gayon, J. 2008. L'Evolution créatrice lue par les fondateurs de la théorie synthétique de l'évolution. In *Annales bergsoniennes IV L'Evolution créatrice 1907–2007: épistémologie et métaphysique*, ed. A. Fagot-Largeot and F. Worms, pp. 59–84. Paris: Presses universitaires de France.

Haldane, J. S. 1917. *Organism and Environment as Illustrated by the Physiology of Breathing.* New Haven: Yale University Press.

Harrington, A. 1996. *Reenchanted Science: Holism in German Culture from Wilhelm II to Hitler.* Princeton: Princeton University Press.

Herring, E. 2016. Des évolutionnismes sans mécanisme: les néo-lamarckismes métaphysiques d'Albert Vandel (1894–1980) et Pierre-Paul Grassé (1895–1985). *Revue d'histoire des sciences*, 69(2): 313–342.

Huxley, J. 1942. *Evolution: The Modern Synthesis.* Cambridge, MA: The MIT Press.

Johnstone, J. 1914. *The Philosophy of Biology.* Cambridge: Cambridge University Press.

Koestler, A. and Smythies, J. R. (eds.). 1969. *Beyond Reductionism: New Perspectives in the Life Sciences.* London: Hutchinson.

Loeb, J. 1912. *The Mechanistic Conception of Life: Biological Essays.* Chicago: University of Chicago Press.

Medawar, P. 1961. Critical Notice. *Mind*, 70(277): 99–106.

Mesarovic, M. D. (ed.). 1968. *Systems Theory and Biology: Proceedings of the 3rd Systems Symposium, Cleveland, Ohio, Oct. 1966.* New York: Springer-Verlag,

Monod, J. 1970. *Le hasard et la nécessité.* Paris: Seuil.

Morange, M. 1994. *Histoire de la biologie moléculaire.* Paris: La Découverte.

Morgan, C. L. 1923. *Emergent Evolution: The Gifford Lectures Delivered at the University of St Andrews in the year 1922.* London: Williams and Norgate.

Needham, J. 1932. Thoughts on the Problem of Biological Organization. *Scientia*, 52: 64–92.

Needham, J. 1936. *Order and Life.* New Haven: Yale University Press.

Noble, D. 2006. *The Music of Life: Biology beyond the Genome.* Oxford: Oxford University Press.

Peterson, E. 2014. The Conquest of Vitalism or the Eclipse of Organicism? The 1930s Cambridge Organizer Project and the Social Network of Mid-Twentieth-Century Biology. *The British Journal for the History of Science*, 47: 281–304.

Peterson, E. 2016. *The Life Organic: The Theoretical Biology Club and the Roots of Epigenetics.* Pittsburgh: University of Pittsburgh Press.

Polanyi, M. 1962. *Personal Knowledge: Towards a Post-Critical Philosophy.* Chicago: University of Chicago Press.

Polanyi, M. 1968. Life's irreducible structure. *Science*, 160(3834): 1308–1312.

Prigogine, I. 1947. *Etude thermodynamique des phenomenes irreversibles.* Liège: Dunod.

Radick, G. 2004. Morgan, Conwy Lloyd. In *The Dictionary of Nineteenth-Century British Scientists*, ed. Bernard Lightman, 4 vols, vol. 3, pp. 1425–1426. Bristol: Thoemmes Press; Chicago: University of Chicago Press.

Radick, G. 2017. Animal Agency in the Age of the Modern Synthesis: W. H. Thorpe's Example. In *Animal Agents: The Non-Human in the History of Science*, ed. Amanda Rees, *BJHS Themes* 2: 35–56. Cambridge: Cambridge University Press.

Ritter, W. E. 1919. *The Unity of the Organism: Or, the Organismal Conception of Life*. Boston: R G Badger. The Gorham Press.

Rosen, R. 1968. Review: A Means toward a New Holism. *Science, New Series*, 161(3836): 34–35.

Russell, E. S. 1924. *The Study of Living Things: Prologomena to a Functional Biology*. London: Methuen and Company Limited.

Schrödinger, E. 1944. *What Is Life? The Physical Aspect of the Living Cell*. Cambridge: Cambridge University Press.

Smuts, Jan C. 1926. *Holism and Evolution*. London: Macmillan.

Spemann, H. and Mangold, H. 1924. Über Induktion von Embryonalangen durch Implantation artfremder Organisotoren. Translated by Viktor Hamburger in 2001 as: Induction of Embryonic Primordial by Implantation of Organizers from a Different Species. *International Journal of Developmental Biology*, 45(1): 599–683.

Stark, J. F. 2016. Anti-Reductionism at the Confluence of Philosophy and Science: Arthur Koestler and the Biological Periphery. *Notes and Records of the Royal Society*, 70: 269–286.

Thompson, D'Arcy Wentworth. 1942. *On Growth and Form*. Cambridge: Cambridge University Press.

Thomson, J. A. 1911. *Introduction to Science*. London: Williams and Norgate.

Weiss, P. A. 1925. Tierisches Verhalten als "Systemreaktion." Die Orientierung der Ruhestellungen von Schmetterlingen (Vanessa) gegen Licht und Schwerkraft. *Biologia Generalis*, 1: 165–248.

Whitehead, A. N. 1920. *The Concept of Nature: The Tarner Lectures Delivered in Trinity College, November 1919*. Cambridge: Cambridge University Press.

Whitehead, A. N. 1925. *Science and the Modern World Lowell Lectures*. Cambridge: Cambridge Univerity Press.

Whitehead, A. N. 1929. *Process and Reality: An Essay in Cosmology: Gifford Lectures Delivered in the University of Edinburgh During the Session 1927–1928*. 1978 edition. New York: The Free Press A Division of Macmillan Publishing Co., Inc.

Whyte, L. 1958. Review: General Systems: Yearbook of the Society for the Advancement of General Systems Theory, Vol. I 1956 by L. von Bertalanffy. *The British Journal for the Philosophy of Science*, 9(34): 170–171.

Wiener, N. 1948. *Cybernetics: Control and Communication in the Animal and the Machine*. Cambridge, MA: The MIT Press.

Woodger, J. H. 1929. *Biological Principles*. London: Routledge & Kegan Paul.

Woodger, J. H. 1930a. The 'Concept of Organism' and the Relation between Embryology and Genetics, Part I. *Quarterly Review of Biology*, 5(1): 1–22.

Woodger, J. H. 1930b. The 'Concept of Organism' and the Relation between Embryology and Genetics, Part II. *Quarterly Review of Biology*, 5(4): 438–463.

Woodger, J. H. 1931. The 'Concept of Organism' and the Relation between Embryology and Genetics, Part III. *Quarterly Review of Biology*, 6(2): 178–207.

29

EMERGENCE IN THE CELL

Michel Morange

Introduction

Consideration of emergence in the cell raises various issues. The first is the widespread and non-rigorous use of the term by biologists. What is true in all branches of science is more evident in biology, where the common language is used for most scientific descriptions. A rapid examination of the occurrence of the word *emergence* in biological articles fully demonstrates this polysemy.

A first use refers to its original meaning: for instance, a larva is said to emerge from the egg, or the adult insect from its pupa. In many cases the verb "appear" could easily replace "emerge," for instance, when new differentiated cell types are described as emerging during development. The most significant uses of the word *emergence* in biological sciences – at least significant for philosophers – are related to evolution. Emergence of life and emergence of consciousness will be discussed in other chapters. But many other, more focussed uses of the word *emergence* also refer to the results of an evolutionary process: emergence of multicellularity, or of new gene families. Emergence of new diseases and of resistant (to antibiotics) forms of microbes is also related to an evolutionary process: emergence of AIDS was the result of mutations in simian immunodeficiency viruses, and resistance to therapeutic treatments is also often the consequence of mutations in the pathogenic agents. Ecologists also discuss of the emergence of cooperativity, which is not surprising since ecological and evolutionary models are closely related.

It is much more difficult to find examples of emergence in cell biology. These less frequent occurrences are related in this case to a different meaning of emergence, which is that the whole is greater than the sum of its parts: emerging behaviours are visible in systems, whereas they are invisible in isolated components.

There is not a well-established list of these emergent cellular behaviours, and in the first and longest part of this chapter, I will focus my study on three different categories of cellular phenomena that have been, at least implicitly, considered as emergent. In the second part, I will briefly describe what has changed since the beginning of the 2000s, with the development of systems and synthetic biology. If phenomena invisible in cellular components can emerge in cells, the same is true at other levels of biological organization: new phenomena may emerge in organs or in the whole organism. Examples will be given in the third part and their significance discussed. Finally, in the concluding remarks, I will discuss the ontological or epistemic nature of the previously described emergent phenomena and their relations with the evolutionary emergence described in another chapter.

The problem of emergence has always been linked with the question of reductionism. Although it is somewhat arbitrary to separate these two issues, I will nonetheless consider what is called emergent in a cell by biologists and leave aside a large part of the abundant literature on reductionism.

Emergence before systems biology

I will describe some of the emergent cellular phenomena studied in molecular cell biology between the 1960s and the 2000s. Limiting the examples that will be discussed to this period is obviously too restrictive, and I will argue in the concluding remarks that some of these studies were anchored in traditions well established already in 19th-century biology. The advantage of choosing a limited period of time is that these emergent phenomena can be classified in a small number of partially dependent categories: the study of oscillations, the chemiosmotic model of Peter Mitchell and the role of membranes within cells and the evidence for "cell decisions."

The importance attributed to oscillations

The first oscillations to be experimentally demonstrated in the 1960s were oscillations in the concentration of components of metabolic pathways, such as the glycolytic pathway. It is rather difficult today to understand why these observations attracted so much interest at the time. One reason is that biologists were already familiar with oscillations, at least since the mid-1920s when Lotka and Volterra proposed a mathematical model of the relations between preys and predators (Lotka 1925). The second is that oscillations have always been seen in science, and in particular in physics, as the result of interactions between different objects, celestial objects, for instance, such as planets, and for this reason were considered obviously emergent phenomena. Finally, in the case of metabolic pathways, it was possible to give a thermodynamic interpretation of the phenomenon in terms of far-from-equilibrium systems at a time when Ilya Prigogine emphasized the importance of dissipative structures in organisms (Prigogine and Stengers 1979).

Metabolic oscillations were not the only cyclic phenomenon to attract the attention of biologists. Some years later, the molecular characterization of the components of the circadian clock had a huge impact (Reppert and Weaver 2002; Goldbeter 1996, 2002).

Another cyclic phenomenon that generated more and more work was the cell cycle. This was not immediate: the first molecular model – the replicon – proposed in 1963 by François Jacob, Sydney Brenner and François Cuzin to explain the mechanism of cell division in bacteria, did not accord a major role to the cyclical nature of the phenomenon (Jacob et al. 1963). But the molecular characterization of the different phases of the much more complex cell cycle of eukaryotic cells (with yeast as a model) – known since the end of the 19th century – attracted more and more researchers. It led to the evidence for the existence of checkpoints in the cell cycle, at which the completion of the previous phase is checked and the decision to pursue the cell cycle taken. I will soon discuss the importance of this notion of cell decision.

The chemiosmotic theory of Peter Mitchell

The late attribution of the Nobel Prize in Chemistry in 1978 to Peter Mitchell for the chemiosmotic model, seventeen years after he proposed it (Mitchell 1961), is often considered a defeat of the reductionist view of biochemists (and molecular biologists). The relation between oxidation and the production of energy (ATP) in mitochondria was shown not to be direct through a molecule called X that was intensively sought but never discovered (later named the 20th-century phlogiston). Rather, it occurred through the formation of a gradient of protons and electric charges between the two sides of the inner membrane of mitochondria that was ultimately converted into the production of ATP.

This discovery was striking for different reasons (Morange 2007). The first was that Peter Mitchell was an outsider; he did not belong to academic circles. The second was the time required to convince the scientific community. Decisive experiments finally came from the study of another phenomenon, photosynthesis in chloroplasts – for which a similar chemiosmotic model had been proposed – through the work of Efraim Racker and Walther Stoeckenius in 1974 (Racker and Stoeckenius 1974). The role of the precursor despised by contemporary scientists apparently perfectly suited Peter Mitchell.

Things, though, were not so simple. Peter Mitchell was not the only one to propose alternatives to the intermediate X, and the model that was finally accepted was not the one he initially proposed: it was not the gradient of protons and electric charges that directly produced ATP, but the transconformational changes induced in ATPase, the enzyme present in the inner membrane of the mitochondria that synthesizes ATP, by the binding of protons to this protein. The chemiosmotic model was more (macro)molecular than supramolecular, more reductionist than holistic.

It is nonetheless true that this model gave membranes, and their organization in cells, a major role in the process. The important role of membranes in cellular activities was emphasized in the 1970s after a new model of membrane structure was proposed by Jonathan Singer and Garth Nicolson in 1972 (Singer and Nicolson 1972): membranes were involved in exchanges of information with surrounding cells and transfer of proteins and information within cells. In this perspective, the significant change was not the chemiosmotic model, but the importance accorded to membranes as a form of organization that cannot be reduced to its component molecules and macromolecules.

The concept of "cell decision"

I mentioned the discovery of the existence of checkpoints in the control of the cell cycle and the ensuing decision of the cell to interrupt (or not) the process. Other forms of cell (cellular) decisions were revealed at the same period. Apoptosis, the controlled cell death that occurs in particular during development to eliminate misplaced cells or to shape embryos, does not result directly from the activation of the apoptotic machinery. There is a decision of the cell to enter (or not) the apoptotic pathway that depends on the balance of pro- and anti-apoptotic signals received by the cell.

The same is true for a neuron that is simultaneously activated or inhibited by the neuromediators secreted by cells in contact with it. The final decision to emit (or not) an action potential depends on an integration of the different signals received and the effects they had on the membrane potential.

Is the decision of the cell an emergent phenomenon, or is it the predictable result of well-defined mechanisms? Are these decisions the result of a computation by the cell and, if so, can we hope to be able to perform this computation in the near future?

What changed with systems and synthetic biology?

Systems biology has deep roots in the past, and its importance was vigorously advocated by von Bertalanffy (von Bertalanffy 1968). Organisms are systems, and their structural and functional characteristics cannot be anticipated from the description of their components.

The influence of this systemic view was limited until the end of the 1990s, when the first genomic sequencing and the development of post-genomic technologies – transcriptomics, proteomics – provided a global view of gene expression and of protein composition in a cell or in an organism. The inventors of these technologies strongly argued that a new view of organisms would emerge from their use, a new "logic of life" (Brown and Botstein 1999). The development of models was necessary to exploit the results generated by the new technologies. Projects to model whole cells or organisms rapidly appeared. Building working cells in silico (e-cells) was initiated at Keio University near Tokyo; similar projects flourished around the world.

But what precisely was expected to emerge? What was this new logic of life? Fifteen years later, it is obvious that the different e-cell projects have not reached their objectives. But the idea that something emerges from the global functioning of systems has not disappeared. There are two different perspectives, one more general and abstract and the other more precise in terms of the nature of what emerges.

In the abstract view, what emerges are cellular states. Different transformations affecting cells, differentiation, death and oncogenic transformation can be seen as the passage from one cellular state to another. Metaphors as attractors or landscapes of energy are used to represent these transitions: they do not palliate the absence of a precise description of these cellular states.

A second group of researchers has considered that what emerges is mainly a certain functional dynamics that the previous static models of cellular and molecular biology totally neglected. The discovery, in 2002, that the activation of NF-κB, a transcription factor essential for the immune response, obeyed very complex kinetics hitherto ignored was emblematic of what was expected to emerge from the global picture provided by the new technologies (Hoffmann *et al.* 2002; Nelson *et al.* 2004). Many studies have since tracked these complex kinetic behaviours, which might play an essential role in the control of cellular processes.

This work has been pursued by synthetic biologists. Since the beginning of the 2000s, they have been particularly interested in the reproduction of emergent phenomena such as oscillations and cellular decisions. Emergence of a specific pattern may be a consequence of stochastic events, of random variations in the level of gene expression. In *Drosophila*, random variations in the expression of the genes encoding *Notch* and its ligand lead through feedback loops to the segregation of neural and epidermal lineages in the epidermis (Heitzler and Simpson 1991). In the same organism, the mosaic eye structure required for colour vision is also generated by random variations in the expression of the *spineless* gene (Wernet *et al.* 2006). In the two previous cases, random variations can lead to cell fate determination. The expression "cellular decision" is now often associated with these noise-controlled processes of differentiation (Balazsi *et al.* 2011). Cells exploit the dynamic interaction of transcriptional pulses to control gene expression and cellular differentiation (Lévine *et al.* 2012, 2013; Lin *et al.* 2015).

Emergence at the organ and organism levels

Emergence can occur at a supracellular level. From the coordinated action of cells emerge global functions that retroactively alter the behaviour of cells. During vertebrate development, oscillations generated by the *Wnt* and *Notch* genes organize the presomitic cells into somites, groups of cells that underpin the organization of the body into segments (Pourquié 2003). The heart has been a paradigmatic example of this type of emerging phenomena since the initial work of Denis Noble in 1960 and his efforts to compute a model of heart functioning based on the famous model proposed by Hodgkin and Huxley in 1952 (Hutter and Noble 1960; Noble 1960). The heart is an integrative structure with multilevel retroactions. For instance, the contraction of the heart modifies the electrical properties of its cells (Noble 2002). The integrated multilevel functioning of the heart may also in pathological conditions generate chaotic processes such as atrial fibrillation, a type of emergent, non-physiological phenomenon. The heart is a model for the study of other more complex organs, such as the brain, in which emergent phenomena are supposed to take place.

But emergence occurs also at the level of the whole organism, through the inter-related functions of the different organs. The most significant emergent phenomena are those occurring in pathological situations such as septic shock, a systemic inflammatory response of the body that in a high percentage of cases results in death.

Concluding remarks

Previous descriptions show the heterogeneous use of the term *emergence*. Is it possible to distinguish different types of emergence? Are these forms of emergence epistemic or ontological? And what is their relation to the use of the notion of emergence to describe evolutionary transformations?

A distinction between synchronic and diachronic emergence is particularly useful. Evolutionary emergence is diachronic: it requires a long period of time and a succession of events to occur; synchronic emergence is not time independent, but it does not require the creative power of time to occur.

The emergent phenomena that I have described in cells and organs are synchronic. But they require for their occurrence a certain form of organization that itself was the result of a long evolutionary history. Consider, for instance, the chemiosmotic model of Peter Mitchell. It requires the characteristic organization of membranes, in this case, of the inner mitochondrial membrane. This is the case in (all) the examples of emergence that I have described previously. The emergent phenomena resulting from the complex dynamics of a system are the consequence of a structural organization and of the multiple intermolecular interactions shaped by evolution. Emergent phenomena studied by 21st-century biologists rely on the same organization that biologists of the 19th century, vitalists or not, Louis Pasteur or Claude Bernard, considered responsible for the characteristic properties of organisms. What is new is the explanation of this organization in terms of a long and complex evolutionary history. The distinction made in the first part of the 20th century by Frederick Gowland Hopkins, the head of the Department of Biochemistry in Cambridge, between biochemistry and chemistry was also based on the existence of this cellular organization.

There is no reason to imagine that most of these emergent phenomena will not be computable in a more or less distant future. In this sense, the dominant conception of emergence in cell biology is epistemic and not ontological. From a sufficient knowledge of the components and of their interactions, it should be possible to deduce the behaviour of the system. The fact that this is not presently possible is the mere consequence of insufficient knowledge and/or limited computational capacity. But from an evolutionary point of view, the properties of systems have emerged (in an ontological sense) during the complex evolutionary history of organisms. The two types of emergence are related in organisms through the process that generated them.

Bibliography

Balazsi G., van Oudenaarden A. and Collins J. J. 2011 Cellular decision making and biological noise: From microbes to mammals. *Cell* 144: 910–925.

Brown P. O. and Botstein D. 1999 Exploring the new world of the genome with DNA microarrays. *Nature Genetics* 21: 33–37.

Goldbeter A. 1996 *Biological oscillations and cellular rhythms* (Cambridge: Cambridge University Press).

Goldbeter A. 2002 Biological rhythms: Clocks for all times. *Current Biology* 18: R751–R753.

Heitzler P. and Simpson P. 1991 The choice of cell fate in the epidermis of Drosophila. *Cell* 64: 1083–1092.

Hoffmann A., Levchenko A., Scott M. L. and Baltimore D. 2002 The IκB-NF-κB signaling module: Temporal control and selective gene activation. *Science* 298: 1241–1245.

Hutter D. F. and Noble D. 1960 Rectifying properties of heart muscle. *Nature* 188: 495.

Jacob F., Brenner S. and Cuzin F. 1963 The regulation of DNA replication in bacteria. *Cold Srping Harbor Symposia on Quantitative Biology* 28: 329–348.

Lévine J. H., Fontes M. E., Dworkin J. and Elowitz M. B. 2012 Pulsed feedback defers cellular differentiation. *PLOS Biology* 10: e1001252.

Levine J. H., Lin Y. and Elowitz M. B. 2013 Functional roles of pulsing in genetic circuits. *Science* 342: 1193–1200.

Lin Y., Sohn C. H., Dalal C. K., Cai L. and Elowitz M. B. 2015 Combinatorial gene regulation by modulation of relative pulse timing. *Nature* 527: 54–58.

Lotka A. J. 1925 *Elements of physical biology* (Philadelphia: Williams and Wilkins).

Mitchell P. 1961 Coupling of phosphorylation to electron and hydrogen transfer by a chemi-osmotic type of mechanism. *Nature* 191: 144–148.

Morange M. 2007 The complex history of the chemiosmotic theory. *Journal of Biosciences* 32: 1245–1250.

Nelson D. E., Ihekwaba A. E. C., Elliott M., Johnson J. R., Gibney C. A., Foreman B. E., Nelson G., See V., Horton C. A., Spiller D. G., Edwards S. W., McDowell H. P., Unitt J. F., Sullivan E., Grimley R., Benson N., Broomhead D., Kell D. B. and White M. R. 2004 Oscillations in NF-κB signaling control the dynamics of gene expression. *Science* 306: 704–708.

Noble D. 1960 Cardiac action and pacemaker potentials based on the Hodgkin-Huxley equations. *Nature* 188: 495–497.

Noble D. 2002 Modeling the heart–from genes to cells to the whole organ. *Science* 295: 1678–1682.

Pourquié O. 2003 The segmentation clock: converting embryonic time into spatial patterns. *Science* 301: 328–330.

Prigogine I. and Stengers I. 1979 *La nouvelle alliance* (Paris: Gallimard).

Racker E. and Stoeckenius W. 1974 Reconstitution of purple membrane vesicles catalyzing light-driven proton uptake and adenosine triphosphate formation. *Journal of Biological Chemistry* 249: 662–663.

Reppert S. M. and Weaver D. R. 2002 Coordination of circadian timing in mammals. *Nature* 418: 935–941.

Singer S. and Nicolson G. L. 1972 The fluid mosaic model of the structure of cell membranes. *Science* 175: 720–721.

von Bertalanffy L. 1968 *General systems theory: Foundations, development, applications* (New York: George Braziller).

Wernet M. F., Mazzoni E. O., Celik A., Duncan D. M., Duncan I. and Desplan C. 2006 Stochastic *spineless* expression creates the retinal mosaic for colour vision. *Nature* 440: 174–180.

30

EVOLUTION, INFORMATION AND EMERGENCE

George Ellis

Introduction

The emergence of life is based on adaptive evolutionary processes that accumulate information about the physical, ecological, and social environment, and on this basis develop appropriate responses to them. Because true complexity is based in modular hierarchical structures (Simon 1996; Booch 2007), this is a process of emergence of successive higher levels of structure and behaviour on evolutionary, developmental, and functional timescales. They each behave according to quite different patterns of causation than the layers below them, on the basis of different kinds of information suited to that level, processed in the way appropriate to that level.

In this chapter, I will attempt to encapsulate these relations and processes in a series of principles that underlie the emergence of autonomous higher levels of function and behaviour arising out of lower levels but not dictated by them. The points raised are the following:

1   Information usage is key to the functioning of life.
2   Complexity such as that of life is based in modular hierarchical structures of both a logical and a physical nature, with different kinds of information occurring at each level.
3   Information use at each level of a logical hierarchy is based on contextually informed, branching logical choices that are realised in the implementation hierarchy.
4   The physically realised information needed for life to evolve and develop intelligence was not present at the beginning of the universe; it came into being over time.
5   Consequently, emergence must take place with genuine causal powers coming into being at each higher level of an implementation hierarchy, allowing the logical processes appropriate to that level to occur largely independent of the lower-level implementing medium.
6   The origin of information is via processes of adaptive selection taking place in a contextual way through adaptation to the physical, ecological, and social environment. This is a multi-level process, with higher-level needs driving lower-level selection.
7   The processes of information origin and usage are therefore only possible via top-down realisation of higher-level requirements at lower levels, with multiple such realisations possible. Each level carries out the kind of work appropriate to that level.

Information use and life

In human beings, incoming information is compared with what is expected to be there on the basis of immediate past experiences and interpreted in this context (Kandel 2012); what we see is not the same as the data flowing in to our visual cortex from the optic nerve; rather, it is what makes sense to us (Frith 2007; Purves 2010). The feedback loops making this possible go from the cortex to the thalamus where incoming sensory data are processed and sent back to the visual cortex (Alitto and Usrey 2003). A crucial need is to sort out information (that is, useful data) from noise (that is, irrelevant data), so billions of bits of unneeded data are discarded each minute. Because living in groups is good for survival, we have a "social brain", with much brain capacity devoted to tracking social relationships (Dunbar 2014) and data associated with social relationships.

The key point is that the logic implemented in decoding our sensory data (which involves solving Helmholz's "inverse problem" (Purves 2010)) and making useful predictions about the future (Hawkins 2004) is logic that makes sense at the psychological and cognitive level,[1] even though it is implemented by physiological structures at lower levels.

It is this high-level information processing capacity, underlying "folk psychology" and logical argumentation, that enables us to survive and prosper in our specific environment. This is there-fore what evolutionary processes have selected for; genes and epigenetic processes are merely the lower-level means selected to enable this to happen.

> **Principle 1: Information usage.** *All biological life is based on function or purpose [Hartwell et al. 1999], and at the higher levels of emergence, intention [Fishbein and Ajzen 1975;Gray 2011]. Deployment of resources to attain specific purposes is dependent on acquisition, filtering, classification, and storage of information, which enables future actions to take place on the basis of reliable prediction of possible future outcomes [Hawkins 2004].*

The way this works out in the case of higher animals, and specifically humans, is shown in Figure 30.1.

The kind of logic processed at this level might include:

1   "If I leave Oxford at 10 am, I can be in London by 11:30 am". (Prediction)
2   "To solve this problem I'll need a Fortran program with six subroutines". (Planning)
3   "Maxwell's equations predict the existence of electromagnetic waves". (Analysis)

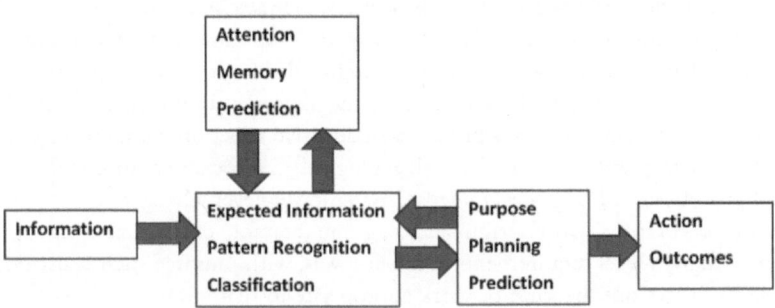

Figure 30.1 **Information use to plan action outcomes.** The logic of these processes is at the psychological level, relying on physical, biological, and social information at that level.

These thoughts are all realised at lower levels by flows of electrons down axons and fluxes of neurotransmitters across synapses (Kandel 2012; Kandel et al. 2013). However, they are all based in information that is meaningful at the macro level, and it is an undeniable fact that thoughts at the psychological level do indeed have causal power in the real world. Thus example 3 led to the existence of radio, television, radar, cell phones and GPS systems, which have had a profoundly transformational effect across all societies.

Emergent complexity and modular hierarchical structures

Generic principles for handling complexity underlie how this happens: one must have a modular hierarchical structure with abstraction and information hiding (Simon 1996; Booch 2007) for complexity to emerge. These principles are derived on the one hand by analysing biological structure and function, and on the other from practical experience in actually constructing genuinely complex systems (in particular, digital computer systems). They apply both to logical functions and to the physical structures and processes that implement that logic. Thus, the logical processes taking place at the psychological level have a modular hierarchical structure, for example, both language (written and oral) and mathematics have such a structure, as do practical plans such as those for constructing an aircraft or making a cake.

These logical processes are implemented by a physical modular hierarchical structure (the brain, based in neurons, proteins, genes, atoms, electrons, and so on) that realises the abstract logic in physical terms (Kandel 2012; Kandel et al. 2013). The underlying physical causation (electronic pulses along axons, voltage gated ion channels allowing ionic flows across membranes, and so on) is quite different from the emergent higher-level logic just discussed, where abstract logical plans can result in physical effects. Both levels (the psychological level and the electron level) are in operation at the same time, each with their own kind of causation, that is, with causation characterised by specific kinds of relations between variables defined at that level.

Principle 2: Complex emergence is based in perpendicular modular hierarchies.

All genuine complexity is based in modular hierarchical structures (Simon 1996;Booch 2007), where a complex task is broken up into simpler tasks that, when done together, coherently enable the complex task to be accomplished. Strongly bound units (modules) accomplish the simpler tasks, hiding their interior processes from the exterior (abstraction). Physically based implementation ("vertical") modular hierarchies enable the functioning of information-based logical ("horizontal") modular hierarchies.

"Abstraction" means that internal variables and detailed logic are hidden from the exterior. Seen from the outside, a module has a set of input–output relations in terms of a small number of interface variables representing an idealised interface that hides the complexity of internal workings.[2]

Logical hierarchies include language (written and spoken), mathematics, pictures, maps, planned processes (tasks), and social roles and functions. Thus, for example,[3] a computer system in a company has a module "Process Customer Order" (Level 1) as part of its overall computer system, which organises physical operations. This has submodules (Level 2) and sub-submodules (Level 3) as shown in Figure 30.2. Each Level 3 module requires further detailed tasks (Levels 4 and on) in order to accomplish its goals. Data flow in to the "Process order" and "Maintain account" submodules, and this drives what happens next. Information is processed at each level of this logical hierarchy, with more and more detailed data at the lower levels ("28 widgets will be dispatched to customer 765432 on 33rd March"). Much of this is hidden from the higher levels, which proceed on the basis of coarse-grained or abstracted information ("2057 units were sold in January", "Warehouse #4 needs 600 more units"). This process hierarchy will be the outcome

1: Process Customer Order			
2: Maintain account	Process order	Ship order	Update Inventory
3: – Retrieve owing – Register payments – Calculate owed – Send bill	– Receive customer order – Process customer order – Fill customer order – Invoice customer	– Retrieve address – Locate stock – Package – Despatch	– Subtract sent – Create order – Register arrivals – Update records

Figure 30.2 **A logical hierarchy.** Three levels are shown. There will be further more detailed levels, coded in computer programs and in operations manuals.

of thoughts in the brains of company officials and computational system designers and will be reflected in the brains of company operatives, as well as in computer programmes (which are abstract entities, even though they are realised in electronic states: they are part of the functional system which is the company as a whole, and quite different pieces of code could play the same role in the company's workings).

There are many other logical hierarchies; an interesting one is social roles in a school, a company, a court, a government, and so on. Note that the role itself (such as being a judge) is causally effective and is distinct from the individual who fills the position (e.g. Judge Mogoeng Mogoeng). Each role is effective through different kinds of information and has different kinds of outcomes.

Implementation hierarchies. The abstract (logical) hierarchy representing abstract logical relations as discussed earlier will be instantiated through an implementation (physical) hierarchy. For example, the logical hierarchy in Figure 30.2 is implemented via a digital computer, which is a physical structure that itself has a modular hierarchical nature (for example, transistors are modules in a central processing unit that is a module in a microprocessor).

For the case of biology in general, this physical hierarchy is given in Campbell and Reece (2005); for the human body, in Rhoades and Pflanzer (1989); and for the brain, in Scott (1995). There is a similar one for natural sciences (Ellis 2016) and for digital computers (Tanenbaum 2006; Ellis 2016). However, here I mainly concentrate on the human sciences hierarchy, indicated in Figure 30.3. Note that the upper levels are emergent entities, and these may be abstract rather than physical, for example, a university is a social institution that is not the same as its buildings or its staff or its students or its charter, but rather is a socially recognised entity binding all these elements in a logical structure, and it has causal powers such as the ability to arrange structured teaching and award degrees. Thus, the hierarchy is a hierarchy of structure and causation, rather than just structure (Ellis 2016).

The Relation between the Two. The relation between the logical and implementation hierarchies is clearest in the case of digital computers (Tanenbaum 2006). Exactly the same logical structure is written in different languages at each level in the logical hierarchy, with compilers or interpreters chaining this logic and related information down a chain of abstract machines via compilers or interpreters until it is coded in binary form (machine code) that matches the instruction set of the particular chip set chosen for the computer, which is implemented in physical terms (electron flows through gates). It can be realised in many different ways at the lower levels to give the identical upper-level logic. This principle of multiple realisability is in essence Turing's great discovery in devising the concept of general-purpose computers (Turing 1936; Hodges 2014): appropriately structured physical systems, whatever their physical details, can do universal computations. Thus, in the brain, logical processes at the psychological level are essentially unrestricted by the details of the lower-level implementation in terms of neural nets.

The Hierarchy of Structure and Causation

Sociology/ecology/economics/politics
Psychology/botany/zoology/physiology
Cell biology/neurology
Genotype/biochemistry/molecular biology
Chemistry
Atomic physics
Particle physics

Figure 30.3 **The human sciences hierarchy of structure and causation** (Ellis 2016). This is an implementation hierarchy. The lower levels are physical, but the higher levels have abstract elements such as values, thoughts, and social roles, which are causally effective in the hierarchy.

The main restrictions are in terms of processing speed and short-term memory capacity; they do not relate to the subject we are considering. We can think of Maxwell's equations or dragons, rockets, or pancakes, holidays, or the political situation, as we wish.

Logical versus physical causation

How does logical argumentation differ from direct causation via the laws of physics? Essentially (Binder and Ellis 2016), it uses information to determine outcomes in a branching logical structure involving logical operations (AND, OR, XOR, NOT) and computations represented by arbitrary functions $T(X)$, whereas physical laws give a specific outcome from any particular initial state, determined purely by particle interactions of a nature determined by fixed physical laws.

> **Principle 3: Logic of information usage.** *Information use at each level of a logical hierarchy is based on contextually informed logical choices of the form {GIVEN CONTEXT C, IF $T(X)$ THEN $F1(Y)$, ELSE $F2(Z)$}, where X, Y, and Z may be the same or different variables and $T(X)$, $F1(Y)$, and $F2(Z)$ are arbitrary functions, including logical operations. This kind of branching logic is realised in the implementation hierarchy via mechanisms such as transistors, voltage, and ligand gated ion channels; molecular recognition via lock-and-key mechanisms; and synaptic thresholds.*

It is this conditional branching of the logical flow of argumentation, determined on a contextual basis, that is the core process whereby information is used to make predictions and so provide the basis for choice selection. It is explicit in computer languages at every virtual machine level and occurs at each level of any logical hierarchy (designing an aircraft, baking a cake, planning organisational processes, and so on). This includes as a special case logical flows that are simply consecutive (do X first, Y second, and so on), which in many cases actually have an implicit conditional nature (do X until it is complete, then do Y until it is complete, etc.).

In the case of life, these logical processes are enabled by the underlying physics, provided complex enough biological systems have emerged that can embody this kind of logic, such as neuronal synapses and membrane allosteric binding sites; but these have to be selected for in order to come into existence. The coming into being of the implementation hierarchy is the outcome of genetic information, read under the control of epigenetic mechanisms based in the underlying molecular biology and physics. In the case of artificial systems, it requires logic gates based, for example, in transistors, but these have to be designed appropriately. The underlying physics and molecular biology allow emergence of the implementation hierarchy, that is then able to allow the logical hierarchy and its use of information as in Principle 3 to come into play.

Information comes into being

Where does all this information come from? There is no law of conservation of information at the micro level, provided one believes that quantum measurements take place, because that is an irreversible process whereby information on the initial state is lost (Ellis 2012). At the macro level, the cosmic context in which this all comes into being shows that biological information (such as that encoded in DNA) and cognitive information (such as that encapsulated in scientific theories) now exists which was not there at the start of the universe (when there was no life and no minds).

> **Principle 4: Information is generated over time.** *The needed information for emergence both of specific forms of life and of intelligent outcomes such as the text of Einstein's 1915 General Relativity paper was not present in the very early universe. This is* inter alia *because of quantum uncertainty at the micro level, which affects macro outcomes. The information needed to determine specific such outcomes is not written there: it has to have come into existence at later times.*

In the inflationary epoch in the very early universe, quantum fluctuations occur that get expanded to macroscopic scales by the exponential expansion at that time through a great many powers of 10, resulting in fluctuations in the interacting matter-radiation system of the Hot Big Bang epoch. These result in seed perturbations on the Last Scattering Surface ("LSS"), which is the set of events in the history of the universe where matter cools so much that it is no longer ionised, hence matter and radiation decouple and radiation propagates freely from then on. The perturbations on the LSS, then, through gravitational attraction are the sources of the astronomical structures we see today.

However, quantum uncertainty is foundational: although we can with certainty predict the statistical outcome of microscopic events, it is not possible even in principle to predict what the specific outcome of the next event will be. These outcomes are determined as they happen, as is demonstrated by the foundational two-slit experiment (Figure 30.4) (Feynman et al. 1963,

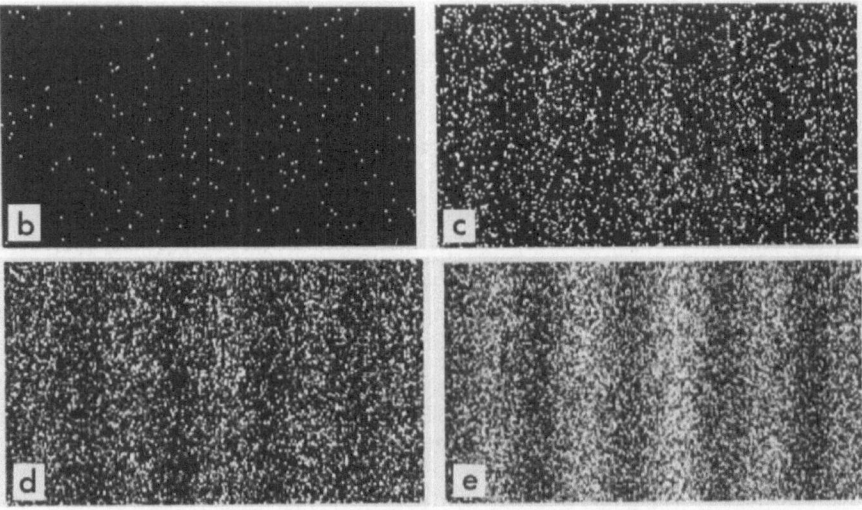

Figure 30.4 **Quantum uncertainty**. Double-slit experiment performed by Dr. Tonomura showing the build-up of an interference pattern of single electrons. The numbers of electrons are (b) 200, (c) 6,000, (d) 40,000, and (e) 140,000. This experiment demonstrates the fact that we cannot even in principle predict the specific outcomes that will occur at the micro level.

Figure 30.5 **The fluctuations on the Last Scattering Surface.** This is what the universe looked like when it was about 300,000 years old. No human thoughts are coded in that data.

[Courtesy of Planck]

1965; Nairz et al. 2002). This will apply also to the quantum fluctuations arising in inflation. Because of the quantum uncertainty at the basis of the process whereby quantum fluctuations in the early universe become classical, we cannot predict what specific galaxies will exist at the present day even if we have complete knowledge of conditions at the start of inflation. This is a cosmic analogue of the uncertainty shown in the previous picture. Additionally, cosmic rays have influenced evolutionary history through their damaging effects on DNA. However, the emission of a cosmic ray is a quantum event, subject to quantum uncertainty: given complete knowledge of the initial state of the system, we do not know when it will occur or where it will go. Thus, existence of human beings as a species is not predictable from complete data about life on Earth 2 billion years ago, even in principle.

In any case, no cosmological transfer function (see Dodelson 2003) can uniquely predict specific intelligent outcomes from the almost Gaussian random fluctuations on the Last Scattering Surface (Figure 30.5). There is no way that data could have been written into the LSS – they would have to specify in detail the entire history of the human race up to now and in the future. This is not remotely possible. And if it were possible, what demi-urge could have written all this into the initial data for the universe?

Occurrence of strong emergence

The implication is that genuine emergence must have occurred in order to enable information and structures that are now causally effective to come into being out of the random kinds of fluctuations that occurred on the Last Scattering Surface in the early universe. These emergent structures and the information they embody were enabled to come into being by that initial data but are not uniquely determined by it. They occur because of their own emergent logic.

> **Principle 5: Strong emergence.** *As a consequence of Principle 4, strong emergence must take place, with genuine causal powers coming into being at each higher level of an implementation hierarchy, allowing the logical processes appropriate to that level to occur independent of the lower-level implementing medium. Logical outcomes occur based on the logic and information at that level.*

Emergence must take place of the physical hierarchy where each level has causal powers appropriate to the relevant laws at that level and of the genetic information underlying the existence of that hierarchy, as well as of logical hierarchies containing specific information such as mathematical representations of the laws of physics, neither being specifically implied by the initial data in the early universe.

Origin of information

How then does the information arise that is embodied in both the implementation and logical hierarchies? In the implementation hierarchy, it arises through processes of natural selection (in the life sciences, see Campbell and Reece 2005) or design (in the artificial sciences, see Simon 1996). In the logical hierarchy, it must also take place as a collective social construction, gradually pieced together by experience, learning, interaction, and scientific investigation (Bronowski 1973). Both take place through processes of adaptive selection.

> **Principle 6: Origin of information.***Information origination is via processes of adaptive selection both in the logical hierarchies (learning and experimentation) and in the implementation hierarchy (via natural selection, engineering design, or a combination of the two, depending on context). In the biological case, this process takes place in a contextual way through adaptation to the physical, ecological, and social environment of developmental systems [Oyama et al. 2003] that inter alia shape proteins, metabolic networks, signal transduction networks, and gene regulatory networks [Wagner 2014]. This is a multi-level process, with higher-level needs driving lower-level selection.*

In more detail:

Creating order out of disorder. Adaptive selection is the only way to create order and associated information out of chaos and disorder. Some process or relation creates an essentially random ensemble $\mathbf{E}$ of entities $\mathbf{X}$, from which a selection process chooses those that satisfy some selection criterion $\mathbf{S}$. These are kept; the rest are deleted or discarded. The result is to create a more ordered system $\hat{\mathbf{E}}$ out of a disordered one, with the outcome depending both on the environmental context $\mathbf{C}$ (which creates niches where entities can flourish) and the selection criteria $\mathbf{S}$ (which attempt to adapt the outcomes optimally to the niche).

The logic of the process is a special case of that discussed in Principle 3:

$$\{\text{IF NOT } S(X{:}C) \text{ THEN DELETE } X\} \tag{9}$$

hence, the effect on the ensemble is

$$\{E(X)\} \rightarrow \{\hat{\mathbf{E}}(X){:} X \ \varepsilon \ S(X{:}C)\} \tag{10}$$

A simple example in the logical hierarchy is deleting emails or files on a computer. In the physical case, Maxwell's demon (Leff and Rex 1990) is the canonical example. It takes place in biology through Darwinian selection (see later) and in manufacturing by rejecting under-specification components. This process of generating a new order cannot be achieved by dynamical systems or unitary transformations, where the information at the end is the same as at the beginning, albeit perhaps transformed in form. It is also not the same as the usual randomising processes in physics which result in a Gaussian distribution which has no information except the mean and standard deviation, and in principle has members $\mathbf{X}$ with all values of the relevant variable (the Gaussian $\mathbf{p(x)}$ is non-zero for all values of $\mathbf{x}$ (Stone 2015:233–234)) and so cannot satisfy (10)).

In any case that distribution is associated with thermal equilibrium, which cannot occur for a living organism, because organisms use free energy to remain alive (Rovelli 2016).

Creating information where there was none. This process generates ordered information about the environment and information on how to make implementation hierarchies, which did not exist before. Thus, on the one hand, it generates genetic information coded in DNA, read out when developmental processes take place and leading to emergence of the implementation hierarchy with genuine causal powers that can generate abstract information on the basis of high-level reasoning (such as (6)–(8) earlier). On the other hand, it generates specific knowledge as well as social and cultural information and individual memories and abilities, all adapted to a greater or lesser extent to the physical, ecological, and social environment. Both of these kinds of information were not there at the start of the universe, nor were they uniquely implied by the data existent at that time (Principle 5).

The implementation hierarchy. With regard to the implementation hierarchy, Darwinian evolution (understood in terms of an extended evolutionary synthesis, see Pigliucci and Müller 2010) leads to emergence of developmental systems (Oyama et al. 2003) which result in brain structures coming into being that can classify information, store what is relevant, reason rationally (Churchland 2012), and predict likely outcomes (Hawkins 2004). This adaptive process is key to the accumulation of useful information in the implementation hierarchy at all levels. It is crucial that there is selection not just for genes but also for:

- Metabolic networks
- Gene regulatory networks
- Signal transduction networks
- Proteins

(see *Arrival of the Fittest*, Wagner 2014). This leads to macro physiological structures such as eyes and lungs that are adapted to the environment (eyes presuppose a transparent atmosphere, lungs that it is largely composed of oxygen). The proteins rhodopsin and haemoglobin are selected for so that eyes and lungs, respectively, can function in that context. They are selected to fulfil a specific purpose.

The logical hierarchy. With regard to the logical hierarchy, the specific knowledge embodied in the operations of our minds is based in learnt information deriving from our experiences, our social environment, and the logical operations of our own minds. Adaptive selection through learning processes (Gray 2011) is the key factor in attaining the information that is useful at these levels, including abstract relationships (Churchland 2012). The detailed knowledge resulting is what it is because of the individual's ongoing experience with the physical, ecological/biological, and social environment. The last factor is particularly important because we are a social species (Dunbar 2014), and our learning processes, starting at a very young age (Gopnik et al. 1999), are exquisitely tuned to adapt to and learn from the subtleties of social interactions – which are not physical entities. Furthermore, this social interaction then allows society to accumulate scientific, engineering, and organisational knowledge that underlies the rise of civilisations (Bronowski 1973) and so greatly enhance our survival prospects, and indeed enabled us to subdue all other species. The language we use for this purpose has to have a modular hierarchical structure because we use it to represent faithfully the modular hierarchical structure of the world in which we live (Deacon 2003).

Bottom-up and top-down processes

The processes at the psychological level involve learning at that level ("A dog might bite me"), involving concepts and information appropriate to that level ("dog", "bite"), but are realised through brain plasticity at the micro level, because the functions of the mind are enabled by the structure of the brain at that lower level. Thus, for example, memory of macro-level events is realised via gene expression in neuronal synapses (Kandel 2006), enabled through epigenetic processes. This is a top-down process in the implementation hierarchy. The high-level information can never have emerged in a purely bottom–up way from the underlying physics or molecular biology because of its contextual nature: it involves information about specific events ("My birthday is 19th March") and places ("We spent a year in London once"). That information is chained down from the psychological level of experienced life to determine details of the synaptic connections in our neocortex that embody our individual memories. These lower-level aspects can also never have emerged in a purely bottom–up way, because they embody specific, contextually-dependent, higher-level data. Synaptic weights are altered to embody memory (Kandel 2006), and synaptic pruning takes place on the basis of experience (Wolpert 2002). Thus, processes at the psychological level – of a logical or factual or emotional kind – affect brain details at the microphysical level (Kandel 1998).

This is a specific case of a more general principle: in order that emergence of truly complex systems can take place, it is essential that as well as bottom–up effects, top–down (contextual) effects take place in both hierarchies (Ellis 2016). In particular, the key role of adaptive selection at every level (see Principle 6) is that it fits the system (structure and function) to the environment in both hierarchies. Hence, this is a top–down process: information about the physical, ecological, or social environment gets embodied in the physical and logical structure and function of the system. In the implementation hierarchies, adaptive selection is the key factor shaping structure and function. This is a top–down process because the organism then in a sense embodies images of the environment in which it exists via its genetic structure (Stone 2015:188–290). In the logical hierarchy at the psychological level, learning processes embody this principle (elements of memories and learnt patterns derive meaning in a contextual way via adaptation to the context) and are then realised in the implementation hierarchy at the micro level (in terms of the patterns and strengths of detailed synaptic connections in the neocortex).

> **Principle 7: Top-down effects in both hierarchies.** *In an implementation hierarchy, the processes of strong emergence and associated higher-level information storage and utilisation are only possible via top-down realisation of higher-level requirements at lower levels (Ellis 2016), with multiple such realisations possible (Auletta et al. 2008). In this way the lower levels carry out the work, but the higher levels decide what will be done (van Gulick 2007; Gazzaniga 2012). In the logical hierarchies, the lower levels always attain their meaning in the context of the higher-level situations, which is what allows information filling in and reliable prediction on the basis of partial information.*

Implementation hierarchy. Bottom-up emergence creates higher levels of structure in the implementation hierarchy, and causation takes place at each level of the hierarchy representing the same high-level logic but with different information and logic at each level. Top-down realisation of the higher-level logic at each of the lower levels enables this to happen. This is a many-to-one relation: in general there are a great many structures and functional ways of implementing the higher-level logic at each lower level. Examples are:

- A thermostat for a room heater has a lever (a macro object) whose setting controls billions of micro states (velocities and positions of molecules in the room).

- Natural selection is a top-down process adapting animals to their environment, thereby altering the details of the base-pair sequence in DNA; thus, "[e]volution is essentially a process in which natural selection acts as a mechanism for transferring information from the environment to the collective genome of the species" (Stone 2015:188).
- In the brain, top-down signalling underlies the way senses function in terms of selective filtering of information and prediction, implemented via downward connections in cortical columns (Bullier 2006) and top-down connections from the cortex to the thalamus (Alitto and Usrey 2003).
- Memory (information storage) is enabled via epigenetic mechanisms controlling gene expression in synaptic connections (Kandel 1998, 2006), a top-down process from the psychological level to the level of synapses.
- In physiology and microbiology, top-down causation takes place via epigenetic mechanisms (Gilbert and Epel 2009, as indicated in Figure 30.6. Higher-level organ and tissue conditions control gene expression; organism and organ and conditions trigger cell signalling; protein and RNA networks select, read, and correct genes (Noble 2012).

Thus, information processing in the implementation hierarchy is a two-way bottom-up and top-down process, and selection through the evolutionary process has ensured that this is so, for this is what enables the kind of adaptive, flexible responses that enhance survival prospects.

Logical hierarchy. In a logical hierarchy the lower levels are always interpreted in the context of a higher-level framework, which is taken for granted and shapes how the lower levels are read or understood (Ellis 2016: Chapter 7). This is the way all the sensory systems work, where expectations of what should be, deriving from experience and understanding of context, shape what is actually seen or experienced (Frith 2007; Kandel 2012), which is not the same as the

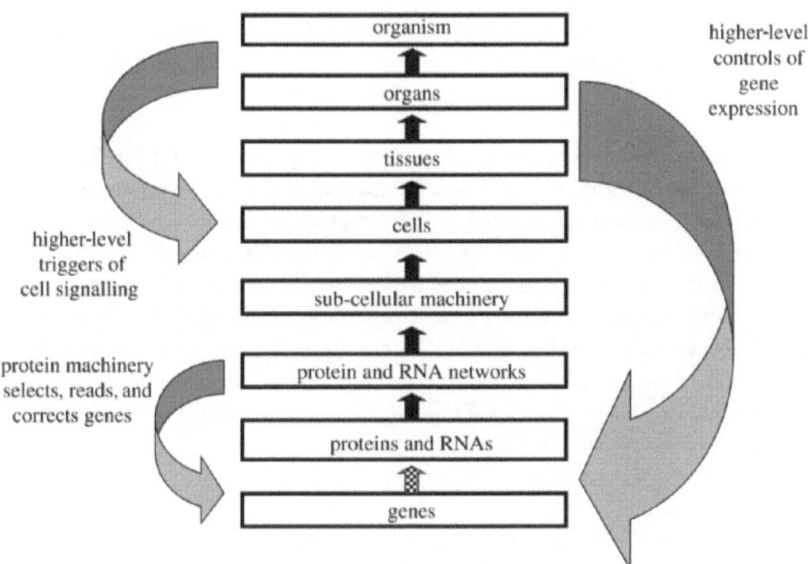

Figure 30.6 **Top-down effects take place as well as bottom-up effects.** This is the case of physiology as presented by Noble (2012), based on his detailed studies of the heart.

[Reproduced with permission]

image projected on the retina by incoming photons (Purves 2010). It is also the way our mental faculties work.

In a mathematical proof, variables in a theorem will have been defined in a previous part of the proof, and the reader will be expected to remember what their definition is. Language is read on the basis of an understanding of context, with the mind interpreting lower-level elements on the basis of that context. For example, the words "they", "she", "then", and "there" in a sentence have no meaning by themselves; they depend on the context of the sentence. The word "plane" has no meaning by itself: it derives its meaning from the context (an airport or a woodwork shop or a geography text). The pronunciation of syllables depends on the words in which they are imbedded ("the tough dough tasted sour") (Strauss 2004). Reading a text or listening to a conversation is a psycholinguistic guessing game in which the reader or listener is all the time filling in unsaid things on the basis of the social and psychological context (Goodman et al. 2016), and phrases are read as a whole rather than as a sequence of individual words. It can be experimentally proven by reading studies that what actually enters the reader's mind is not the same as what is written: it is a text with the same broad meaning, but with some words substituted for others (Flurkey et al. 2008): thus, it is not a bottom-up process of reading each symbol in turn and assembling the results together. Overall, the lower levels and component variables in a logical hierarchy always attain their meaning in a contextual way.

Interaction between them. In the implementation hierarchy in a living system, emergence processes (evolutionary and developmental) create higher levels of structure, and causation takes place at each level of that hierarchy, with the lower levels performing supporting functions for the higher-level operations. There is a different functional language and syntax operational at each level, storing and processing different kinds of information through different kinds of emergent physical structures.[4] In the logical hierarchy, the lower levels represent decompositions of high-level logic into smaller units, read together in a contextual way to create the higher-level function and meaning. Top-down realisation of the higher-level logic at each of the lower levels enables this to happen. The emergence of the higher levels occurs via contextual constraints so that they are adapted to the environment.

In summary:

- Higher levels of the implementation hierarchy emerge from lower levels; each level has its own kinds of variables and·information formalised in a language appropriate to that level.
- Same-level logic structures causation at each level of the implementation hierarchy, according to the laws and information appertaining to that level; these also have a hierarchical structure, with realisation of the higher logical levels through combinations of lower-level modules, with multiple lower-level realisations of the same higher-level logic being possible.
- The higher-level logic decides what will be done, but this is implemented by the lower-level implementation structures in terms of the logic at their level, with supporting operations being carried out at each lower level that enable the higher-level properties to emerge, even when they have a completely different character.

This is particularly clear in the case of digital computers (Tanenbaum 2006), where the chaining down of logical structure from higher- to lower-level languages in the hierarchy of virtual machines takes place via compilers and interpreters. The application program–level logic gets compiled in terms of virtual machines such as a Java virtual machine; this chains down to the level of assembly language, and then to machine code (binary), which can be interpreted directly by a suitable chip. The flow of electrons through gates does not resemble in any qualitative sense the appearance of images on a screen that it enables, even though they encode the same logical operations, which are represented in different syntaxes and variables at every level of the logical hierarchy represented by the tower of virtual machines.

More generally, as shown in Figure 30.7,[5] the relation is:

1   Causation is a relation involving two events at different times.
2   Realisation is a relation involving two things or collections of things which exist at the same time but on different scales or explanatory levels.
3   A natural way to describe the phenomena is to say that there is a vertical relation of realisation between entities at different levels (with the lower-level entities realising the higher-level entities) and parallel layers of "horizontal" causation at each level.
4   All the levels of causal explanation are equally valid and "real". They complement each other. They take place at the same time and don't compete. For example, lots of atoms are bonded together to form the lever in the thermostat. We can say that the movement of the lever causes the air to warm, or that the movement of the giant constellations of atoms that compose the lever causes the atoms that compose the air to move faster.
5   These are parallel and complementary causal stories at different levels of explanation. The latter story abstracts away from the mindbogglingly complex details of the former story, and is much more general. The latter story could remain true even if the former story were somewhat different (e.g., because the thermostat lever was a slightly different shape). The latter story is in this respect more general. The entities it posits (e.g., the "lever") are "multiply realizable" at lower levels.

The multiple-level nature of such processes can be illustrated by a specific example (Figure 30.8). Human bodies are composed of a huge number of cells. Because the kinds of living cells out of which our bodies are made need oxygen for metabolism, which need is based in turn on the underlying thermodynamics, there is a lower-level need for oxygen to be provided to each cell in the body. Hence, there is a macroscopic emergent need for us to breathe oxygen, which can then be circulated to the cells. Evolution has found that a good way to do this is via lungs and a circulatory system (Rhoades and Pflanzer 1989). But in order for this work, blood must have a way of transporting oxygen; hence, it needs haemoglobin molecules at the micro level to make this possible. Thus, natural selection brought into existence, through adaptive Darwinian processes, both the circulatory system (lungs, heart, veins, and so on) at the macro level and haemoglobin molecules at the micro level. When the system is functioning, at the macro level the heart is beating in a way controlled by a combination of bottom-up and top-down processes indicated in

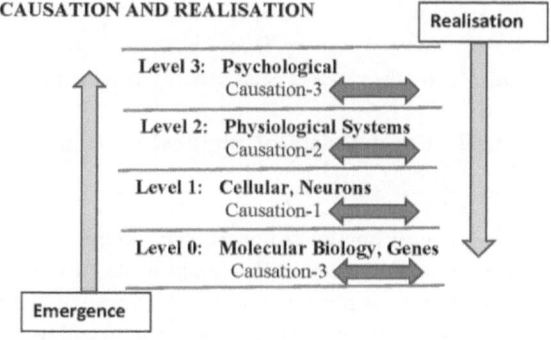

Figure 30.7   Multiple levels of causation related by emergence and realisation.

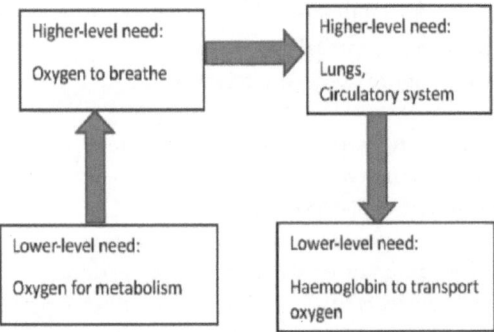

Figure 30.8   Multiple levels and the circulatory systems

Figure 30.6 (Noble 2012), and thereby transporting oxygen to all the tissues in the body. At the micro level, haemoglobin is taking up oxygen in the lungs and then releasing it in bodily tissues. These are the kinds of parallel processes indicated in Figure 30.7.

Issues regarding top-down causation

1   **What about supervenience?** Suppose one could reproduce in every detail the structure of a computer or brain and its current state of excitation. Would the identical higher-level behaviour/effects emerge? Yes – maybe. So the higher-level behaviour is determined in a bottom-up way by the lower level; for example, the brain works out the effects of Maxwell's equations in a bottom-up way. There is thus no need for top-down effects to occur; this is the argument from supervenience (McLaughlin and Bennett 2011). But what it omits is that you can't get to that initial state in a purely bottom-up way – the brain or computer can't self-assemble. How did the electrons in our cortex know about Maxwell's equations? They exist in a context of neural connections that over time have been taught about those equations through a social process of education, based in the fact that mathematicians have discovered the nature of the underlying mathematics through the pattern recognition properties of the neural nets in the cortex (Churchland 2012), and Maxwell then realised they accurately described the nature of electromagnetic interactions. The relevant neuronal patterns are shaped by learning processes at the psychological level, incorporating eternal, unchanging abstract patterns of mathematical relationships that have been explored by humans over some hundreds of years of conceptual exploration. You can't get the final result without top-down causation, even if the final causal step is purely bottom up. Thus, while synchronic emergence of complex systems from lower-level structures and information is possible ("supervenience" can take place at each instant), diachronic emergence (getting the information there) is not.

2   **What about the causal completeness of physics?** A claim often made is that emergence of causally effective higher-level variables and processes is not possible, because the lower-level physics is causally closed and determines what happens at all higher levels (Vicente 2007). But lower-level physics is not causally complete! Philosophers and many physicists seem not to have taken on board the implications of the quantum revolution of last century (Figure 30.4). Physics is not causally closed because of quantum uncertainty: real wave function collapse alters the chances of subsequent events, and so physical laws alone cannot determine the chances of events, only physical laws plus the outcomes of wave function

collapses, which are not deterministic. And it cannot be closed, because information has come into being over time (for example, that in all the references at the end of this chapter) that was not there at the start of the universe (see Principle 4). This proves conclusively that causal closure cannot be true.

3 **Role of chance**: The occurrence of chance does not mean the outcome is random; rather, it provides the basis for the selection of outcomes on the basis of higher-level selection criteria, thus creating order out of disorder (Principle 6). Indeed, microbiology thrives on randomness (Hoffmann 2012), as does brain function (Glimcher 2005; Rolls and Deco 2010). Together with statistical randomness between levels, quantum uncertainty provides the material on which selection processes can operate.

4 **Is it top-down causation, or something else?**
Various labels have been used for the top-down effects discussed here:

- *Top-Down Constraint*? The macrostates are described by the values of the constraints, and they affect lower-level states. This is indeed the case, for example, in digital computers, where wiring constrains the flow of electrons. But it seems a rather passive name to describe the way that my choice of software decides which patterns of electron flows will take place through lower-level gates. Each microstate satisfies the constraints, and a key question is how tightly do they constrain lower levels and hence higher levels that then emerge?
- *Top-Down Causation*? That seems a good description of what happens when I decide to get up and walk across the room, and this results in electrons flowing in my muscles according to this choice. But there are philosophical objections to this. Douglas Campbell, for example, suggests there are no bottom-up or top-down causal relations in Figure 30.7.
- *Top-Down Realisation*? There are a huge number of microstates corresponding to one macrostate that realise the higher-level function or purpose. One of them will be selected for. This may well be the best description of the way higher levels affect lower-level states.

5 **Multiple realisation as a key feature.** It is the equivalence class at the lower level that realises a higher-level variable that has the real causal power, rather than the specific lower-level member of the class that is realised (Auletta et al. 2008; Ellis 2016). But how to quantify this? Some formulation in terms of mutual or relative information and associated relative entropy may be appropriate (Stone 2015; Rovelli 2016). But it must take seriously the full hierarchy of structure and causation. The integrated information theory of Tononi and colleagues seems very promising in this regard (Tononi 2008, 2015), particularly because it can indicate how causal power can be stronger at macro rather than micro levels (Oizumi et al. 2014; Hoel et al. 2014, 2016).

6 **Greatest causal power.** Given that all levels have causal power, can we nevertheless characterise some of them as having greater causal power than others? It would appear that this is indeed the case: for example, in the case of the digital computer, it is the high logical level where algorithmic structures are set out that determines what happens when this logical structure is instantiated at lower levels via compilers or interpreters. Hoel et al (2014, 2016) show how this can be made rigorous through use of the idea of integrated information.

Issues regarding emergence

- Do some composite objects have properties and behaviour that go beyond those of their basic constituents? Yes indeed they do: psychological-level properties and logic are quite different from pure physical causation. Principles 3 and 6 detail the kinds of causation not implied by Newton's laws, Maxwell's equation, Schrödinger's equation, or the Dirac equation.

- Is it the case that, underlying the apparent complexity, there is unity: just a few kinds of things, governed by a few simple laws, which provide the ultimate basis for everything that happens? Yes indeed. All ordinary matter is made out of electrons, protons, and neutrons interacting by the equations just mentioned.

- Is emergence a coherent possibility? It has to be because it manifestly happens on evolutionary timescales, developmental timescales, and functional timescales. Laughlin (2000) emphasises that one cannot even in principle predict the equations of superconductivity from the underlying lower-level physics of interactions between electrons and protons, because the derivation depends on higher-level properties of crystal lattices that cannot be described at the lower level. The same is true for all topological effects in physics (Hasan and Kane 2010), for example, the quantum Hall effect, and Scott (1995) emphasises it applies also to the derivation of the Hodgkin-Huxley equations for action potential propagation along axons at the neuron level of description of the brain. At an even higher level, the psychological level comes into being with its own causal powers that are realised through the lower-level molecular biology and physics.

- It is genuinely interesting because life and mind – the emergence of the psychological level – have quite different properties from inanimate matter. Evidence for this is the fact that the chapters in this book exist, containing complex patterns of thought and analysis that cannot have been uniquely predicted even in principle from data describing in full detail everything existing in the very early universe (Principle 4), *inter alia* because that data do not imply that the Earth will necessarily exist.

Issues regarding information

- Detailed information is lost in the upward process of coarse graining, and information hiding takes place when macro variables emerge out of micro variables; entropy is a measure of how much detail is hidden (Stone 2015). However, not all higher-level information is coarse-grained, lower-level data: for example, geographical and historical data as well as logical relationship and scientific knowledge are intrinsically higher-level information. They get encoded somehow in lower-level structures (synaptic connection patterns and weights), but that is the opposite process to coarse-graining: it is a top-down process of physical realisation.

- New information is created when selection processes delete or discard unneeded information and so create order that did not exist before (Principle 6). This takes place in the implementation hierarchy (through Darwinian evolution) and in the logical hierarchy (through learning). This is a key top-down process (because it involved adaptation to the physical, ecological, and social environment) without which complex entities cannot come into being. Randomness or noise is needed to make this work by providing an ensemble of entities to choose from.

- Information is lost when such a selection process takes place, because the final state has lost the full information that was in the initial ensemble. Sometimes this is to do with absorption and irreversible processes at various levels; sometimes it is discarded to the environment. Part of it may be retrievable (for example, dinosaur skeletons), but much is not (we have difficulty reconstructing the processes by which language emerged). It might be claimed that we never lose information in principle, because dissipational processes result from coarse-graining microscopic processes that are, in fact, reversible, even if we lose it in practice. However, the view that I will put here is that this is not the case: in the end, loss of information is irretrievable because it is associated with the quantum measurement process (Ellis 2012). Physics is not in fact reversible at the micro level. When a wave function collapses from a superposition to an eigenstate, one loses the information of what that superposition was.

In summary: information is not conserved. It is created and destroyed at all emergent levels in the implementation hierarchy. It is coming into existence is the basis of the emergence of life.

Notes

1 For the notion of levels, see Figure 30.3.
2 See Wikipedia. Abstraction (software science). https://en.wikipedia.org/wiki/Abstraction_(software_engineering).
3 See Borysowich, C. Data Driven Design: Other Approaches. *Toolbox*. http://it.toolbox.com/blogs/enterprise-solutions/data-driven-design-other-approaches-16350.
4 They are always emergent because we do not know what the base level is, if it exists (perhaps it is super-strings/M theory, perhaps not). For practical purposes, the standard model of particle physics (Oerter 2006) is the lowest level.
5 I thank Douglas Campbell (Christchurch) for this formulation.

References

Alitto H. J. and W. M. Usrey (2003) "Corticothalamic feedback and sensory processing" *Current Opinion in Neurobiology* 13:440–445.
Auletta G., G. F. R. Ellis, and L. Jaeger (2008) "Top-down causation: From a philosophical problem to a scientific research program" *Journal of the Royal Society Interface* 5:1159–1172 [arXiv:0710.4235].
Binder P. M. and G. F. R. Ellis (2016) "Nature, computation and complexity" *Physica Scripta* 91:064004.
Booch G. (2007) *Object Oriented Object Oriented Analysis and Design with Applications* (Boston, MA: Addison-Wesley).
Bronowski J. (1973) *The Ascent of Man* (London: BBC Books) (reprint 2011).
Bullier J. (2006) "What is fed back?" In J. L. van Hemmen and T. J. Sejnowski (eds.), *23 Problems in Neuroscience* (Oxford: Oxford University Press).
Campbell N. A. and J. B. Reece (2005) *Biology* (San Francisco: Benjamin Cummings).
Churchland P. (2012) *Plato's Camera: How the Physical Brain Captures a Landscape of Abstract Universals* (Cambridge, MA: MIT Press).
Deacon T. (2003) "Universal grammar and semiotic constraints." In M. Christiansen and S. Kirby (eds.), *Language Evolution* (Oxford: Oxford University Press), pp. 111–139.
Dodelson S. (2003) *Modern Cosmology* (Cambridge, MA: Academic Press).
Dunbar R. (2014) *Human Evolution* (London: Pelican Books).
Ellis G. F. R. (2012) "On the limits of quantum theory: Contextuality and the quantum-classical cut" *Annals of Physics* 327:1890–1932 [arXiv:1108.5261].
Ellis G. F. R. (2016) *How Can Physics Underlie the Mind? Top-Down Causation in the Human Co Text* (Springer). Available at http://link.springer.com/book/10.1007/978-3-662-49809-5
Feynman R. P., R. B. Leighton, and M. Sands (1963) *The Feynman Lectures on Physics: Mainly Mechanics, Radiation, and Heat* (Reading, MA: Addison-Wesley).
Feynman R. P., R. B. Leighton, and M. Sands (1965) *The Feynman Lectures on Physics: Quantum Mechanics* (Reading, MA: Addison-Wesley).
Fishbein M. and I. Ajzen (1975) *Belief, Attitude, Intention, and Behaviour: An Introduction to Theory and Research* (Reading, MA: Addison-Wesley).
Flurkey A. D., E. J. Paulson and K. S. Goodman (2008) *Scientific Realism in Studies of Reading* (Mahwah, NJ: Lawrence Erlbaum).
Frith C. (2007) *Making Up the Mind: How the Brain Creates Our Mental World* (Malden: Blackwell).
Gazzaniga M. (2012) *Who's in Charge?: Free Will and the Science of the Brain* (London: Hachette).
Gilbert S. F. and D. Epel (2009) *Ecological Developmental Biology* (Sunderland, MA: Sinhauer).
Glimcher P. W. (2005) "Indeterminacy in brain and behaviour" *Annual Review of Psychology* 56:25–56.

Goodman, K. S., P. Fries, S. Strauss and E. Paulson (2016) *Reading: The Grand Illusion. How and Why Readers Make Sense of Print* (London: Routledge).

Gopnik A., A. N. Meltzhoff and P. K. Kuhl (1999) *The Scientist in the Crib* (New York: Harper Collins).

Gray P. (2011) *Psychology* (New York: Worth Publishers).

Hartwell L. H., J. J. Hopfield, S. Leibler and A. W. Murray (1999) "From molecular to modular cell biology" *Nature* 402 (Supplement):C42–C52.

Hasan M. Z. and C. L. Kane (2010) "Topological Insulators" *Reviews of Modern Physics* 82:3045.

Hawkins, J. (2004). *On Intelligence* (New York: Times Books).

Hodges A. (2014) *Alan Turing: The Engima* (New York: Vintage).

Hoel E. P., L. Albantakis, and G. Tononi (2014) "Quantifying causal emergence shows that macro can beat micro" *Proceedings of the National Academy of Science* 110:19790–19795.

Hoel E. P., L. Albantakis, and G. Tononi (2016) "Can the macro beat the micro? Integrated information across spatiotemporal scales" *Neuroscience of Consciousness* 1:1–13.

Hoffmann P. M. (2012) *Life's Ratchets: How Molecular Machines Extract Order from Chaos* (New York: Basic Books).

Kandel E. (1998) "A new intellectual framework for psychiatry" *American Journal of Psychiatry* 155:457–469.

Kandel E. (2006) *In Search of Memory: The Emergence of a New Science of Mind* (New York: W W Norton).

Kandel E. (2012) *The Age of Insight: The Quest to Understand the Unconscious in Art, Mind, and Brain, from Vienna 1900 to the Present* (New York: Random House).

Kandel E., J. H. Schwartz, T. M. Jessell, S. A. Siegelbaum, and A. J. Hudspeth (2013) *Principles of Neural Science* (New York: McGraw Hill Professional).

Laughlin R. B. (2000) "Fractional quantisation" *Reviews of Modern Physics* 71:863.

Leff H. S. and A. F. Rex (eds.) (1990) *Maxwell's Demon: Entropy, Information, Computing* (Bristol: Adam Hilger).

McLaughlin B. and K. Bennett (2011) "Supervenience." In E. N. Zalta (ed.), *The Stanford Encyclopedia of Philosophy*. Available at http://plato.stanford.edu/archives/win2011/entries/supervenience/

Nairz O., M. Arndt, and A. Zeilinger (2002) "Quantum interference experiments with large molecules" *American Journal of Physics* 71:319–325.

Noble D. (2012) "A theory of biological relativity: No privileged level of causation" *Interface Focus* 2:55–64.

Oerter R. (2006) *The Theory of Almost Everything: The Standard Model, the Unsung Triumph of Modern Physics* (New York: Plume).

Oyama S., P. E. Griffiths, and R. D. Gray (2003) *Cycles of Contingency: Developmental Systems and Evolution* (Cambridge, MA: MIT Press).

Oizumi, M., L. Albantakis and G. Tononi (2014) "From the phenomenology to the mechanisms of consciousness: Integrated information theory 3.0." *PLoS Computational Biology* 10(5):e1003588.

Pigliucci M. and G. B. Müller (2010) *Evolution: The Extended Synthesis* (Cambridge, MA: MIT Press).

Purves D. (2010) *Brains: How They Seem to Work* (Upper Saddle River, NJ: FT Press Science).

Rhoades R. and R. Pflanzer (1989) *Human Physiology* (Fort Worth: Saunders College Publishing).

Rolls E. T. and G. Deco (2010) *The Noisy Brain: Stochastic Dynamics as a Principle of Brain Function* (Oxford: Oxford University Press).

Rovelli C. (2016) "Meaning = Information + Evolution". arXiv:1611.02420v1.

Scott A. (1995) *Stairway to the Mind* (New York: Springer).

Simon H. A. (1996) *The Sciences of the Artificial* (Cambridge, MA: MIT Press).

Stone J. V. (2015) *Information Theory: A Tutorial Introduction* (Sheffield: Sebtel Press).

Strauss S. L. (2004) *The Linguistics, Neurology, and Politics of Phonics: Silent E Speaks Out* (Abingdon, UK: Routledge).

Tanenbaum, A. S. (2006) *Structured Computer Organisation* (Englewood Cliffs: Prentice Hall).

Tononi G. (2008) "Consciousness as integrated information: A provisional manifesto" *Biological Bulletin* 215:216–242. Available at www.jstor.org/stable/25470707

Tononi, G. (2015) "Integrated Information Theory" Scholarpedia. Available at www.scholarpedia.org/article/Integrated_Information_Theory*

Turing A. (1936) "On computable mumbers, with an application to the entscheidungsproblem" *Proceedings of the London Mathematical Society* 2(42):230–265.

van Gulick R. (2007) "Who's in charge here? And whose doing all the work." In Nancey Murphy and William R. Stoeger, SJ (eds.), *Evolution and Emergence: Systems, Organisms, Persons* (Oxford: Oxford University Press), p. 74.

Vicente A. (2006) "On the causal completeness of physics" *International Studies in the Philosophy of Science* 20:149–171.

Wagner A. (2014) *Arrival of the fittest* (Penguin: Random House).

Wolpert L. (2002) *Principles of Development* (Oxford: Oxford University Press).

31

A-MERGENCE OF BIOLOGICAL SYSTEMS

Raymond Noble and Denis Noble

Introduction

We argue that (1) emergent phenomena are real and important; (2) for many of these, causality in their development and maintenance is necessarily circular; (3) the circularity occurs between levels of organization; (4) although the forms of causation can be different at different levels, there is no privileged level of causation a priori: the forms and roles of causation are open to experimental investigation; (5) the upward and downward forms of causation do not occur in sequence, they occur in parallel (i.e. simultaneously); (6) there is therefore no privileged direction of emergence – the upper levels constrain the events at the lower levels just as much as the lower levels are necessary for those upper-level constraints to exist; and (7) to emphasize this point, we introduce the concept of a-mergence, which expresses the lack of causal directionality. We illustrate these points with a major test case: Schrödinger's distinction between physics and biology in which he proposed that physics is the generation of order from molecular disorder, while biology is the generation of order from molecular order. This characterization of biology is physically impossible. Modern biology has confirmed both that this is impossible and that, on the contrary, organisms harness stochasticity at low levels to generate their functionality. This example shows in fine detail why higher-level causality can, in many cases, be seen to be more important than lower-level processes. The chapter highlights a number of further examples where a-mergence seems to be a more appropriate way of describing what is happening than emergence.

Emergent phenomena are real and important

Biological reductionism can be seen to have originated with Descartes in the seventeenth century, while relying heavily on Newtonian mechanics later in the century, and in later centuries on the mathematical genius of Pierre-Simon Laplace. Descartes laid the foundation by arguing that animals could be regarded as machines in some way, comparable to the ingenious hydrostatic robots that had become popular among the aristocracy in their gardens. Newtonian mechanics cemented the foundation with the laws of mechanical motion, and Laplace systematized the ideas with his famous statement that a supreme intelligence could use mathematics to predict the future completely and retrodict the past as well. Everything that has or will happen would be clear to such a being. Descartes even foresaw one of the central ideas of Neo–Darwinism:

> If one had a proper knowledge of all the parts of the semen of some species of animal in particular, for example of man, one might be able to deduce the whole form and configuration of each of its members from this alone, by means of entirely mathematical and certain arguments, the complete figure and the conformation of its members.
>
> (On the Formation of the Fetus)[1]

which is essentially the idea that there is a complete mathematical "program" there in the semen, prefiguring Jacob and Monod's "genetic program". Complete, because he writes "from this alone". The causation, in this view, is entirely one way.

It is therefore significant that the first clear statement of the opposite view can be traced back to Descartes' main philosophical opponent. In 1665, just two years after the foundation of The Royal Society, Benedict de Spinoza, working in Holland, was in extensive correspondence with the first secretary of that society, Henry Oldenburg, working in London.

Oldenburg had just returned from meeting Spinoza in Holland and had been fascinated by his discussions with him on "the principles of the Cartesian and Baconian philosophies". Spinoza was opposed to the dualism of mind and body espoused by Descartes. This was necessary in Descartes' view of animals as automata, since he wished to exclude humans from this view and so attributed their free will to a separate substance, the soul, which could interact with the body. Spinoza was in the process of seeking to publish his great work (*The Ethics: Ethica ordine geometrico demonstrata*) in which he proposes an alternative philosophy. Spinoza did not publish in *Philosophical Transactions*, but this correspondence includes an important letter from Spinoza which could form a text for the systems approach and the concept of biological relativity (Noble 2012 2016). The original letter in Latin is still kept in the Royal Society library. He writes: "every part of nature agrees with the whole, and is associated with all other parts" and "by the association of parts, then, I merely mean that the laws or nature of one part adapt themselves to the laws or nature of another part, so as to cause the least possible inconsistency". He realized therefore some of the problems faced in trying to understand what, today, we would characterize as an open system. An open system is one that freely exchanges energy and matter with its surroundings. By definition, in a closed system each part must be influenced only by rules governing the behavior of the parts within it. If those parts behave deterministically, then the whole must also do so. But when parts of wholes are considered as sub-systems, they are necessarily open in the context of the whole. As we will explain Figure 31.1, even the equations used to describe the behavior of parts require initial and boundary conditions provided from outside the system. Thus, biological systems are open in relation to their environment.

Spinoza therefore appreciated the difficulty in working from knowledge of minute components to an understanding of the whole:

> Let us imagine, with your permission, a little worm, living in the blood, able to distinguish by sight the particles of blood, lymph etc, and to reflect on the manner in which each particle, on meeting with another particle, either is repulsed, or communicates a portion of its own motion. This little worm would live in the blood, in the same way as we live in a part of the universe, and would consider each particle of blood, not as a part, but as a whole. He would be unable to determine, how all the parts are modified by the general nature of blood, and are compelled by it to adapt themselves, so as to stand in a fixed relation to one another.[2]

This paragraph could stand even today as a succinct statement of one of the main ideas of biological relativity. He doesn't use a mathematical medium to express his idea, but this could be so expressed as the aim to understand how the initial and boundary conditions of a system constrain the parts to produce a particular solution to the differential equations describing their motions. We need then to move to the complete system (with whatever boundary we choose to use to define that) in order even to understand the behavior of the parts.

The essence of Spinoza's argument, to use modern language, is that organisms are open systems. This must be so since they can survive only by exchanging matter and energy with their environment. If an organism, or a part of an organism, is treated as a closed system by experimentally preventing those exchanges, it will die. The great majority of biochemical and molecular biological experiments are performed on dead and dying organisms, or their parts, such as cells and molecules. To understand how they operate as complete organisms, it is completely necessary to take into account the exchanges of matter and energy with their environment. It is through those interactions that organisms can be alive.

Causality in the development and maintenance of emergent processes is necessarily circular

Since the environmental influences arise from a higher level, circular causality must occur, downwards as well as upwards. "Down" and "up" here are metaphors and should be treated carefully. The essential point is the more neutral statement: there is no privileged scale of causality, which is the *a priori* principle of biological relativity. One of the consequences of the relativistic view is that genes, defined as DNA sequences, cease to be represented as active causes. They are templates and are passive causes, used when needed to make more proteins or RNAs. Active causation resides in the networks, which include many components for which there are no DNA templates. It is the interactive relationships of those dynamic networks which determine what happens. No single component or single mechanism can do so.

This view of organisms can be formalized mathematically as shown in Figure 31.1. Many models of biological systems consist of differential equations for the kinetics of each component. These equations cannot give a solution (the output) without setting the initial conditions

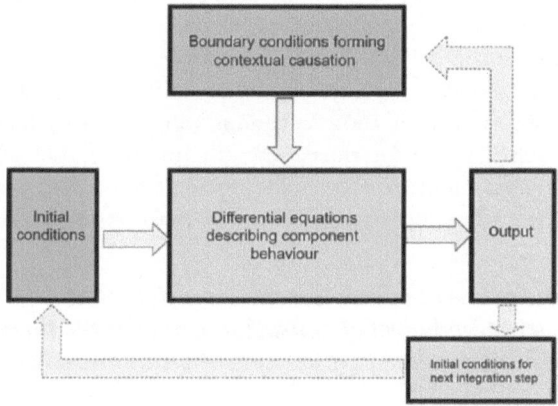

Figure 31.1 Diagram of causal sequences involved in integrating differential equation models. Description in text.

[From Noble (Noble 2012)]

(the state of the components at the time at which the simulation begins) and the boundary conditions. The boundary conditions define what constraints are imposed on the system by its environment and can therefore be considered a form of contextual causation from a higher scale. This diagram is highly simplified to represent what we actually solve mathematically. In reality, boundary conditions are also involved in determining initial conditions, and the output parameters can also influence the boundary conditions, while they in turn are also the initial conditions for a further period of integration of the equations. The arrows are not really unidirectional. The dotted arrows complete the diagram to show that the output contributes to the boundary conditions (although not uniquely) and determines the initial conditions for the next integration step.

Several important conclusions follow from this analysis. First, the equations used in modeling biology cannot even be solved if we do not specify the boundary and initial conditions. Second, those conditions necessarily require causal information about the environment of the system we are modeling. Third, this conclusion is true irrespective of whether we consider the world to be determinate. Even a Laplacian determinist would have to accept this. Recall that Spinoza also was a determinist. We can, of course, introduce stochasticity into the modeling to produce a non-determinate model. In fact, this is necessary to formulate the complete principles of biological relativity (Noble 2016, chapter 6), but this does not change the essential need for input from the environment of any open system.

The circularity occurs between levels of organization

Consider as a concrete example the regularity of the normal heartbeat and how it is disturbed in life-threatening arrhythmias. The normal heartbeat is an attractor caused by a circular form of causality in which both the cell potential and the individual proteins are entrained by their interaction. Once the rhythm begins, it can continue indefinitely. Even large perturbations in the individual proteins or their genes can be resisted (Noble 2011). This is precisely what is meant by an attractor. If you represent the parameters as a multidimensional space, there are large volumes within this space representing possible parameter sets, from which the system will automatically move towards the attractor.

Now consider what happens when a different kind of attractor is established. This happens in the heart when abnormal spiral waves of excitation arise at the level of the whole heart. The individual molecules in each cell are now constrained to dance to a different and more chaotic rhythm. Viewed from the level of the individual molecules, both of these influences from the higher levels of the cell or the whole organ will seem inexplicable. The molecules are like boats tossed around in a storm beyond their own control. Yet the storm also depends on their activity. Indeed, it can be modeled using the equations for that activity (Carro et al. 2011). But as explained in the previous section, those equations will necessarily represent the circularity between the causal levels. Each of the three views, the molecular, the cellular and the organ, are valid, but only from the higher levels can we provide a full account of what is happening, including the lower levels whose behavior is in need of explanation.

Although the forms of causation can be different at different levels, there is no privileged level of causation a priori: the forms and roles of causation are open to experimental investigation

The principle that there is no privileged scale of causality can easily be misinterpreted. It is important now to introduce some clarifications.

First, we must distinguish between its conceptual status and its practical implications. It is an *a priori* statement, that is, a statement about what we should or should not assume in advance of doing the experiments. We should not assume that causation *necessarily* resides at a particular (e.g. molecular) scale. That is the mistake made by naïve reductionism in biology. The reduction to molecular-level events is treated as a methodological necessity, whereas it should emerge, if at all, from the experiments themselves. Before we do those experiments, we cannot know which parts of a system are involved in its behavior, nor attribute any privileged position to them.

But that does not mean that all scales must be involved in any given example. The circles of causal networks may span particular ranges of scales, which may be more or less limited in extent. And there may be particular levels that act as important hubs. Those facts are for us to discover as empirical observations. For example, many biologists regard the cell as a central level of integration in much of biology. That conclusion is a result of extensive experiments on cells showing their functional integrity and that many physiological functions cannot be ascribed to entities lower than the cell. Cells contain the main metabolic networks, circadian and various other rhythm networks, cell cycle networks and so on. Moreover the great majority of living organisms are single cells.

The genome also has a unique position, but it is not the one most often ascribed to it as a program dictating life. As the American cell biochemist Franklin Harold puts it in his book *In Search of Cell History*, "The genome is not the cell's command center but a highly privileged databank, something like a recipe or a musical score, yet for the purpose of parsing evolution, genes have a rightful claim to center stage" (Harold 2014). Parsing is the analysis of strings of symbols, usually with guidance from some rules of grammar. In the case of DNA, the start and stop sequences and those for binding transcription factors, among other features, provide those guidelines. Analysis of this kind has indeed been exceptionally useful in the interspecies DNA sequence comparisons that now form the basis of much of our understanding of evolutionary history.

The genome sequences are therefore comparable to a formal cause[3] in Aristotle's classification of the forms of causation, while the causation from the networks operating at higher levels than the genome can be regarded as an efficient cause (Noble 2016, pp. 176–181). The sequences are a formal cause since they form templates to enable ribosomes to construct the proteins specified by those sequences, while those proteins form part of the dynamic networks that form the efficient cause[4] necessary for the attractor to exist. This distinction is particularly clear in the example of cardiac rhythm discussed earlier. The attractor doesn't even require the involvement of DNA or RNA sequences until the cell requires more proteins to be made. Some other rhythmic attractors do involve DNA sequences in the cycle. Circadian rhythm is a good example. One form of the attractor includes feedback from the level of the protein involved to inhibit the formation of the protein (Hardin et al. 1990; Foster and Kreitzman 2004). But even in this case, the genome is not completely necessary. So-called "clock" genes in mice can be knocked out without affecting circadian rhythm (Debruyne et al. 2006).

Each feature of organisms at the various levels may therefore have unique causal properties. The principle of biological relativity should not be taken to require that all forms of causation involved are equivalent.

The upward and downward forms of causation do not occur in sequence; they occur in parallel (i.e. simultaneously)

It is important to understand that the processes represented in Figure 31.1 all occur as a process. It is merely a convenience of representation that the integration step is represented as coming after setting the initial conditions, which then precedes the formation of the output. In a computer program representing the sequence, we do indeed write the code in precisely this sequence. But this is a mathematical fiction arising from the fact that we solve the equations in finite steps.

Differential equations themselves do not express finite steps. On the contrary, the differential symbol "d/dt" represents a vanishingly small step. In reality also, all the processes represented in the equations proceed simultaneously. Our "difference" equations actually solved by the computer are simply approximations. The test we use for whether they are accurate enough is precisely to reduce the integral step length until the solution converges to an arbitrarily high degree of accuracy. In principle, for infinitely good accuracy, we would have to reduce the step length to zero, which is exactly what differential equations themselves represent. In this respect the equations are better representations of what we are modeling than any particular computer simulation. In the rare cases in nonlinear differential equation models where we can solve the equations analytically (Hunter et al. 1975; Jack et al. 1975), the solution is revealed as a complete solution as a function of time, with no sequence of causation. This is an important reason for which we will introduce the concept of *a*-mergence at the end of this chapter.

There is therefore no privileged direction of emergence; the upper levels constrain the events at the lower levels just as much as the lower levels are necessary for those upper-level constraints to exist

It follows that it is simply a matter of convenience that we often talk of the higher-level functions arising *from* the interactions of the components. It would be just as correct to say that the constraints on the lower-level components arise *from* the existence of the higher-level function. Best of all, we should conclude that they necessarily co-arise.

To emphasize this point, we introduce the concept of a-mergence, which expresses the lack of causal directionality

We have developed our argument with examples from cardiac and circadian rhythms. We will now illustrate all these points with a central test case: Schrödinger's distinction between physics and biology in which he proposed that physics is the generation of order from molecular disorder, while biology is the generation of order from molecular order (Schrödinger 1944). This is a central test case because, as both Watson and Crick acknowledged, the formulation of the central dogma of molecular biology was greatly influenced by Schrödinger's ideas. It is also hard to think of a more concrete example where the directionality of causation is widely accepted to be one way. The central dogma has been interpreted to mean that the genome sequences cause the organism but are not themselves affected by the organism. This view has been taken to deny the existence of emergent properties, and it is implicit in versions of evolutionary theory that equate the central dogma to the Weismann barrier or, at least, claim that the Weismann Barrier is now "embodied by" the central dogma. We develop this final section of our argument in four stages.

It is a mistake to interpret Crick's statement to mean one-way causation

The relevant statement is:

> The central dogma of molecular biology deals with the detailed residue-by-residue transfer of sequential information. It states that *such information* cannot be transferred back *from protein* to either protein or nucleic acid.
>
> *(Crick 1970, p. 561)*

We have italicized *"such information"* and *"from protein"* since it is evident that the statement does not say that *no* information can pass from the *organism* to the genome. In fact, it is obvious that it must do so to produce many different patterns of gene expression, which enable many different phenotypes (e.g. many different cell types in the same body) to be generated from the same genome.

This information from organisms is conveyed to their genomes by patterns of transcription factors, genome marking, histone marking and many RNAs, which in turn control the patterns of gene expression. These controls are exerted through preferential targeted binding to the genome or histone proteins. For example, methylation of cytosines preferentially occurs at CpG sites. Binding to histones preferentially occurs at the histone tails. Even though these are the targeted molecular mechanisms by which the functional control is exerted, there is no guarantee that the functionality will be evident at the molecular level. It would require many correlations between the *patterns* of binding and the functional processes at a higher level to identify the functionality involved. Without that correlation, the binding patterns will appear random.[5] Yet it is those patterns that control the expression of individual genes. Those patterns are phenotypes, not themselves genotypes. A good example comes from the study of the evolution of hemoglobins in many avian species to adapt to altitude. Natarajan et al. show that "predictable convergence in hemoglobin function has unpredictable molecular underpinnings" (Natarajan et al. 2016).

That first point establishes that the same genotype can be used to create an effectively unlimited number of phenotypes (Feytmans et al. 2005). That demonstration does not, however, exhaust the role of the phenotype in determining the functioning of the genome. Not only is it true that the same genotype can be used to generate many different phenotypes, *it is the phenotype that enables it to be the same (or even a different) genotype.*

It is the phenotype that enables the genome to be the same genotype

If correct, this statement would completely reverse the direction of causality assumed in reductionist explanations of living organisms. How, then, was the currently accepted one-way genotype − › phenotype explanation ever thought to be correct? The answer lies in Schrödinger's book *What Is Life?* (Schrödinger 1944). That book makes one spectacularly correct prediction and a second necessarily incorrect prediction. The correct prediction was that the genetic material would be found to be what he called a nonperiodic crystal. Remember that this was in 1942 before it had been shown that genetic information is found in DNA sequences. If one thinks of a linear polymer as a crystal that does not endlessly repeat itself, then nonperiodic (or a-periodic) crystal is quite a good description of what molecular biology subsequently discovered. Remember too that the book was written at a time when x-ray crystallography had come into use to "read" the molecular structure of organic molecules. This enabled Dorothy Hodgkin to determine the structure of cholesterol in 1937, penicillin in 1946 and vitamin B_{12} in 1956. These were spectacular achievements. What was more natural than to conclude that if the genetic material is a form of crystal, it could also be "read" in a determinate way? That was indeed the conclusion Schrödinger drew in his book.

But he was too good a physicist not to notice, initially at least, that this conclusion was "ridiculous":

> We seem to arrive at the ridiculous conclusion that the clue to understanding of life is that it is based on a pure mechanism, a "clock-work".
>
> *(1944, p. 82)*

"Ridiculous", because how could biological molecules not show the extensive stochasticity that would arise from their possession of kinetic energy?[6] That was precisely why he had, earlier in his book, concluded that physics was the generation of order, for example, the laws of thermo-dynamics, from disorder, that is, molecular-level stochasticity.

But he had difficulty harmonizing the two insights. Confusingly, he wrote:

> The conclusion is not ridiculous and is, in my opinion, not entirely wrong, but it has to be taken "with a very big grain of salt".

(1944, p. 82)

He then explains the "big grain of salt" by stating that even clockwork is, "after all statistical" (p. 103).

Schrödinger clearly realized that something is far from right but was struggling to identify what it might be.

It is the phenotype that enabled the genome to be a different genotype

We would now say that the molecules involved (DNA) *are* subject to frequent statistical varia-tions (copying errors, chemical and radiation damage, etc.), which are then corrected by the cell's protein and lipid machinery that enables DNA to become a highly reproducible molecule. This is a three-stage process that reduces the copy error rate from 1 in 10^4 to around 1 in 10^{10}, which is an astonishing degree of accuracy. In a genome of 3 billion base pairs, this works out as fewer than one error in copying a complete genome, compared to millions of errors without error correction. The order at the molecular scale is therefore actually created *by the system as a whole*, including lipid components that are not encoded by DNA sequences. This requires energy, of course, which Schrödinger called negative entropy. Perhaps therefore this is what Schrödinger was struggling towards, but we can only see this clearly in retrospect. He could not have known how much the genetic molecular material experiences stochasticity and is constrained to be highly reproducible *by the organism itself.* The order at the molecular (DNA) level is actually imposed by higher-level constraints. If we ever do synthesize from scratch the complete genome of a living organism, we would have to give it this cellular environment in which to function accurately. Otherwise, any genome longer than about 10,000 bases would fail to be preserved reliably at the first copying process.

The central dogma was originally formulated by Crick in 1958 (Crick 1958) in a very hard form: DNA −› RNA −› protein. This formulation would have completely protected the genome from alteration of its sequence by the organism. No changes in proteins or their relative expression patterns could conceivably have altered a genome that was isolated in such a way. By 1970, how-ever, the dogma had to be modified after the discovery of reverse transcriptase (Temin and Mizu-tani 1970) to become DNA ‹−› RNA −› protein, and even to become as shown in Figure 31.2.

Reverse transcription enables the white upward arrow to occur to allow RNAs to be back-transcribed into DNA, while the upper curved arrow enables DNA sequences to be pasted directly into the genome. The two together completely counter the central dogma since they enable sequences of any length to be moved around the genome, either directly as DNA or indirectly via RNA. Way back in the 1930s and 1940s Barbara McClintock had shown that such transfers do occur in plants in response to environmental stress. This was why, on winning the Nobel Prize for mobile genetic elements in 1983, she referred to the genome as a

> highly sensitive organ of the cell, monitoring genomic activities and correcting com-mon errors, sensing the unusual and unexpected events, and responding to them, often by restructuring the genome.

(McClintock 1984, p. 800)

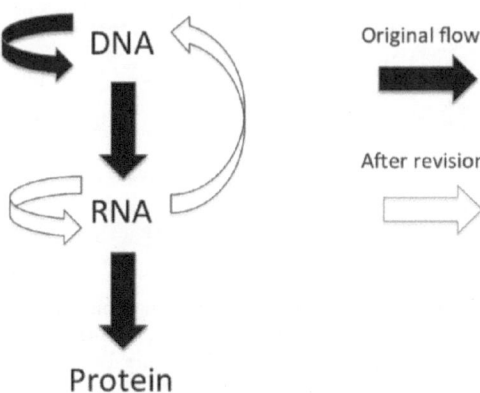

Figure 31.2   The central dogma of molecular biology

Did this happen in the evolution of genomes? The answer is yes – it must have. The evidence comes from the comparative sequencing of genomes from many different species, ranging from yeast to man, reported in the 2001 *Nature* report on the first full draft of the human genome sequence (International Human Genome Mapping Consortium 2001). The gene sequences for both transcription factor proteins and chromatins show precisely this kind of massive genome reorganization (Shapiro 2014).

This process has also been recorded in real-time experiments performed on bacteria evolving in a nutrient medium that does not provide what was an essential metabolite. By periodically allowing conjugation with bacteria that metabolize the new chemical and gradually removing the usual essential metabolite, the bacteria succeeded in weaning themselves completely off their usual nutrient. Sequence analysis showed that conjugation had allowed the shuffling of gene domains during the periodic conjugations. Significantly, the authors entitle their article with reference to "directed evolution" (Crameri et al. 1998). In a recent article we have shown why this kind of process should be characterized as "directed" since it arises from circular causation that represents a form of organism intelligence. Hosseini et al. (2016) have confirmed such findings, which they characterize as "phenotypic innovation through recombination".

It is the phenotype that enables the genome to be a different genotype

Notice the small difference in tense compared to statement (c). In this section we ask whether the phenotype can be observed to alter the genome in real-time observations on the evolution of cells and organisms.

It is in fact well known already that cells can harness stochasticity to generate specific functions, since cells of the immune system show the phenomenon of highly targeted somatic hypermutation. Figure 31.3 summarizes what we know. Faced with a new antigen challenge, the mutation rate in the variable part of the genome can be accelerated by as much as 1 million times. So far as we know, the mutations occur randomly. But the location in the genome is certainly not random. The functionality in this case lies in the precise targeting of the relevant part of the genome. The mechanism is directed, because the binding of the antigen to the antibody itself activates the proliferation process.

This example from the immune system shows that functionally significant targeted hypermutation can occur in the lifetime of an individual organism. There is no reason why this kind of mechanism should not be used in evolutionary change, and it is.

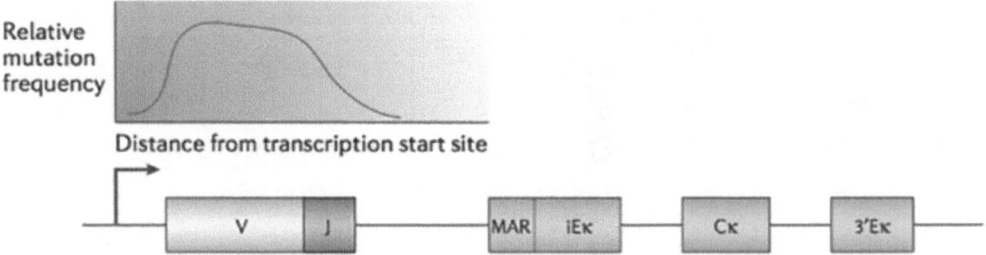

Figure 31.3 Schematic diagram of gene-specific targeted hypermutation in immunoglobulin gene loci. The mutation rate is greatly increased only in the variable part of the genome, which is a ~1.5-kilobase region in each of the three immunoglobulin loci. In this figure, the graph above the rearranged variable (V) and joining (J) gene segments that form the variable region of Igκ depicts the mutation domain in the κ-light chain (Igκ) locus. 3′Eκ, Igκ 3′ enhancer; Cκ, Igκ constant; iEκ, Igκ intronic enhancer; MAR, matrix attachment region

[From Odegard and Schatz 2006]

A well-known functionally driven form of genome change is the response to starvation in bacteria. Starvation can increase the targeted reorganizations of the genome by five orders of magnitude (i.e. by a factor of over 100,000). This is one of the mechanisms by which bacteria can evolve very rapidly and in a functional way in response to environmental stress. It would be important to determine whether such targeted reorganization occurs in experiments on conjugating bacteria discussed earlier. The question would be whether bacteria sensing deprivation trigger higher frequencies of conjugation and shuffling of domains. This "sensing" of the environment, as in the immune system, is precisely what constitutes the feedback necessary for the process to be characterized as directed (see Noble and Noble 2017 for the relevant definitions of agency and directionality in the evolution of organisms and their populations).

A similar targeting of location where genomic change can occur has been found in experiments on genetically modified fruit flies. One of the common ways in which genetic modification is achieved is to use a particular kind of mobile genetic element that can move around the genome using a cut-and-paste mechanism that does not require an RNA intermediate. Most often the insertions occur in a random way. But when DNA sequences from certain regulatory regions are used, they get inserted preferentially near the gene from which the sequence was derived (Bender and Hudson 2000). This process targets the changes in a way that is clearly not random with respect to possible function.

There are many more examples in bacterial evolution (Noble 2017).

Conclusions

It will be evident that we have a concern about the word "emergence". The "e" naturally leads to us asking "what emerges *from* what?" Our position makes it obvious that this is often the wrong question. Of course, there may be "emergence" in the sense that, *on a temporal time scale*, atoms emerged from fundamental particles, stars emerged from condensations of matter, life emerged from the formation of suitable planets and so on. But at each stage a new level of organization takes over. Once an attractor has formed, the components are constrained *by the attractor itself.* The direction of causality then changes. The term "a-mergence" tries to make that clear. There is no privileged level of causality in the sense that all levels can be causal. But it is clear from what we have written that the nature of causality changes with the level and that the higher levels can be said to be directed functionally, and in that sense they are privileged.

Moreover, we doubt whether any directionality of causation should be assumed, whether sequential or in parallel. Once an attractor has formed, the best description would be to say that this condition of the system is followed by that condition. There is no need to isolate any components, at any level, as the primary cause. The condition is the a-mergent state, and this condition is causative. Moreover, the organization of the state is precisely what defines the level at which it can be said to occur. Thus, we refer to atomic, molecular, cellular, tissue, organ, organism, niche, habitat, etc., each with a dynamic of causative, functional (goal-directed) organization.

Does the concept of goal-directed processes lead to a better understanding? It is difficult to understand causality without knowing this logic. We understand "the function" of a thermostat better by understanding that it operates *to* maintain temperature within a certain range – it is the logic of the thermostatic system. We understand better the function of baroreceptors in the circulatory system by knowing they are part of a system to maintain pressure within a given range and facilitate blood flow round the body – it is the logic of the system. Knowing or understanding the logic of a system is as important as any reductionist detail about the system. It is an organizational logic – no one part of the system has primacy in that logic. How each part behaves is influenced by its arrangement within the system. It is the situational logic of the system.

The thoughts and discussion we have had in writing this chapter are part of the dynamic organization at a social level, where action can be identified not only as purposive but also intentional. It is only at this level that behavior can be described, for example, as "selfish" or "altruistic", as only at this level can there be reasoned choice, and reasoned logic, or where we can talk of motivation and emotion, hopes, desires, fears and anxieties.

Science by method will hold variables constant to study the effect of changing a given one – clamping a voltage, for example. What we know is that this is artificial and establishes an artificially fixed sequence of events. The a-mergent state is in continuous flux, but biological processes maintain such states within a functional range. Life as an a-mergent state is in this sense autopoietic, or self-maintaining (Maturana and Varela 1980; Luisi 2016).

The conclusions we have drawn in this chapter are firmly based on experimental findings on the ways in which organisms harness stochasticity to generate functionality. Our reinterpretation of Schrödinger's ideas to produce a conclusion diametrically opposite to the one he himself drew, and which has dominated biological science ever since, is clearly based on experimental findings on the mechanism of reproducibility of DNA copying in whole cells, which could not occur without the integrative activity of cells as a whole. And the cardiac rhythm and targeted hypermutation examples are also firmly based on experimental findings, requiring the integrative action of whole cells.

Our conclusions also strongly support the philosophical approaches developed by, for example, Nancy Cartwright and John Dupré. In his book *The Disorder of Things* ((Dupré 1993), p 101) Dupré writes "causal completeness at one particular level is wholly incredible. By contrast with even the weakest versions of reductionism, the pluralism I have in mind precludes the privileging of any particular level". This statement accurately reflects the metaphysical position adopted in our work and its empirical basis. Dupré's work focuses on biology. Cartwright (Cartwright 1999, p. 31) comes to very similar conclusions in her study of causality in physics and economics: "nature is governed in different domains by different systems of laws not necessarily related to each other in any systematic or uniform way; by a patchwork of laws". This nicely expresses our analysis that the constraints exerted by higher-level systems on lower-level components depend on the rules being followed by the system, not the highly stochastic behavior of the molecular components.

Modern philosophers of science arrived at these conclusions at least twenty years ago on the basis of careful recognition of the significance of experimental work already achieved at that time. But the silo-ization of disciplines has meant that there has been very little cross-fertilization back from these philosophical works and the scientific community. The veritable flood of experimental work now appearing (Noble 2017) that makes the conclusions even more convincing may, we hope, now have its impact in the strategy of experimental biological science. It is high time to escape the limited metaphysical straightjackets of purely gene-centric interpretations.

Acknowledgements

Some parts of the text have been developed from our previous articles and books. We thank Nancy Cartwright and John Dupré for valuable criticisms.

Notes

1   The French text reads "Si on connoissoit quelles sont toutes les parties de la semence de quelque espece d'Animal en particulier, par exemple de l'homme, on pourroit déduire de la seul, par des raisons entierement Mathematiques et certaines, toute la figure & conformation de ses membres"; (de la formation du fœtus, para LXVI p. 146).
2   The Latin text is "Concipiamus iam, si placet, vermiculum in hoc fluido, nempe in sanguine, vivere, qui visu ad discernendas particulas lymphae et chyli etc. valeret, et ratione ad observandum, quomodo unaquaequae particula ex alterius occursu vel resilit, vel partem sui motus alteri communicat. Ille quidem in sanguine, ut nos in hoc universi parte viveret, et unamquamque sanguinis particulam ut totum non vero ut partem consideraret nec scire posset, quomodo partes omnes ab universali natura sanguinis moderantur, et invicem prout universalis natura sanguinis exigit accomodari coguntur ut certa ratione inter se consentiant".
3   A formal cause exists when it is the geometrical arrangement of something that influences the outcome. It is the formal arrangement of nucleotides in sequences that gives the genome the power to determine amino acid sequences in proteins and nucleotide sequences in RNAs.
4   An efficient cause exists when it is the motion of something that affects the outcome, as in billiard balls colliding. The billiard balls may all have the same form, but their movements are different.
5   The stochasticity is therefore epistemological. In principle, once the higher-level constraints are known, a bottom-up computation becomes conceivable. However, given the effectively unlimited combinations and the associated requirement for unlimited computer power, it is extremely unlikely that such computations could be successful. And they almost certainly could not be performed without the insights provided by the higher-level functional information. As we emphasized earlier, the initial and boundary conditions are essential for the computation to be performed.
6   The only way known to modern physics would be for the molecules to form a Bose-Einstein condensate. But molecules can only do this at extremely low temperatures near absolute zero (Whitfield 2003). Nevertheless, some biologists have speculated along these lines (Ho 2008). Whether or not this happens, it is not needed as an explanation, since we already know that the stochasticity is present, even in copying DNA.

References

Bender, W. and A. Hudson (2000). "P element homing to the Drosophila bithorax complex." *Development* 127: 3981–3992.

Carro, J., J. Rodríguez, P. Laguna and E. Pueyo (2011). "A human ventricular cell model for investigation of cardiac arrhythmias under hyperkalaemic conditions." *Philosophical Transactions of the Royal Society A: Mathematical, Physical and Engineering Sciences* 369(1954): 4205–4232.

Cartwright, N. (1999). *The Dappled World: A Study of the Boundaries of Science.* Cambridge, Cambridge University Press.

Crameri, A., Raillard, S.-A., Bermudez, E., Stemmer, W. P. C. (1998). DNA shuffling of a family of genes from diverse species accelerates directed evolution. *Nature* 391: 288–291.

Crick, F. H. C. (1958). "On protein synthesis." *Symposia of the Society for Experimental Biology* 12: 138–163.

Crick, F. H. C. (1970). "Central dogma of molecular biology." *Nature* 227: 561–563.

Debruyne, J. P., E. Noton, C. M. Lambert, E. S. Maywood, D. R. Weaver and S. M. Reppert (2006). "A clock shock: Mouse clock is not required for circadian oscillator function." *Neuron* 50(3): 465–477.

Dupré, J. (1993). *The Disorder of Things.* Cambridge, MA, Harvard University Press.

Feytmans, E., D. Noble and M. Peitsch (2005). "Genome size and numbers of biological functions." *Transactions on Computational Systems Biology* 1: 44–49.

Foster, R. and L. Kreitzman (2004). *Rhythms of Life.* London, Profile Books.

Hardin, P. E., J. C. Hall and M. Rosbash (1990). "Feedback of the *Drosophila* period gene product on circadian cycling of its messenger RNA levels." *Nature* 343: 536–540.

Harold, F. (2014). *In Search of Cell History.* Chicago, University of Chicago Press.

Ho, M.-W. (2008). *The Rainbow and the Worm: The Physics of Organisms.* London, World Scientific Publishing.

Hosseini, S.-R., Martin, O. C. and Wagner, A. (2016). Phenotypic innovation through recombination in genome-scale metabolic networks. *Proceedings of the Royal Society B 283.* doi:10.1098/rspb.2016.1536.

Hunter, P. J., P. A. McNaughton and D. Noble (1975). "Analytical models of propagation in excitable cells." *Progress in Biophysics and Molecular Biology* 30: 99–144.

International Human Genome Mapping Consortium (2001). "A physical map of the human genome." *Nature* 409: 934–941.

Jack, J. J. B., D. Noble and R. W. Tsien (1975). *Electric Current Flow in Excitable Cells.* Oxford, Oxford University Press.

Luisi, P. L. (2016). *The Emergence of Life: From Chemical Origins to Synthetic Biology.* Cambridge, Cambridge University Press.

Maturana, H. and F. Varela (1980). *Autopoiesis and Cognition.* Dordrecht, Reidl.

McClintock, B. (1984). "The significance of responses of the genome to challenge." *Science* 226: 792–801.

Natarajan, C., F. G. Hoffmann, R. E. Weber, A. Fago, C. C. Witt and J. F. Storz (2016). "Predictable convergence in hemoglobin function has unpredictable molecular underpinnings." *Science* 354: 336–339.

Noble, D. (2011). "Differential and integral views of genetics in computational systems biology." *Interface Focus* 1: 7–15.

Noble, D. (2012). "A theory of biological relativity: No privileged level of causation." *Interface Focus* 2: 55–64.

Noble, D. (2016). *Dance to the Tune of Life: Biological Relativity.* Cambridge, Cambridge University Press.

Noble, D. (2017). "Evolution viewed from physics, physiology and medicine." *Interface Focus,* 7: 20160159.

Odegard, V. H. and D. G. Schatz (2006). "Targeting of somatic hypermutation." *Nature Reviews Immunology* 8: 573–583.

Schrödinger, E. (1944). *What Is Life?* Cambridge, Cambridge University Press.

Shapiro, J. A. (2011). *Evolution: A View from the 21st Century.* Upper Saddle River, NJ, Pearson Education Inc.

Temin, H. M. and S. Mizutani (1970). "RNA-dependent DNA polymerase in virions of Rous sarcoma virus." *Nature* 226: 1211–1213.

Whitfield, J. (2003). "Molecules form new state of matter." *Nature* doi:10.1038/news031110-16.

32

EMERGENCE IN THE SOCIAL SCIENCES

Julie Zahle and Tuukka Kaidesoja

Introduction

In the social sciences, discussions of emergence mainly focus on social phenomena as they emerge from individuals. Social phenomena are commonly taken to be exemplified by universities, states, traffic jams, wealth distributions, declarations of war, firms' firing of employees and norms. Social scientists who invoke the notion of emergence typically maintain that social phenomena are emergent insofar as they arise from individuals and possess certain additional features such as being novel, irreducible, unexplainable and unpredictable relative to individuals. Among social scientists, there is no consensus as to which features should be regarded as the additional features constitutive of emergence. Moreover, the same features are sometimes characterized in divergent ways. Accordingly, diverse notions of emergence are being advocated in the social sciences.

Émile Durkheim's work from around the turn of the 19th century is often regarded as containing one of the earliest social scientific discussions of emergence, even though he does not use the term "emergence" (Sawyer 2005: 100). In the following decades, the notion of emergence is rather sporadically brought up. For instance, it is cursorily mentioned by social scientists such as Talcott Parsons, George Homans and Peter Blau (Parsons 1968[1937]; Homans 1950; Blau 1964). It is not until the latter part of the 20th century that the idea of emergence begins to receive more sustained attention in social theorizing. During this period, a number of influential approaches that appeal to emergence came into being, including the school of critical realism (see Bhaskar 1998[1979]), systems theory à la Niklas Luhmann (see Luhmann 1995[1984]) and agent-based computational modeling (see Epstein and Axtell 1996). Today, the idea of emergence continues to be explored and debated in the social sciences. One way in which to characterize current discussions is to note that some social scientists defend an epistemic notion of emergence, that is, they view emergence as a feature that social phenomena have relative to our limited knowledge of them, whereas others opt for an ontological notion of emergence, that is, they regard emergence as a feature of social phenomena independently of our knowledge about them. Ontological notions of emergence tend to be invoked in the context of discussions about how social phenomena should be explained and whether properties of social entities can somehow be reduced to properties of individuals. It is ontological notions, rather than epistemic ones, that have received the most attention in recent social scientific debates.

In this chapter, we examine the theories of ontological emergence advanced by two proponents of critical realism: Roy Bhaskar and Dave Elder-Vass. The school of critical realism

400

is arguably the social scientific approach in which the notion of emergence has played – and continues to play – the most central role. Moreover, it is currently a highly influential social scientific movement with many followers, its own journal, viz. *Journal of Critical Realism*, yearly conferences and so on. Roy Bhaskar's two seminal works, *A Realist Theory of Science* from 1975 and *The Possibility of Naturalism* from 1979, laid the foundation for the movement (Bhaskar 1978[1975], 1998[1979]). Subsequently, his views have been further elaborated by critical realists such as Margaret Archer, Andrew Collier, Dave Elder-Vass, Tony Lawson, Douglas Porpora and Andrew Sayer (see, e.g., Archer 1995; Collier 1989; Elder-Vass 2010; Lawson 1997; Porpora 2015; Sayer 2010[1984]). Among these, it is Elder-Vass who has offered the most systematic exposition and elaboration of the notion of emergence. We begin by considering Bhaskar's account of emergence and then move on to consider Elder-Vass' position.

Bhaskar on emergence

Bhaskar (1978, 1998) develops his views on causal powers, emergence and social ontology in his first two books that established the critical realist movement in the social sciences. In these works, he construes the notion of emergence in synchronic terms, that is, he takes emergent phenomena to *co-occur* with the more basic phenomena from which they emerge. When presenting his philosophy of the social sciences, Bhaskar discusses emergence in relation to social structures (a subset of social phenomena). However, he does not provide a detailed analysis of what he means by social structures and emergence in this context. For this reason, it is necessary to do some reconstructive work to describe his views on these matters. Bhaskar introduces the notion of emergence in his philosophy of the natural sciences and then discusses it in the context of his philosophies of mind and social sciences. Similarly, we begin with a brief discussion of Bhaskar's account of emergence as applied to the entities studied in the natural sciences before addressing his views on emergence with respect to social structures.

Bhaskar (1978) argues that the natural sciences as we know them are possible only if nature consists of entities (or things) with causal powers whose interactions generate the observable patterns of events. Apart from ultimate entities that have no structure (if there are any), Bhaskar (e.g. ibid: 51, 88) thinks that causal powers are emergent properties of complex entities, such as chemical substances and living organisms, and that complex entities possess these powers in virtue of their intrinsic structures (or essential natures). Further, he holds that the emergent powers of an entity are possessed by it conceived of as a whole and make the entity "capable of acting back on the materials out of which" it was formed (ibid: 114). Causal powers of this kind can be characterized as dispositional properties of entities to generate specific types of effects (or engage in specific types of activities) in suitable conditions.

Bhaskar (ibid: 181) maintains, too, that, on the condition that an emergent power of an entity and its parts have been identified and described, a reductive explanation (or explanatory reduction) of the emergent power may be provided. A reductive explanation specifies how the emergent power of an entity synchronically depends on, or arises from, the interrelated parts of the entity. Bhaskar emphasizes that reductive explanations do not eliminate emergent powers from the scientific ontology because they leave "the reality of higher order entities intact" (ibid). For example, the emergent powers of liquid water to put out fires and to dissolve sodium chloride (i.e. table salt) can be explained by reference to its molecular structure within a certain range of temperature and air pressure. The emergent powers of liquid water remain intact in spite of the fact that chemists have provided reductive explanations of them. Hence, this view of emergent powers of natural entities is compatible with these powers being reductively explainable.

In his works on the philosophy of the social sciences, Bhaskar (e.g. 1986, 1998) stresses the differences between natural entities and social structures. For our purposes, it is enough to mention that, according to Bhaskar (1998: 38–39, 174–175), there are two features that differentiate social structures from (most) natural entities: (i) social structures are ontologically dependent on the activities of human individuals (considered as self-conscious persons) and their (true or false) conceptions of these activities and that (ii) social structures can be geographically quite specific and are prone to historical transformations that may sometimes take place quite fast. Despite their differences from natural entities, Bhaskar thinks that social structures have emergent powers relative to individuals.

How should we then understand Bhaskar's notion of social structures? Though he does not clearly articulate the sense in which he uses the notion, we think that he most often construes social structures as consisting of the internally related social positions that are relatively enduring and distinct from the individuals occupying these positions and their activities. The idea is that internally related social positions, such as teacher and pupil, capitalist and worker or husband and wife, can only be understood in relation to each other. According to Bhaskar (ibid: 41–43), this entails that the social positions are constituted by the internal relations that connect them, meaning that these positions would not be what they essentially are unless they were related to each other in the particular way they are. In addition to internal social relations of this kind, there are external social relations, such as the relation "between two cyclists crossing on a hilltop" (ibid: 42). The latter are relations that do not define social positions.

In Bhaskar's view, social structures have emergent powers relative to those individuals whose social actions and interactions they enable, motivate and constrain. He argues that individuals occupy positions in social structures that are autonomous and (at least typically) precede them and that they reproduce or transform these structures by their activities (ibid: 33–35, 40–41). In addition, he suggests that individuals mostly reproduce social structures over time via the unintended consequences of their intentional actions (ibid: 35, 39). For example, the positions of teacher and pupil precede each new generation of teachers and pupils who enter these positions and whose reasons for engaging in various activities and practices in school classes usually have quite little to do with the reproduction of the social structure governing their school work. Bhaskar (e.g. ibid: 25–26, 36, 40–42) ascribes emergent powers to social structures of this kind because he thinks that they (i) affect the available resources and opportunities of individuals occupying the positions in these structures and (ii) define the rights, duties, tasks and informal norms that these individuals have to take into account in their social activities.

These reflections can be illustrated by considering a capitalist entrepreneur whose firm designs and manufactures products, such as clothes or technological devices. It can be argued that the entrepreneur is able to make profit on the condition that her business uses the latest production technologies. Further, she must avoid paying too high wages to her workers because otherwise she will not succeed in the competition with other firms in the market. In her entrepreneurial activities, she also has to take into account the relevant legislation about ownership rights, minimum wages and working conditions in order to avoid legal sanctions. According to Bhaskar (e.g. ibid: 42–44, 51), opportunities and constraints of these kinds are due to the capitalist relations of production that constitute the social positions of the capitalist and worker. Though he grants that the effects of the emergent powers of capitalist relations of production are always mediated through the actions of individuals, Bhaskar (ibid: 33–37, 49–54) tends to hold that social structures of this kind should be strictly separated from the activities of individuals and concrete social groups. This view is supported by his claims that structures (typically at least) precede individuals occupying positions in them and that structures may continue to exist once particular individuals in the relevant positions have been replaced by others. Bhaskar nevertheless believes that, at the large scale, the capitalist relations of production are reproduced, among other things, by people's daily activities as capitalists and workers even though none of them is necessarily intending to accomplish this with their actions. Furthermore, a class of people occupying the social position of worker (or capitalist) in a capitalist society may also unite and seek to transform the relations of production according to their vested interests.

Bhaskar (ibid) provides his views on emergent powers of social structures as contributions to the agency–structure and individualism–holism debates in the social sciences. The agency–structure debate concerns the extent to which individuals and social structures, respectively, causally influence what happens in the social world, as well as the relation between individuals' social actions and social structures. Bhaskar maintains that because individuals and social structures are distinct and have irreducible emergent powers, they both causally contribute to social events. And, as was indicated earlier, he thinks that social structures are dependent on the activities of many individual agents and that individuals always reproduce and transform social structures through their activities.

The individualism–holism debate focuses on explanation rather than causation. It concerns the question of the extent to which social scientific explanations should focus on individuals and social phenomena, respectively. Bhaskar (e.g. ibid: 27–31) rejects methodological individualism, that is, the view that all social scientific explanations should revolve around individuals only. He maintains that sociologists, or maybe even social scientists in general, should focus on offering explanations that include reference to social structures with emergent powers.

Bhaskar's views on emergent powers of social structures and the relations between individuals and structures have generated a lively debate that has continued to the present date. Here we mention three critiques that are relevant to our topic.

First, Margaret Archer (1995, chapter 5) and Anthony King (1999) have argued that there is a tension between Bhaskar's emphasis on the activity and concept dependence of social structures and his account of social structures as internally related social positions that are autonomous from, and pre-exist, those individuals who currently occupy them. However, the critics differ in their view as to how this tension could be resolved.

Second, Charles Varela and Rom Harré (1996) and Tuukka Kaidesoja (2013: 72–76, 129–133) have argued that Bhaskar illegitimately ascribes emergent powers to internally related social positions because these positions are taxonomic categories rather than entities with causal powers.

Third, Kaidesoja (2013: 179–187) has indicated that Bhaskar uses at least two different concepts of emergent power without specifying their differences. The first concept of emergent power focuses on compositional relations between entities and their parts and is therefore compatible with the possibility of reductive explanations of emergent powers. This concept is mostly used in the context of Bhaskar's (1978) philosophy of the natural sciences. The second concept of emergent power ascribes emergent powers to internal relations between social positions instead of compositionally organized entities and, accordingly, these emergent powers cannot be reductively explained in terms of interrelated parts. This concept is used in the context of Bhaskar's (1998) philosophy of the social sciences.

Elder-Vass on emergence

Elder-Vass has published extensively on the topic of emergence while offering the most concentrated discussion of the issue in his book *The Causal Power of Social Structures. Emergence, Structure and Agency* (2010). In his writings, Elder-Vass offers a general theory of emergence and, drawing on this theory, he then shows that social phenomena sometimes qualify as emergent too. In the following, we present his account of emergence as applied to social phenomena. In many respects,

Elder-Vass' theorizing builds on Roy Bhaskar's work. Roughly speaking, Elder-Vass develops Bhaskar's specification of emergence in relation to the natural world, while arguing, *pace* Bhaskar, that this concept is equally applicable to the social world.

The starting point of Elder-Vass' discussion is the notion of a social entity, which he defines as a relatively enduring whole composed of interrelated parts (see ibid: 17). Notable types of social entities are organizations (like a firm, a household, a school and a religious association) and norm circles, that is, social groups, each sustaining a particular social norm. In most of his writings, Elder-Vass mainly identifies a social entity's parts with individuals. Moreover, he proposes that individuals who compose such an entity are first and foremost interrelated via their beliefs and dispositions; it is only sometimes that their physical or spatial relations matter too (ibid: 200). We concentrate on his position thus specified. Yet it should be noted that in a more recent paper Elder-Vass stresses that nonhuman material objects are often parts in social entities too (Elder-Vass 2017).

Social entities have various properties that Elder-Vass also refers to as causal powers. To begin with, social entities have resultant properties or powers relative to individuals. These are the properties of social entities that their parts, viz. individuals, also have in isolation or as elements in an unstructured collection of parts. Examples of these properties are the power to scream or walk as ascribed to a kindergarten; these are resultant properties since the children also have these properties independently of being, at that moment, parts of the kindergarten.

Furthermore, social entities have emergent properties specified as the properties of social entities that their parts do *not* have in isolation or as elements in an unstructured collection of parts. In this fashion, the emergent causal powers of social entities are *novel* properties relative to the properties possessed by their parts, viz. individuals. Social entities, Elder-Vass explains, have these emergent causal powers in virtue of at least some of their parts standing, at that moment, in certain relations to each other (see Elder-Vass 2014: 7). Here, "at that moment" is meant to capture that the emergent properties co-occur with the interrelated individuals from whom they emerge. In other words, Elder-Vass' notion of emergence is synchronic.

Elder-Vass specifies that there are two types of emergent social properties. One is the emergent properties ascribed to social entities as wholes and exemplified by a firm's power to adopt a new sales strategy and a barbershop quartet's "ability to produce [. . . a] harmonized performance" (Elder-Vass 2010: 154). The other is emergent properties of social entities that are exercised by their parts. Examples of these properties are the power to fire, to hire, to grade and to vote. At first sight, it may perhaps seem puzzling that these properties should be regarded as emergent properties of social entities rather than individuals. Yet Elder-Vass contends that they are properties of social wholes because individuals would not possess these properties if they were not, at that moment, interrelated parts of social entities. As an illustration of this point, he mentions the power to fire an employee:

> [a] manager could not dismiss an employee unless both were parts of an organization of a certain kind, thus the causal power is a power of the organization, exercised on its behalf by the manager, and not a power of the manager as an individual.
>
> *(ibid: 74)*

The fact that the emergent properties of social entities occur as a consequence of some or all of their parts being suitably interrelated means that it is always, in principle, possible to offer reductive explanations of the emergent properties of social entities. This point, Elder-Vass emphasizes, is compatible with the recognition that social entities have irreducible causal powers in the sense of powers that are not identical with the causal powers that their parts, viz. individuals, also have in isolation or as elements in an unstructured collection of parts (Elder-Vass: 2010: 54).

As noted earlier, Elder-Vass' account of emergence applies to other phenomena than social ones. Accordingly, individuals may not only be considered from the perspective of being parts of social wholes. Individuals also may be regarded as wholes, composed of parts in terms of cells and possessing both resultant and emergent properties. A similar perspective may be adopted in regard to cells: these, too, may be viewed as entities composed of parts, and so on, until some basic parts that are not themselves composed of parts are perhaps reached.

Like Bhaskar, Elder-Vass regards his account of emergence as making a contribution to the agency–structure debate. In Elder-Vass' view, social structures have causal powers only insofar as they are identified with social entities; the ascription of causal powers to structures otherwise understood is simply wrong. Accordingly, his theorizing is not only a corrective to positions in the debate that deny that social structures have causal powers. Also, it goes against accounts that agree that social structures are causally effective yet fail to identify these structures with social entities. The kinds of accounts he opposes have been defended by Bhaskar and other critical realists. To avoid mistaken imputations of causality to social phenomena, Elder-Vass recommends that social scientists should carry out ontological analyses in which they identify the entity said to have the causal powers, analyze how its emergent properties arise from the entity's interrelated parts and so on.

In his writings, Elder-Vass also makes it clear that his ontological analysis has implications for the individualism–holism debate. According to him, explanations of events are always partial: they never describe all the multiple interacting causal powers that contributed to bringing about a particular event. Moreover, he holds, when methodological individualists claim that social scientific explanations should solely focus on individuals, this amounts to the view that social scientists should only offer explanations which lay out how individuals, in virtue of the proper-ties they also have in isolation or as elements in an unstructured whole, partially brought about particular events.

On this basis, consider an explanation which states that a social entity, in virtue of its *resultant* social property, partially produced some event. An explanation along these lines may be rephrased by specifying how individuals who are parts of this entity each possess this property and together contributed to the generation of the event in question. By contrast, an explanation which states that a social entity, in virtue of its *emergent* social property, partially produced some event cannot be replaced by an explanation that focuses on individuals only. The properties that individuals also have in isolation or as elements in an unstructured collection of parts do simply not add up the emergent property that a social entity has in virtue of at least some of the individuals being suitably interrelated. By implication, these properties of individuals are unable, on their own, to generate the same effect as the emergent social property. Therefore, explanations that specify how social entities, in virtue of their emergent social properties, contributed to bringing about various specific events are indispensable: on pain of leaving aspects of these events unexplained, it will not do solely to offer explanations that focus on individuals. In this fashion, Elder-Vass maintains, his account shows that methodological individualists are wrong to insist that social scientists should confine themselves to providing explanations that revolve around individuals only.

A number of objections to Elder-Vass' position have been raised. For example, King has argued that Elder-Vass' account misrepresents the social world by holding that it "includes an ontolog-ically distinct and causally efficacious layer: structure" (King 2007: 214). There is nothing more to the social world, King contends, than individuals' social relations and practices.

Coming from a different direction, Douglas Porpora, himself a critical realist, has complained that Elder-Vass' discussion fails to acknowledge that the relations between individuals who occupy various social positions sometimes have distinct causal effects. To illustrate this point, Porpora mentions the power relation between a superordinate and a subordinate. A relation of this sort, he states, "is something analytically distinct from both a behavior and a rule and has its own distinct causal effects. A formal relation of power, for example, can give rise to informal power, the power, say, to harass" (Porpora 2007: 196). The focus of analysis should be these relations, or relational properties, Porpora continues, rather than the emergent properties of social entities. It may be noted that Porpora's claim that relations are causally efficacious is in alignment with Bhaskar's view on this matter.

Finally, Julie Zahle has criticized Elder-Vass' position on the ground that he identifies explanations that focus on individuals with ones that state how individuals, in virtue of the properties that they also have in isolation or as elements in an unstructured collection of parts, brought about particular events. Methodological individualists typically work with a broader notion of these explanations according to which they may also appeal to properties that individuals have in virtue of being, at that moment, suitably interrelated parts of a social entity. This being the case, methodological individualists are likely to reject Elder-Vass' argument against their position on the ground that it trades on his narrow conception of explanations that focus on individuals (Zahle 2014).

Conclusion

The examination of Bhaskar's and Elder-Vass' positions has brought out that there are differences as to how they each spell out the notion of emergence and apply it to the social world. The most significant difference is that Elder-Vass rejects Bhaskar's doctrine of internal relations and attribution of emergent powers to abstract relations between social positions. In contrast to Bhaskar, Elder-Vass stresses that social structures with emergent powers have to be identified with social entities that are composed of interrelated individuals. Though Elder-Vass' view on emergent powers of social entities is not accepted by all critical realists, the conceptual clarity of his views and the thoroughness of his arguments have surely raised the standards by which new accounts of emergence in critical realism should be evaluated.

Though their accounts vary, critical realists in general converge on the importance of offering ontological analyses (that appeal to emergence) of the social world. In their absence, they think, it is impossible to carry out social science in a satisfactory manner; to be successful, social scientific practice must rest on a firm ontological footing. In this sense, ontology precedes methodology and explanation. This approach seems to be one important reason why ontological discussions in critical realism tend to concentrate on rather simple and stylized examples of emergent phenomena instead of examples drawing on empirical research about social phenomena. As a result of this focus, it is reasonable to wonder about the following: may the theories of emergence, offered by the critical realist, serve as the basis for satisfactory in-depth analyses of the more complex social phenomena that current explanatory practices and theories in the social sciences typically deal with? Also, since social scientists are concerned with enormously heterogeneous social phenomena, may these phenomena all be adequately understood within the framework of critical realist accounts of emergence? Finally, it is worth noting that many social scientists oppose the view that getting the ontology straight is a precondition for doing good social science. In their opinion, there is really no need to provide ontological accounts of the social world before engaging in social research. Also, they often continue, the critical realist focus on ontology means that important epistemological and methodological issues tend not to be sufficiently addressed. These points raise another issue: How exactly, and to what extent, is the critical realist concern with ontology (and emergence) justifiable? These are all questions that are currently being discussed and that do indeed deserve careful consideration.

References

Archer, M. 1995. *Realist Social Theory: The Morphogenetic Approach.* Cambridge: Cambridge University Press.

Bhaskar, R. 1978[1975]. *A Realist Theory of Science.* Brighton: Harvester Press.

Bhaskar, R. 1986. *Scientific Realism and Human Emancipation.* London: Verso.

Bhaskar, R. 1998[1979]. *The Possibility of Naturalism: A Philosophical Critique of the Contemporary Human Sciences.* London: Routledge.

Blau, P. 1964. *Exchange and Power in Social Life.* New York: John Wiley.

Collier, A. 1989. *Scientific Realism and Socialist Thought.* Hemel Hempstead: Harvester Wheatsheaf.

Elder-Vass, D. 2010. *The Causal Power of Social Structures: Emergence, Structure and Agency.* Cambridge: Cambridge University Press.

Elder-Vass, D. 2014. "Social Emergence: Relational or Functional?" *Balkan Journal of Philosophy* 6(1): 5–16.

Elder-Vass, D. 2017. "Material Parts in Social Structures." *Journal of Social Ontology* 3(1): 89–105.

Epstein, J. M. and Axtell, R. 1996. *Growing Artificial Societies: Social Science from the Bottom Up.* Washington, DC: Brookings Institution Press.

Homans, G. C. 1950. *The Human Group.* New York: Harcourt, Brace & World.

Kaidesoja, T. 2013. *Naturalizing Critical Realist Social Ontology.* London: Routledge.

King, A. 1999. "The Impossibility of Naturalism: The Antinomies of Bhaskar's Realism." *Journal for the Theory of Social Behavior* 29(3): 267–288.

King, A. 2007. "Why I Am Not an Individualist." *Journal for the Theory of Social Behaviour* 37(2): 211–219.

Lawson, T. 1997. *Economics and Reality.* London: Routledge.

Luhmann, N. 1995[1984]. *Social Systems.* Stanford, CA: Stanford University Press.

Parsons, T. 1968[1937]. *The Structure of Social Action.* New York: The Free Press.

Porpora, D. 2007. "On Elder-Vass: Refining a Refinement." *Journal for the Theory of Social Behaviour* 37(2): 195–200.

Porpora, D. 2015. *Reconstructing Sociology: The Critical Realist Approach.* Cambridge: Cambridge University Press.

Sawyer, K. 2005. *Social Emergence: Societies as Complex Systems.* Cambridge: Cambridge University Press.

Sayer, A. 2010[1984]. *Method in Social Science: A Realist Approach.* London: Routledge.

Varela, C. and Harré, R. 1996. "Conflicting Varieties of Realism: Causal Powers and the Problem of Social Structure." *Journal for the Theory of Social Behavior* 26(3): 313–325.

Zahle, J. 2014. "Holism, Emergence, and the Crucial Distinction." In J. Zahle and F. Collin (eds.). *Rethinking the Individualism-Holism Debate: Essays in the Philosophy of Social Science.* New York: Springer, 177–196.

INDEX

Note: Page numbers in italics indicate a figure on the corresponding page.